2nd International Conference on Mechanical Engineering and Applied Composite Materials (MEACM2018)

IOP Conference Series: Materials Science and Engineering
Volume 544

Harbin, China
7 – 8 December 2018

ISBN: 978-1-5108-9381-8
ISSN: 1757-8981

Printed from e-media with permission by:

Curran Associates, Inc.
57 Morehouse Lane
Red Hook, NY 12571

Some format issues inherent in the e-media version may also appear in this print version.

Printed by Curran Associates, Inc. (2019)

For permission requests, please contact the Institute of Physics at the address below.

Institute of Physics
Dirac House, Temple Back
Bristol BS1 6BE UK

Phone: 44 1 17 929 7481
Fax: 44 1 17 920 0979

techtracking@iop.org

Additional copies of this publication are available from:

Curran Associates, Inc.
57 Morehouse Lane
Red Hook, NY 12571 USA
Phone: 845-758-0400
Fax: 845-758-2633
Email: curran@proceedings.com
Web: www.proceedings.com

Table of contents

Volume 544

2nd International Conference on Mechanical Engineering and Applied Composite Materials

7–8 December 2018, Harbin, China

Accepted papers received: 29 April 2019
Published online: Pl Online DATE

Preface

Papers

IOP Conf. Series: Materials Science and Engineering **544** (2019) 011001 doi:10.1088/1757-899X/544/1/011001

Preface

Dear Distinguished Authors and Guests,

We are pleased to present the proceedings from the 2nd International Conference on Mechanical Engineering and Applied Composite Materials (MEACM2018), which was held on December 7-8, 2018 in Harbin, China.

MEACM2018 aims to present the latest research and results of scientists (professors, students, PhD Students, engineers, and post-doc scientist) related to Mechanical Engineering and Materials topics. This conference provides opportunities for the different areas delegates to exchange new ideas and application experiences face to face, to establish business or research relations and to find global partners for future collaboration.

After the peer-review process, the submitted papers were selected on the basisof originality, significance, and clarity for the purpose of the conference. Theselected papers and additional late-breaking contributions to be presented aslectures will make an exciting technical program. The conference program isextremely rich, featuring high-impact presentation. We hope that the conference results constituted significant contribution to the knowledge inthese up to date scientific field.

The proceeding records the fully refereed papers presented at the conference.The main conference themes and tracks are Mechanical Engineering and Applied Composite Materials and other related topics. Hopefully, all participants and other interested readers benefit scientifically from the proceedings and also find it stimulating in the process.

We would like to thank all the authors who have contributed to this volume and also to the speakers, reviewers, chairpersons, organising committee and all participants for their support to MEACM 2018.

Finally we wish all the authors and attendees of MEACM 2018 a unique, rewarding and enjoyable memory at MEACM 2018 in Harbin, China.

With our warmest regards,

Peng-Sheng Wei

Conference Organizing Chair

January 9th 2019

Committees

Conference Chair

Professor Peng-Sheng Wei, National Sun Yat-Sen University, Kaohsiung, Taiwan

Program Committee

Prof. Steven Y. Liang, Georgia Institute of Technology, USA

Prof. Dr. Farid Nasir bin Ani, Universiti Teknologi Malaysia, Malaysia

PROF. JACOB OPADEYI, The University of the West Indies, Trinidad and Tobago

PROF. DR. MOHAMMAD NAZRI BIN MOHD. JA'AFAR, Universiti Teknologi Malaysia, Malaysia

Dr. Norhayati Binti Ahmad, Universiti Teknologi Malaysia, Malaysia

PROF. DR. MD MUSTAFIZUR RAHMAN, Universiti Malaysia Pahang, Malaysia

ASSOC. PROF. JIYUN ZHAO, City University of Hong Kong, Hong Kong

ASSOC. PROF. YONGSHENG CHEN, The Chinese University of Hong Kong, China

ASST. PROF. YU (WILL) WANG, Washington State University, USA

ASST. PROF. ZHIQUAN SHU, Washington State University, USA

ASST. PROF. MOHAMMAD H. RAHMAN, University of Wisconsin-Milwaukee, USA

ASST. PROF. KOW KIEN WHO, University College Sedaya International, Malaysia

DR. MOHD FARID BIN MUHAMAD SAID, Universiti Teknologi Malaysia, Malaysia

DR. NORHAYATI BINTI AHMAD, Universiti Teknologi Malaysia, Malaysia

DR.M.K.MARICHELVAM ASSISTANT PROFESSOR (SR. GRADE), Mepco Schlenk Engineering College, India

DR. M. SREENIVASAN, National Technical University of Ukraine, Ukraine

Peer review statement

All papers published in this volume of *IOP Conference Series: Materials Science and Engineering* have been peer reviewed through processes administered by the proceedings Editors. Reviews were conducted by expert referees to the professional and scientific standards expected of a proceedings journal published by IOP Publishing.

Research on the typical case of industrial and commercial capital promoting the increase of the farmer's income

Ying Li[1] And Zhuohuai Guan[2]*

Research Assistant, Nanjing Research Institute for Agricultural Mechanization Ministry of Agriculture, Nanjing, China
Research Assistant, Nanjing Research Institute for Agricultural Mechanization Ministry of Agriculture, Nanjing, China
E-mail:675290487@qq.com

Abstract. The farmer's income in China is still low and increasing slowly. The entry of industrial and commercial capital into "three rural" areas in China is an important way to promote the increase of the farmer's income, and there are some successful cases of this entry existing in different areas of China which give us lots of useful experience. Some policies and measures should be implemented to improve the effect of the entry, mainly including changing the local government investment, correctly guiding industrial and commercial capital to participate in the development of agricultural industrialization, improving the investment environment of industrial and commercial capital, and optimizing the business model of industrial and commercial capital entering "three rural" area have become to be the policy focus of the government's support the industrial and commercial capital promoting the doubling of the farmer's income in China.

Key Word Industrial and commercial capital; The famer's income; Income doubled; Policy advice

1. The typical case of industrial and commercial capital promoting the increase of the farmer's income

1.1. Jiangsu "Dai zhuang mode"

Dai zhuang is a village in Jiangsu province, and it became famous for Yafu Zhao creating the agricultural cooperative "Dai zhuang mode". It is a form of cooperation that make the scientists, village committees and farmers back to the land together, regarded by Japanese experts as efficient and in line with the realities of rural in China."Daizhuang mode" give full play to the advantages of the industrial and commercial capital, taking the way of secondary industry with agriculture-oriented. Through the guidance of cooperative business capital and farmers to establish a closer producing and selling link mechanism, that is the famer in charge of producing, processing, sale and distribution to industrial and commercial capital, finally let the business capital and the farmer can achieve the value-added profit in the agricultural industry. Farmer's specialized cooperative organization shall be encouraged to establish their own leading enterprises by the way of mutual fund assistance and joint stock cooperation. In promoting the construction of professional cooperative organizations for farmers, the cooperative organizations should play a leading role in increasing farmers' income, and give priority to supporting the cooperative organizations in undertaking efficient agricultural projects, processing and distribution enterprises of agricultural products, constantly expanding the scale of the cooperative organizations so that farmers can get more benefits. "Dai zhuang mode" began in 2001 when the per capita net income of the village in Dai zhuang was less than 2000 yuan. Through 10 years of exploration and practice, the organic agriculture professional cooperatives in Dai zhuang have developed nearly 800 households, accounting for more than 90% of all village households, with the per capita income of the farmers reaching 13006 yuan. The demonstration has led to the positive development of the surrounding administrative villages.

1.2. Shandong "Xisen mode"

"Xisen mode" is made by industrial and commercial capital entering into village renovation, improving the farmer's living condition and environment, and establish enterprise company in the replacement of old curtilage base and abandoned fields for agricultural industrialization management, taking the form of the farmer buying the share with village land to comprehensive develop and utilize the land, cultivated land, making experiment and popularization of agricultural science and technology project, and achieve the development of ecological agriculture circular economy mode. In 2001 and 2004, liang xisen, a farmer entrepreneur, invested 122 million yuan to renovate two villages, Liangzhui village and Xujia village, Huangjia town, Leling city, Shandong province, and invested 1 billion yuan to build the xisen sanhe group and xisen zhonglian potato industry group on the base of vacated old basement field, so as to develop ecological agriculture and circular economy. "Xisen mode" is a typical representative of the business capital participating in the new socialist rural construction, creating good economic and social benefits. Local villagers bought the share with their own land in the enterprise, gaining at least 500 yuan a year per acre of guaranteed income, and they can also take part in enterprises' year-end dividend. The other villages' residents that 5000 meters around Liangzhui village also gained more than before. In 2005, for example, enterprises spent more than 10 million yuan on purchasing corn stalks for feed. That is to say, nearby villagers gained more than 10 million yuan simply by selling straw. Among them, the detoxification and planting potato project area not only directly brought an increase of 1000-1500 yuan to the income of farmers, but also could provide employment opportunities for 100,000 farmers and settle a large number of rural surplus labor force, so as to achieve the improvement of living conditions, income of farmers and the unification of enterprise benefits, social benefits and ecological benefits.

2. The revelation that the typical cases bring out

These typical cases show us some important revelation. Firstly, Chinese farmers are not conservative and backward. On the contrary, they have a strong desire to accept industrial and commercial capital for common development and income increase. They are willing to cooperate with industrial and commercial capital whether they take the form of buying share with land or directly participating in industrial production. Secondly, the successful cases of adopting industrial and commercial capital mode have strong demonstration effect and can drive the imitative participation of surrounding areas. Thirdly, Industrial and commercial capital entering "three rural" have intense effect on getting rich. The gathering of production factors and the modernization of agricultural organization can accelerate the process of increasing the farmer's income and getting rich. Fourthly, the timely improvement of external environment plays an important role in promoting the implementation of the mode. At present, the participation rate and success rate of industrial and commercial capital entering the new rural construction are relatively low. The important reason is that there are still some problems and risks in the entry of industrial and commercial capital, such as the farmland leasing access system is not sound, local government functions are absent, and relevant supporting facilities and investment platforms are not perfect.

3. Policy advice

*3.1. **Transform local government input and correctly guide industrial and commercial capital to participate in the development of agricultural industrialization***

It is far from enough for industrial and commercial capital to enter only the planting link of "three rural" areas. It should be encouraged to enter the serving, processing and selling of the whole agriculture production. The range of guiding industrial and commercial capital entering "three rural" should include the following aspects. First, public welfare agriculture area should be included. We will give priority to encouraging capital investment in new rural development, mainly in such areas as public services, infrastructure for rural living and social programs that offer universal benefits. In addition, it also includes the public welfare part of agricultural socialization service system, such as animal and plant epidemic prevention and control, agricultural technology service promotion station and so on which have the public welfare characteristics. Second, the area that suitable for the enterprise management of farming industry should be included. It mainly includes the fields where the industrial chain is relatively long, the industrialization degree is relatively high, and the agricultural industrial chain can be relatively extended, the agricultural multi-function can be expanded, and the value chain of agricultural products can be improved, such as facility agriculture, circular agriculture, three-dimensional agriculture and large-scale breeding industry. Third, the agricultural industrial chain pre - and post-natal links should be included. Industrial and commercial capital should be encouraged to invest in the upstream and downstream areas suitable for industrialization, scale and intensification of the agricultural industrial chain, extending the agricultural industrial chain and increasing the added value of agricultural products, and let farmers share the value-added benefits in the processing and circulation of agricultural products.

3.2. Improve the environment for industrial and commercial capital investment

In terms of improving the investment environment of industrial and commercial capital, it is mainly aimed at keeping up with the follow-up services provided by the government and departments after industrial and the sector of commercial capital entering the agricultural. Therefore, it is necessary to strengthen the transformation of functions, strengthen service awareness, and establish an effective platform for agricultural services and industrial and commercial capital entering the agricultural sector. First, information service. We will establish and improve the information network for the production, marketing and sale of agricultural products, and the land transfer information network. We will intensify the publicity of policies concerning agriculture, rural areas and farmers, provide consulting services on policies, laws and regulations for enterprises, issue investment guidelines for agricultural industries regularly, and prevent and control the investment risks of enterprises. Second, technical services. Departments involved in agriculture and agro-technical popularization should contact enterprises with specialized personnel on a regular basis and guide enterprises to solve technical problems in the production process in a timely manner. We will help industrial and commercial enterprises to strengthen ties with agricultural colleges and universities and agricultural research institutes, ensure the supply of specialized personnel, speed up the transformation of scientific research results, and intensify personnel training. Third, management services. The government's construction projects should be inclined to the grain and other planting and breeding enterprises, and the industrial and commercial capital subject with high efficiency of output per unit of land, large number of local employment, and remarkable ecological function should be exempted from taxes or compensated by awards, and the agricultural product processing enterprises should be given preferential policies on water and electricity use. Fourth, we will focus on improving financial services. We will provide financial products and services to industrial and commercial enterprises through various channels, explore the security guarantee of facilities and equipment of enterprises, ground attachments of production bases, evaluation of agricultural products discount, and solve the financing problems of agricultural and industrial enterprises.

3.3. Optimize the business model of industrial and commercial capital entering "three rural"

In terms of optimizing industrial and commercial capital entering agricultural mode, local governments should give reasonable guidance so that the progress of capital going to rural areas is organically integrated with the process of industrialization and urbanization, and is compatible with the scale and speed of land circulation. At the same time, industrial and commercial capital should be guided to invest in the relatively weak infrastructure construction in rural areas, and leading enterprises should be encouraged to focus on agricultural infrastructure construction projects with small externalities and clear beneficiaries. The government should issue relevant policies to regulate the investment mode and investment behavior of industrial and commercial capital. First of all, we should advocate and encourage the way of "company + base + farmer" mode of agricultural capital combination, namely after choosing base, the transfer of the base of the farmer's right to the contracted management of land does not occur, the company signed with the farmers crops purchase orders, and coordinate the farmland shall supply planning, varieties, technology, information, provide loan guarantees, develop high-quality, high-yield agriculture, realize the scale production in some area, unified purchase, unified processing, the sale, maintain the farmer's right to the contracted management of land, and guarantee the enterprise the source of agricultural products, agricultural materials confrontation. Industrial and commercial capital is prohibited from entering "real estate development, rural tourism and other fields whose main purpose is large-scale enclosure, which may endanger food security and social stability". And take up agricultural land in the way of "low price, long time, large scale", affecting produce supply.

Reference

[1] Huang Caiying and Su Yaran 2003 Rational thinking of industrial and commercial capital boosting agricultural industrialization construction Agricultural Economy vol 7 p 20–21

[2] Jia Guangjun 2008 Financial system arrangement of industrial and commercial capital participating in new rural construction -- a survey study based on "Xisen mode" in Shandong province. Financial Research vol 8 p 186–191

[3] He Junwei and Wang Zhonghai 2013 Thoughts and Suggestions on industrial and commercial capital entering agriculture China development observation vol 7 p 47–50

[4] Jiang Yongmu Zhang Zunshuai 2014 Research on the generation and realization of the guiding catalogue of industrial and commercial capital investment in agriculture Modern economic research vol 5 73–77

Acknowledgments

Project Name: Study on brand construction of journal of agricultural engineering in new period, Project Number: 151814-43.

IOP Conf. Series: Materials Science and Engineering **544** (2019) 012002 doi:10.1088/1757-899X/544/1/012002

Preparation of iron tailings glass-ceramics by sol-gel method

Changwei Xu1, a and Jianxin Zhao2,b

1 Professor, Shenyang JianZhu University, Shenyang, China
2 Postgraduate, Shenyang JianZhu University, Shenyang, China
E-mail: [a]1301663317@qq.com, [b]1282693098@qq.com

Abstract. For the preparation of glass-ceramics by iron tailings, the sintering and melting methods have been used in the past, but the two methods are short of high energy consumption. In this test, the sol-gel method has been adopted. Because the sol-gel method requires high purity to prepare raw materials, and there are many impurities in iron tailings, which is inconsistent with the preparation requirements of sol-gel method. Therefore, in this experiment, useful elements in iron tailings were extracted. Iron tailings were obtained by thermal-alkali reaction, and Fe was obtained by acid reaction, so as to prepare Ca-Mg-Al-Si system glass-ceramics. After the preparation of sol-gel, the basic glass was obtained, and after nucleation and crystallization, the microcrystalline glass whose main phase was diopside was obtained.

1. Introduction

Microcrystalline glass, also known as glass ceramics, is a new kind of inorganic non-metallic material[1-2]. It is a new material obtained by making full use of the advantages of glass in thermodynamics. It is a new composite phase material composed of micro crystal and residual glass.[3-4]

According to incomplete statistics, more than 10 billion tons of tailings and waste rock are discharged every year around the world. At present, the comprehensive utilization rate of China's tailings is less than 10%, and the amount of iron tailings deposited is as high as over 1 billion tons. There are problems such as occupying a large amount of cultivated land, wasting resources, polluting soil, water and air. Therefore, the comprehensive recycling and utilization of iron tailings has been widely concerned by the whole society[5]. The raw materials for the preparation of glass-ceramics can be pure chemical reagent, tailings, steel slag, quartz sand, glass waste, etc. The application of industrial wastes not only reduces the cost of preparation, but also reuses waste such as tailings, reduces the pressure of waste accumulation and protects the ecological environment[6]. The preparation of sol-gel requires high purity of raw materials and high purity of chemical reagent but high cost. In this experiment, tailings were selected to neutralize the raw material cost. Ethyl orthosilicate, aluminum nitrate, calcium oxide and magnesium oxide were selected as the basic materials. Ethanol as a solvent, nitric acid as a catalyst and reagent for removal of impurities. Sodium hydroxide is used as raw material for thermal-alkali reaction.

As there are many impurities in iron tailings, sol-gel method is not suitable. Generally speaking, sintering and melting methods are mainly used in the preparation of glass-ceramics from iron tailings. In this experiment, some treatment of iron tailings was carried out to purify the iron tailings, so that they could be used in the sol-gel method [7]. The microcrystalline glass structure is compact, but there is cracking and shrinkage in sol gel process. In order to prevent this adverse factor, dry cracking control agent is commonly used. In this experiment, the method of grinding basic glass into powder is adopted[8].

2 Extraction of useful elements from iron tailings

In this experiment, Ca-Mg-Al-Si microcrystalline glass was prepared by sol-gel method, and the specific process is shown in figure 1.

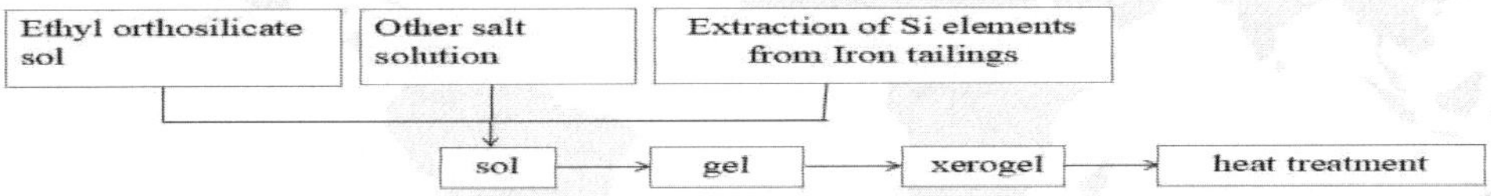

Fig.1 Experimental procedure

The main components of iron tailings are SiO_2, Fe_2O_3 and Al_2O_3, CaO and MgO.The content of SiO_2 was the highest. According to the designed Ca-Mg-Al-Si microcrystalline glass, the maximum amount is Si element. Fe^{3+} can be used as a nucleating agent. Considering the purity of raw materials required by sol-gel method, the elements in iron tailings should be extracted first to realize the efficient use of iron tailings[9-10].The sodium silicate solution was prepared by reaction with acid and alkali, and used as the silicon source to prepare the glass.

2.1 Iron tailings react with acid

The main chemical composition of the tailing raw material selected in this experiment is shown in table 1.

Table 1 Chemical composition of iron tailings

Components	SiO_2	Al_2O_3	Fe_2O_3	CaO	MgO	Na_2O
Mass percent/%	68.58	2.69	13.81	3.68	4.98	0.36

2.1.1 Different acids react with iron tailings. HF, HCl, H_3PO_4, HNO_3 and H_2SO_4 were mainly considered in this experiment.After passing through 200 mesh sieves, the iron tailings were weighed and added with over acid. After reaction for a period of time, filter, wash, dry, weigh the quality of filter residue, and conduct feasibility analysis.It is found that the dissolution of SiO_2 in iron tailings by HF is feasible in thermodynamics, but the actual reaction rate is extremely slow, which is not feasible from the point of view of kinetics.Other acids can only react with metal oxides in iron tailings, not with SiO_2.It is found that nitric acid is easy to be decomposed, and the decomposition product is poisonous, and has little contribution to the actual reaction.Although sulfuric acid can react, SO_4^{2-} is an impurity ion in the preparation of sol and gel, and is difficult to remove impurities. Other acids can only react with metal oxides in iron tailings, not with SiO_2. As $FeCl_3$ can be used as crystal nucleating agent, hydrochloric acid is finally used to acidify iron tailings.

2.1.2 Reaction time with acid. First, 5g iron tailings were weighed and added with excessive dilute hydrochloric acid to react at room temperature for different times. Then the quality of filter residue was determined by filtration, washing and drying. Finally, formula (1) is used to calculate the dissolution rate. The experimental results are shown in table 2.

It can be seen from table 2 that dissolved percent of iron increases with the prolonged reaction time with acid at room temperature. However, when the time exceeds a certain limit, the dissolved percent starts to grow slowly, which can be regarded as the total dissolution. In this experiment, Fe^{3+} in iron tailings was extracted by reacting 2days with acid at room temperature.

$$X = \frac{m_0 - m_1}{m_0} \times 100\% \tag{1}$$

m_0 - iron tailings mass before reaction with acid;

m_1- iron tailings mass after reaction with acid.

Table 2 Dissolved percent of Fe reacted with acid at room temperature

Time/h	2	4	6	8	24	48
Dissolved percent/%	47.6	60.1	68.4	74.5	78.6	80.3

2.2 Iron tailings react with alkali.

In this experiment, the reaction of high temperature with alkali was used to extract Si elements.Firstly, proper amount of iron tailings (mainly SiO_2) after reaction with acid was weighed, and a certain amount of NaOH was added.And then mixed evenly, the mixture placed into the muffle furnace heated to 500 ℃, and the heat preservation for a period of time.Next, the mixture needs to be poured into a certain amount of distilled water and heated to stir for a while. At last, the sodium silicate solution was obtained by filtration and washing.

2.2.1 Effect of holding time on the dissolution rate of SiO_2. Taking 5g iron tailings after reacting with acid and 10gNaOH ,and mixing them thoroughly.Put them at the 500℃,heated respectively in 20,30,40,50, 60 minutes, immediately pouring 190 ml water.We will put the mixture at 70℃, heated and mixed 30min, and then filtered.Equation(2) is used to calculate the influence of the heat preservation time on dissolved percent of Si .The experimental results are shown in table 3 and figure 2.

$$X = \frac{c_0 - c_1}{c_0} \times 100\% \tag{2}$$

c_0-iron tailings mass before alkali reaction;

c_1-iron tailings mass after alkali reaction.

Table 3 Dissolved percent of silicon in iron tailings at different time of heat preservation

Time of heat preservation /min	20	30	40	50	60
Dissolved percent /%	26.17	30.15	34.47	39.68	39.93

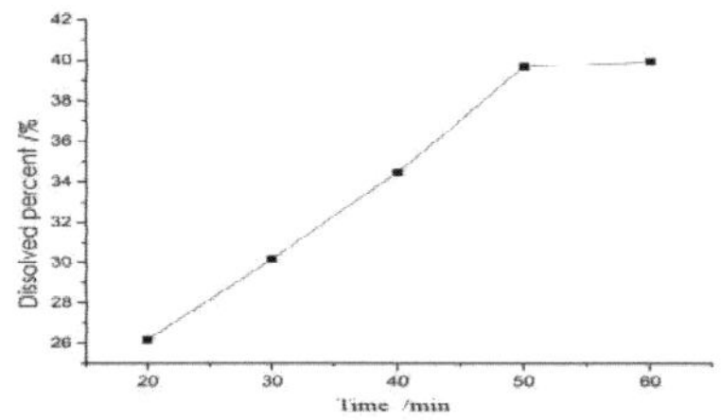

Fig. 2 Dissolved percent curves of silicon in iron tailings at different heat preservation times

It can be seen from the figure that when the heat preservation time is relatively short, Dissolved percent of silica increases with the heat preservation time.However, dissolved percent of silica does not change much with the prolonged heat preservation time, and tends to be stable.

2.2.2 The influence of water consumption (NaOH concentration) on dissolved percent of SiO_2. Firstly, taking 5g iron tailings after reacting with acid,and mixing well with 10gNaOH.Secondly, heated 50 min at 500 ℃, and were respectively put into the 240, 190, 90, 57, 40ml(NaOH concentration were 4%, 5%, 10%, 15%, 20%)water.Thirdly, the mixture was stirred and heated 30min at 70℃,and then filtered.Finally, it was found that the difference of water consumption resulted in dissolved percent of Si.The experimental results are shown in table 4 and figure 3.

Table 4 Effect of different NaOH concentration on dissolved percent of silicon

The concentration of NaOH /%	4	5	10	15	20
Dissolved percent/%	26.69	39.69	39.36	38.15	35.21

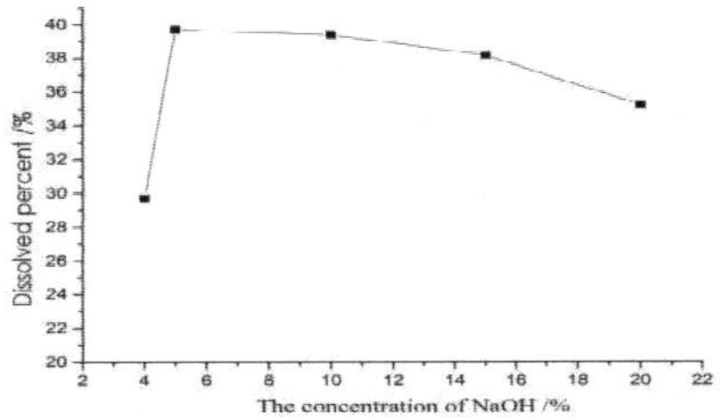

Fig. 3 Dissolved percent curves of silicon with different NaOH concentrations

According to the figure, when sodium hydroxide concentration is low,Si dissolution percentage increases with the increase of mass fraction of NaOH concentration.When sodium hydroxide concentration is high, Si dissolution percentage decreases with the increase of mass fraction of NaOH concentration.When the concentration of sodium hydroxide solution is large, the viscosity of the solution system increases, thus reducing the reaction rate.

2.2.3 The influence of temperature on dissolved percent of SiO_2 during stirring. Firstly, taking 5g iron tailings after reacting with acid, and mixing well with 10gNaOH.Secondly, heated 50 min at 500 ℃, and were respectively put into 190ml water.Thirdly, the mixture was stirred and heated 30min at 30,50,70,90,100℃,and then filtered.Finally, dissolved percent of Si can be calculated at different temperatures.The experimental results are shown in table 5 and figure 4.

Table 5 Dissolved percent of SiO_2 at the different temperature when stirred

Temperature/℃	30	50	70	90	100
Dissolved percent/%	33.39	35.92	39.68	39.23	35.21

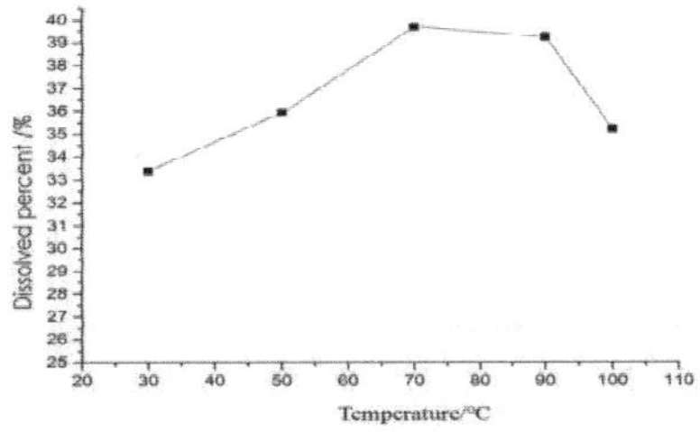

Fig. 4 Dissolved percent curves of SiO2 at different temperatures when stirred

The temperature of solution system has great influence on the reaction rate and process.Basically, as the temperature of the solution increases, the speed of the reactant molecules in the solution increases. The number of effective collisions per unit time increases and the reaction rate increases.By the figure that, as the temperature increases, Si dissolution percentage reached the maximum at 70℃.But when the temperature exceeds 70℃, dissolution percentage decline.The reason is that, although the reaction speed increases from a thermodynamic perspective, the viscosity increases as the temperature increases.From the kinetic point of view, the viscosity inhibits the reaction rate, and dissolved percent decreases.

2.2.4 Influence of different stirring time on dissolved percent of SiO2. Firstly, taking 5g iron tailings after reacting with acid, and mixing well with 10gNaOH.Secondly, heated 50 min at 500 ℃, and were respectively put into 190ml water. Thirdly, the mixture was stirred and heated 30,40,60,80min at 70℃,and then filtered.Finally,dissolved percent of Si at different stirring time can be calculated.The experimental results are shown in table 6 and figure 5.

Table 6 Dissolved percent of SiO_2 at different stirring time

Stirring time/min	20	30	40	60	80
Dissolved percent/%	34.48	39.68	39.76	40.13	40.22

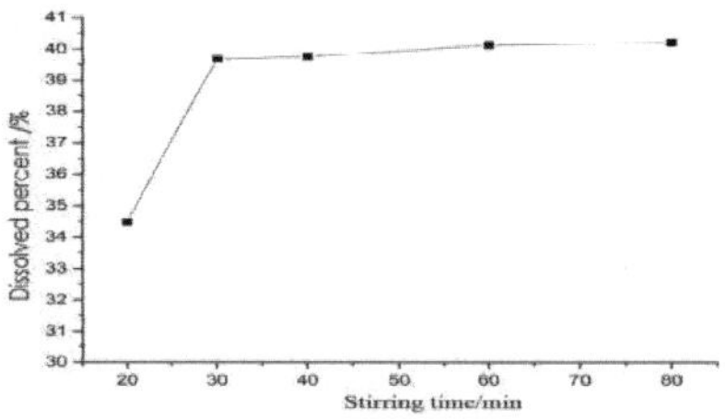

Fig. 5 Dissolved percent curves of SiO_2 at different stirring time

The dissolved percent of silicon increased with the increase of stirring time.It did not change significantly after 60min.The reason was that with the increase of reaction time, the concentration of NaOH decreased, the content of SiO_2 decreased, the reaction speed decreased, and the reaction was basically completed.The reason is that with the increase of reaction time, the concentration of NaOH decreases and the content of SiO_2 decreases.

The reaction rate is reduced and the reaction is basically completed.In summary, when iron tailings react with alkali.Under the following conditions: 500 ℃ heating 50 min, pour 190 ml water, heated at 70℃ and stirred 60 min,.Then you can filter. The dissolved percent of SiO_2 in iron tailings is higher.

3 Basic glass was prepared by sol-gel method

According to table 7, the CaO,$FeCl_3$ and MgO needed can be weighed.Add the nitric acid to dissolve and form the calcium nitrate solution and magnesium nitrate solution.Since Al_2O_3 did not react thoroughly with acid at room temperature, $Al(NO_3)_3{\cdot}9H_2O$ was substituted for Al_2O_3. $Al(NO_3)_3{\cdot}9H_2O$ is highly soluble in water, and the weighed $Al(NO_3)_3{\cdot}9H_2O$ was slowly added to the solution.

Table 7 Base glass chemical composition

Components	SiO_2	Al_2O_3	CaO	MgO	$FeCl_3$
Mass percent/%	55	5	22	15	3

As it is difficult to prepare the gel from a single sodium silicate, ethyl orthosilicate is used instead of 30% sodium silicate.According to the molar ratio of TEOS: H2O=1:1, TEOS: ethanol =1:3 and TEOS: HNO_3=1:0.04, H_2O, ethanol, nitric acid and TEOS were respectively weighed.After mixing them, they

IOP Conf. Series: Materials Science and Engineering **544** (2019) 012002 doi:10.1088/1757-899X/544/1/012002

were stirred for 1h at room temperature to complete the pre-hydrolysis of ethyl orthosilicate [11].In the prepared sodium silicate solution, ethanol was added according to the molar ratio of 1:3, and HNO_3 was added to adjust the pH value to about 10.5.Let it sit for a while.The prehydrolyzed ethyl orthosilicate was added to the salt solution.After that, sodium silicate solution was added under the effect of magnetic stirring, and the pH value of the solution system was maintained at 8-9 to form the gel.Place the gel in 80 ℃ oven aging for 6 h.A clear homogeneous dry gel was obtained.DSC detection was performed and the DSC figure was obtained as shown in figure 6.

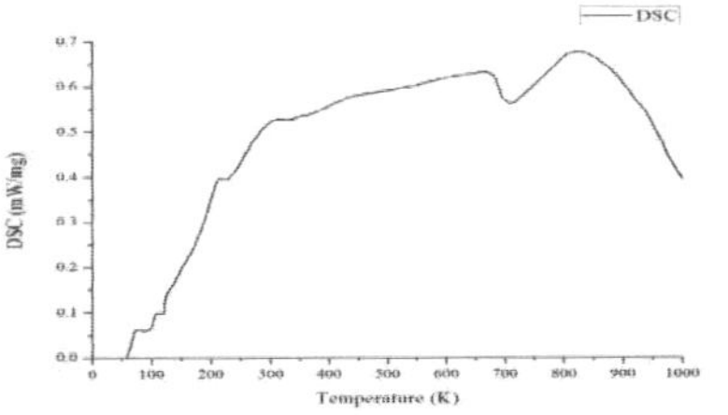

Fig. 6 The DSC test chart of the sample

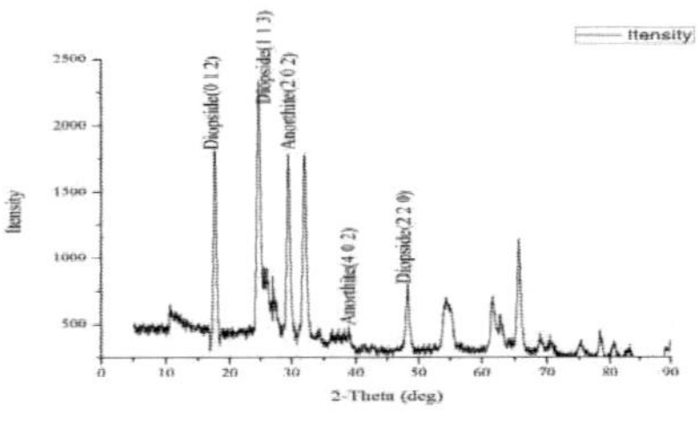

Fig. 7 XRD diffraction pattern of the sample

4 Heat treatment

It can seen from the graph that a tiny endothermic peak at 80 ℃. The ethanol in the dry gel may evaporate from the system.There is a larger endothermic peak appears at 120 ℃. It is mainly the elimination of the combined water in the system.A heat absorption valley appears at 720 ℃. It could be that the glass phase begins to nucleate.A larger exothermic peak appears at 850 ℃.This crystallization exothermic peak is significant. However, there is a large heat absorbing valley before the peak, which is not nucleated heat absorbing peak.It is caused by the softening deformation, microstructure rearrangement and heat absorption of the products in the heat treatment process.So the actual crystallization temperature needs to be higher than that.

Therefore, in the process of heat treatment need to slowly heat up to 120 ℃, and then heat preservation for a period of time. Ensure all organic solvents and water are drained.It quickly heats up to 830 ℃ and keeps 2 h.Then, it needs to raise the temperature to 950 ℃, and wait for 2h.The glass samples can be obtained by cooling them to room temperature.

5 Microscopic detection

After the sample is cooled,the samples were taken out and analyzed by XRD and SEM.

1 X-ray diffraction analysis (XRD)

The XRD diffraction pattern of the sample is shown in figure 7.

2 Scanning electron microscope analysis(SEM)

The sample of microcrystalline glass was corroded with 1% HF solution for 2min.Then wash it, dry it and spray gold. Its microstructure was observed by scanning electron microscopy .

The SEM diagram of the sample is shown in figure 8.

Fig. 8 SEM diagram of the sample

It can be seen from Fig 8 that it meets the requirement of microcrystal size.Diopside crystal is short columnar, belongs to monoclinic system, complete crystal is rare, cross section is similar to square, aggregate can be columnar, dense block or radial shape.The morphology of the precipitated crystals was observed to be columnar or lamellar, which was consistent with the morphology of diopside crystals.At the same time, the analysis results are consistent with the previous XRD analysis results.

6 Conclusion

(1)When iron tailings reacted with acid, made it react at room temperature for 2d.The iron elements in iron tailings were dissolved more under this condition.

(2)When iron tailings reacted with alkali, the way to dissolve more silicon is that the mixture heated at 500 ℃ for 50min, then immediately poured 190 ml water,finally heated and stirred at 70 ℃ for 60 min.

(3)Through XRD analysis of the glass, it can be known that the samples are mainly diopside.The microscopic morphology seen by SEM is consistent with XRD analysis.

References:

[1]Jinshu Cheng,Hong Li,Liying Tang,Feng He.Glass-ceramics [M].Chemical Industry Press, 2006.
[2]Bing Liao,Zhenguo An,Jingjie Zhang.Research Progress of Magnetic Glass-Ceramics [J].New material industry,2014,(7)： 717-719.
[3]Jing Wang.MgO-Al2O3-SiO2 Study on structure and Properties of transparent Glass-Ceramics: [doctoral thesis].Wuhan: Wuhan University of Technology, 2013.
[4]Tuowen Li,Jialu Zhang,Xiaozhong Zjao.Glass-ceramics [J]. Chinese glass.1992(1):28-31.
[5]Xingzhong Guo,Hui Yang,et al. Compound nucleating agent CaO-MgO-Al2O3-SiO2 Crystallization and properties of glass-ceramics [J].Silicate Journal,2013,41(4)： 454-458.
[6]Jing Wang.Study on preparation of Glass-Ceramics for Li-Zn-Al-Si system by Sol-Gel method [D].Wuhan: Wuhan University of Technology,2012.
[7]Ritzberger C,Apel E,Höland W,et al. Properties and Clinical Application of Three Types of Dental Glass-Ceramics and Ceramics for CAD-CAM Technologies[J]. Maral,2010,(4)： 667-674.
[8]Caishi Lai,Xianzhi Hu,Futing Zi. Present situation and Countermeasures of Comprehensive Utilization of Mine tailings Resources in China [J].Comprehensive utilization of mineral resources, 2011, (4):31-37.
[9]Mingxi Zhang,Fengfeng Li,Chaoxia Lu.Study on purification of Nano-silica from Iron tailings.New chemical materials,2013,41： 33-35.
[10]Zhi Li.Study on iron removal and aluminum removal by acid leaching of iron sulphide tailings.Guizhou Chemical Industry.2006,31(3)： 8-15.
[11]Hai Wang,Feng Ren,Chunbo Huang,et al. Progress in preparation of carrier silica gel by sol-gel method.Aging and Application of synthetic Materials.2018,47(2)： 111-113.

XPS analysis of char residue from an intumescent flame retardant rigid PU composite foam

Ting Zheng[1,2], Linying Yang[1], Peibang Dai1*, Chang Qin[1], QichenTang[1]

1. School of Materials Science and Engineering, Guilin University of Electronic Technology, Guilin 541004, P.R.China,
2. Guilin institute of intelligence science and technology, Guilin 541004, P.R.China
*E-mail: daipeibang@163.com

Abstract: A flame retarded rigid PU foam was prepared using polyols and diisocyanates with a modified intumescent flame retardant (MIFR) via polyoxyethylene nonylphenol ether (surfactant TX-10). The MIFR included a new carbon source, calcium 3-hydroxy-2, 2-bis(hydroxymethyl) propyl phosphate. The carbon residue from the flame retardant rigid PU foams after UL-94 test was investigated with XPS. The data showed that the addition of MIFR into PU foams could promote formation of pure carbon char and P-C bond.

Keywords: Polyurethane; Intumescent flame retardant; Composite

1. Introduction

A traditional intumescent flame retardant (IFR) system contains generally three components that are the basic elements to form a char: acid donor (ammonium polyphosphate, APP), blowing agent (melamine, MA) and carbon donor (pentaerythritol). Material containing such flame retardant system will form a char layer at the surface during combustion to protect the material [1-4]. IFR is normally used as fire proof coating. In recent years, it has been added into polymer materials as flame retardant due to its superior flame retardant effect. A great amount of papers have reported flammability of intumescent flame retardant polymer composites, i.e. flame retarded polypropylene, polyethylene, poly[ethylene-co-(vinyl acetate) and ABS, etc., which were prepared by incorporating the components of intumescent flame retardant into the polymers directly.

However, high loading level of IFR is needed in flame retardant polymer composites for achieving superior flame retardancy. When high content of IFR is used, the mechanical and environmental durability of the materials can be severely degraded. It also causes an increase to the viscosity of the polymer melt which make processing more difficult. IFR systems are particularly sensitive to processing conditions and exhibit a limited processing window. Therefore, surface-modification is needed for intumescent flame retardant. The study in this paper aims to prepare flame retarded a rigid PU foam using polyols and diisocyanates with a

IOP Conf. Series: Materials Science and Engineering **544** (2019) 012003 doi:10.1088/1757-899X/544/1/012003

modified intumescent flame retardant (MIFR) via surfactant TX-10 and investigate the carbon residue from the flame retardant rigid PU foams after UL-94 test. The MIFR included a new carbon source, calcium 3-hydroxy-2, 2-bis(hydroxymethyl) propyl phosphate.

2. Experimental

The MIFR included calcium 3-hydroxy-2, 2-bis(hydroxymethyl) propyl phosphate, APP and melamine and was modified with TX-10 via water solution method.

The MIFR were added into the compound of polyols and diisocyanates(the ratio is 1.0:1.0) and stirred for 1 minutes at high speed. After mixing, the mixture was molded under ambient temperature for 8 minutes to sheets of suitable thickness.

The UL-94 vertical burning tests were carried out using a CZF-1 type instrument (made in China) on the sheets of 127×12.7×3.0mm3 according to the standard UL-94 test ASTM D635.

The XPS characterization of the residue composition and chemistry was carried out on a XSAM800 System (Kratos Campany, UK) using Al Kα radiation.

3. Results and discussion

3.1 Flame retardancy

Table1 shows the effects of MIFR on the flame retardancy of the PU flame retardant composite foams. The formulation containing more than 22.5wt% MIFR was V-0 rating for UL-94 test while the formulation with 20.0wt% MIFRC could only pass V-2 rating of UL-94 test. The formulation containing less than 15.00wt% MIFR as well as the formulation no MIFR could not pass UL-94 test. When the formulations contained MIFR more than 20.0%, the flame retardancy of the flame retardant PU composite foams were improved effectively.

Table 1. Flame retardancy of flame retardant PU foams.

Sample	MIFR content(%)	UL-94 rating
PU	0	No Rating
PU/MIFR10	10.0	No rating
PU/MIFR20.0	20.0	V-2
PU/MIFR22.5	22.5	V-0

3.2 Char Residue Analysis

The chemical composition of residues from PU10.0 and PU/MIFR22.5 was quantitatively analyzed by XPS measurement. The composition of the residue is summarizes in Table2 and Figure1 illustrates single XPS spectra. For the two samples an additional broad phosphorus signal at between 137.0-131.0eV was detected and could be interpreted as an overlapping of peaks from phosphorus-oxygen species, such as metaphosphate($\approx$134.5eV), pyrophosphate($\approx$133eV) and phosphate($\approx$132.5eV). The carbon signal with a peak at 285eV identified the benzene hydrocarbon. The nitrogen peak at between 397eV and 403eV identified –CN and $-NH_2$ groups. Analysis of the phosphorus signal showed that different phosphorus species existed in PU/MIFR22.5 compared with in PU/MIFR10. The position of the peak between 131.5-137.5eV could be interpreted as an overlapping of peaks from phosphorus-oxygen species. The position of the shoulder between 129.0-132.0eV fits well with XPS features of phosphorus-carbon species ($\approx$131eV). Analogous to PU/MIFR10, the carbon spectrum identified the benzene hydrocarbon and carbon impurity. A shoulder peak between 283eV and 288eV indicated pure carbon char in good correspondence to the inhomogeneous black and grey color of the residue [4]. The nitrogen peak at between 397eV and 403eV identified –CN and $-NH_2$. Another nitrogen signal with a peak at 406eV indicated $-NO_2$.

Table 2. Chemical composition of the residue by XPS measurements.

Sample	Carbon (%)	Phosphorus (%)	Oxygen (%)	Nitrogen (%)	Calcium (%)
PU/MIFR10	59.72	14.24	23.85	2.19	
PU/MIFR22.5	58.31	19.66	15.69	4.05	2.29

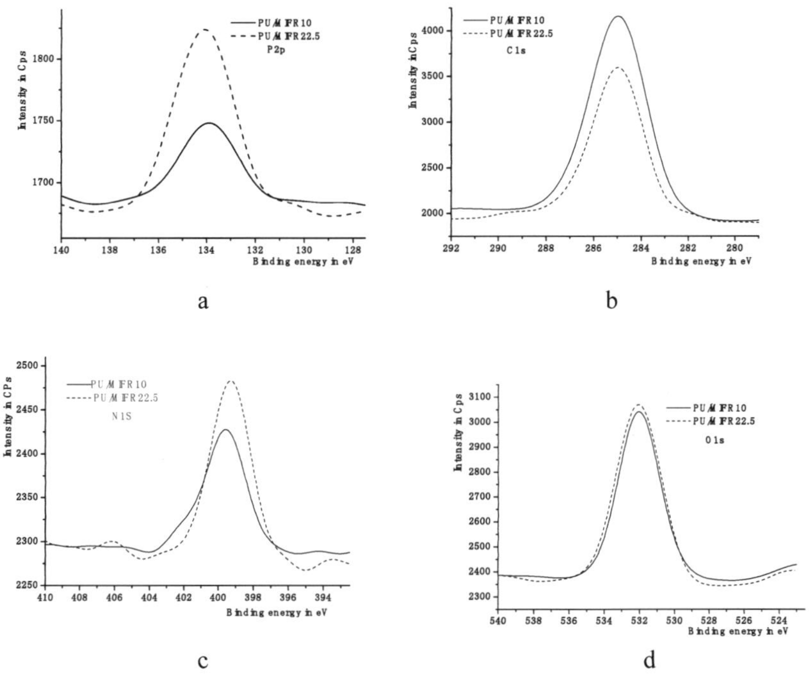

Fig 1. XPS spectra of residue char obtained from PU/MIFR10 and PU/MIFR22.5.

4. Conclusion

A surface-modified intumescent flame retardant (MIFR) was prepared by compounding ammonium polyphosphate, melamine and Calsium 3-hydroxy-2, 2-bis(hydroxymethyl) propyl phosphate with TX-10 surfactant. The carbon residue from the flame retardant rigid PU foams after UL-94 test was investigated with XPS. The XPS characterization of the residue composition and chemistry revealed that a suitable amount of MIFR into PU foams could promote formation of pure carbon char and P-C bond.

Acknowledgements

This work was supported by the Scientific Research and Technology Development Project of Guilin city and Guangxi Key Laboratory of Information Materials (Guilin University of Electronic Technology, China).

Reference

[1] Liping Li, Gang Wang and Chuigen Guo. Influence of intumescent flame retardant on thermal and flame retardancy of eutectic mixed paraffin/polypropylene form-stable phase change materials 2016 *Applied Energy* **162(15)** 428-34
[2] Ma HY, Tong LF, Xu ZB, Fang ZP, Jin YM and Lu FZ. A novel intumescent flame retardant: Synthesis and application in ABS copolymer 2007 *Polym Degrad Stab* **92 (4)** 720-26.
[3] Xia Y, Jiang XG, Li JF, Wang XH and Xu YY. Synergistic Effect of Montmorillonite and Intumescent Flame Retardant on Flame Retardance Enhancement of ABS 2007 *Polym-Plast. Technol* **46(3)** 227-32.
[4] Maria WP and Ryszard K.. The Thermal Characteristics of Different Intumescent Coatings 1999 *Fire Mater* **23** 33-43.

IOP Conf. Series: Materials Science and Engineering **544** (2019) 012004 doi:10.1088/1757-899X/544/1/012004

The theoretical studies on the adsorption and migration of hydrogen atom on the two-dimensional nanocrystal surface

L J Zhou[1], X X Yao[2] and J G Guo[2]

[1] Department of Mechanical Engineering, Tianjin University of Technology and Education, Tianjin 300222, China
[2] Tianjin Key Laboratory of Modern Engineering Mechanics, School of Mechanical Engineering, Tianjin University, Tianjin 300072, China

E-mail: zhoulj_tj@163.com

Abstract. An analytical model is presented to study the adsorption and migration of atoms/ions on the surface of two-dimensional nanocrystal. An atom/ion interacts with every atom of the two-dimensional nanocrystal through a pairwise potential which can be approximated by the Lennard-Jones (L-J) potential. Using the Fourier expansion of the interaction potential, the total interaction energy between the adsorption atom/ion and the two-dimensional nanocrystal is derived. Adsorption energy and adsorption stability of a hydrogen atom are analyzed on three special sites for three kinds of typical crystal surfaces. The interaction potential of the hydrogen atom migrating along specific paths on the surface of three kinds of typical nanocrystals is illustrated and analyzed.

1. Introduction

Two-dimensional nanocrystals have excellent mechanical, thermal and electrical properties, and they have potential applications in many fields. The adsorption and migration of atoms/ions on the surface of two-dimensional nanocrystal is significant in the fields of high sensitivity sensors, atomic-scale friction, rechargeable lithium-ion batteries, energy storage and drug delivery system (DDS) and so on, which has attracted much attention in recent years and some researches have been made [1-10]. However, there are still some unsolved scientific questions for its applications.

Amft *et al.* [1] made a systematic density functional study (DFT) on the adsorption of Cu, Ag and Au adatoms on pristine graphene. According to the differences of the total energies at the three adsorption sites, they found that the migration of Cu and Au took place along carbon-carbon bonds, while the Ag adatoms could move almost unrestrictedly on graphene sheet. By the same methods, Xian *et al.* [2] studied the potential energy variations and migrations of Si and C atoms on a single layer of graphene and between graphene sheets, and they found that Si atoms could move almost freely on graphene and between graphene layers, while C atoms had much larger migration barriers. Ji *et al.* [3] investigated the adsorption of lithium atom on graphene, and three different adsorption sites were taken into consideration. They found that the hollow site was found to be the most stable, followed by the bridge and top sites. What's more, Xia *et al.* [4] studied the adsorption, migration and desorption of hydrogen on the graphene surface by DFT method.

Apart from DFT method, Zhou *et al.* [5] studied the adsorption and migration of lithium on the graphene with divacancy and Stone-Wales defects using the first-principles calculations to understand the effect of point defects on the lithium adsorption on graphene, and the results showed that the presence of point defects could enhance the lithium adsorption on graphene. Tachikawa *et al.* [6]

studied the migration processes of the lithium-ion on a fluorinated graphene surface by means of a direct molecular orbital-molecular dynamics (MO-MD) method. The interaction of Mg atom with graphene was also researched by using MD simulations [7]. Uthaisar *et al.* [8] studied the influence of edge effects of graphene on the migration of lithium-ion on the graphene surface. Yu *et al.* [9, 10] made a theoretical investigation on adsorption and migration of atoms/ions on graphene surface based on an analytical continuous model, and they also discussed the influence of edge and size effects of graphene on the migration of atoms/ions.

In this work, a theoretical model will be presented to investigate the adsorption stabilities of atoms/ions on the surface of two-dimensional nanocrystal, and the total interaction energy between the adsorption atom/ion and two-dimensional nanocrystal will be derived. Moreover, adsorption energy and adsorption stability of a hydrogen atom will be analyzed on three special sites for three kinds of typical crystal surfaces. The interaction potential of a hydrogen atom moving along specific paths on the surface of three kinds of typical nanocrystals is illustrated and discussed.

2. Theoretical model

A two-dimensional nanocrystals with periodic lattice structure is assumed to be perfect, flat and infinite compared to atomic scale (as shown in Fig. 1).

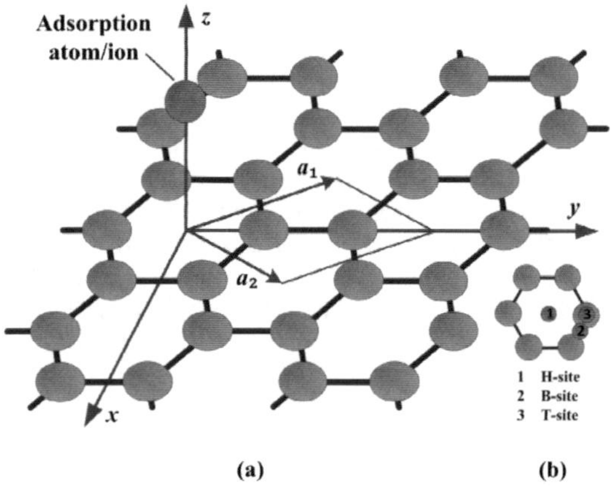

Figure 1. Schematics of (a) an adsorption atom/ion and a two-dimensional nanocrystal, (b) lattice structure with three particular adsorption sites: hollow (on top of a lattice, H-), top (on top of an atom, T-), and bridge (on top of a bond, B-).

An adsorption atom/ion is on top of two-dimensional nanocrystal. The interaction between the atom/ion and nanocrystal can be defined by van der Waals force. It is assumed that a 6-12 L-J potential is applied to determine the van der Waals interaction potential between the adsorption atom and each atom in the nanocrystal. What's more, due to the periodic surface lattice structure of nanocrystal, the total interaction potential u(r) between the adsorption atom and nanocrystal surface can be expressed via a Fourier series [9, 10].

$$u(r) \approx w_0(z) + \sum_{g>0} \sum_{k=1}^{q} w_g(z) \exp[i\boldsymbol{g} \cdot (\boldsymbol{m}_k + \boldsymbol{\tau})] \tag{1}$$

where,

$$w_0(z) = \frac{2\pi\varepsilon q}{A}\left(\frac{2\sigma^{12}}{5z^{10}} - \frac{\sigma^6}{z^4}\right) \text{ and } w_g(z) = \frac{2\pi\varepsilon}{A}\left[\frac{\sigma^{12}}{30}\left(\frac{g}{2z}\right)^5 K_5(gz) - 2\sigma^6\left(\frac{g}{2z}\right)^2 K_2(gz)\right] \tag{2}$$

where q is the number of the nanocrystal atoms in the unit cell and A the area of unit cell (as shown in Figure 1), ε and σ denote the potential well depth and the distance when the interaction potential is equal to zero, respectively, K_2 and K_5 are the modified Bessel function of the second kind, and g is dependent on the surface lattice structure of nanocrystal.

For nanocrystals with hexagonal, square and triangular lattice, the dimensionless total interaction potential between the adsorption atom/ion and the nanocrystal can be derived in the Cartesian coordinate system as shown in Figure 1, respectively.

$$u^*(r)=\lambda q\left(\frac{2\sigma^{*12}}{5z^{*10}}-\frac{\sigma^{*6}}{z^{*4}}\right)-\lambda(\alpha-\beta)\cdot\left[2\cos\left(\frac{2\pi}{\sqrt{3}}y^*\right)\cos(2\pi x^*)+\cos\left(\frac{4\pi}{\sqrt{3}}y^*\right)\right] \quad (3)$$

$$u^*(r)=\lambda q\left(\frac{2\sigma^{*12}}{5z^{*10}}-\frac{\sigma^{*6}}{z^{*4}}\right)-\lambda(\alpha-\beta)\cdot\left[\cos(2\pi x^*)+\cos\left(2\pi y^*\right)\right] \quad (4)$$

$$u^*(r)=\lambda q\left(\frac{2\sigma^{*12}}{5z^{*10}}-\frac{\sigma^{*6}}{z^{*4}}\right)+\lambda(\alpha-\beta)\cdot\left[2\cos\left(\frac{2\pi}{\sqrt{3}}x^*\right)\cos(2\pi y^*)+\cos\left(\frac{4\pi}{\sqrt{3}}x^*\right)\right] \quad (5)$$

where dimensionless parameters have following forms, $A^*=A/a^2$, $z^*=z/a$, $g^*=ga$, $\sigma^*=\sigma/a$, $x^*=x/a$, $y^*=y/a$, $u^*=u/\varepsilon$, $\lambda=2\pi/A^*$, $\beta=2\sigma^{*6}\left(g^*/2z^*\right)^2K_2(g^*z^*)$, $\alpha=\sigma^{*12}\left(g^*/2z^*\right)^5K_5(g^*z^*)/30$.

3. Results and discussions

The Eqs. (3)-(5) describe the interaction potentials between a single adsorption atom/ion and the nanocrystal surface with different lattice structure (hexagonal, square and triangular lattice), respectively. For the convenience of discussion, three typical two-dimensional nanocrystals (graphene, caesium chloride (CsCl) and molybdenum disulfide (MoS2)) are selected. Their lattice structures are hexagonal, square and triangular, respectively. It can be found from the Eqs. (3)-(5) that the values of the interaction potential are dependent on the position of the adsorption atom/ion on crystal surface and the parameters of L-J potential which is varied for different nanocrystal. A hydrogen atom (H atom) is chosen as the adsorption atom. The L-J potential parameters can be specified for H atom and the surface atoms of three typical two-dimensional nanocrystals. Then the interaction potentials between the H atom and two-dimensional nanocrystals can be calculated via the Eqs. (3)-(5) when it is at the different positions of crystal surface.

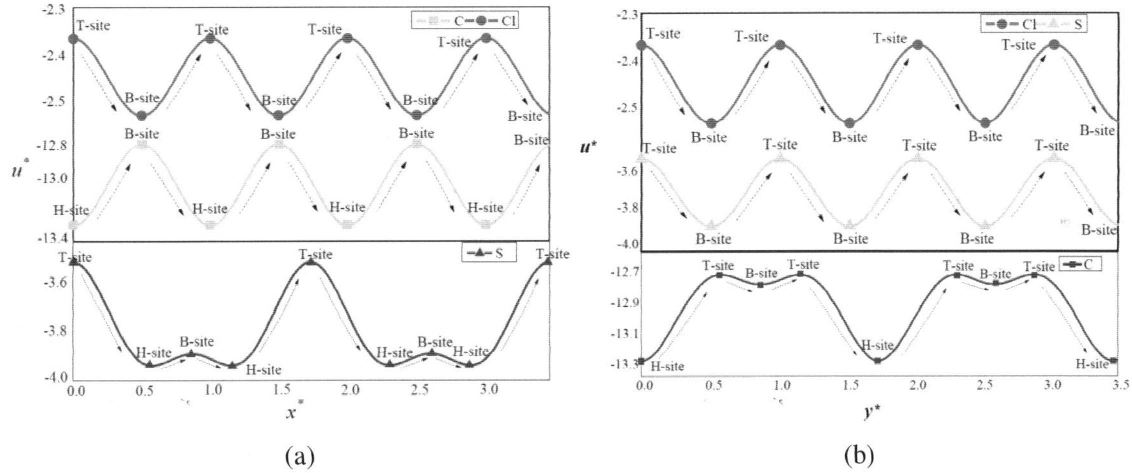

(a) (b)

Figure 2. The variations of interaction potential between H atom and three kinds of nanocrystals in the (a) x-direction and (b) y-direction, respectively.

The variations of interaction potential between the H atom and three two-dimensional nanocrystals are illustrated in the Figure 2 and 3 with respect to the positions in the x, y and z directions, respectively. It can be seen form Figure 2 that the potential curves are periodic in the x and y directions, and there are three periodic equilibrium positions, corresponding to three particular adsorption sites of H atom on crystal surface. They are hollow (on top of a lattice, H-), top (on top of an atom, T-), and bridge (on top of a bond, B-) sites (as shown in Figure 1(b)), respectively. At three adsorption sites, the equilibrium height and interaction potential at the equilibrium which can be called binding energy are different. Moreover, via the equilibrium equations in the z direction, the equilibrium height and binding energy at three adsorption sites can be calculated for three kinds of nanocrystals, respectively. For different nanocrystals, the equilibrium height and binding energy at the each adsorption site are different. However, for three nanocrystals, the equilibrium height and binding energy at the H-site are the lowest, followed by those at the B-site, and those at the T-site are the highest. Thus, H-site is the most stable adsorption site, and T-site is the most unstable. Figure 3 shows the variations of interaction potential of the H atom at the T-sites of three nanocrystals along the z direction. Although the variation trends are almost identical for the three nanocrystals, the equilibrium height and binding energy are also different.

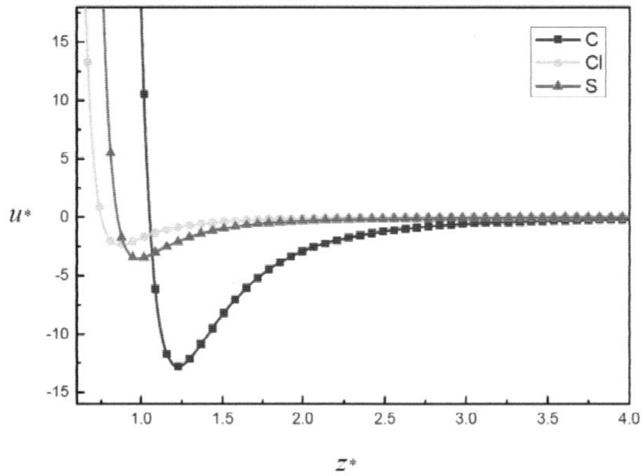

Figure 3. The variations of interaction potential of H atom and three kinds of nanocrystals at T-site along the z-direction

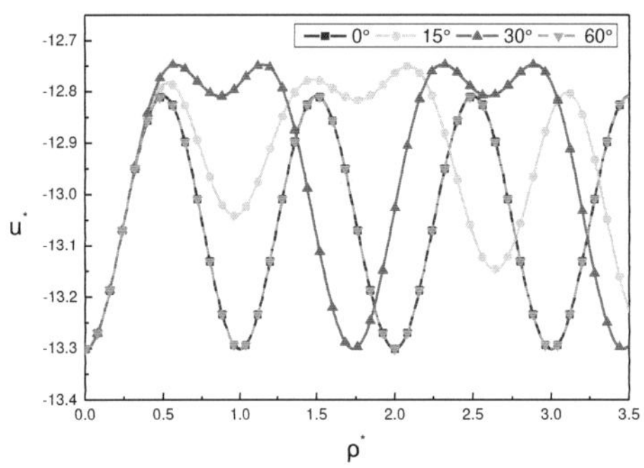

(a)

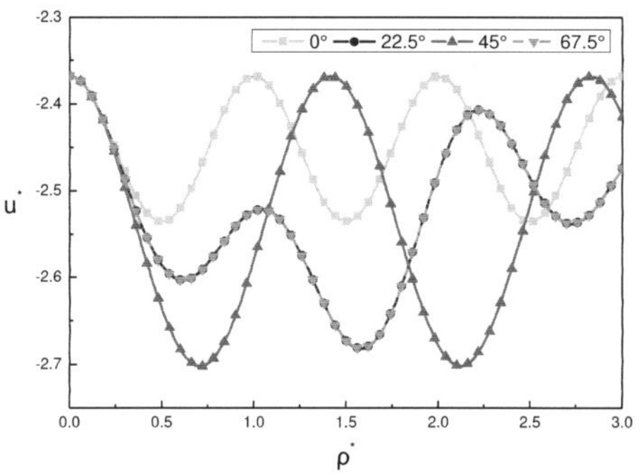

(b)

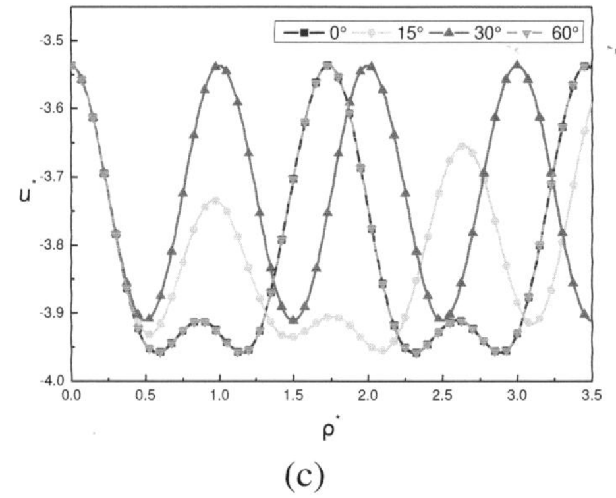

(c)

Figure 4. The dimensionless interaction potential is path-dependent and the path-dependence is periodic when a hydrogen atom migrates (a) along 0°, 15°, 30°, 60° angles from the x-axis on the graphene surface, (b) along 0°, 22.5°, 45°, 67.5° angles from the x-axis on the CsCl surface, (c) along 0°, 15°, 30°, 60° angles from the x-axis on the MoS_2 surface.

Since the binding energy of the H atom at three adsorption sites is different, the H atom has to get over an energy barrier when it migrates from an adsorption site to another one. The energy to drive the H atom is defined as migration energy. Obviously, the migration energy is dependent on the migration path. It is assumed that a hydrogen atom migrates along the specified paths, and its initial adsorption site is assumed to be H-site. The dimensionless interaction potentials between the H atom and three kinds of nanocrystals for different migration directions are illustrated in the Figure 4. It can be found from these figures that the periods of interaction potential are varied for different migration directions, and the minimum migration energy are also varied., i.e., the migration of the H atom on nanocrystal surface is path-dependent. What's more, the path-dependence exhibits periodic. It can be found along the same lattice orientation, such as 0° and 60° angles for graphene, the variations of interaction potential are same.

4. Conclusion

An analytical model is presented to investigate the interaction potential when an atom/ion is adsorbed and migrates on the surface of two-dimensional nanocrystals, the mechanism of adsorption and migration are discussed. The equilibrium height and binding energy at three particular adsorption sites are illustrated and analyzed for three kinds of typical nanocrystals, and the adsorption stability are discussed. The order from low to high for both equilibrium height and binding energy is H-site, B-site and T-site, and H-site is the most stable adsorption site. When an adsorption atom/ion migrates on the surface of nanocrystal, there are energy barriers it has to get over, and the migration is path-dependent, and the path-dependence exhibits periodic.

References

[1] Amft M, Lebegue S, Eriksson O and Skorodumova NV 2011 Adsorption of Cu, Ag, and Au atoms on graphene including van der Waals interactions *J. Phys.: Condens. Matter* **23** pp 395001

[2] Ji Z, Contreras-Torres F F, Jalbout A F and Ramírez-Treviño A 2013 Surface diffusion and coverage effect of Li atom on graphene as studied by several density functional theory methods *Appl. Surf. Sci.* **285** pp 846-52

[3] Xian L and Chou M Y 2012 Diffusion of Si and C atoms on and between graphene layers *J. Phys. D: Appl. Phys.* **45** pp 455309

[4] Zhou L, Hou Z and Wu L 2012 First-principles study of lithium adsorption and diffusion on graphene with point defects *J. Phys. Chem. C* **116** 21780-7

[5] Xia Y, Li Z and Kreuzer H J 2011 Adsorption, diffusion and desorption of hydrogen on graphene *Surf. Sci.* **605** pp 70-3

[6] Tachikawa H 2008 A direct molecular orbital-molecular dynamics study on the diffusion of the li ion on a fluorinated graphene surface *J. Phys. Chem. C* **112** pp 10193-9

[7] Tachikawa H, Iyama T and Kawabata H 2009 MD simulation of the interaction of magnesium with graphene *Thin Solid Films* **518** pp 877-9

[8] Uthaisar C and Barone V 2010 Edge effects on the characteristics of Li diffusion in graphene *Nano Lett.* **10** pp 2838-42

[9] Yu Y Z, Guo J G and Zhou L J 2016 Theoretical investigation on the adsorption and diffusion of lithium-ion on and between graphene layers with size and defect effects *Adsorpt. Sci. Technol.* **34** pp 212-26

[10] Yu Y Z, Guo J G and Kang Y L 2015 An analytical model for adsorption and diffusion of atoms/ions on graphene surface *J. Nanomater.* **2015** pp 382474

Acknowledgments

The work is supported by the National Natural Science Foundation of China (Grant no. 11502167) and the Development Fund of Tianjin University of Technology and Education (Grant no. KJ14-52).

IOP Conf. Series: Materials Science and Engineering **544** (2019) 012005 doi:10.1088/1757-899X/544/1/012005

Analysis of the Key Structures of RV Reducer Based on Finite Element Method

Lei Song[1,2], Liang Shunke[3], Zhenxing Zheng[2], Feixin Chen[2]

1 South China Robotics Innovation Research Institute, Foshan 528315, China
2 School of Mechatronic Engineering, Guangdong Polytechnic Normal University, Guangzhou 510665, China
3 Guangzhou College of South China University of Technology, Guangzhou 510800, China
E-mail: s_forrest@126.com

Abstract. In this paper, key structures of the RV reducer were calculated by finite element method respectively, including the stress and deformation of cycloidal-pin wheel and eccentric shaft. The results show that the contact stress and equivalent stress of cycloidal-pin wheel and eccentric shaft are far less than the strength of materials. It means the strength is not the main consideration of design and manufacture. The stiffness of eccentric shaft is less than the cycloidal-pin wheel, which had a great influence on the transmission accuracy of the machine. In order to improve the rigidity so as to improve the transmission accuracy of gear reducer machine, the modification methods and modification should be chosen to design by the deformation.

1. Introduction

Cycloidal-pin wheel reducer was invented by L. Braren[1], which was improved by the Nabtesco company. Then the RV (Rotate Vector) gear reducer was invented. The RV reducer is composed of planetary gear transmission and cycloidal-pin wheel transmission, which is widely used in industrial robots, machine tools, assembly equipment, handling equipment and other fields. It has more meshing gear at the same time and is small, lightweight, high stiffness, high overload etc. Due to small clearance, rotational vibration and inertia, it has good speed performance, which can realize smooth running and get the correct location accuracy[2-3].

In theory, cycloid gear and pin gear are completely tangent and there is no gap between each other. However, because of the manufacturing error and installation error, in order to install the cycloid gear and pin gear smoothly, the teeth profile have been modified, which causing the clearance and idle and decreasing the torsional stiffness and transmission accuracy of the RV reducer. In view of the transmission accuracy problem of RV reducer, domestic and international scholars have much in-depth research. Naren Kumar[4]have established the single-stage elastic torsional flexibility of cycloidal-pin wheel transmission and torque transfer efficiency by static experiment and finite element analysis, which helps to realize the dynamic optimization design of cycloidal-pin wheel transmission and standardization. Blanche[5-6] studied the single rotary precision of the cycloid-pin gear planetary reducer with pure geometry method, including torsional vibration causing by backlash and fluctuating ratio of speed, deduced several kinds of calculation formulas about machining error and backlash caused by installation error, and used the CAD method to find out the relationship between the tooth gap, the fluctuating ratio of speed and the torsional vibration. Although this method is rigorous and the conclusion has guidance, it only considers the single stage and single cycloid gear transmission accuracy, which does not apply to the RV reducer composing of planetary gear transmission, cycloidal-pin wheel transmission and several cranks. Ta-Shi Lai[7-8] used the finite element method to calculate the contact force and contact deformation about the theory of cycloid tooth profile. There is large error about the calculation result because the actual tooth profile was modified.

In view of the present study, the structure of the RV reducer was analyzed through the finite element method in this paper, especially about the cycloidal-pin wheel and the eccentric shaft. Base on the analysis of stress and deformation calculation results, the influence of accuracy has been clear between the key components and the whole RV reducer machine, providing technical reference for precision gear reducer design, manufacture and application.

2. Finite Element simulation analysis of the RV reducer

2.1 The construction of a finite element model

The RV-40E is a kind of RV reducer widely used. In this paper, it was selected to be the research object. The 3D model of the RV-40E was established in the Pro/E. Then import the model into ANSYS15.0. Because of the irregular structure components, the RV reducer cannot be divided by using hexahedron mesh. Therefore, the tetrahedron elements were used to divide the model. Finally, the model was divided into 1055948 nodes and 1055948 elements. The material parameters of components of the RV reducer are shown in Table 1.

Table 1 The material parameters of components of the RV reducer

	Material	Young's modulus(GPa)	Poisson's ratio	Yield limit(MPa)
Cycloid gear	20CrMo	211	0.292	700
pin	20CrMo	211	0.292	700
Pin wheel	20CrMo	211	0.292	700
Eccentric shaft	GCr15	219	0.3	518
Eccentric bearing	20CrMo	211	0.292	700
others	20CrMo	211	0.292	700

The output flange works as the output. So attach the fixed constraint to the pin wheel. Pin and pin wheel were attached cylindrical constraint, restraining it rotate only along the circumferential direction. The input and output flanges connected through the pin hole. Then attached constraints on them to restrain they rotate only along the center axis of rotation. Cycloidal gear and pin, cycloid gear and eccentric bearing, eccentric bearing and eccentric shaft, planetary gear and sun gear, planetary gear and eccentric shaft are all set as the frictional contact relationship between each other. The friction coefficient was set as 0.15. In this paper, the selected type of RV reducer's output torque was 572 Nm. Converting to the input shaft, it is 4.77 Nm. Converting to the cycloid wheel, it is 14.3 Nm. Converting to the eccentric shaft, it is 7.15 Nm. The established finite element model is shown in Figure 1.

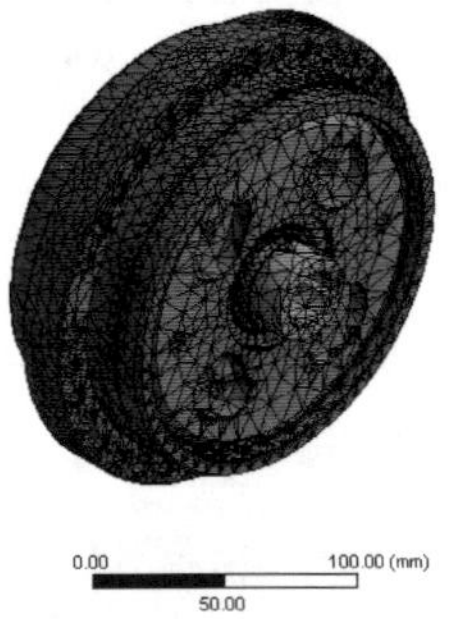

Figure 1 The finite element model of RV reducer

2.2 Cycloidal-pin wheel and eccentric shaft mechanism simulation Results

The first level planetary gear mechanism in the RV reducer is the traditional involute gear. There are much research about involute gear has been relatively mature. The related research results can be directly applied to the analysis of the first transmission mechanism, which was no longer to be analyzed in this paper. The output flange's stiffness is large. Some literature has already carried out comparatively detailed analysis. So it was no longer to be analyzed in this paper too. Submit the finite element model of the RV reducer to the ANSYS solver to calculate and get the contact stress nephogram, equivalent stress nephogram and equivalent deformation nephogram of the RV reducer cycloid-pin wheel as shown in Figure 2, Figure 3 and Figure 4. The maximum contact stress between cycloid wheel and pin is 166.22 MPa according to the contact stress nephogram and each cycloidal gear only has half area existing contact stress. The maximum equivalent stress is 279.48 MPa between cycloidal gear and pin, which is bigger than the contact stress of 113.26 MPa. It shows that the cycloidal-pin wheel mechanism loads the main driving force. The contact stress and internal stress all have important influence to cycloidal-pin wheel mechanism. The maximum contact stress and equivalent stress of cycloidal-pin wheel mechanism is far less than the yield limit. So the strength is enough. The maximum deformation of cycloid gear is 0.0365 mm according to Fig.4, while the deformation of pin is approximate to zero. It is mainly due to the pin and the pin gear connected through the pin hole and the large stiffness. The deformation is not easy to happen when it would be loaded.

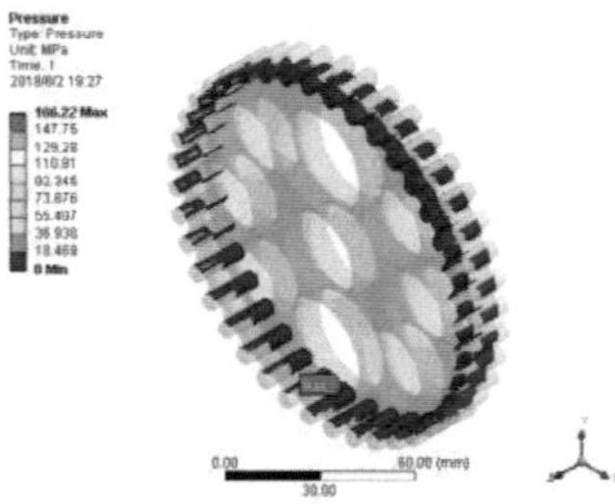

Figure 2 The contact stress nephogram of cycloid-pin wheel

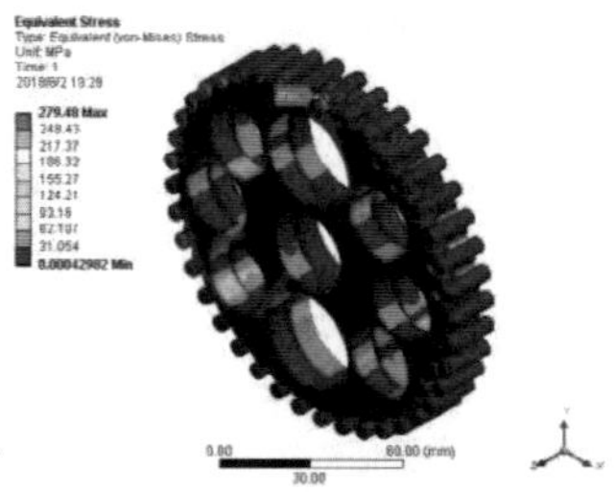

Figure 3 The equivalent stress nephogram of cycloid-pin wheel

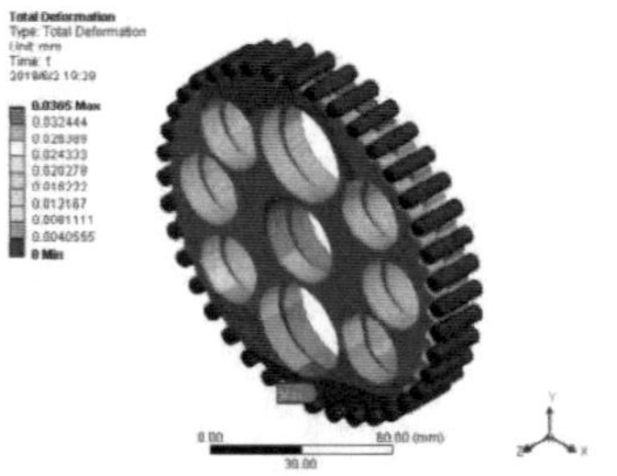

Figure 4 The equivalent deformation nephogram of cycloid-pin wheel

The contact stress, equivalent stress and deformation calculation results of the eccentric shaft mechanism are shown as Figure5-7 respectively. The maximum contact stress of the RV reducer eccentric shaft mechanism is 113.99 MPa and the maximum equivalent stress is 51.46 MPa. The contact stress is bigger than the equivalent stress, meaning the eccentric shaft mechanism is mainly composed of contact stress. Both the maximum contact stress and the maximum equivalent stress of the eccentric shaft mechanism are far less than the yield strength. So the strength is enough. The deformation calculation results of RV reducer eccentric shaft are as shown in Fig.7. The deformation in mechanism is mainly distributed on the eccentric shaft, which is similar to the stress distribution. The maximum deformation is 0.045 mm. The deformation also presents obvious annular, meaning the torsion deformation is the main deformation form of eccentric shaft. The larger deformation areas are mainly distributed on the two eccentric shaft sections. Under the condition of rated load, the eccentric shaft deformation is larger. It will be one of the main factors affecting the whole transmission stiffness.

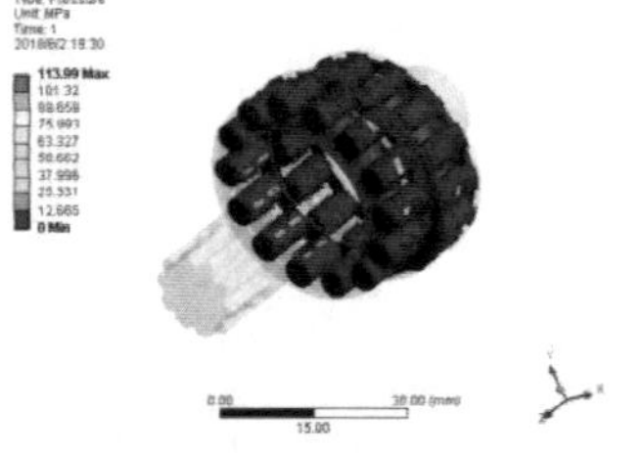

Fig.5 The contact stress nephogram of eccentric shaft mechanism

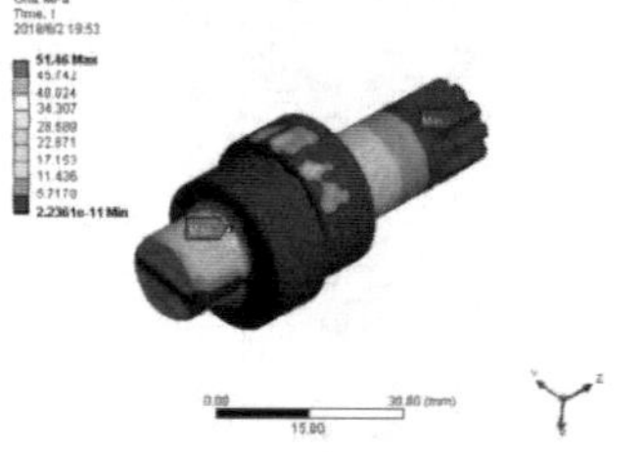

(a) The equivalent stress nephogram of eccentric shaft

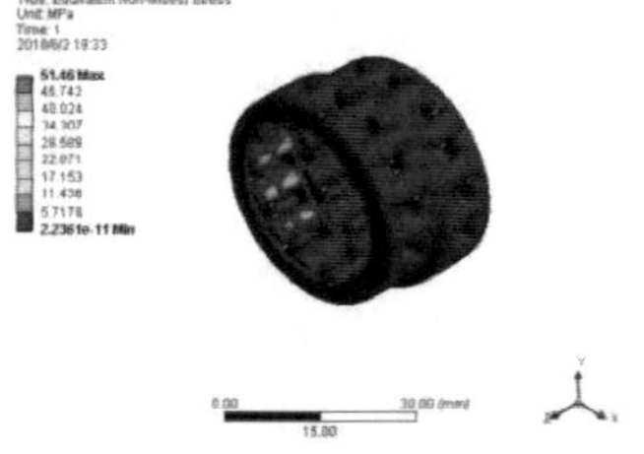

(b) The equivalent stress nephogram of eccentric shaft bearing

Fig.6 The equivalent stress calculation results of eccentric shaft mechanism

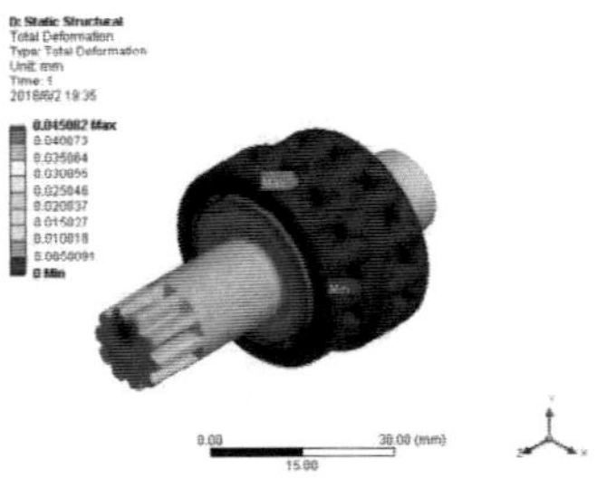

(a) The deformation nephogram of the eccentric shaft mechanism

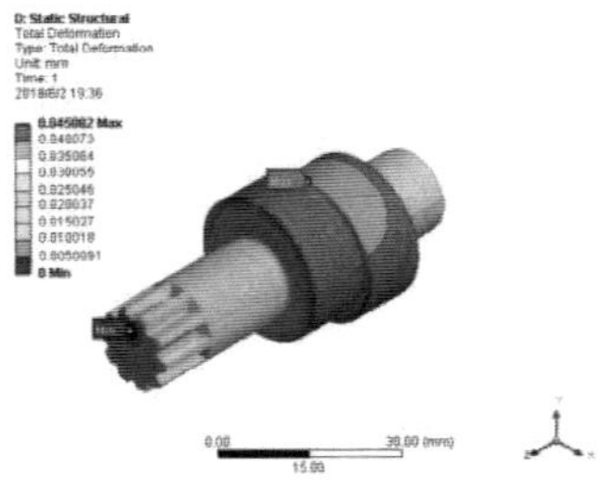

(b) The deformation nephogram of eccentric shaft

Fig.7 The deformation calculation results of eccentric shaft mechanism

2.3 The analysis of calculation results

The finite element calculation results show that the cycloidal-pin wheel mechanism transmit the main force. The stress status of eccentric shaft and eccentric bearing mechanism is mainly contact stress. The internal stress of components is smaller and far less than the corresponding material strength. It suggests that the strength is not the main consideration of the RV reducer components in design and manufacture. In order to ensure the accuracy in the process of the RV reducer running, stiffness of the components is the key problem in the process of design and manufacture. The above calculation results show that the deformation of cycloidal-pin wheel mechanism is less than that of the eccentric shaft mechanism. In order to ensure the transmission precision of the reducer machine, the stiffness of eccentric shaft mechanism should be ensure firstly in the process of design and manufacture. The deformation of the eccentric shaft and cycloid-pin wheel are roughly along the radial direction from inside to outside increasing gradually. In the process of manufacture, the materials and manufacture processes of the eccentric shaft should be rational chosen to improve its stiffness. In the process of the cycloid gear tooth profile modification, the reasonable modification method and modification should be chosen.

3. Conclusion

In this paper, the finite element method were used to analyze the key structures of the RV reducer, such as the cycloidal-pin wheel mechanism and the eccentric shaft mechanism. The results are shown as follow:

(1) The stress status of cycloidal-pin wheel mechanism is mainly equivalent stress, while that of the eccentric shaft mechanism is mainly contact stress. The internal stress of the components is smaller and far less than the corresponding material strength. It suggests that the strength is not the consideration of RV reducer components in design and manufacture.

(2) The stiffness of eccentric shaft is smaller than the cycloidal-pin wheel, which has a great influence on the transmission accuracy of the whole machine. Therefore, the materials and manufacturing processes should be reasonable selected to improve its stiffness. As for the cycloid gear, it needs to choose the modification methods and modification according to the deformation reasonably so as to improve its rigidity. Finally, the transmission accuracy of reducer will be improved.

4. References

[1] He Weidong, Shan Lijun 2016 Status and Development of RV Reduce *Journal of DaLian JiaoTong University* 37 13.

[2] Mirko Blagojevic, Nenad Marjanovic, Zorica Djordjevic, et al. 2011 A New Design of a Two-Stage Cycloidal Speed Reducer *Journal of Mechanical Design* 133 085001

[3] Chiu-Fan Hsieh. 2014 Dynamics Analysis of Cycloidal Speed Reducers With Pinwheel and Nonpinwheel Designs *Journal of Mechanical Design* 136 091008

[4] Naren Kumar, Vladis Kosse, Adekunle Oloyede 2016 A new method to estimate effective elastic torsional compliance of single-stage Cycloidal drives *Mech. Mach. Theory* 105 185.

[5] J. G. Blanche, D. C. H. Yang. 1989 Cycloid Drives With Machining Tolerances *ASME J. Mech. Transm. Automn Des* III 337.

[6] D. C. H. Yang, J. G. Blanche. 1990 Design and Application Guidences for Cycloid Drives with Machining Tolerances *Mech. Mach. Theory* 25 487.

[7]Hong-Sen Yan, Ta-Shi Lai. 2002 Geometry design of an elementary planetary gear train with cylindrical tooth-profiles *Mech. Mach. Theory* 37 757.

[8] Ta-Shi Lai. 2006 Design and machining of the epicycloid planet of cycloid drives *Int. J. Adv. Manuf. Technol* 28 665.

Acknowledgements
This work was financially supported by the Guangdong Natural Science Foundation (2018A030313509) and Major Special Project of Guangdong Provincial Science and Technology Plan (2017B090910007).

IOP Conf. Series: Materials Science and Engineering **544** (2019) 012006 doi:10.1088/1757-899X/544/1/012006

Dynamic Properties for Laminate Composites in the Marine Environment

Luo Zhong and Zhu Libao
Department of Naval Architecture Engineering, Naval University of Engineering, Wuhan, China
E-mail: luo_zhong@163.com

Abstract. Dynamic properties of high strain rate and absorbing energy for laminate composites in the marine environment were studied in this paper. The high strain rate dynamic compressive tests were carried out on Split Hopkinson Pressure Bar(SHPB) experimental facility at strain rate 1800/s, 1530/s, 1230/s and 950/s. The front surface failure modes were Analyzed by optical micrographs. A fractured fibres mode was discussed in high rate compression. The stress-strain rate curves showed that, the specimen dynamic properties were better for the samples immersed in marine environment for 12 days. The energy absorption properties were test by ballistic trajectory experiment. Experimental results and penetration faces showed that, the ballistic trajectory properties with 60days of hygrothermal environment was better.

1. Introduction

Laminate composite has excellent impact resistance, low density and stable chemical performance, and is one of the ideal materials for ship protection structures. Under the marine environment, the protective laminate composite may be affected by environmental loads such as water, temperature and salt, resulting in the degradation of its performance. Therefore, it is of certain significance to study the effects of marine environment on the mechanical properties of the ultra high molecular weight polyethylene (UHMWPE) composites. The high-speed impact behavior of composites is influenced by the dynamic compressive properties of the material[1-2]. Extensive research has been conducted on the quasi-static and dynamic compressive behavior of composites and the split Hopkinson pressure bar (SHPB) was usually used to study the dynamic properties[3-7]. Song Bo[8] investigated the compressive behavior of the S-2 glass/SC15 composite. The peak stress along the through-thickness direction was found to be much higher than those along the in-plane direction. Khubab Shaker[9] investigated the high strain rate compressive properties of Dyneema H62, Dyneema H5T and UHMWPE/aramid hybrid composite and peak stress was found to increase with higher strain rate. The goal of this paper is to investigate the effects of marine environment on the dynamic compressive properties of laminate composites.

2. Experimental work

2.1. SHPB experiment

Dynamic properties were test for the UHMWPE laminate composite in out-of-plane direction. The high strain rate dynamic compressive tests were carried out on a split Hopkinson pressure bar (SHPB) experimental facility. Schematic of the SHPB equipment is shown in Figure 1.

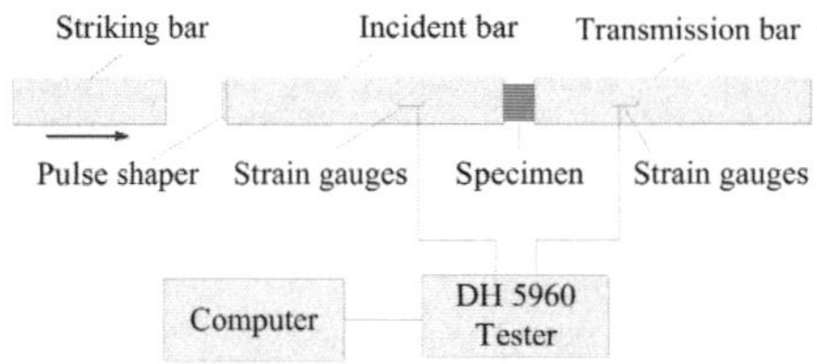

Figure 1. Schematic of the SHPB equipment

The material of the SHPB facility bars are $60Si_2Mn$ and had density of 7850 kg/m^3, elastic modulus of 206GPa. The 20 mm-diameter striking, incident and transmission bar had lengths of 300, 2000 and 1500mm, respectively. Bar wave velocity $c=\sqrt{E/\rho}$ = 5123m/s was verified by standardized experiments. Strain gauges, with sensitivity of 2.13 were attached to the incident and transmission bar. Distances of the strain gauges to the specimen/bar contact surfaces were approximately 1000 and 745 mm for gauges on the incident bar and transmission bar, respectively. The strain signals were measured using a DH5960 dynamic strain tester (sampling rate 20MHz) by half bridge connection to the strain gauges. During the test, the speed of the bullet was controlled by adjusting the driving pressure in the range of 0.2 - 0.45MPa. Annealed copper disks with thickness of 1mm and diameters ranging from 8 – 14mm were used to smooth and control the incident pulse. Molybdenum disulfide was used at the specimen/bar contact surface to reduce the friction. At least three replicates were conducted by SHPB for each driving pressure.

The SHPB experiment is based on the no-dispersion one-dimensional stress wave theory and dynamic stress equilibrium assumption in the specimen. A parameter R was considered to evaluate the stress equilibrium.

$$R(t)=\left|\frac{\Delta\sigma(t)}{\sigma_{avg}(t)}\right|=2\left|\frac{F_1-F_2}{F_1+F_2}\right| \quad (1)$$

where, F_1, F_2 represent the specimen axial force history in the incident bar/specimen end and transmission bar/specimen end, respectively, and can be calculate according to Eq.(2)

$$\begin{aligned} F_1 &= AE(\varepsilon_i+\varepsilon_r) \\ F_2 &= AE\varepsilon_t \end{aligned} \quad (2)$$

Where, A represent the equilibrium parameter; E represent the modulus of elasticity; ε_i, ε_r represent the strain.

Figure 2 shows a typical pulse received from the incident bar and transmission bar and the equilibrium parameter calculated according to Eq.(2), indicating that the specimen was in dynamic stress equilibrium from 5-130s. Thus, the strain rate, engineering strain and engineering stress of the specimen can be calculated according to Eq.(3)

$$\begin{aligned} \varepsilon(t) &= \frac{2c_0}{L}\varepsilon_r(t) \\ \varepsilon(t) &= \frac{2c_0}{L}\int_0^t \varepsilon_r(t)d\tau \\ \sigma(t) &= \frac{S_B E}{S_s}\varepsilon_t(t) \end{aligned} \quad (3)$$

where, c_0 is the longitudinal wave velocity in the bar, S_B, E are the cross sectional area and elastic modulus of the bar, L and S_S are the length and cross sectional area of the specimen.

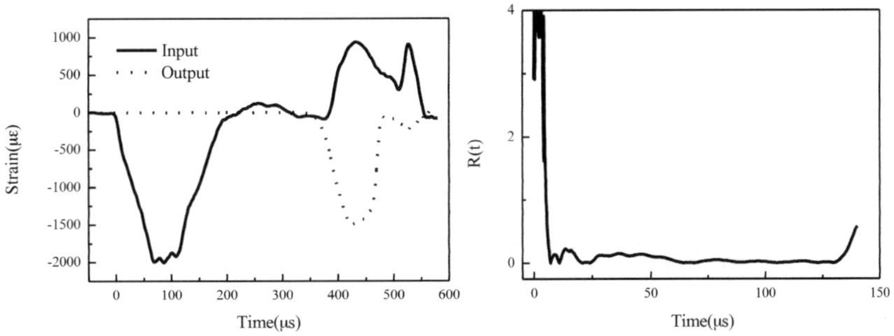

(a) (b)

Figure 2. (a) A typical pulse received from the incident bar and transmission bar (b) Force equilibrium parameter in the specimen

2.2. Ballistic trajectory experiment

The ballistic trajectory experiment of UHMWPE composite was done , on dry samples and samples subjected to hygrothermal aging for 30 and 60 days. The bullets were eradiated by ballistic trajectory gun with 14.5mm caliber, which was drive by powder. A laser equipment was used to measure the velocity of bullets in front and behind of the drone. The material of bullets was 45# steel with quenching, which was enwrapped by aluminum with 13.5g fragment simulated projectiles. The size of composite target shell was 300×300×10mm, which were fixed on shelves. The properties of resist bullets of composite were described by Eq.:

$$E = \frac{1}{2}m\left(v_i^2 - v_r^2\right)/\rho \tag{4}$$

Where m was the mass of bullets, v_i and v_r were the initial and final velocity, ρ was the plane density.

3. Results and discussions

3.1. SHPB properties

High strain rate compression tests were conducted on the virgin dry samples, 12days and 24 days hygrothermal exposed samples, respectively. Stress-strain curves of the 12 days hygrothermal treated samples at strain rate 950/s, 1230/s, 1530/s and 1800/s are shown in Figure 3. The stress increases with strain until reaching the maximum stress and strain softening occurs at large strain. Stress plateau was found around the maximum stress for the 1530/s and 1800/s curves. The strain softening and stress plateau may indicate the compression-induced damage such as matrix cracking and delamination[8]. For dynamic compression of UHMWPE composite, the high speed striking induced damages and the inner defects cannot be extended immediately, resulting in the lager loading capacity and striking energy absorption capacity. Thus, maximum stress increases with higher strain rate.

The failure modes of UHMWPE composite under high strain rate compression were quite different from the static compression failure modes. Figure 3 shows the failure patterns for the 12 and 24 days hygrothermal treated specimens at different strain rate. Damage appears on the specimen surface near the incident bar and we reference it as "front surface".

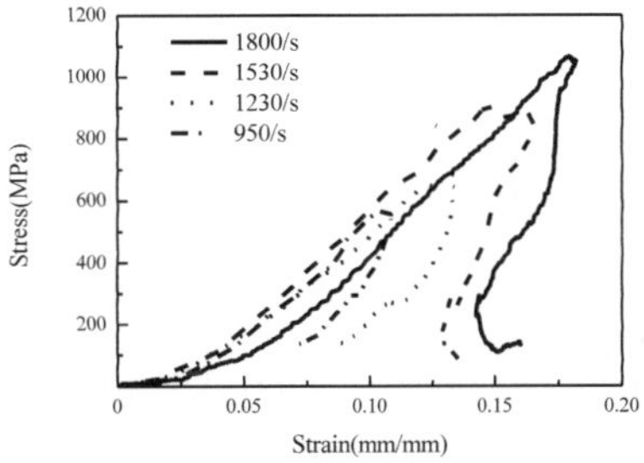

Figure 3. Stress-strain curves of the 12 days hygrothermal treated specimens

For the 12 days treated specimens, (1) fibres were knocked off on the front surface at 1230/s, as shown in Figure 4(a); (2) several layers delaminated and the outer region of the delaminated layers warped outwards the surface at strain rate 1530/s, as shown in Figure 4(b); (3) the periphery delaminated fibres were carbonized at strain rate 1800/s as shown in Figure 4(c). There were no obvious differences in failure modes for the 12 and 24 days treated specimens at strain rate 1800/s, as shown in Figure 4(c) and (d).

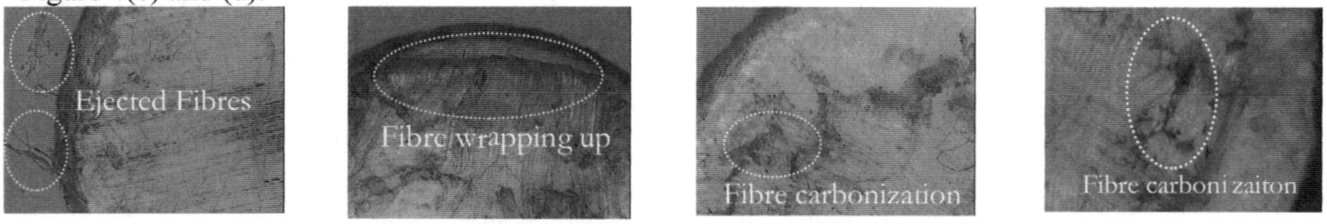

(a) (b) (c) (d)

Figure 4. Optical micrographs of 12 days wet specimens at strain rate (a) 1230/s (b) 1530/s (c) 1800/s and the 24 days wet specimens at strain rate 1800/s(d)

During the high strain rate compression, a small amount of fractured fibres were ejected out from the front delaminated plies, as shown in Figure 5. High strain rate compression is a transient process during which a large amount of energy is transferred to the specimen, resulting in a temperature rise in the specimens[8]. The outer delaminated fibres will be fractured and carbonized at higher strain rate, resulting in different damage patterns comparing with the quasi-static specimens.

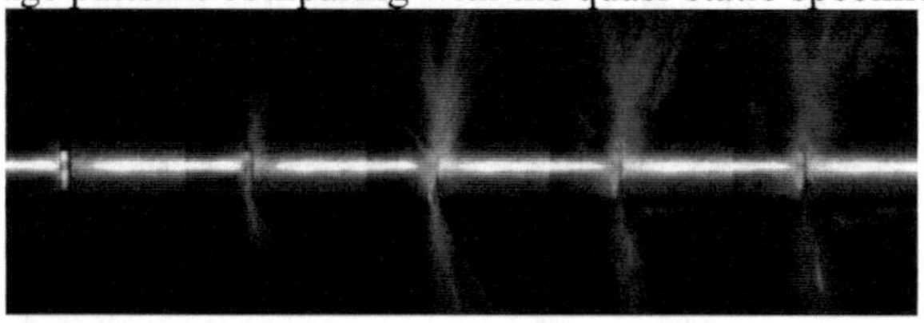

0.167ms 0.334ms 0.501ms 0.668ms 0.835ms

Figure 5. The high speed images of specimens at strain rate 1530/s

The compressive stress-strain curves of both dry and wet specimens at dynamic strain rate around 1500/s and 1800/s are shown in Figure 6. Hygrothermal treatment improves the maximum stress (wet 12 days) and then decreases the maximum stress (wet 24 days) while the corresponding strain of the maximum stress were decreased firstly and then increased. The strain softening effects was degraded firstly and then upgraded as well.

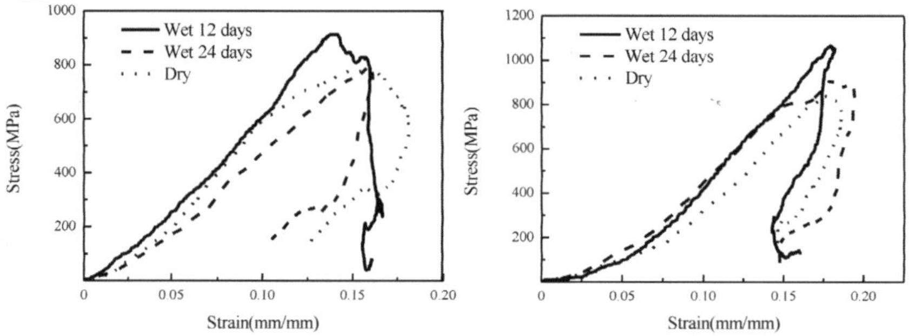

Figure 6. (a)Stress-strain curves at strain rate1500/s (b) Stress-strain curves at strain rate1800/s

Figure 7 shows the maximum stress-strain rate curves of both the dry and wet specimens. Twelve days hygrothermal treatment increased the maximum stress and the improvement was declined in 24 days treatment especially at high strain rate above 1500/s. The fit curves of the maximum stress as a function of strain rate are as follows:

$$y = 941.30552 - 1518.5046 * 0.99851^x \quad \text{Dry}$$
$$y = 3187.38925 - 3325.51795 * 0.99975^x \quad \text{Wet 12 days} \tag{5}$$
$$y = 1084.55607 - 2300.84035 * 0.99861^x \quad \text{Wet 24 days}$$

Matrix plays an important role in the out-of-plane compression and plasticization of the matrix improves the dynamic compressive properties while fibre/matrix interface degradation and voids expansion degrades the compressive properties. For the 12 days hygrothermal treated specimens, the effects of plasticization dominate the compressive properties and thus improve the maximum stresses. However, under the treatment of moisture and long-term high temperature, the degradation of fibre/matrix interface and expansion of the inner voids dominates the dynamic compressive properties for 24 days treated specimens. The dynamic compressive properties thus degraded while still stronger than the virgin dry specimens.

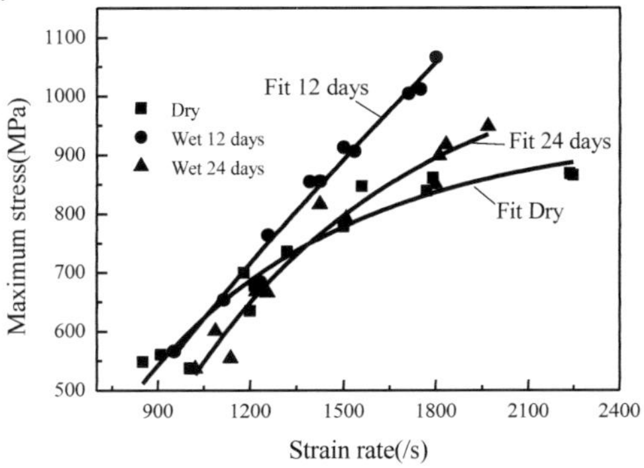

Figure 7. Effects of hygrothermal treatment on the maximum stress of out-of-plain compression

3.2. *Ballistic trajectory properties*

The results of ballistic trajectory experiment were showed by table 1, which were dry and hygrothermal aging in 30 days and 60 days. The results show that average of absorbing energy were 186.3, 109.5 and182.7 J·m^2/kg. The uptrend of absorbing energy is initially increasing and then decreasing, and the dates of hygrothermal aging in 60 days were close to the dry composite.

Table 1 Experimental results of ballistic trajectory

Status	Designation	Thickness (mm)	Area density (kg/ m^2)	Mass of bullet (g)	Initial velocity (m/s)	Residual velocity (m/s)	E (J·m^2/kg)
Dry	O-1	10.0	10.2	13.4	546.1	No Perforation	> 195.9
	O-2	10.1	10.3	13.4	554.0	188.3	176.6
Hygrothermal 30 days	HT30-1	10.0	9.8	13.5	487.9	288.0	107.1
	HT30-2	10.2	9.8	13.5	503.6	310.6	108.7
	HT30-3	10.1	9.8	13.4	409.6	60.7	112.6
Hygrothermal 60 days	HT60-1	10.1	9.8	13.4	497.2	No Perforation	> 169.6
	HT60-2	10.0	9.8	13.4	512.1	No Perforation	> 179.9
	HT60-3	10.0	9.8	13.4	616.2	300.4	198.6

The process of bullet ballistic trajectory include steps of compress upsetting, cut compress, and drawing distortion. Front faces and back faces of the HT30-2 and HT60-2 composite plates after penetration show in Figure 8. The Initial velocity of the two case were close, which front face of the composite were cut compress destroyed. The ballistic trajectory properties was depressed in 30 days of hygrothermal environment, which fiber was destroyed after margin drawing distortion. The ballistic trajectory properties with 60days of hgrothermal environment was better, which can not penetrate by bullets.

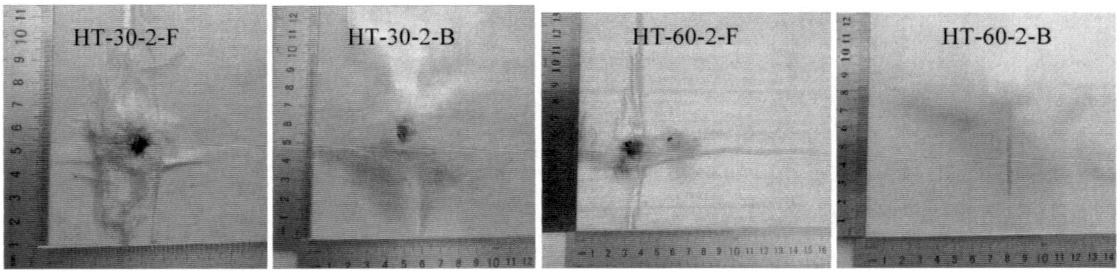

Figure 8. Front faces and back faces of the HT30-2 and HT60-2 composite plates after penetration

4. Conclusions

The conclusion can be summarized as follows, which may be useful for engineering application.

(1) The Laminate UHMWPE composites have a relatively large strain rate sensitivity in the out-of-plane direction under dynamic compression. The peak dynamic compressive stress of the original dry samples increases by nearly 2 times as strain rates increasing from 950/s to 1800/s. Hygrothermal treatment has no effects on the elastic modules for dynamic compression.

(2) The ballistic trajectory properties was decreased in 30 days, and increased in 60 days, which is closed to the dry composite.

5. References

[1] O'Masta, M. R., Crayton, D. H., Deshpande, V. S., & Wadley, H. N. G. (2016). Indentation of polyethylene laminates by a flat-bottomed cylindrical punch. *Composites Part A Applied Science & Manufacturing, 80*, 138-147.

[2] O'Masta, M. R., Deshpande, V. S., & Wadley, H. N. G. (2014). Mechanisms of projectile penetration in dyneema ®; encapsulated aluminum structures. *International Journal of Impact Engineering, 74*, 16-35.

[3] Attwood, J. P., Khaderi, S. N., Karthikeyan, K., Fleck, N. A., O'Masta, M. R., & Wadley, H. N. G., et al. (2014). The out-of-plane compressive response of dyneema® composites. *Journal of the Mechanics & Physics of Solids, 70*(1), 200-226.

[4] O'Masta, M. R., Deshpande, V. S., & Wadley, H. N. G. (2015). Defect controlled transverse compressive strength of polyethylene fiber laminates. *International Journal of Solids & Structures, 52*, 130-149.

[5] Luo, H., Roy, S., & Lu, H. (2012). Dynamic compressive behavior of unidirectional IM7/5250-4 laminate after thermal oxidation. *Composites Science & Technology, 72*(2), 159-166.

[6] Pankow, M., Salvi, A., Waas, A. M., Yen, C. F., & Ghiorse, S. (2011). Split Hopkinson pressure bar testing of 3D woven composites. *Composites Science and Technology*, 71(9), 1196-1208.

[7] Shi, B. B., Sun, Y., Chen, L., & Li, J. L. (2010). Energy absorption of ultra-high molecular weight polyethylene fiber-reinforced laminates at high strain rates. *Applied Mechanics & Materials, 34-35*, 1532-1535.

[8] Song, B., Chen, W., & Weerasooriya, T. (2003). Quasi-static and dynamic compressive behaviors of a S-2 glass/SC15 composite. Journal of Composite Materials, 37(19), 1723-1743.

[9] Shaker, K., Jabbar, A., Karahan, M., Karahan, N., & Nawab, Y. (2016). Study of dynamic compressive behavior of aramid and UHMWPE composites using split Hopkinson pressure bar. *Journal of Composite Materials, 51*(1).

IOP Conf. Series: Materials Science and Engineering **544** (2019) 012007 doi:10.1088/1757-899X/544/1/012007

Quasi-static Characters for UHMWPE Composites after Hygrothermal Aging

Luo Zhong and Zhu Libao
Department of Naval Architecture Engineering, Naval University of Engineering, Wuhan, China
luo_zhong@163.com

Abstract. This paper studied the Moisture absorption behaviors, DMTA temperature spectrum, Infra-red spectrogram, Quasi-static characters of the dry and hygrothermal UHMWPE composites. The material was UHMWPE fiber reinforced polyurethane matrix composite manufactured by hot press method, which was firstly kept in a heating oven, and then put in a beaker of distilled 70°C water bath for 24 days and 60 days. The results show that there were two plateaus for the moisture absorption curve of the UHMWPE composites. The coupled effects of moisture and temperature reduce the composite glass transition temperature(Tg) by matrix plasticization and degrades the fiber/matrix interface. The storage modulus and loss modulus of dry and 24 days hygrothermal were gradually decreased in the initiatory calefaction, which drop sharply in Tg. The tensile strength was lossing in the beginning, and comeback to the initial at last in the hygrothermal environment. The quasi-static compressive strength is reduced by about 8% after 24 days treatment. Hygrothermal treatment has no effects on the quasi-static characters.

1. Introduction

Due to the need of protection and the lightweight consideration, the fiber reinforced composite material is gradually applied to the ship protection structures to resist the penetration of high-speed fragments. UHMWPE composite has excellent impact resistance, low density and stable chemical performance, and is one of the ideal materials for ship protection structures. Under the hygrothermal environment, the protective UHMWPE composite may be affected by environmental loads such as water, temperature and salt, resulting in the degradation of its performance. Hygrothermal environment may lead to the plasticization of composite matrix and degrade the fiber/matrix interface, resulting in a decrease of the quasi-static mechanical properties, especially the matrix-dominant mechanical properties. However, researchers have come to different conclusions about the influence of moisture on the static characters of composites. Woldesenbet's study shows that the dynamic compressive peak stress of the graphite/epoxy composite was firstly increased and then decreased with the increase of moisture absorption[1]. However, HAQUE[2] and SNWOSU[3] found that moisture absorption will reduce the dynamic compressive peak stress of composites. The effects of hygrothermal treatment on the compressive behavior of composites needs further research[4-8]. This paper studied the effects of hygrothermal treatment on quasi-static characters of UHMWPE composites. Quasi-static characters conducted by a MTS machine, and DMTA test, and infra-red spectrogram experiment. The moisture absorption and its effects on the composite compressive properties were studied in detail.

2. Experimental work

IOP Conf. Series: Materials Science and Engineering **544** (2019) 012007 doi:10.1088/1757-899X/544/1/012007

2.1. Material description

The material studied in this paper was UHMWPE fiber reinforced polyurethane matrix composite manufactured by hot press method. UHMWPE fibers, coated with polyurethane, were stacked by 0/90° and then kept at pressure of 15MPa, temperature below 120°C for approximately 30 minutes in a hot press machine to fabricate the plates measuring 300×300mm with a thickness of 4 or 10mm. The single ply of the plates had a thickness of about 0.13mm and the mass fraction of the polyurethane matrix was ~15%. Figure 1 shows the surface of the as-manufactured UHMWPE composite, proving a good compatibility of the UHMWPE fibres and the PU matrix. The composite panels were cut into quasi-static compressive specimens of 10 × 10 × 10mm cubes, and dynamic compressive specimens of Φ × h = 12×4mm cylinders by water-jet method. For the dynamic compressive specimens, diameter Φ was the in-plane direction while height h was the through thickness direction. The basic material description is listed in Table1.

Table 1. Material description of the [0/90]n UHMWPE/polyurethane composites

Composite Density g/cm^3	Resin Mass Fraction	Ply Thickness / mm	T_g / °C	Tensile Strength / MPa	Tension Modulus / GPa	Short-beam Shear Strength / MPa
0.95	0.1-0.15	0.13±0.02	118	541	28.5	6.65

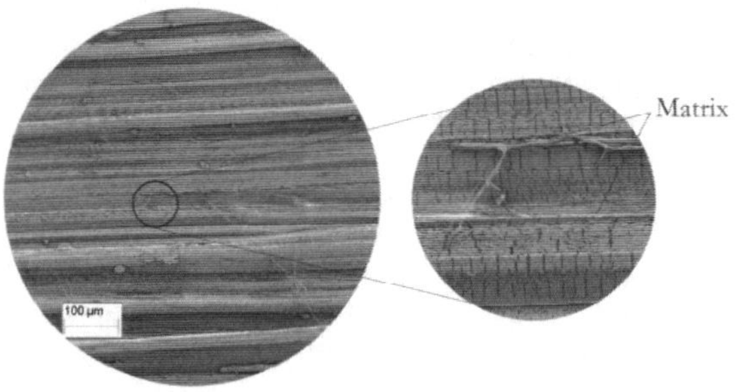

Figure 1. SEM images of the UHMWPE composite surface

2.2. Specimens of Hygrothermal Aging

Specimens were firstly kept in a heating oven to reach the engineering dry condition until the daily mass loss less than 0.02%. Heating temperature was set at 50°C which is below the composite glass transition temperature T_g = 118°C. Subsequently, the dry specimens were put in a beaker of distilled water and the beaker was at kept in a 70°C water bath for 24 days. Five specimens were numbered and taken out regularly for weighing using an OHAUS analytical balance. Hot, wet specimens were cooled for a short time to reach the room temperature and surface water was removed before weighing. Moisture absorption rate of the composite under hygrothermal environment is calculated according to formula (1)

$$W_t = \frac{M_t - M_0}{M_0} \times 100 \quad (1)$$

where, W_t, M_t and M_0 represent the moisture absorption rate, weight of the hygrothermal treated specimens and weight of the dry specimens, respectively. Average moisture absorption rate of the numbered five specimens was regarded as the moisture absorption rate of the UHMWPE composite.

2.3. DMTA Tests

The DMTA tests of dry specimens and hygrothermal specimens with 15 days, 30days, and 60days were measured by METTLER TOLEDO 1/236 machine. The specimens size were 40×4×2mm with long, wide, and thickness. Every experiment is done for 3 specimens. The method of experiment was three point bending test of frequency 1Hz, loading 5N, temperature 30-140 °C, speed of heating 3K/min.

2.4. *Infra-red spectrogram and SEM Tests*

Infra-red spectrogram test were done by Nicolet IS5 spectrum machine. The specimens consist of dry, hygrothermal 30 days, and hygrothermal 60 days. The method of test was ATR, which scan scope was 400-4000cm^{-1}. The SEM test of fracture area for drawing specimens were done by Zeiss field emitter electron microscope SUPRA™ 55.

2.5. *Quasi-static Experiments*

The drawing and interceded cutting tests were conducted. The specimens consist of dry, hygrothermal 4 days, 8 days, 15 days, 30 days and 60 days. A test method of drawing was chosen from reference [8]. The specimens were dumbbell shape, which was 240 × 4mm, and the middle wide was 4mm. The specimens were connected with tongs by M4 bolt, which solved the specimens delaminating by traditional methods. The schematic of the tension specimen and test system were showed in Figure 2. A test method of cutting test was choosing from JC/T 773-2010, which was short beam test method. The specimens were 30×8×4mm. The span of loading was 20mm. The test machine was MTS E45, whose loading speed was 2mm/min. Each test case was carrier out for 4 specimens.

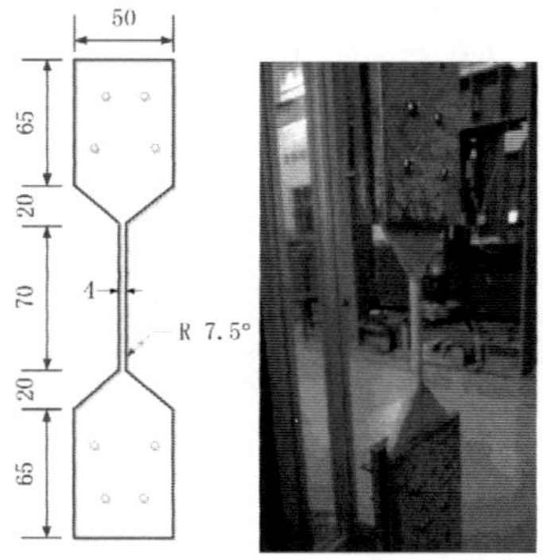

Figure 2. Schematic of the tensile specimen and test system

3. Results and discussions

3.1. *Moisture absorption*

Moisture absorption rate of the hygrothermal treated UHMWPE material is shown in 0. The moisture absorption rate rises proportionally to the square root of the time initially and then reaches two plateaus of 1.92% and 2.47% on day 6 and day 12, respectively. The moisture absorption process may be explained as follows. Moisture firstly diffused into the composite matrix and occupied the inner voids, reaching the first plateau. The inner voids and cracks expanded under the long time treatment of moisture and high temperature, leading to more moisture absorption and then reaching the second plateau.

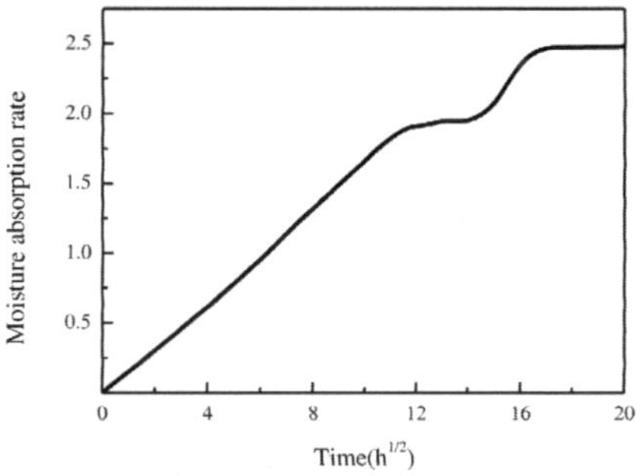

Figure 3. Moisture absorption curve of the hygrothermal treated specimens

The moisture absorption property of UHMWPE/PU was showed in Figure 4, which was hygrothermal in 70±3°C, 95±5% R.H. The moisture absorption rate was faster in the initiatory phase. The case was the water molecule attaching the sample surface, and pervade in PU. The pervasion process of water in PU was slowness. The absorption rate gradually turn slowly, and mitigation ultimately. UHMWPE/PU moisture absorption rate was low, because the UHMWPE fiber was non-polarity material, which absorption rate blow 0.01%[4], and the content of PU was lesser, which showed that the influence was little to whole composite sample.

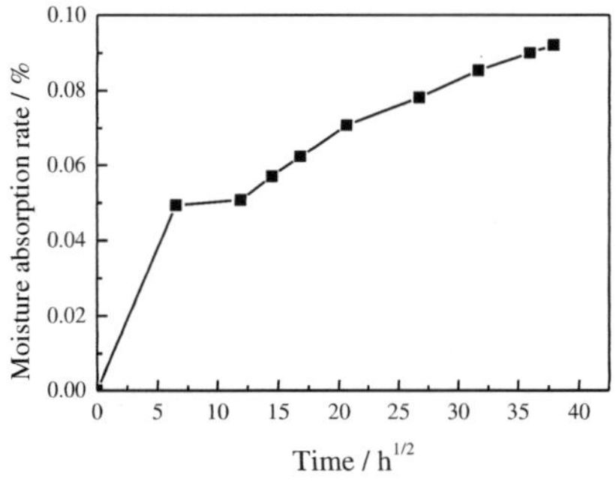

Figure 4. Moisture absorption rate curve of the UHMWPE/PU composite

3.2. DMTA

The DMTA temperature spectrum of the dry UHMWPE/PU composite showed in Figure 5. The storage modulus (*E'*) and loss modulus (*E''*) were gradully decreased in the initiatory calefaction, which fastly reduce in T_g. The loss factor (tan δ) gradully hoist, and form a pinnacle in T_g, which suddenly surge after T_g. If the first pinnacle of the composite was the T_g , the T_g of dry sample was 120.5°C.

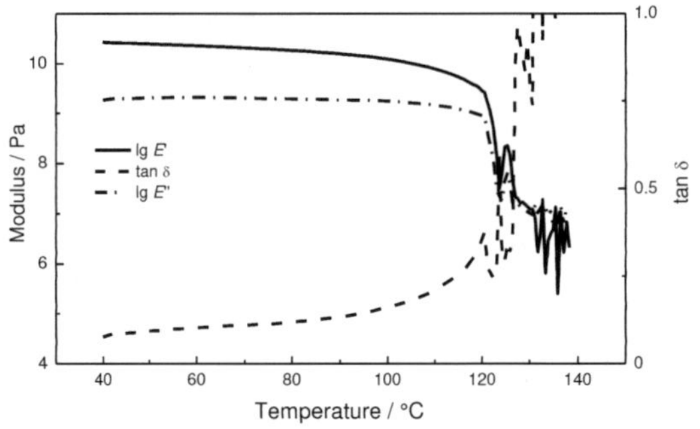

Figure 5. DMTA temperature spectrum of the dry UHMWPE/PU composite

The storage modulus, and loss modulus, and tan δ curves of the UHMWPE/PU composite with dry and in different hygrothermal times sample were showed in Figure 6. The modulus didn't deal with logarithm, which showed the change of storage modulus with different hygrothermal time. The storage modulus firstly reduced along with the increasing of hygrothermal time, then rising between the tempeture 40-100°C. The storage modulus curves were superposition of dry and hygrothermal with 60 days. The trend of loss modulus is similar to the storage modulus. The T_g of UHMWPE/PU were 121.4°C, 120.4°C, and 120.3°C, which hygrothermal with 15, 30, and 60 days.

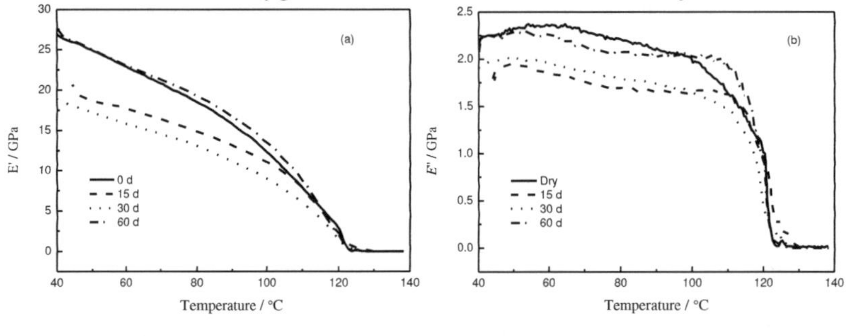

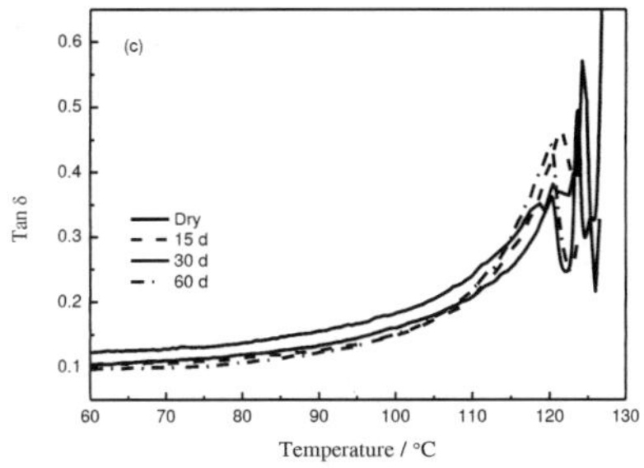

Figure 6. (a) Storage modulus, (b)loss modulus and (c) tan δ curves of the UHMWPE/PU composite in hygrothermal environment with time

The DMTA tests results of the dry and 24 days hygrothermal treated UHMWPE composites are shown in 0. Hygrothermal treatment seems to decrease the storage modulus(E') and to reduce the glass transition temperature T_g from 118°C to about 115°C (the first peak of tan Delta). When the temperature exceeds T_g, the storage modulus and the loss modulus drop sharply, leading to the irregular change of the tan Delta. The decrease of T_g may indicate the plasticization of the specimen matrix because plasticization usually degrades the glass transition temperature[7-8]. Plasticization also leads to lower buckling stress of composite materials and this is consistent with the quasi-static compression results described later.

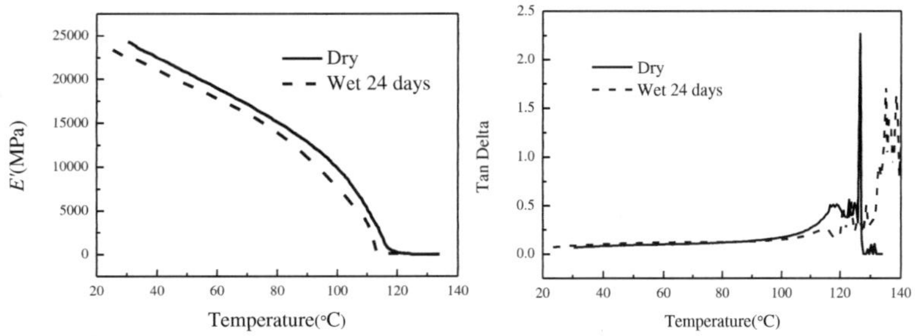

Figure 7. DMTA tests results of the dry 24 days wet specimens

3.3. Infra-red spectrogram

The Infra-red spectrograms of UHMWPE/PU composite at different aging time were showed in Figure8. There were not new diagnostic peak in IR spectra after hygrothermal enviroment. The location and relatively size of peak were the same with the dry samples. These showed thate there were not new matter engender between the process of hygrothermal.

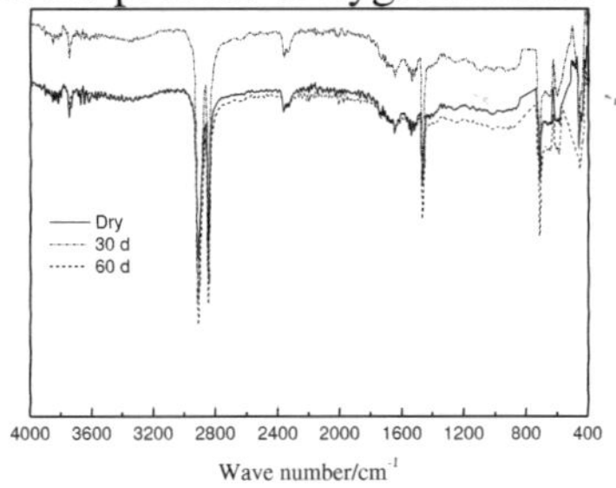

Figure 8. Comparison of IR spectra of UHMWPE/PU composite at different aging time

3.4. Quasi-static characters

The SEM photographs of the fractured the UHMWPE / PU composite was showed in Figure 9. The UHMWPE fiber extend to rupture after neck shrink, which produced the nicer extend property. The PU attaching in the face of the fiber was little, because of the min content of PU, and non- polarity of fiber.

The variation curves of tensile strength and short-beam strength of UHMWPE/PU composite in hygrothermal environment with time were showed in Figure 10. The initial tensile strength was 541.5 MPa. The tensile strength was losing in the beginning, and comeback to the initial at last in the hygrothermal environment. The reason maybe the hygrothermal environment reinforce the tensile strength. The short-beam strength was decreased following the hygrothermal time. The reserve rate of short-beam strength was 69.7% after 60 days hygrothermal.

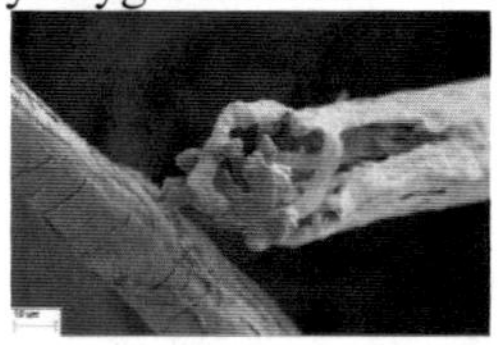

Figure 9. SEM photographs of the fractured the UHMWPE / PU composite

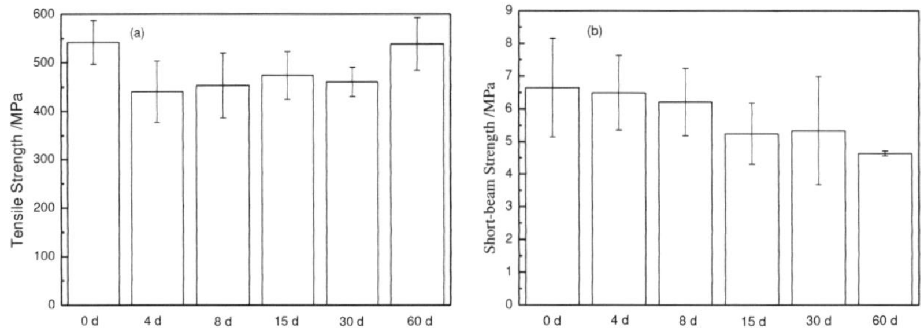

Figure 10. Variation curves of (a)tensile strength and (b)short-beam strength of UHMWPE/PU composite in hygrothermal environment with time

The out-of-plain quasi-static compressions at strain rate 0.001/s and 0.01/s were conducted on the virgin dry samples and 24 days hygrothermal treated samples. Hygrothermal treatment decreased the compressive strength by about 8%, as depicted in Figure 11 (a). The typical compressive stress-strain curves at strain rate 0.001/s are shown in Figure 11 (b). Stress had initially increased nonlinearly with strain before 0.15 and then increased linearly until a catastrophic failure occurred with a very loud burst sound and fractured plies ejecting from the edges of the samples. Fibres extruding out transverse to the local fibre direction can be seen over the side faces of the samples, as shown in Figure 12.

Ravichandran have studied the out-of-plane compression behaviour of Dyneema UHMWPE composites. The existence of shear-lag zone was proved at the periphery of specimens. Lower matrix shear strength, and void-like defects resulting from missing groups of fibres, could degrade the out-of-plane compressive strength[6]. The degradation effects of the hygrothermal treatment on compressive strength in our tests thus can be explained as follows. Water absorption usually causes plasticization of the composite matrix, inconsistent swell of matrix and fibres, and expansion of the inner voids. The lower matrix quasi-static shear strength caused by plasticization thus decreases the compressive strength. Fibre/matrix interface degradation and inner voids expansion caused water absorption have the same effects as lower the matrix shear strength.

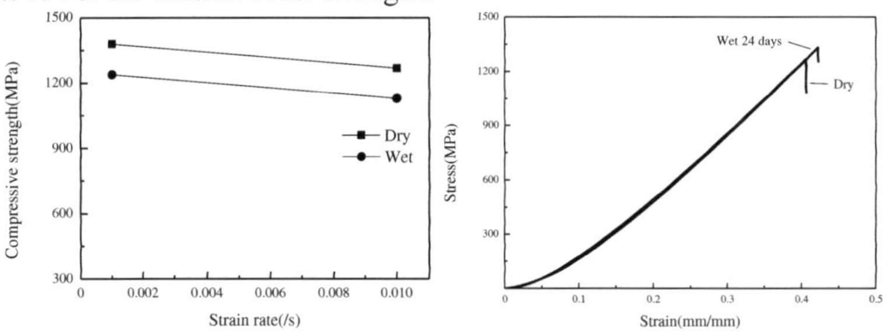

(a) (b)

Figure 11. (a) Quasi-static strength and (b) stress-strain curves of dry/wet composite

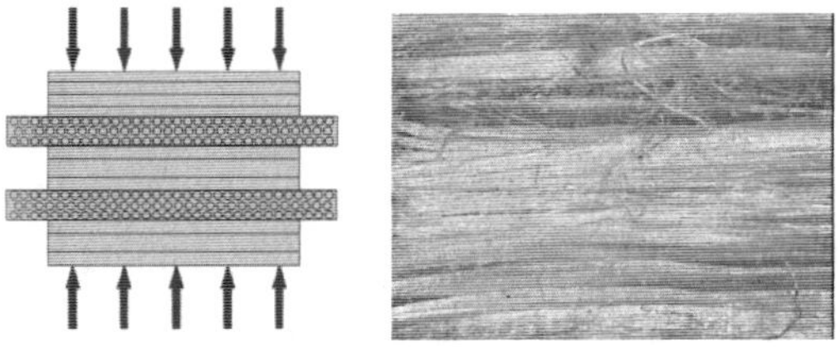

Figure 12. Optical micrograph of the quasi-static compression specimen

4. Conclusions

In this paper, we investigated the quasi-static characters of hygrothermal treatment UHMWPE/PU composites. The conclusion can be summarized as follows:

(1) There were two plateaus for the moisture absorption curve of the UHMWPE composites. The coupled effects of moisture and temperature reduces the composite T_g by matrix plasticization and degrades the fiber/matrix interface.

(2) The storage modulus and loss modulus of dry and 24 days hygrothermal were gradully decreased in the initiatory calefaction, which drop sharply in T_g.

(3) There were not new diagnostic peak in IR spectra after hygrothermal enviroment.

(4) The quasi-static compressive strength is decreased by about 8% after 24 days treatment.

5. References

[1] Woldesenbet, E., Gupta, N., & Vinson, J. R. (2002). Determination of moisture effects on impact properties of composite materials. *Journal of Materials Science, 37*(13), 2693-2698.

[2] Haque, A., & Hossain, M. K. (2003). Effects of moisture and temperature on high strain rate behavior of s2-glass-vinyl ester woven composites. *Journal of Composite Materials, 37*(7), 627-647.

[3] Wosu, S. N., Hui, D., & Daniel, L. (2012). Hygrothermal effects on the dynamic compressive properties of graphite/epoxy composite material. *Composites Part B Engineering, 43*(3), 841-855.

[4] Frew, D. J., Forrestal, M. J., & Chen, W. (2001). A split hopkinson pressure bar technique to determine compressive stress-strain data for rock materials. *Experimental Mechanics, 41*(1), 40-46.

[5] Frew, D. J., Forrestal, M. J., & Chen, W. (2005). Pulse shaping techniques for testing brittle materials with a split hopkinson pressure bar. *Experimental Mechanics, 45*(2), 186-195.

[6] Ravichandran, G., & Subhash, G. (1994). Critical appraisal of limiting strain rates for compression testing of ceramics in a split hopkinson pressure bar. *Journal of the American Ceramic Society, 77*(1), 263-267.

[7] Musto, P., Galizia, M., Scherillo, G., & Mensitieri, G. (2014). *Water Sorption Thermodynamics in Polymer Matrices. Durability of Composites in a Marine Environment.* Springer Netherlands.

[8] Colin, X., & Verdu, J. (2014). Humid ageing of organic matrix composites. *Solid Mechanics & Its Applications, 208*, 47-11.

IOP Conf. Series: Materials Science and Engineering **544** (2019) 012008 doi:10.1088/1757-899X/544/1/012008

Influence of additives on hydration heat of mass concrete

Y Li[1], M Li[1,2], H Chen[1]

1. College of Architecture and Civil Engineering, Beijing University of Technology, 100124, China

liyue@bjut.edu.cn

Abstract. In this paper, the development of early hydration heat and the performance of large temperature difference of mass concrete structures are studied. Five mass concrete samples with two strength grades and five mixing ratios were made. The size of samples is 1000mm*1000mm*1400mm. Temperature sensors are embedded in the component while the components are constructed. Temperature of early hydration heat temperature rise and instant heating on one side are tested by the sensors. It is found that the early hydration heat temperature rise of self-made crack resistance concrete with internal curing agent or shrinkage reducing agent is lower than that of common commercial concrete. The results of single-side instantaneous temperature rise test show that the temperature changes least on the heating surface of concrete and most in the middle position.

1. Introduction

Due to the large amount of hydration heat generated during hydration of cement, cracks will occur in the process[1, 2]. Therefore, in order to ensure the safety and durability of mass concrete structure[3], it is particularly important to test and analyze the temperature changes during the design and construction stages[4]. For concrete, besides water cement ratio, cement dosage and mineral admixture[5, 6], admixture also has certain influence on the hydration heat of concrete. In the early stage of hydration, the superplasticizer can inhibit the hydration of cement[7, 8]. At the same time, the expansion agent, shrinkage reducing agent and internal curing agent can also reduce the hydration heat of cement in varying degrees[9].

It can be seen from the above that most of the research on the hydration heat of concrete is still focused on the water cement ratio, cement dosage, mineral admixtures and other fields. The existing research on the influence of admixtures on the hydration heat of cement is mostly confined to the level of the hydration heat of cement. At present, there is little research on the effect of admixtures on the hydration heat of mass concrete. This paper makes experimental measurements and theoretical

[2] Corresponding author: M. Li, E-mail:limin.bjut@foxmail.com

calculations on the temperature changes of large-scale concrete members with different proportions, so as to study the development process of early hydration heat of mass concrete structures. In addition, by analyzing the temperature gradient in the single-side instantaneous heating test, the large temperature difference resistance of concrete members with different proportions is studied.

2. Test conditions

2.1 Mix ratio and materials

The cement is P.O42.5R ordinary Portland cement with a density of 3100kg/m^3. The fly ash is F class II grade. The slag powder is S95 grade high performance slag powder. The gradation of gravel is 5-20mm, the sand is II grade medium sand. The superplasticizer is JK-5 naphthalene series superplasticizer with a 23% water reduction rate. Shrinkage reducing agent is SBT-SRAI concrete shrinkage reducing agent. Internal curing agent is a kind of composite polymer water-absorbing and water-retaining composite material made by China Academy of Railway Sciences. The mix proportions are shown in table 1.

Table 1. Mix proportions (Kg/m^3)

Mix type	Cement	Fly ash	Sand	Gravel	Water	Super plasticizer	Shrinkage reducing agent	Internal curing agent
C50-S	280	88	582	1237	144	6	0	0
C40-N	280	88	582	1237	180	4	0	32
C50-N	280	88	582	1237	144	6	0	32
C40-J	280	88	582	1237	180	4	6	0
C50-J	280	88	582	1237	144	6	6	0

2.2 Design of concrete specimen

In order to simulate the actual situation and test the internal temperature of concrete, rectangular members with length of 1000mm, width of 1000mm and height of 1400 mm were made. Components of this size basically conform to the definition range of mass concrete size in 《Specification for mix proportion design of ordinary concrete (JGJ55-2011, J 64-2011) 》.

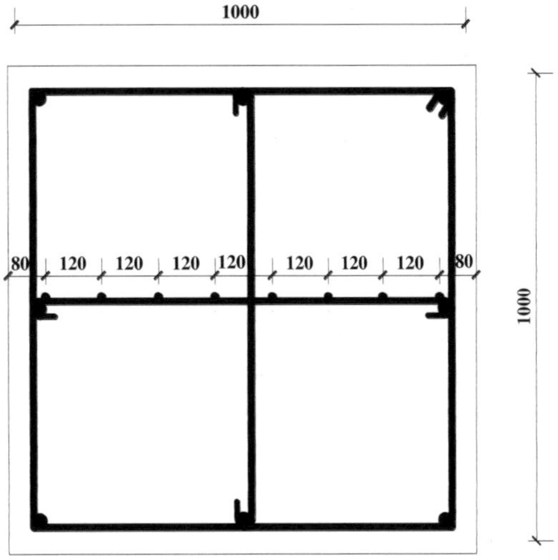

Figure 1. Temperature sensors **Figure 2.** Test of adiabatic temperature rise

As shown in figure1, the sensors are arranged from the outside of the concrete member to the inside according to the spacing of 80 mm, 7*120 mm, 80 mm, and all the sensors are at a plane height. The sensor wire is fixed on the steel bar of the specimen. When the concrete is constructed the sensor is wrapped in concrete and fixed in place according to the design.

2.3 Experimental programme

The flash concrete and the specimen model are formed at the same time, and the concrete is vibrated while constructing to ensure the compactness of the specimens. The test of the hydration temperature begins after the specimens are constructed. The hydration heat is tested after modeling. The adiabatic temperature rise at the center of the concrete specimen was measured at 0.1d, 0.2d, 0.3d, 0.4d, 0.5d, 0.6d, 0.7d, 0.8d, 0.9d, 1d, 2d, 3d and 5d. Respectively, the environmental temperature was recorded. The field test is shown in figure2. The hydration heat test is completed on the 5 day of the specimen formation, and after the test sample was dismantled, the temperature gradient under the simulated temperature condition was tested. When the surface temperature of each specimen reaches 55℃, the temperature is kept constant and the internal temperature is recorded.

3 Results and discussion

3.1 Early hydration heat

3.1.1 Calculation of early hydration heat. The adiabatic temperature rise of mass concrete specimens is estimated by the double exponential equation of cement hydration heat and the adiabatic temperature rise estimation formula of concrete. The maximum adiabatic temperature rise of each proportion of concrete 1d~3d is calculated by equation (1) and (2).

$$Q(\tau) = Q_0(1 - e^{-a\tau^b}) \qquad (1)$$

Where : a , b — Calculating coefficients

Q_0 — Total amount of hydration heat of cement (kJ/kg)

τ — Age of concrete (d)

for ordinary Portland cement：a=0.69, b=0.56 , Q_0=330

$$\theta(\tau)=\frac{Q(\tau)(W+kF)}{c\rho} \qquad (2)$$

Where：$\theta(\tau)$ — Adiabatic temperature rise of concrete (^{0}C)

$Q(\tau)$ — Cumulative hydration heat of cement (kJ/kg)

W — Cement consumption per cubic metre of concrete (kg/m^3)

c — Specific heat of concrete (kJ/kg·^{0}C)

ρ — Concrete density (kg/m^3)

k — Reduction coefficient ，for concrete with fly ash $k=0.25$

F — Mineral admixture per cubic metre of concrete (kg/m^3)

3.1.2 Test results and analysis of hydration heat. The hydration heat of mass concrete specimens with different proportions are tested, the measured results are shown in figure 3. At the same time, the calculation results are also placed for comparison.

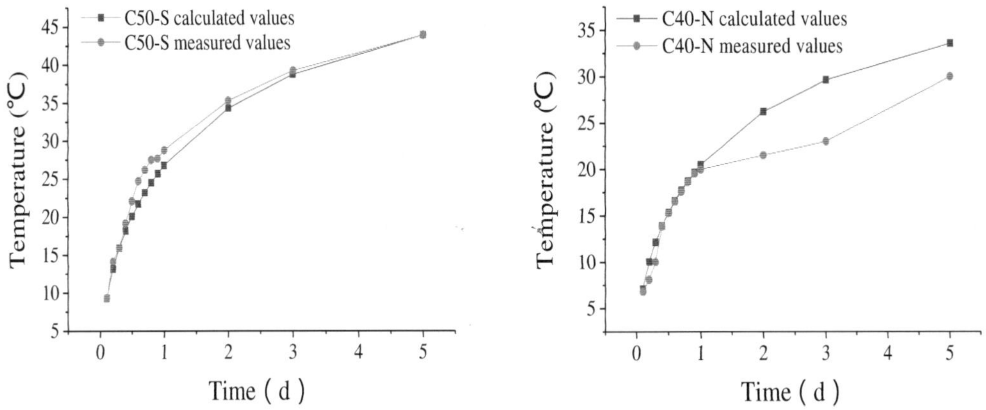

(a) Adiabatic temperature rise of C50-S (b) Adiabatic temperature rise of C40-N

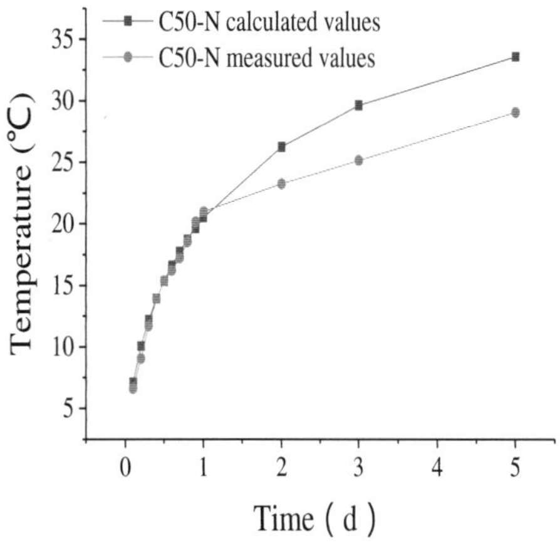

(c) Adiabatic temperature rise of C50-N

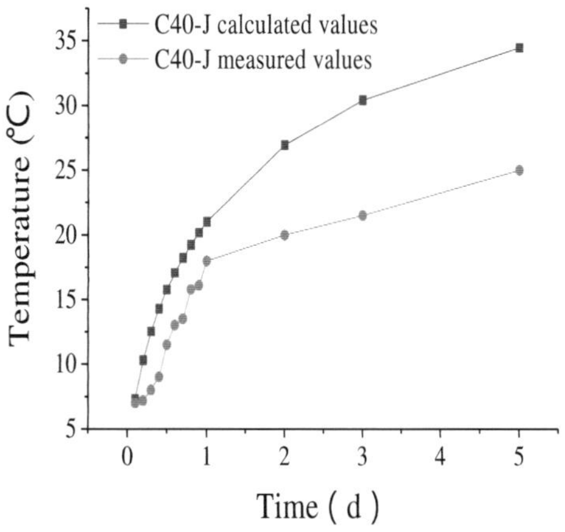

(d) Adiabatic temperature rise of C40-J

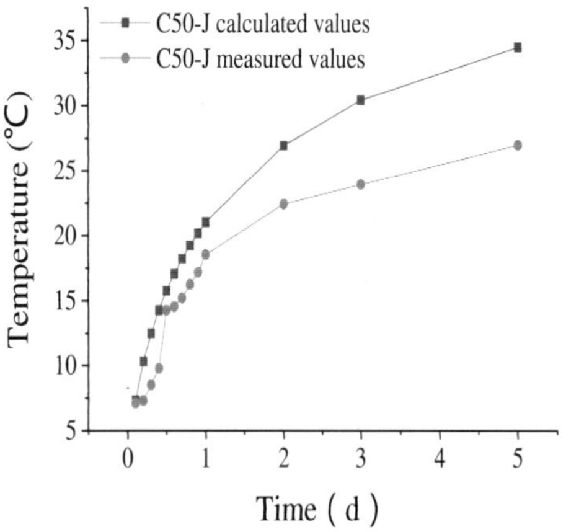

(e) Adiabatic temperature rise of C50-J

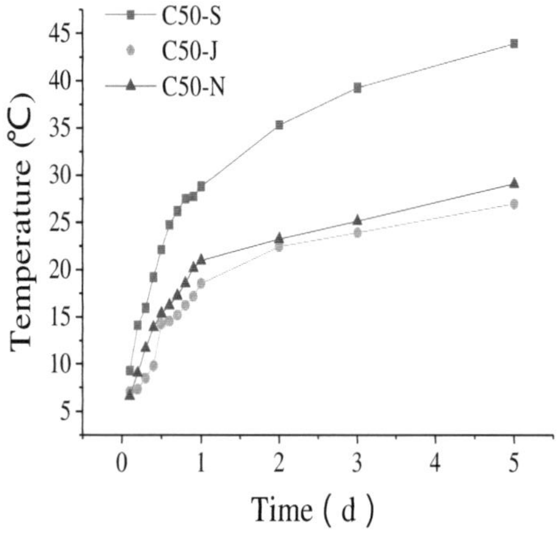

(f) Measured adiabatic temperature rise

Figure 3. Adiabatic temperature rise

By comparing the internal heat of hydration between commercial concrete and mass concrete specimens with internal curing agent and shrinkage reducing agent, it is found that the temperature rise of internal heat of hydration is positively related to the age of concrete. That is, with the advancing of age, the hydration heat and temperature of concrete will gradually increase. At the same time, it is found that the hydration heat temperature rise of mass concrete specimens with internal curing agent and shrinkage reducing agent in 5 days is lower than that of ordinary mass concrete specimens in 5 days. Comparing the test results with the calculation results, it is found that the test results of the hydration heat rise of the ordinary mass concrete specimens are higher than the calculation results, while the test results of the hydration heat rise of the mass concrete specimens with internal curing agent and shrinkage reducing agent are lower than the calculation results.

Shrinkage reducer can reduce the surface tension of porous solution, cause the change of ion concentration in porous solution, and then slow down the transformation of ettringite to calcium sulfonate monosulfate, which has a certain retardation effect on hydration reaction[10]. Internal curing agent can react with water to form hydrogel, which can hinder the contact between cement and water, thus inhibiting the hydration reaction of cement[11].

3.2 Results and analysis of the single side momentary heating test

After form removal, the temperatures of the specimens' surface are simulated. The simulation uses a one-sided instantaneous heating method, while the internal temperature of the specimen is tested. The test results are shown in figure 4. It can be seen that the temperature change is the smallest in the concrete heating surface and the largest change occurs in the middle position. At the same time, there is little difference between the temperature gradients of different proportions of concrete at the same measuring point. And the influence of the change of measuring point position on the temperature difference of concrete with different proportions is similar. The results of remote monitoring on the heating surface show that the temperature variation trend of different test points is similar, the temperature difference is small, the temperature difference between the inside and outside of the concrete sample is small too, which means the volume stability of the concrete is good, and there is little possibility of cracking of the concrete. The temperature difference between the surface temperature and the ambient temperature of the mega-cylinder is large, which agrees with the calculation results and accords with the law of heat conduction.

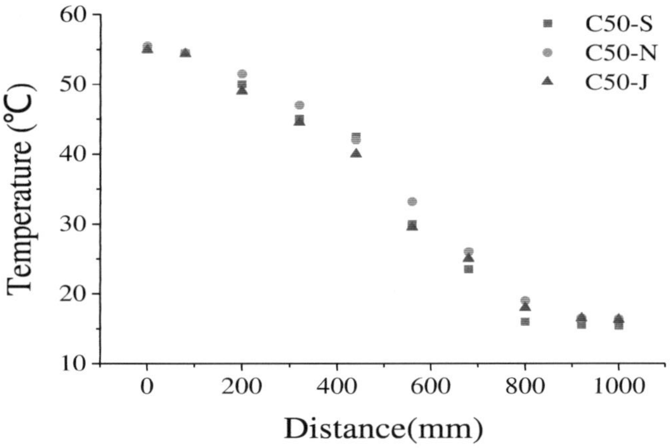

Figure 4. Temperature distribution of specimens

Conclusions

(1) The hydration heat temperature rise of concrete is positively related to the early age of concrete. The smaller the age is, the faster the hydration heat of concrete develops.

(2) The test results of the temperature rise of hydration heat of mass concrete with internal curing agent or shrinkage reducing agent are lower than the ordinary mass commercial concrete.

(3) The temperature rise of hydration heat in 5 days of mass concrete with internal curing agent or shrinkage reducing agent is lower than that in 5 days of ordinary mass concrete.

(4) In the single-side instantaneous heating test, the temperature change is the smallest on the concrete heating surface and the biggest on the middle part. There is little difference between the temperature gradient of concrete with different proportions at the same survey point.

Acknowledge

The authors would like to acknowledge the financial support provided by National Key R&D Program of China-Key materials and preparation technology of high crack resistant ready-mixed concrete(2017YFB0310100), National Natural Science Foundation of China (51678011); Beijing Natural Science Foundation (8162005).

IOP Conf. Series: Materials Science and Engineering **544** (2019) 012008 doi:10.1088/1757-899X/544/1/012008

References

[1] Xie Z, Fu S, Wang S 2018 *J. Adv Mater Sci Eng.* **2018**:1-12.
[2] Zhu B F 2014 *M. Thermal Stresses & Temperature Control of Mass Concrete.* 83-103.
[3] Mehta P K, Burrows R W 2001 *J. Concrete Intl.* **23**(3):57-63.
[4] FU S J, WANG G J. ZHANG S H 2011 *J. Sci China Technol Sc.* **54**(3):565-572.
[5] ZHANG Y, SUN W, LIU S 2002 *J. Cement Concrete Res.* **32**(9):1483-1488.
[6] LANGAN B W, WENG K, WARD M A2002 *J.Cement Concrete Res.* **32**(7):1045-1051.
[7] PEI M, WANG Z, LI W 2008 *J. Constr Build Mater.* **22**(12):2382-2385.
[8] YOUSUF M, MOLLAH A, PALTA P, 1995 *J. Cement Concrete Res.* **25**(3):671-682.
[9] GE C C 2015 *D. Hefei University of Technology.*
[10] DU Y J 2001 *D. Hohai University.*
[11]WANG Z, XING X L ,SHI L 2013 *J. Concrete.* **2013**(12):110-112.

The Elasticity of Calcium Phosphate MAO Coatings Containing Different Concentrations of Chitosan

E. Soldatova[1], E. Bolbasov[1], A. Kozelskaya[1], V. Ignatov[2], V. Smirnov[2], K. Kholodov[3] and S. Tverdokhlebov[1]

[1]School of Nuclear Science & Engineering, Tomsk Polytechnic University, Tomsk, 634050, Russia
[2]School of Advanced Manufacturing Technologies, Tomsk Polytechnic University, Tomsk, 634050, Russia
[3]School of Earth Sciences & Engineering, Tomsk Polytechnic University, Tomsk, 634050, Russia
E-mail: tverd@tpu.ru

Abstract. Calcium-phosphate coatings were formed on the commercially pure titanium by the method of microarc oxidation (MAO) from the electrolyte, which additionally contains 2.5-15 wt.% solution of chitosan. The surface morphology, wettability and elasticity of the formed coatings were investigated. It is shown that the addition of chitosan to the electrolyte composition increases porosity and elasticity of the coatings.

1. Introduction

In traumatology, implants made on the basis of titanium and its alloys, capable of withstanding mechanical loads during operation, are widely used. However, low wear and corrosion resistance of metal implants in a corrosive liquid medium of the human body, contribute to the release of metal ions into the body [1] . To prevent the release of metal ions into the human body, as well as to increase the biocompatibility and reduce the time of treatment, various coatings are deposited to the surface of titanium implants. The latter must combine biological activity and high mechanical strength. As a rule, calcium phosphate materials, in particular hydroxyapatite, are used as biocompatible coatings. Hydroxyapatite has a chemical composition similar to that of bone tissue and provides a high biocompatibility of the implant with the body.

Another important property of coatings for implants is their developed porosity. With the introduction of such implants into the body, the bone tissue grows into the pores of the coating. This ensures the strength and long-term fixation of an implant, its normal functioning in the human body. As presented in a number of studies, porous bioceramics has a high biocompatibility, supporting bone growth in the implant [2,3].

One of the most commonly used methods for deposition of porous calcium phosphate coatings on titanium implants is the microarc oxidation method (MAO), which is a process of anodic oxidation with high plasma voltage [4,5]. One of the drawbacks of calcium phosphate coatings formed by the MAO method is their low elasticity. In addition, with increasing coating thickness, adhesion strength significantly decreases [6], which causes the destruction and detachment of the coating as a result of significant flexural deformations to which the elastic implant undergoes, both during implantation and during operation.

To increase elasticity of the electrolytically deposited calcium-phosphate coatings, the authors of this work used an electrolyte containing chitosan solution. Chitosan is a biodegradable, biocompatible and non-toxic polysaccharide, which in recent years has been considered as one of the most promising materials for biomedical applications [7–10].

Thus, the purpose of this work is to study the influence of chitosan on elasticity of the calcium phosphate coatings formed by the MAO method.

2. Materials and methods

The samples of the commercially pure titanium with sizes 60×20×0.5 mm^3 were used as a material of substrates. Preparation of the surface of the samples before coating included cleaning in an ultrasonic bath in distilled water and chemical etching in an aqueous solution of nitric and hydrofluoric acids taken in volume ratios of HN:HF:H_2O = 1:2.5:2.5, at temperature of 15-20ºC for 10-15 seconds, followed by neutralization in a 1% aqueous solution of sodium hydroxide and repeated washing with distilled water.

The formation of calcium phosphate coatings by the microarc oxidation method using the setup created in National Research Tomsk Polytechnic university was carried out in a saturated solution of CaO in 10% H_3PO_4 with the addition of hydroxyapatite with a dispersion phase with a particle size of up to 70 μm, additionally containing 2.5-15 wt.%, by weight of chitosan solution obtained when dissolving dry powder chitosan in acetic acid with the concentration of 4.5%. The coating was formed at voltage of 190 V with a voltage rise rate of 3 V/s, with a pulse repetition rate of 50 Hz and a pulse duration of 9.7 ms for 20 minutes. The process was conducted at a temperature of 15°C.

The thicknesses and the roughness of the CaP coatings was determined using Talysurf 5-120 contact profilometer (Tailor-Hobson, England).

The study of the surface morphology of the coatings was performed using scanning electron microscopy (SEM) using a JEOL-6000 electron microscope.

The diameter measurement and counting of the number of pores in the obtained images was carried out using the Image J 1.38 software package (National Institutes of Health, USA).

The elasticity of the coatings was investigated using Bend instrument (Gradient Techno, Russia) according to ISO 1519:2011. The smallest diameter of the rod, bending around which did lead to mechanical destruction or detachment of a coating was taken as the value of elasticity index. The values obtained from the results of 5 measurements for each of the studied groups were taken as the value of the elasticity index.

The wettability of each CaP coating was studied using an EasyDrop device (Krüss, Germany) with the 'sit' drop method by measuring the contact angle of a 3 μL liquid drop one minute after placing it on the sample surface. To avoid surface contamination, contact angle measurements were carried out immediately after surface modification.

3. Results and discussion

The studies of morphology of coatings using SEM showed that the surface of the calcium phosphate coating formed by the MAO method without the addition of a polymer has a complex spatial organization. On the surface, there are numerous craters at the top of which pores with an average diameter (0.3 ± 0.2) μm are located (Fig. 1, a). The surface roughness of such coating is Ra = 0.8 ± 0.2 μm. Adding a polymer to the electrolyte composition leads to an slight increase in the number of pores in the coatings, while the pore size does not change (Table 1). It should also be noted that the average roughness also does not change (Table 2).

The thickness of the coatings is up to 15 microns μm for all coatings.

The results of measuring wettability of coatings showed that addition of a polymer to the electrolyte composition leads to an increase in the wetting angle. Moreover, with an increase in the mass content of the polymer from 2.5 wt.% to 15 wt.%, hydrophobicity of the coatings increases (Table 3).

The studies of the elasticity of coatings showed that the coatings formed in an electrolyte containing a polymer are characterized by the increased elasticity compared to coatings formed in a conventional electrolyte. It should be noted that elasticity of coatings does not depend on the mass content of the polymer (Table 1).

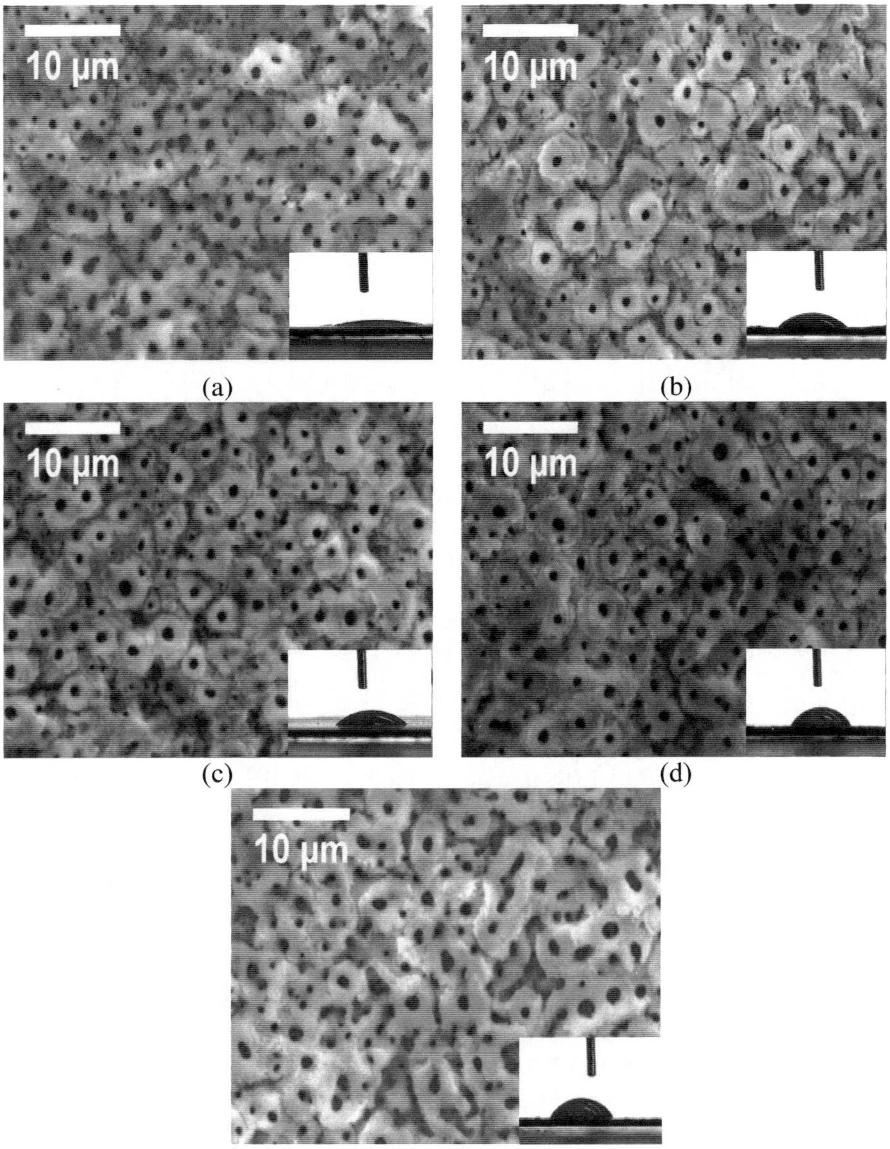

(e)

Figure 1. SEM images of calcium phosphate coatings formed on Ti substrates by MAO methods in electrolytic solution without chitosan addition (a) and with the addition of 2.5% wt. chitosan (b), 5% wt. chitosan (c), 10% wt. chitosan (d) and 15% wt. chitosan (e). The contact angles of CaP coatings are shown in the inserts.

Table 1. Elasticity, number and diameter of pores of the CaP coatings (means ± standard deviations).

	Elasticity, mm	Pores/μm^2	Diameter of pores, μm
CaP	7	2.4±0.2	0.3±0.2
CaP + 2.5 wt.% chitosan	1	2.6±0.4	0.6±0.3
CaP + 5 wt.% chitosan	1	3.5±0.1	0.7±0.3
CaP + 10 wt.% chitosan	1	3.6±0.3	0.7±0.3
CaP + 15 wt.% chitosan	1	2.8±0.3	0.8±0.3

Table 2. Roughness of the CaP coatings (means ± standard deviations).

	CaP	CaP+2.5 wt.% chitosan	CaP+5 wt.% chitosan	CaP+10 wt.% chitosan	CaP+15 wt.% chitosan
R_a	0.8±0.2	0.9±0.2	0.8±0.2	0.9±0.2	0.8±0.1
R_z	2.5±0.5	2.5±0.3	2.8±0.5	3.1±0.5	2.9±0.4

Table 3. Contact angles of CaP coatings (means ± standard deviations).

	CaP	CaP+2.5 wt.% chitosan	CaP+5 wt.% chitosan	CaP+10 wt.% chitosan	CaP+15 wt.% chitosan
Contact angle, °	14.7±2	39.2±1	41.1±0.7	53.3±0.5	59.7±2

4. Conclusions

The calcium phosphate coatings are formed on titanium substrates by MAO using electrolytes containing various chitosan concentrations. It is shown that the addition of chitosan to the calcium phosphate electrolyte contributes to an increase in the elasticity and slight increase of the coating porosity. All coatings formed by MAO from electrolytes containing a polymer are characterized by the same elasticity. With increasing of a polymer concentration in the electrolyte, an increase in the wetting angle of the surface of the formed coatings is observed.

5. Acknowledgments

This work was supported by the Ministry of Education and Science of the Russian Federation, Federal Target Program (agreement #14.575.21.0140, unique identifier RFMEFI57517X0140).

6. References

[1] Reddy B T, Patel J B, Powell D L and Michaels A D 2009 Interatrial shunt closure devices in patients with nickel allergy *Catheter. Cardiovasc. Interv.* **74** 647–51

[2] Hench L L 2005 Bioceramics *J. Am. Ceram. Soc.* **81** 1705–28

[3] Takemoto S, Kusudo Y, Tsuru K, Hayakawa S, Osaka A and Takashima S 2004 Selective protein adsorption and blood compatibility of hydroxy-carbonate apatites *J. Biomed. Mater. Res.* **69A** 544–51

[4] Sankara Narayanan T S N, Park I S and Lee M H 2014 Strategies to improve the corrosion resistance of microarc oxidation (MAO) coated magnesium alloys for degradable implants: Prospects and challenges *Prog. Mater. Sci.* **60** 1–71

[5] Pan Y K, Wang D G and Chen C Z 2014 Effect of negative voltage on the microstructure, degradability and in vitro bioactivity of microarc oxidized coatings on ZK60 magnesium alloy *Mater. Lett.* **119** 127–30

[6] Sharkeev Y P, Legostaeva E V., Eroshenko Y A, Khlusov I A and Kashin O A 2009 The Structure and Physical and Mechanical Properties of a Novel Biocomposite Material, Nanostructured Titanium–Calcium-Phosphate Coating *Compos. Interfaces* **16** 535–46

[7] Luo Y and Wang Q 2014 Recent development of chitosan-based polyelectrolyte complexes with natural polysaccharides for drug delivery *Int. J. Biol. Macromol.* **64** 353–67

[8] Cheung R, Ng T, Wong J and Chan W 2015 Chitosan: An Update on Potential Biomedical and Pharmaceutical Applications *Mar. Drugs* **13** 5156–86

[9] Muxika A, Etxabide A, Uranga J, Guerrero P and de la Caba K 2017 Chitosan as a bioactive polymer: Processing, properties and applications *Int. J. Biol. Macromol.* **105** 1358–68

[10] Wang J, de Boer J and de Groot K 2004 Preparation and Characterization of Electrodeposited Calcium Phosphate/Chitosan Coating on Ti6Al4V Plates *J. Dent. Res.* **83** 296–301

IOP Conf. Series: Materials Science and Engineering **544** (2019) 012010 doi:10.1088/1757-899X/544/1/012010

Study on Preparation and thermal reflective properties of energy saving pigments with selective solar reflection

Yunlong Li [1, a] and Yuanquan Yang [2,b]

1 Postgraduate, Shenyang JianZhu University, Shenyang, China
2 Doctoral student, Dalian University of Technology, Dalian, China
E-mail: [a] yunlong_lee921@126.com, [b]aquarius0109@163.com

Abstract. Currently, the color of common reflective insulation coating is mainly white, but the white color coating is monotonous and its resistance is poor, moreover, the reflective performance tends to drop after pollution. In addition, color coating has the advantages of anti-dirty and beautiful appearance, so study on colored paint is of great significance. Based on the principle of three-tone color, the effect of thermal reflective pigments on the near-infrared reflectance of color reflective heat-insulating coatings was studied by blending three kinds of color reflective heat-insulating coatings of orange, green and purple with composite inorganic heat-reflective pigments as raw materials. The results show that the thermal insulation effect of heat-reflective pigments is significantly improved, and the near-infrared reflectance of green pigments can be increased by about 25%. The thermal reflection thermal insulation coating of the green system is 2.78 times higher than that of the same color common thermal insulating coating.

1. Introduction

Thermal reflective insulation coating is the most widely used building materials on exterior wall in the direction of energy conservation. In recent years, it has become a hotspot of research[1]. The exterior surface of the building receives solar radiation can be divided into three parts: ultraviolet region (200-400nm), visible light region (400-720nm), infrared light region (720-2500nm). The visible light region and infrared light region occupy 95% energy that the solar radiation to the Earth surface [2]. Due to the absorption of inorganic pigments and organic pigments in the visible light area, the thermal insulation properties of heat-reflective heat-insulating coatings are adversely affected by common pigments. In order to obtain a better visual effect, the multi-color thermal reflective heat-insulating coating is obviously superior to the white paint, so the study of heat-reflective pigments with both reflective insulation effect and good visual effect is the focus of current research.

There are many research results of near infrared reflective heat insulation pigments based on cooling materials of roof and exterior wall in foreign countries[3]. In 1940s, some foreign researchers began entertaining air with low thermal conductivity into multi-layer aluminum film, and then preparing heat insulation coatings after compounding with it, which opened the prelude of heat reflection heat insulation coatings. The Lawrence Berkeley National Laboratory (LBNL) and Oak Ridge National Laboratory (ORNL) began to study the solar reflectance and hemispherical emissivity of pigments in the 1990s, and the pigments were divided into several categories and the effects of the properties, composition, roughness and purity of the materials on the solar reflectance were studied [4]. In 1998, the American Cold Roofing Association (CRCC) was established to develop accurate, credible evaluation, labeling methods and the dissemination of relevant knowledge for solar and thermal radiation properties of roofing products [5]. Domestic research on the influence of heat reflective pigments on the heat insulation performance of heat reflective insulation coatings is however less .In the 1990s, thermal insulation coatings was introduced into China, and domestic researchers began to conduct in-depth research and development on them [6]. The effect of the optical properties of different color heat-reflective pigments on the thermal insulating properties of the coatings was studied, and the color reflective heat-insulating coating was the same as that of the common reflective heat-insulating coating, and the thermal reflective heat-insulating property was significantly improved.

2 Raw materials and mix ratio

2.1. Materials and mix ratio

Ferric oxide (Fe_2O_3) was provided by Hebei Luquan blue color pigment factory. Zirconium red iron (Zr-Si-Fe) was provided by Jingdezhen wonderful Ceramic Pigments Co., Ltd. Titanium nickel Yellow (Ti-Sb-Ni-O) was supplied by Heng Ming New Materials Co., Ltd. PR yellow (ZrSiO4-Pr) from Jingdezhen wonderful Ceramic Pigments Co., Ltd was used here. Phthalocyanine Blue ($C_{32}H_{16}CuN_8$) comes from Yancheng color Tatsu Pigment Co., Ltd. Titanium dioxide (TiO_2) was produced by Jinzhou Peng Titanium Dioxide Manufacturing Co., Ltd. Self-crosslinking high-gloss acrylic emulsion was supplied by Jiangyin National Union Chemical Co., Ltd. Silicone modified resin was from Shanghai Summer soil New Materials Co., Ltd. Defoaming agent, hardener, thickener were also used.

2.2 Test instrument.

UV/Vis/NIR spectrophotometer Lambda950, perkinelmer; Automatic Film Drying Machine Afa-iii, Shanghai Worship Force; Coating Thickness gauge QUJ, Tianjin Material Testing Machine Factory; X-ray Diffractometer Uitima-iv, Japan Science Company; Chromatic Aberration Instrument NR110 Shenzhen 3nh Co., Ltd.

Table 1 Basic formulations of white thermal reflective heat-insulating coatings

Raw material	m/g
water	X
packing	85~125
Titanium dioxide	appropriate
latex	Y
Curing agent	3
dispersant	3
defoaming agent	2.5
Thickening agent	appropriate

The amount of titanium dioxide added in the basic formula of white reflective energy-saving paint can be appropriately adjusted according to the needs of color matching. The effect of heat reflective pigments on thermal insulation performance was studied in this experiment. The white base paint with 7% titanium dioxide was not clearly indicated.

2.3 Test Methods

2.3.1 Preparation of White Heat Reflective Heat Insulation Coatings. Paint was prepared according to the mixing proportion in Table 1 for the following process.Sample preparation process: Weigh a proper amount of white powder pigment before and after the heat treatment, add film-forming substances such as emulsion, dispersant, defoamer, water and additives, fully mix for 15 mins under the action of a magnetic stirrer with a rotating speed of 200 r, then add curing agent, and then continue to mechanically stir for 5 min. Cleaning the transparent polyester film with alcohol and placing it flat on the coating base of the automatic film dryer.

2.3.2 Optical Performance Test. Color test: Using colorimeter to test the surface color parameters of pigments and powders L *, a *, b * values, The chromatic aberration value of different colors △E. D65 light source is used for measurement. Colors can be classified according to their lightness. When the value of L is less than 40, it is a dark color. L values between 40 and 80 are medium colors, and L values greater than 80 are light colors [7]. Solar reflectance and near infrared reflectance tests: refer to **GJB 2502.2 - 2006** spacecraft thermal control coating reflectance and emissivity test methods part 2: solar absorptivity test, open the vacuum pump to fix it., adjust blade height is 1.00mm, place the mixed pigment in front of the coating device. The coating rate is 6mm/s. After coating is stopped, the heating and drying system is kept at 40 degrees Celsius for 10min. Heat reflective coating was prepared.

3 Results and discussion

3.1 Testing of Optical Properties of Titanium Dioxide

Rutile titanium dioxide is the white pigment with the best heat reflection effect, and it is the key factor affecting the heat insulation performance of the white reflective heat insulation coating base paint. Due to the different surface treatment process of the manufacturer, it will bring about the difference in its performance. The solar reflectance and L (representing brightness, i.e. black and white), A (representing red and green) and B (representing yellow and blue) values of different brands and manufacturers of titanium dioxide pigments were tested. The test results are shown in Table 2 and Figure 1. As can be seen from the data in Table 2, the lightness value L of titanium dioxide pigments is all around 88, and the appearance color is basically the same.The reflectivity curve of rutile-type titanium dioxide pigments to solar light is basically the same except for 2 # domestic samples.

TSR is relatively low, and other titanium dioxide pigments have high solar reflectance (TSR). In the visible region, the solar reflection of 1# and 4# samples is relatively high, about 90% . The 3 # and 4 # samples have relatively high near infrared reflectance in the long-wave near infrared region (1100 ~ 2500 nm). In this study, 4 # titanium dioxide samples were selected to prepare white heat reflective insulating base coating slurry, with titanium dioxide dosage of 7 % and PVC of 60 %.

Table 2 Optical Performance Data of titanium dioxide pigments

Serial number	model	L^*	a^*	b^*	TSR/%	NIR/%
1	PR601	88.75	-0.57	1.72	88.61	90.36
2	PR511	86.54	-0.09	2.64	78.32	77.80
3	PR521	89.99	-0.41	3.67	88.98	91.71
4	PD909	88.65	-0.81	3.21	88.54	92.01·

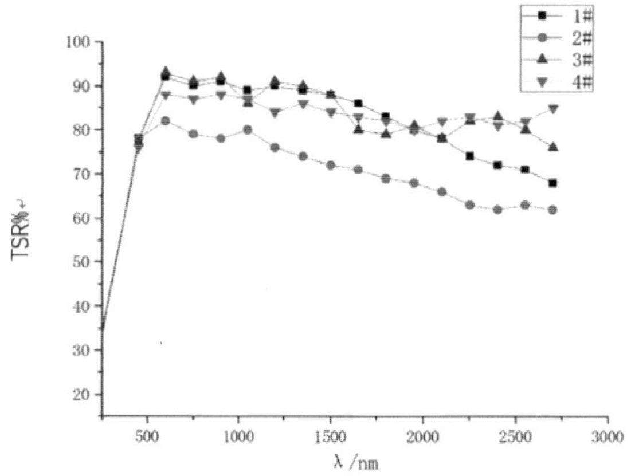

Fig. 1 Solar reflectance of titanium dioxide

3.2 Optical Properties of Heat Reflecting Pigments

Heat-reflective pigments with near-infrared reflectivity are mostly composed of several metal mixtures calcined at high temperatures.The optical data of orange, green and purple heat reflective pigments were tested and shown in Table 3.

Table 3 Proportion Scheme of Secondary Color Pigment

Powder pigment	mass ratio			
Fe_2O_3: Ti-Sb-Ni-O -- (Orange 1#)	1:1	1:2	1:3	1:4
	1:5	1:6	1:7	1:8
Fe_2O_3: ZrSiO4-Pr -- (Orange 2#)	1:1	1:3	1:6	1:9
	1:12	1:15	1:18	1:21
$C_{32}H_{16}CuN_8$: Ti-Sb-Ni-O -- (green1#)	1:1	1:2	1:3	1:4
	1:5	1:6	1:7	1:8
$C_{32}H_{16}CuN_8$: ZrSiO4-Pr -- (Green 2#)	1:2	1:4	1:6	1:8
	1:10	1:12	1:14	1:16
$C_{32}H_{16}CuN_8$: Zr-Si-Fe -- (purple1#)	1:1	1:2	1:3	1:4
	1:5	1:6	1:7	1:8
Fe_2O_3: $C_{32}H_{16}CuN_8$-- (purple 2#)	1:1.2	1:2	1:3	1:4
	1:5	1:6	1:7	1:8

Table 4 Optical Data Of Different Color Heat Reflecting Pigments

Serial number	Color	L^*	a^*	b^*	TSR/%	NIR/%
1	Orange 1#	42.54	20.06	14.32	73.8	74.0
2	Orange 2#	39.01	22.21	15.29	71.4	62.6
3	Green 1#	48.36	-5.06	-3.32	62.1	64.6
4	Green 2#	36.33	-6.55	-4.79	51.9	53.4
5	Purple 1#	28.74	1.43	-1.67	47.5	48.6
6	Purple 2#	29.29	1.59	-1.20	37.7	43.5

As shown in Figure 2 below, in comparison with orange and green pigments, the TSR effect of orange 1 # and green 1 # is better, the TSR effect of purple pigment is average, and iron red and titanium nickel yellow are better pigment choices. As can be seen from fig. 2, in the near infrared region, the NIR effect of orange 1 # and green 1 # is better, and the near infrared reflection effect of purple pigment is generally worse. In contrast to orange and green, iron red and titanium nickel yellow are obviously better pigments because they have better L * and covering ability.

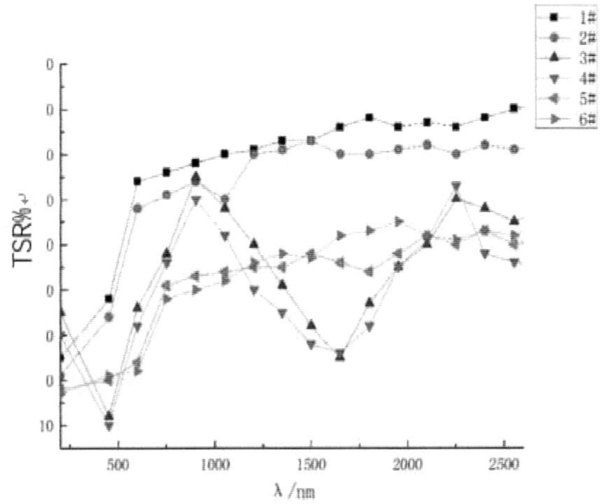

Fig. 2 TSR of different heat reflecting pigments

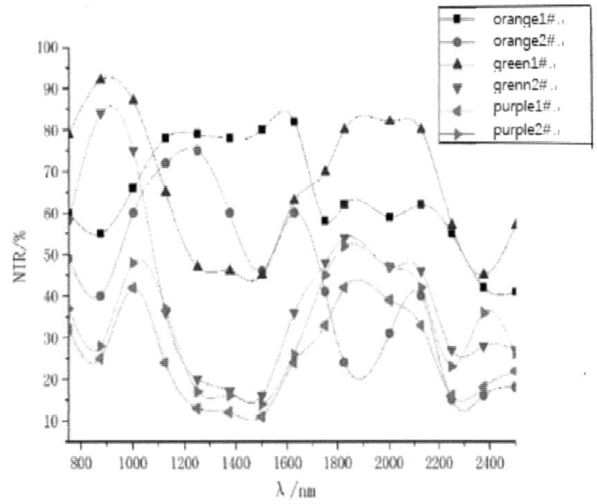

Fig. 3 NIR of different heat reflecting pigments

3.3 Comparative Study on Heat Reflective Performance of Heat Reflective Thermal Insulation Coatings and Ordinary Thermal Insulation Coatings

Color reflective heat-insulating paint is blended in the heat-insulating paint base paint, and water-based general color paste is added to the white base paint to prepare common color heat-insulating paint of the same color. The minimum value of color difference that human eyes can distinguish is 1.0. The optical performance data of the color difference △E≤3.2 kinds of color reflective thermal insulation coatings allowed in this study are shown in Table 5.

Table 5 Optical Performance Data of Heat Reflective Thermal Insulation Coatings and Common Coatings

Color classification	Thermal reflective thermal insulation coatings			Common heat insulation coatings			ΔE
	Serial number	L^*	NIR/%	Serial number	L^*	NIR/%	
Orange	1#	42.54	74.0	1#	40.42	68.1	1.58
	2#	39.01	65.6	2#	38.16	59.9	2.64
Green	1#	48.36	74.6	1#	48.18	50.1	2.98
	2#	36.33	53.4	2#	35.07	26.3	1.15
pueple	1#	28.74	38.6	1#	28.91	17.4	0.59
	2#	29.29	43.5	2#	29.23	24.9	3.15

As can be seen from table 5, color-reflective thermal insulation coatings compared to common thermal insulation coating, NIR has different degrees of improvement, the improvement of green system pigments are particularly obvious, with the increase of the color paste content of the coating color, the value of l* decreased. The lower the lightness is, the greater difference between the two. Because of the shallowness of orange pigment and the increase of orange pigment amount, the brightness value of orange pigment is not obvious. L * between 39~42, near infrared reflectance is relatively high, about 65% to 74% . The addition of heat reflective pigment has little effect on the NIR of the coating.

4 Conclusions

(1) White pigment titanium dioxide has the highest visible and near-infrared thermal reflection performance. Yellow heat-reflective pigments have a high thermal reflectance and are formulated as light and medium-color reflective heat-insulating coatings that act the same as ordinary yellow pigments.
(2) Considering the effect of chromatic aberration factor, it can be considered that the orange thermal reflective pigment has the same effect as ordinary pigments. Red and blue thermal reflective pigments, due to their excellent thermal reflectance in the near-infrared light region, can significantly improve the thermal insulation of dark-coloured heat-reflective insulation coatings (generally up to about 10).
(3) Through the rational formulation design and the addition of heat-reflective pigments, color reflective heat-insulating coatings with rich color and high thermal reflectance can be prepared.

Acknowledgement

The financial support from the University Innovation Team Project of Liaoning Province (No. LT2015024) and the Key Technology Research and Development Project of Shenyang (No.17-48-2-00) are also gratefully acknowledged.

References

[1] Yongxiang Xu, Yunde Li, Hua Shi, et al. Progress in solar heat reflective insulation coatings [J].Coatings Industry, 2010 (1): 70-74.
[2] Gong Xiao Zheng. Infrared reflective pigment and its application [J]. Shanghai coatings, 2008, 46 (12): 19-23.
[3] Berdahl P, Bretz S E. Preliminary survey of the solar reflectance of cool roofing materials [J]. Energy & Buildings, 1997, 25(2):149-158.
[4] Levinson R, Berdahl P, Akbari H, et al. Methods of creating solar-reflective nonwhite surfaces and their application to residential roofing materials[J]. Solar Energy Materials & Solar Cells, 2005, 91(4):304-314.
[5] Levinson R, Akbari H, Reilly J C. Cooler tile-roofed buildings with near-infrared-reflective non-white coatings[J]. Building & Environment, 2007, 42(7):2591-2605.
[6] JOUDIA, SVEDUNG H, CEHLIN M, etal. Reflective coatings for interior and exterior of Building sand improving thermal performance [J]. Applied Energy. 2013, 103: 562—570.
[7] LEVINSONR, BERDAHL P, AKBARI H. Solar spectral.optical properties of pigments, part I: model for deriving scattering and absorption coefficients from transmittance and reflectance measurements [J]. Solar Energy Materials &Solar Cell, 2005, 89 (4): 319—320.

Preparation and properties of silane-treated banana fiber/ poly(lactic acid) biocomposites

T Insung, N Chaiyut, B Ksapabutr and M Panapoy

Department of Materials Science and Engineering, Faculty of Engineering and Industrial Technology, Silpakorn University, Nakhon Pathom 73000, Thailand
E-mail: chaiyut_n@su.ac.th

Abstract. Poly(lactic acid) (PLA) reinforced with silane-treated banana fiber (BFSi) at various fiber contents was studied. Banana fiber surfaces were firstly treated by sodium hydroxide and subsequently by 3-aminopropyl triethoxysilane. The biocomposites were then prepared by melt blending method in a twin-screw extruder. The effect of BFSi fiber contents on the mechanical and thermal properties of the biocomposites were investigated. Scanning electron microscope (SEM) was used to reveal the cryo-fractured surfaces of the composites. Mechanical properties, thermal characteristics and thermal stability were carried out by universal testing machine, differential scanning calorimeter (DSC) and thermogravimetric analyzer (TGA), respectively. Interfacial adhesion between PLA matrix and BFSi fibers was significantly improved even fiber content up to 40% by weight. There were few fiber pull-outs or voids appeared. With increasing fiber contents, tensile modulus and strength increased while elongation at break decreased. Moreover, glass transition temperature (T_g) and cold crystallization temperature (T_{cc}) decreased, whereas, crystallinity (X_c) significantly increased, however, melting temperature (T_m) was almost intact when fiber contents increased. The composites with higher amount of fibers exhibited faster thermal decomposition and higher residue contents at 550°C.

1. Introduction

Biodegradable plastics have been received much attention to be used as green materials. Poly(lactic acid) (PLA) is one of these plastics which can be synthesized from renewable resources. Recently, many countries are making every effort to develop natural fibers to replace synthetic fibers. Natural fibers are not only inexpensive, but also hypoallergenic. Moreover, the advantages of such fibers over man-made fibers include low density, low cost, recyclability, and biodegradability. Fiber from the banana stem is a natural fiber derived from plants. Natural fibers compose of cellulose which enhances the strength and uniformity of the fiber; hemicellulose which increase strength to the fibers; and lignin which increases the flexibility of the fiber. It has been discovered that a natural fiber derived from banana, the amount of cellulose, hemicellulose and lignin ranges in 60-65, 6-8 and 15-10%, respectively. Fibers from banana stem have density, tensile stress and Young's modulus of 1.3 g/cm^3, 791 MPa and 30 GPa, respectively. The amount of cellulose from banana trees is high then the fibers from banana trees show high toughness. The fibers are also suitable to be used to produce textile products [1]. In this study, we actively manufactured high performance banana fiber reinforced PLA biocomposites. Banana fiber surfaces were modified in order to improve the compatibility between banana fibers and PLA in order to enhance the mechanical strength, to reduce the cost of the materials, and to meet the request of eco-friendly composites. Effect of BFSi contents on the properties of biocomposites was investigated.

2. Experimental

2.1. Materials
Poly(lactic acid) was purchased from NatureWorks®,USA (product name – NatureWorks® PLA Polymer 7000D), USA. Banana fibers were obtained from "Musa Sapientum Linn" by Agriculture Women's Club in Hua-Hin, Prachuap Khiri Khan, Thailand. 3-Aminopropyltriethoxysilane (APS) was purchased from Merck, Germany. Sodium hydroxide (NaOH) was purchased from Better Syndicate, Thailand.

2.2. Banana fiber surfaces modification
The chopped banana fibers (6 mm) were soaked in 10%w/w NaOH solution for 1 h at room temperature together with stirring by mechanical stirrer. The fibers were washed with distilled water containing small amount of acetic acid until neutralized and were dried in a hot-air oven at 60°C for 48 hr. Banana fibers treated with NaOH (BFNa) were then undergone silane treatment. 5 wt% APS was dissolved in a mixture solvent of water:ethanol (40:60 %v/v). The pH of solution was adjusted to 4 by acetic acid and stirred continuously for 1 hr. Next, the BFNa fibers were soaked in the solution for 3 hr. NaOH treated Banana fibers followed by silane treatment (BFSi) were washed several times in distilled water. Finally, BFSi fibers were dried in hot-air oven at 60°C for 48 hr.

2.3. BFSi reinforced PLA biocomposites preparation
Silane-treated banana fibers (BFSi) were molten mixed with PLA in a twin screw extruder. BFSi/PLA biocomposites with three different amounts of the fiber at 20, 30 and 40 wt% were prepared. Finally, an injection molding machine was used to produce testing specimens.

2.4. Morphological study
Scanning electron microscope (CamScan, Bruker, UK) was used for the morphological characterization of banana fiber surfaces and cryo-fractured surfaces of BFSi/PLA biocomposites.

2.5. Thermal properties
Differential scanning calorimetry (DSC) measurement was taken on the Mettler Toledo DSC 1 using nitrogen as the purge gas at a scanning temperature ranges from 25-200°C at a heating rate of 10°C/min. Thermogravimetry analysis (TGA) measurement was taken at a scanning temperature ranges from 35-550°C whilst maintaining a nitrogen airflow with a constant heating rate of 10°C/min.

2.6. Mechanical properties
The tensile specimens were tested according to the ASTM D638 standard using an Instron-5969 universal testing machine. Load cell of 50 kN and crosshead speed of 50 mm/min were set for testing. At least 10 specimens were tested.

3. Results and discussion

3.1. Morphological studies
Scanning electron micrographs of each banana fiber and BFSi/PLA biocomposites at different BFSi contents were shown in Figure 1. It can be seen that surfaces of the banana fibers can be altered after the alkali and silane treatments. The untreated fiber (BF) surface (Figure 1a) is found to be slightly smooth due to the presence of oils and waxy substances including lignin that makes the fiber surface shiny and smooth [2]. The morphology of BFSi fiber shows different structures. Smoother surface can be observed in Figure 1b which is due to the deposition of the silane compound on the fiber surface [3]. Figure 1c-f describes the effect of fiber contents on the fractured surface morphology of the composites. It can be seen from Figure 1c that neat PLA normally shows smooth surface. With increasing the fibers contents from 20-40 wt% (Figure 1d-f), the fibers become closer together but don't show any sign of agglomeration. There are few fiber pull-outs or voids appeared on the surfaces. However, there is no gap observed at the interphase between the fiber and the matrix. This indicates that the interfacial adhesion between PLA matrix and BFSi fibers is significantly improved even fiber content up to 40 %

by weight [4]. Moreover, the fiber breakages can be seen in Figure 1f which supporting the interfacial adhesion improvement.

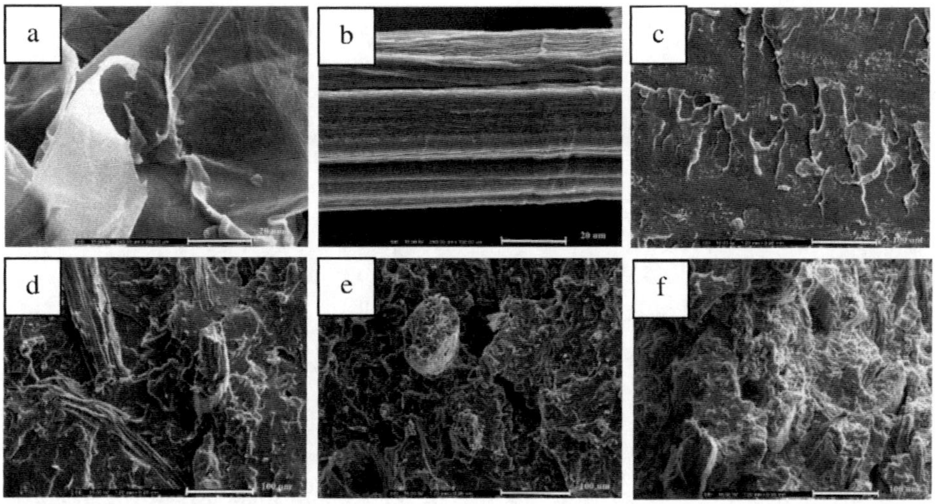

Figure 1. SEM micrographs of (a) BF fiber, (b) BFSi fiber, (c) neat PLA, and BFSi/PLA composites at (d) 20, (e) 30 and (f) 40% by weight of BFSi fiber.

3.2. Thermal characteristics

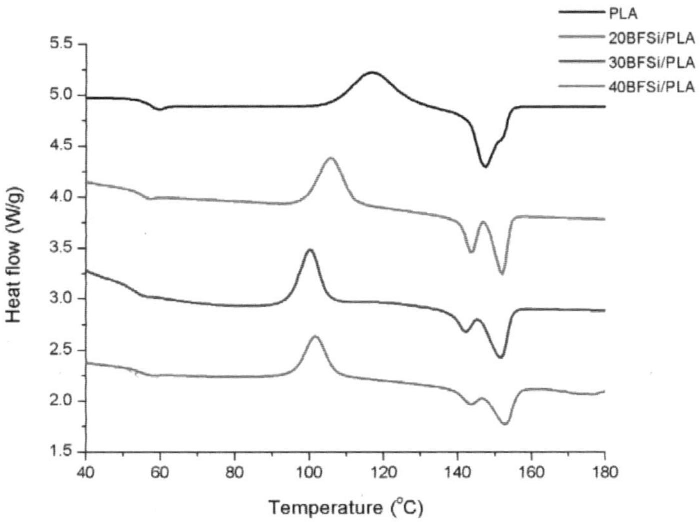

Figure 2. DSC thermograms of neat PLA, BFSi fiber and BFSi/PLA biocomposites at different fiber contents.

The thermal characteristics obtained from DSC experiment are shown in Figure 2 and summarized in Table 1. It can be seen that T_g and T_{cc} decrease with the addition of BFSi to the PLA which resulting from density of crystalline state [5]. Moreover, the addition of BFSi to PLA results in an increase in crystallinity (X_c) of the PLA matrix. This can be explained by the nucleating ability of BFSi which promoting the crystallization of PLA. As a result of the silane modification on the fiber surfaces, bonding between the fiber and PLA matrix is improved due to the increasing in interfacial interaction by potential hydrogen bonding and mechanical interlocking between them [4]. The better the interfacial interaction between BFSi fibers and PLA matrix the more the fiber surfaces act as nucleation sites for the crystallization of PLA [6]. However, T_m seems not significantly affected by BFSi fiber contents. It can be seen that peak intensity at T_{m2} is higher than that of T_{m1} when fiber contents increase. This indicate that more crystallites nucleated with more BFSi fibers loading.

Table 1. Thermal properties obtained from DSC study of neat PLA and BFSi/PLA biocomposites at different fiber contents.

Sample	T_g (C°)	T_{cc} (°C)	T_{m1} (°C)	T_{m2} (°C)	X_c (%)
Neat PLA	55.62	116.8	147.3	-	30.94
20BFSi/PLA	55.95	106.6	144.9	153.0	35.77
30BFSi/PLA	53.59	100.3	142.4	151.4	37.40
40BFSi/PLA	54.51	101.9	143.3	152.6	37.74

3.3. Thermal stability

TGA thermograms of neat PLA, BFSi fiber and BFSi/PLA composites are shown in Figure 3. PLA matrix, BFSi fiber, BFSi/PLA composites at the fiber content of 20, 30 and 40wt% show maximum decomposition temperature (T_d) at 385.8, 342.8, 343.7, 322.0 and 316.0°C, respectively. The BFSi fiber and the composites with different fiber contents clearly show two-step decomposition. The first step during 40-100°C is due to the loss of water molecules and the second step during 250-375°C is due to the decomposition of cellulose in the fiber together with PLA because they decompose in the same temperature range. T_d of the composites becomes lower as the fiber content increases. This is due to BFSi fiber has less thermal stability than neat PLA. In addition, the composites with higher BFSi loading shows the higher residue content at 550°C toward the BFSi fiber itself [7].

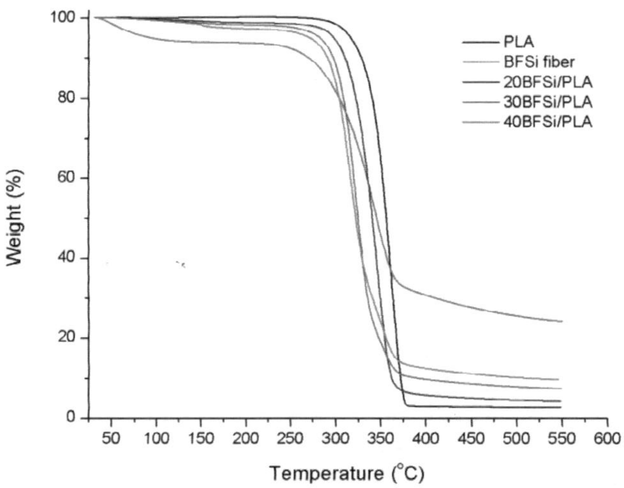

Figure 3. TGA thermograms of neat PLA, BFSi fiber and BFSi/PLA biocomposites at different fiber contents.

3.4. Mechanical properties

Table 2 displays the summary results of mechanical properties from the tensile testing of PLA and BFSi/PLA composites. Tensile modulus of the composite increases with increasing amount of the BFSi fibers because the transcrystallization at the surfaces of fibers enhances the degree of crystallinity in the

IOP Conf. Series: Materials Science and Engineering **544** (2019) 012011 doi:10.1088/1757-899X/544/1/012011

composites [8]. The tensile strength of the composites increases when the loading of BFSi fibers in the composites increase. There is interfacial adhesion enhancement between PLA matrix and BFSi fibers which produces the good force transfer along the fibers. Moreover, the orientation of the fibers in the composites produce positive effect of tensile strength [9]. However, the elongation at break of the composites becomes decreased with the fiber loading contents. Because the fibers hinder the conformation of the PLA molecules, subsequently, slippage between PLA molecules hardly takes place.

Table 2. Tensile properties of neat PLA and BFSi/PLA biocomposites at different fiber contents.

Sample	Tensile modulus (MPa)	Tensile strength (MPa)	Elongation at break
Neat PLA	1,127±90	67.39±0.97	9.82±1.32
PLA/20BFSi	1,367±41	67.83±1.01	6.86±0.64
PLA/30BFSi	1,684±133	68.94±1.74	5.41±0.21
PLA/40BFSi	1,892±75	76.16±2.00	5.21±0.44

4. Conclusions

This work demonstrated that the eco-friendly composites with good mechanical and thermal properties can be successfully developed using banana fibers as a reinforcement and PLA as a matrix. Silane treatment improves the compatibility between banana fibers and PLA resin. Thermal characteristic slightly changes compared to neat PLA, though crystallinity obviously increases with fiber contents. Thermal stability of the composites becomes more unstable with increasing amount of the fiber. Mechanical properties of BFSi/PLA biocomposites was significantly higher than those of the PLA matrix.

5. References

[1] Deepa B, Abraham E, Cherian B M, Bismarck A, Blaker J J, Pothan L A, Leao A L, Ferreira de Souza S and Kottaisamy M 2011 *Bioresour. Technol.* **102** 1988
[2] Ibrahim M M, Dufresne A, El-Zawawy W K and Agblevor F A 2010 *Carbohydr. Polym.* **81** 811
[3] Jandas P J, Mohanty S and Nayak S K 2013 *J. Clean. Prod.* **52** 392
[4] Goriparthi B K, Suman K N S and Roa N M 2012 *Compos.* A. **43** 1800
[5] Islam M S, Pickering K L and Foreman N J 2010 *Compos.* A. **41** 596
[6] Paul S A, Joseph K, Mathew J D C, Pothen L A and Thomas S 2010 *Compos.* A. **41** 1380
[7] Monteiro S N, Calado V, Rodriguez R J S and Margem F M 2012 *Mater. Sci. Eng.* C. **557** 17
[8] Joseph P V, Joseph K, Thomas S, Pillai C K S, Prasad V S, Groeninckx G and Sarkissova M 2003 *Compos.* A. **34** 253
[9] Huda M S, Drzal L T, Mohanty A K and Misra M 2008 *Compos. Sci. Technol.* **68** 424

Acknowledgement
The authors would like to thank Department of Materials Science and Engineering, Faculty of Engineering and Industrial Technology, Silpakorn University and Center of Excellence on Petrochemical and Materials Technology Chulalongkorn University, Bangkok, Thailand for financial supports.

IOP Conf. Series: Materials Science and Engineering **544** (2019) 012012 doi:10.1088/1757-899X/544/1/012012

Study of contents ratio of cellulose, hemicellulose and lignin on the mechanical properties of sisal fibers reinforced polylactic acid (PLA) composites

Z H Zhu [1], B H Mo [2] and M Y Hao [3]

[1,2] School of advanced manufacturing technology, Guangdong Mechanical & Electrical Polytechnic, Guangzhou, China;

[3]The Key Laboratory of Polymer Processing Engineering, Ministry of Education, South China University of Technology, Guangzhou, China

Email: 184823174@qq.com

Abstract. Untreated sisal fibers (USFs) with the highest tensile strength and alkali-treated SFs (ASFs) with the maximum interfacial strength with polylactic acid (PLA) matrix were mixed to make hybrid SFs (HSFs) according to different mass ratios and then compounded with PLA to prepare PLA/HSF composites. It was found that PLA/HSFs had the best mechanical property when the mass ratio of USFs and ASFs was 4:6. and the content of cellulose, hemicellulose and lignin at this moment were 71.4%、16.6% and 12.0% respectively.

1. Introduction

The interest of using natural fibers as reinforcement in polymer composites has increased in recent years due to their lightweight, non-toxicity, low cost, and biodegradable properties. [1-2] Natural fibers consist of cellulose, hemicellulose and lignin. They can be considered as composites of hollow cellulose fibrils held together by a lignin and hemicellulose matrix. [4] Lignin can be considered as a structural supporting material in natural fibers. In our previous work, SFs were alkali treated with different time to prepare ASF samples. The changes of chemical compositions of SFs were investigated, and their influences on the microstructure, tensile strengths of SFs and the interfacial strength of SFs and PLA were studied. It was found that USF had the best tensile strength and ASF1h (be alkali-treated for 1h) had the maximum interfical strength with the PLA matrix. It could also be observed that suitable treatment intensity can reserve appropriate content of lignin and hemicelluloses, which had some positive effect on the performance of SF and SF-reinforced PLA composites. [6] Therefore, USFs was mixed with ASF1h according to different mass ratios to make HSFs in this work. Then different HSFs were compounded with PLA to make PLA/HSF composites. The aim is to obtain the PLA/HSFs composite with the best mechanical properties, the optimal hybrid ratio of USFs and ASF1h and the contents of cellulose, hemicellulose and lignin at this time.

2. Experiment

2.1. Materials

PLA polymer 3051D, Nature Works™, was provided by Shenzhen Bright China Industrial Company (China). SF bundles (GB/T 15031-94) were provided by Guangxi Sisal Company of Guangxi Province, China. 10% w/v aqueous solution of NaOH was commercial product.

2.2. Preparation of USFs and ASF1h

Firstly, SF bundles were cut short and submerged into 10% w/v aqueous solution of NaOH for 1h. Then alkali-treated SFs (ASFs) were washed with distilled water and dried in the open air for 48h. Finally, the USFs and ASF1h were dried in an oven at 80℃ for 4h.

2.3. Preparation of PLA/HSF composites

Firstly, PLA, ASFs and USFs were dried at 80℃ for 4h. Then, USFs were manually mixed with ASF1h to fabricate eleven kinds of HSFs, which mass ratios of USF and ASF1h were listed in Table 1. After that, PLA were melted and compounded with different HSFs respectively to make into eleven kinds of PLA/HSF composites according to a mass ratio of 4: 1 using a two-roll plastic mill . At last, they were compressed into sheets with a thickness of 1 mm or 4 mm by a hot press and cut into standard samples (GB/T 1039-1992) for mechanical tests. Five specimens were used for each test.

Table 1. Ratios of USFs and ASFs of PLA/HSFs composites

	C1	C2	C3	C4	C5	C6	C7	C8	C9	C10	C11
Mass ratio of USF to ASF1h	0:10	1:9	2:8	3:7	4:6	5:5	6:4	7:3	8:2	9:1	10:0

2.4. Scanning electron microscopy (SEM)

The specimens were fractured in liquid nitrogen to obtain unaffected fractured surfaces. The fractured surfaces of the samples were then sprayed with gold powder to prepare for scanning electron microscopic (SEM) observation by a SEM (S-3700N, Hitachi, Tokyo, Japan).

3. Results and discussion

3.1. Mechanical property of SFs reinforced PLA composites with different SFs

In our previous work, it can be found that the tensile strength of USF was the highest while the interfacial bonding between ASF1h and PLA was the strongest. It was generally considered that the mechanical properties of fiber-reinforced composites primarily depends on three factors: (1) the strength and modulus of the fibers, (2) the strength and toughness of the matrix, and (3) the effectiveness of fiber–matrix bonding in transferring stress across the interface. [5] If the processing parameters are the same, the performance of PLA as a matrix is constant. Therefore, the mechanical properties of PLA composites mainly depend on the properties of fibers and the interface of fiber and matrix. Therefore, USF was mixed with SF withASF1h according to different mass ratios to make HSFs. The aim is to make cellulose, hemicellulose and lignin play their roles adequately and thus prepare the PLA/HSFs composites with the best mechanical properties.

Figure 1 showed the mechanical properties of PLA, PLA/USFs, PLA/ASFs and PLA/HSF composites. It was found that the mechanical properties of PLA/USFs composite and PLA/ASFs composite were worse than those of pure PLA, which was mainly because of the bad interfacial compatibility of PLA and USFs and the poor mechanical property of ASFs.

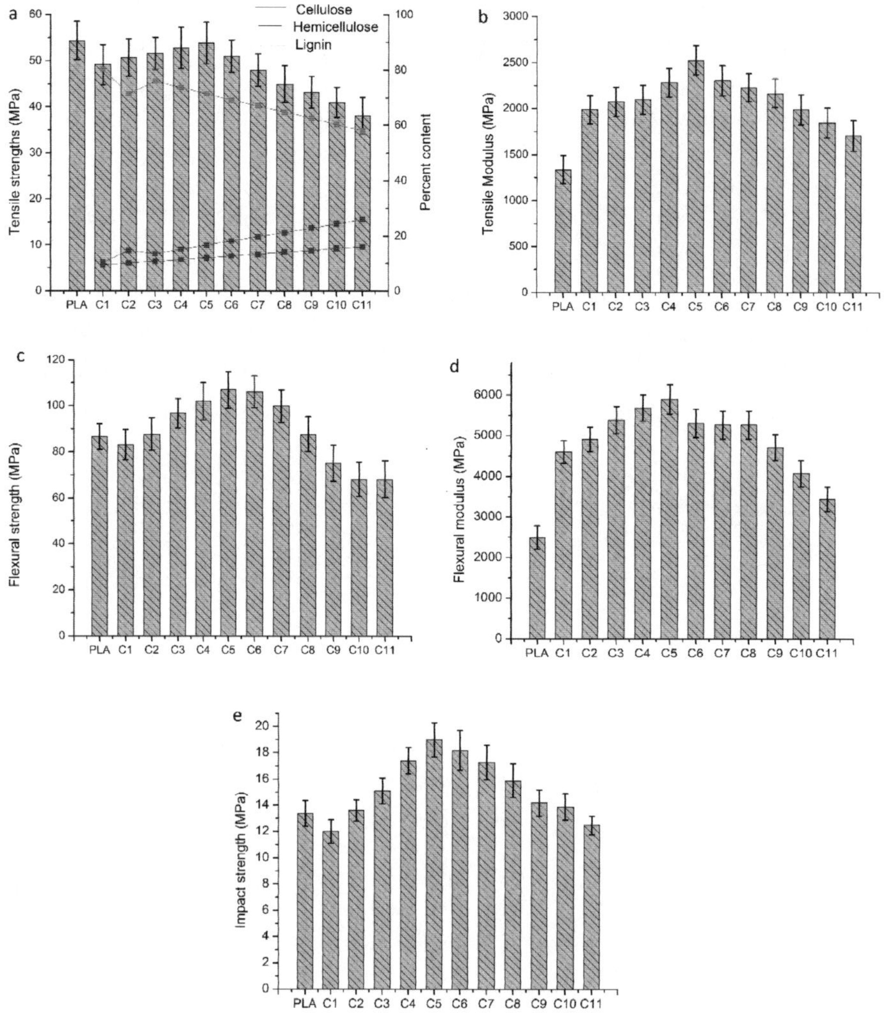

Figure 1. Mechanical properties of PLA and PLA composites

However, the addition of HSFs could improve the flexural and impact properties of PLA effectively. For example, the flexural strength, flexural modulus and impact strength of C5 composite were 23.7, 137.3 and 41.7% higher than those of pure PLA. Comparing PLA/HSFs with PLA/USFs and PLA/ASFs, it can be observed that the strength and Young´s modulus of PLA/HSFs were better than those of PLA/USFs and PLA/ASFs except for C9 and C10 composites. For example, the tensile strength and modulus of C5 were 53.9MPa and 2526.7MPa, which were 41.5% and 48.1% higher than those of PLA/USFs, and 10.0% and 27.1% higher than those of PLA/ASFs. The impact strength of C5 was 19 KJ/m^2, which was 48% and 58.3% higher than that of PLA/USFs and PLA/ASFs.

Therefore, it can be concluded that HSFs could strengthen PLA matrix more effectively than USFs and ASFs. There may be the following several reasons. Firstly, the interface bonding of PLA and ASFs was better and the mechanical property of USFs was higher. When USFs and ASFs were compounded, they could play their synergistic effect and strengthened PLA matrix effectively. Secondly, appropriate contents of hemicellulose and lignin are good for SFs and the interface bonding of PLA and SFs, which had been proved by above experiment. Thirdly, hemicellulose and lignin can make SFs well disperse in the polymer matrix and hence enhance the dispersion uniformity of HSFs [6] At last, some substance existing in the USFs can induce the formation of β-crystal, which had been proved in our previous work.[7] β-crystal higher impact strength and better thermal properties than α-crystal, which may especially explain why PLA/HSFs and PLA/USFs had better impact strength than PLA/ASFs.

3.2. Influence of hybrid ratios of USF and ASF on the strength of PLA/SF composites

The mechanical properties of PLA/HSF composites changed with the hybrid ratios of USFs and ASF1h. When the mass ratio of USFs and ASF1h was 0:10 and 1:9, the mechanical properties of PLA/HSFs composites were relatively worse, which was mainly due to the higher content of ASF1h. As the content of USF increasing, the mechanical properties of PLA/HSFs composites became better and better. When the mass ratio of USFs and ASF1h was 4:6, the mechanical properties of PLA/HSFs composite was the best and the content of cellulose, hemicellulose and lignin at this moment were 71.4%、16.6% and 12.0% respectively. However, with the content of USFs further increasing, the mechanical properties of PLA/HSFs composites reduced. Especially with the USFs content increasing, the reduction amplitude in mechanical properties increased. It presented that the interfacial strength was important for the mechanical properties of fiber reinforced PLA composite. When the interface bonding is too weak, it is possibly difficult for SFs to play their reinforcing roles.

From the above investigation and analysis, it can be found there existed doubts about the pretreatment of natural fibers. Although pretreatment promoted the interface bonding of fibers and matrix, it also caused the loss of the mechanical properties of fibers. By mixing untreated fibers with pretreated fibers to prepare hybrid fibers and fill them in the polymer matrix, the contradiction can be preferably solved and then the reinforcement of fibers for matrix can be achieved effectively. Therefore, different pretreatments can be employed to prepare natural fibers with different surface morphology, compositions and properties. If these natural fibers with different properties are filled in the polymer matrix, composites with excellent properties will perhaps obtained through synergistic effect of different natural fibers.

3.3. SEM analysis

Figure 2 was the SEM graphs of the impact sections of eleven PLA composites obtained under magnification of 100X. It can be found from Figure 2(a) that the impact section of PLA/ASF composite was smooth and fewer cavities were observed, which showed better interface bonding of PLA matrix and ASF.

Meanwhile, many torn SFs were visible in Figure 2(a). It was mainly because that the strength of SF declined obviously after alkali treatment and better bonding between ASF and matrix made SFs absorb more impact energy. It was also found in the following Figure 2(b-k) that pulled-out SFs became more and more with the USF content increasing. This phenomenon is more obvious in Figure 2(g-k). For example, in the 5*4 mm^2 range, the number of holes in Figure 2(a) is 10 and those in Figure 2(g) and Figure 2(k) reached 26 and 32, respectively. Those presented that the interface bonding became weaker with the USF content increasing. Moreover, it can be observed that lengths of pulled-out SFs in Figure 2(f-k) were longer than those in Figure 2(a-e) obviously. In combination with experiment conditions, the length of SFs was 1cm in the beginning, and became shorter and shorter after alkali treatment and processing. It is mainly because alkali treatment removed some compositions, the diameter and length of sisal fiber were greatly reduced to about 30% of its origin. The fibers also became brittle. Then in the process of mixing with the matrix, the fibers broke and get shorter after turning and twisting. Thus, it may be concluded that a large number of SFs are pulled out directly and didn't bear external force when composites suffering from external force. These stated that fibers can't play an effective role as reinforcement if the interface bonding of fiber and matrix is too weak.

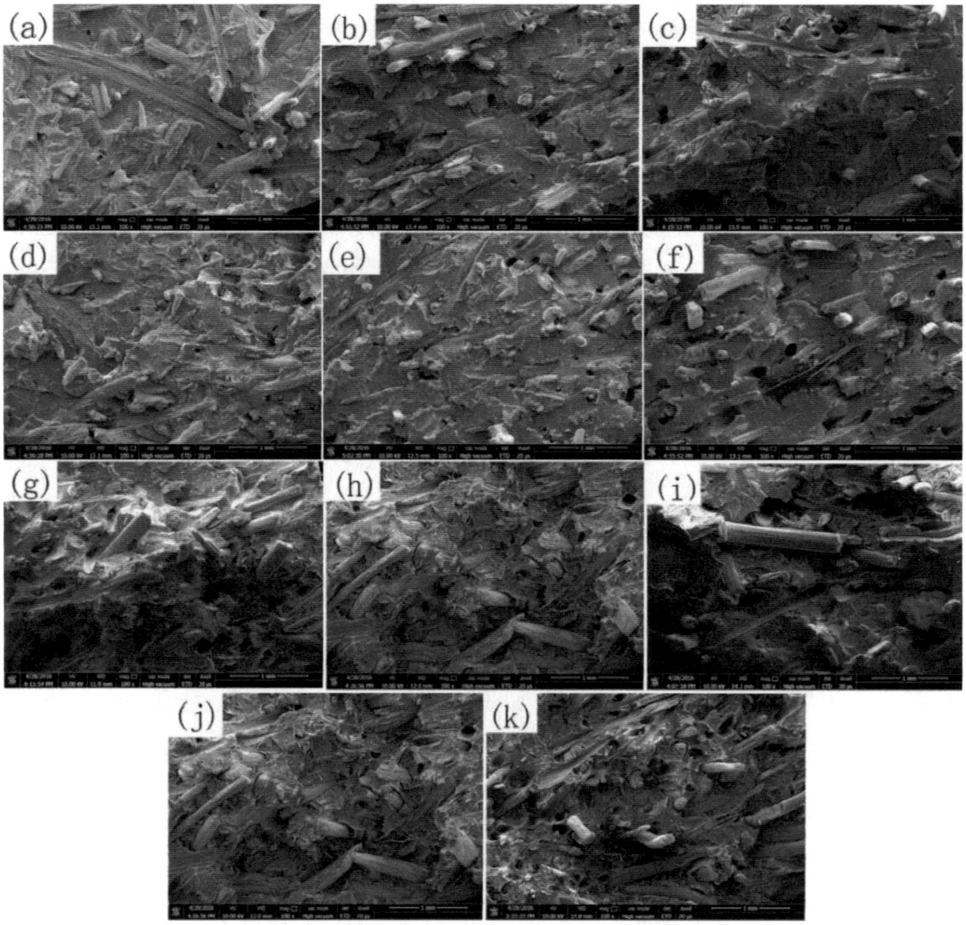

Figure 2. SEM graphs of eleven PLA composites. (a): C1; (b): C2; (c): C3; (d): C4; (e): C5; (f): C6; (g): C7; (h): C8; (i): C9; (j): C10; (k): C11.

4. Conclusions

It was found the reinforced effect of HSFs for PLA composites is better than that of USFs and ASFs in this work. The mechanical properties of PLA/HSFs were better than those of PLA/USFs and PLA/ASFs in this work. PLA/HSFs composite could reach the best mechanical properties when the hybriding ratio of USFs and ASF1h was 4: 6, and the content of cellulose, hemicellulose and lignin at this moment were 71.4%、16.6% and 12.0% respectively. Based on the above analysis, it can be found the reason why treated fibers are difficult to strengthen polymer composites effectively. Though various pretreatments improve the compatibility between fiber and matrix, they can also decline the mechanical properties of fibers themselves. By using hybrid fibers, the high strength of untreated fibers and the good compatibility of treated fibers can play a synergistic role to strength the polymer matrix effectively. Furthermore, we can use different pretreatments to obtain fibers with different morphology, composition and properties. Then these fibers can be mixed and filled into polymer matrix to prepare composites with high performance.

5. References

[1]. Hassan, E., Wei, Y. Plant fibers reinforced poly(lactic acid)(PLA) as a green composites, review. Int. J. Eng. Sci. 2012, 4, 4429-4439.

[2]. Ramamoorthy S.K., Bakare, F. Performance of biocomposites from surface modified regenerated cellulose fibers and lactic acid thermoset bioresin. Cellulose. 2015, 22, 2507-2528.

[3]. Shahin, M.M. Structure-property relationships in polyester fibres. Polym Test. 1995, 14, 243-251.

[4]. Zhu, Z.Z., Hao, M.Y. Influence of contents of chemical compositions on the mechanical property of sisal fibers and sisal fibers reinforced PLA composites. J. Nat. Fiber. DOI: 10.1080/15440478.2018.1469452.

[5]. Tragoonwichian, S., Yanumet, N. Effect of fiber surface modification on the mechanical properties of sisal fiber-reinforced benzoxazine/epoxy composites based on aliphatic diamine benzoxazine. J. Appl. Polym. Sci. 2007, 106, 2925-2935.

[6]. Rozman, H.D., Tan, K.W. The effect of lignin as a compatibilizer on the physical properties of coconut fiber-polypropylene composites. Euro. Polym. J. 2000, 36, 1483-1494.

[7]. Zhu, Z.Z., Ye, C.L. Improvement in mechanical and thermal properties of Polylactic Acid (PLA) biocomposites due to the addition of hybrid sisal fibers and diatomite particles. Int. J. Polym. Anal. Charact. 2016, 3, 365-377.

Acknowledgment

This research was supported by National Key Basic Research Program of China (973 Program) under Grant <number 2012CB025902>.

IOP Conf. Series: Materials Science and Engineering **544** (2019) 012013 doi:10.1088/1757-899X/544/1/012013

Effect of various factors on extraction efficiency of total anthocyanins from Butterfly pea (*Clitoria ternatea L.* Flowers) in Southern Vietnam

Tri Nhut Pham[1,2], Tri Duc Lam[1,3], Minh Tien Nguyen[1], Xuan Tien Le[4], Dai-Viet N. Vo[5], Tran Quoc Toan[6], Thanh Sang Vo[1,*]

[1]NTT Hi-Tech Institute, Nguyen Tat Thanh University, Ho Chi Minh City, Vietnam
[2]Faculty of Chemical Engineering and Food Technology, Nguyen Tat Thanh University, Ho Chi Minh City, Vietnam
[3]Graduate School of Biotechnology, Kyung Hee University, South Korea
[4]Department of Chemical Engineering, HCMC University of Technology, VNU-HCM, Ho Chi Minh City, Vietnam
[5]Faculty of Chemical and Natural Resources Engineering, Universiti Malaysia Pahang, Malaysia
[6]Institute of Natural Products Chemistry, Vietnam Academy of Science and Technology, Ha Noi, Vietnam
*Corresponding author: vtsang@ntt.edu.vn; labams2013@gmail.com

Abstract. Nowadays, anthocyanins have become an interesting topic of much scientific research and act as potential sources of eco-friendly natural food dyes and their helpful effects on human health.. The purpose of the present article was to apply conventional extraction methods of pigments from Butterfly pea (Clitoria ternatea L.Flowers) in Southern Vietnam were tested to extract anthocyanins. The effects of various factors, including the solvent concentration, extraction time, temperature, and solvent/material ratio on the extraction total anthocyanins from Butterfly pea were investigated. In this study, the concentration of monomeric anthocyanins was completed by the spectrophotometric pH differential method.The highest recovered anthocyanin level was achieved with 50 % ethanol (143.49 mg/L), while the lowest one was obtained with water (90.9 mg/L). The extraction yield was significantly affected by temperature and the excellent condition was 60°C. The extraction yield raised with increasing of solvent/material ratio until equilibrium has arrived at the optimal ratio of 25:1 (mL/g). The extraction yield increased first and then decreased with an extension in time values.

1. Introduction

Nowadays, Herbs become more and more popular in many topic of scientific researches as natural products due to their chemical composition, following with two main reasons: applications as natural options for food ingredients and helpful influences on human health based on their antioxidant characteristics [1-6]. Anthocyanins play a vital function on pharmacological against difference human diseases, such as allergies, cancer, cardiovascular disease and so on. [7,8]. Extraction is the first major step in the purification and recovery of active ingredients from plant materials. the purpose of an extraction method in order to provide for the maximum yield which consists of the high concentration of target compounds.

There are many developed techniques which applying to extract anthocyanins, such as microwave-assisted, conventional solvent extraction, supercritical fluid extraction and ultrasound-assisted, however, the solvent extraction technique (liquid-liquid and solid-liquid extraction techniques) is the most generally applied, and has demonstrated to be an efficient and reliable method [9-11]. The difference extraction conditions affect on extraction yields of anthocyanins from plant tissues. The efficacy of solvent extraction depended on many factors such as solvent concentration, temperature, time, pH and so on. Each single factor analyses can provide vital information on the varieties for major extraction parameters on the extraction of anthocyanins compounds from plant materials [12,13]. Clitoria ternatea belong to the family Fabaceae and sub-family Papilionaceae, also known as Butterfly pea.,. Butterfly Pea flower contain high amount of anthocyanins compared to Roselle and dragon fruit [14]. The anthocyanins base found in this flower is delphinidinas. Anthocyanins found in natural sources were useful to food colorant industry when it could give vibrant colors to the food product [15]. Clitoria ternatea (L.) native come from tropical Asia acts as a perennial, twining herbal medicinal plant, has a long history of applying as anxiolytic agent and a memory enhancer. In considering the increasing interest in estimating the antioxidant potential of herbal medicine. Therefore, this research was proposed to determine the best conditions including solvent concentration, type of solvent, liquid-to-solid ratio, extraction time and temperature in order to maximize the yield of total anthocyanins level from Butterfly pea by individual factor experiments.This research contributes a chance to evaluate the potential of Butterfly pea (Clitoria ternatea L. Flowers) in Southern Vietnam as an available source of natural antioxidants and a potential supplement in the pharmaceutical and food industry.

2. Material and methods

Dried petals of butterfly pea were ground using a commercial grinder and were weighed to 10g, put in the two neck round bottom flask and was extracted by Ethanol 50° solutions. Ethanol (C_2H_5OH) is purchased from Sigma Aldrich (US). For extraction parameter study, 10g of butterfly pea powder was placed in the two neck round bottom flask as and was extracted by ethanol with concentration at 50. The liquid/solid ratio in this experiment ranges from 10:1 to 30:1 (mL/g) and. The extraction temperature is adjusted from 40 to 80 (°C) and time varies from 15to 75 (min). Then, centrifugation took place at 4000 rpm for 15 min by high speed centrifuge Model LACE16 (from COLO lab expert). The supernatant was collected and the extract, after being filtered with filter paper, was transferred into plastic bottle for yield estimation. The pH scanning of supernatant ranges from 400 nm to 700 nm. Monomeric anthocyanin was calculated as cyanidin-3-glycoside using a pH differential method [16].

3. Result and discussion

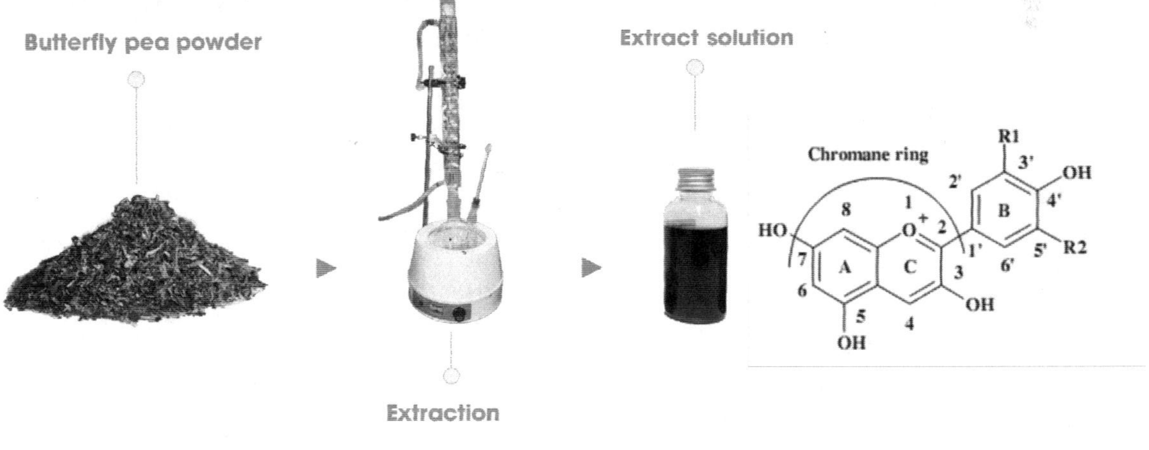

Figure 1. The process of extraction anthocyanins from Butterfly pea (Clitoria ternatea L.Flowers)

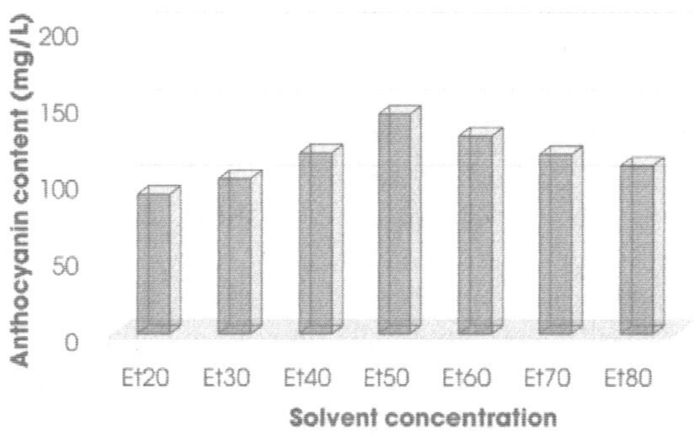

Figure 2. Effect of solvent concentration on anthocyanin content from Butterfly pea

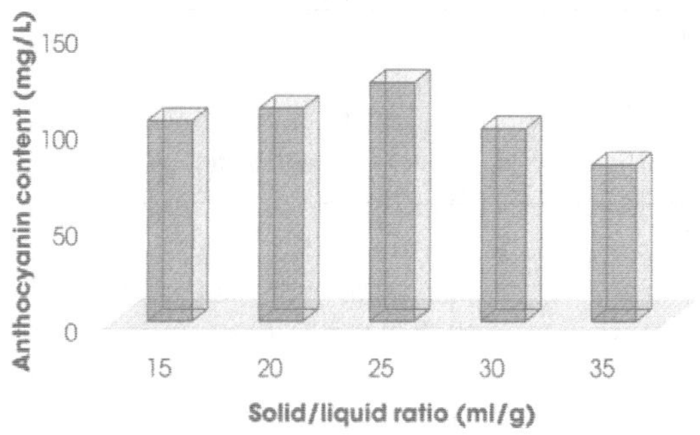

Figure 3. Effect of liquid-solid ratio on anthocyanin content from Butterfly pea

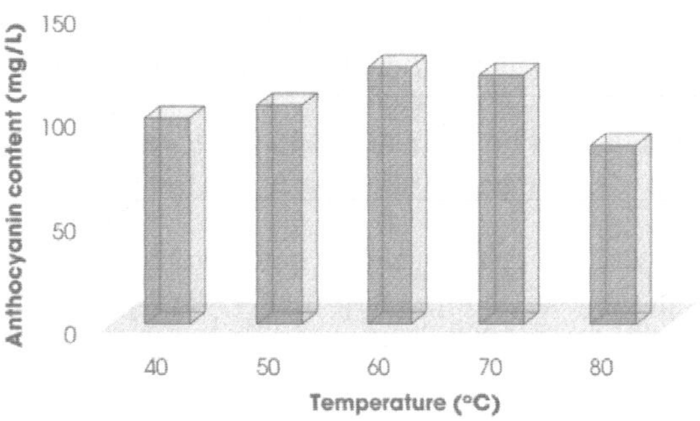

Figure 4. Effect of temperature on anthocyanin content from Butterfly pea

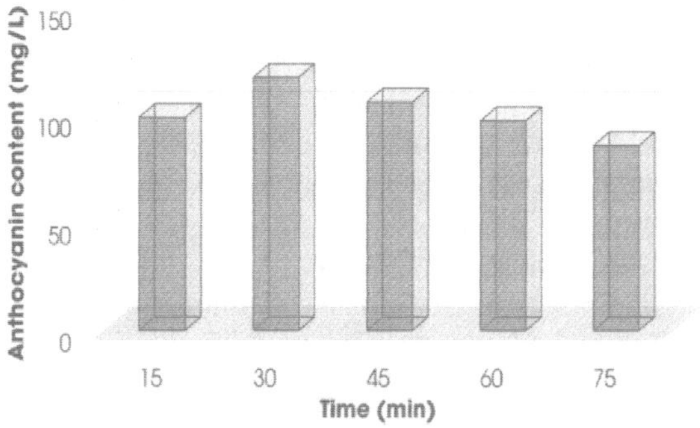

Figure 5. Effect of time on anthocyanin content from Butterfly pea

3.1. Effect of solvent concentration on anthocyanin content

Solvent extraction plays an important common method of extraction of anthocyanins compounds. In overall, the larger the dielectric loss and dielectric constant, the higher is the ability of the disseminate microwave energy and solvent to absorb.. Anthocyanins, which have hydrophobic hydrocarbons, are only slightly soluble in organic solvents, but have functional polyphenols well soluble in polar solvent. Therefore, to extract anthocyanin, organic solvent or polarizer (usually water) are suitable solvent systems. Alike Figure 2, the anthocyanin content increased steadily from 90.9 mg/L to critical point at 143.49 mg/L when exchange sequence solvent concentration. Therefore, we selected ethanol/water with the ratio of 50:50 as the solvent used in the extraction for this investigation.

3.2. Effect of liquid-solid ratio on anthocyanin content

In many cases, a large volume of solvent is also used for extraction and recuperation of extracts, which can be a major problem in terms of environmental considerations. Figure 3 demonstrated the effect of liquid-solid ratio on anthocyanin content. By increasing the liquid-solid ratio from 15:1 to 25:1, the extracted anthocyanin content increased from 103.526 mg/L to 121.58 mg/L. However, from 30:1 to 35:1 of liquid-solid ratio, the anthocyanin content decreased sharply. This can be explained as follows. When the ratio of materials/solvent is insufficient to fill up the material, the hypertonic environment could not be created, detaining the color in the vacuole of the material. However, when the solvent/material ratio reaches a certain value (depending on the characteristics of the material), the cell, owing to its capability to rapidly absorb water, swells to the maximum and bursts out simultaneously, releasing the color within the vacuole. Therefore, based on the graph we choose the liquid/solid ratio of 25/1.

3.3. Effect of temperature on anthocyanin content

High extraction temperature generally also leads to increased solubility. This can be demonstrated by the fact that higher temperature leads to reduced intermolecular interactions within the solvent, giving increase to higher molecular motion, and rose solubility of the target compounds in the extraction solvent. The increasing temperature can also cause destruction of cell matrix, resulting in increased availability of components for extraction. Moreover, at higher temperature levels, solvent viscosity decreases, increasing the solubility, and thus it also increases the efficiency of extraction. But in some cases, it has been remarked that the extraction efficiency rises with the increase in temperature up to a certain temperature level and then the extraction yield begins falling with the further increase in temperature levelThis also differences with the type of target compounds to be extracted as the temperature of degradation for every compound. As showed in Fig. 4, the anthocyanin content in the fluid as well as the effect of color separation from the material increased with increasing temperature and reached the highest level at the temperature of 60°C. However, from 60 to 70°C, the anthocyanin content in the extract did not change significantly and rising temperature past 70°C caused the content to diminish due to decomposition. Therefore, in practice, based on the combined effects of the good extraction yield of anthocyanins, an extraction temperature of 60 °C could be applied.

3.4. Effect of time on anthocyanin content

As shown in Figure 5, the level of anthocyanin in the fluid tends to decrease with prolonged extraction time. The extraction yield of anthocyanin from Butterfly pea raised as the time values rose from 15 to 30 minute, but it declined as the time values higher than 45 minutes were used, which showed that the extraction time value significantly influenced the extraction yield associated with other factors. Anthocyanin content was highest at 30 minutes, but as the time increased past 45 minutes, the content decreases markedly due to anticyclonic decomposition caused by the long exposure to high temperature. Therefore, the appropriate time to extract anthocyanin is 30 minutes.

4. Conclusion

In this study, the single factor experiments were determined for recognizing the optimum condition of each independent variable influencing the total anthocyanins content extraction efficiency of Butterfly pea (Clitoria ternatea L. Flowers), namely liquid-solid ratio, the solvent concentration, extraction time and temperature.Commonly, high extraction yield was achieved using aqueous acetone (about 50%) as the solvent, the liquid/solid ratio of 25/1 and the extraction yield could further be improved using a prolonged time of 30 min and a higher incubation temperature of 60°C. Under these optimized conditions, the experimental maximum yield of total anthocyanins content of 143.49 mg/L was taken. The optimal ranges (minimum and maximum values) acquired for all extraction parameters in the present research can also be served as fundamental information for the scale-up extraction of antioxidant compounds from Butterfly pea.

Acknowledgments: This work was supported by grants from Nguyen Tat Thanh University, Ho Chi Minh City, Viet Nam.

References

1. T H Tran, H H H Nguyen, D C Nguyen, TQ Nguyen, H Tan, L T H Nhan, D H Nguyen, L D Tran, S T Do and T D Nguyen 2018 *Processes* **6** 206
2. A N Q Phan, L G Bach, T D Nguyen, N T H Le 2019 *Journal of Nanoscience and Nanotechnology* **19** 974-978
3. N P T Nhan, T T Hien, L T H Nhan, P N Q Anh, L T Huy, T C T Nguyen, D T Nguyen, L G Bach 2019 *Solid State Phenomena* **279** 235-239
4. T S Vo, D H Ngo, L G Bach, D N Ngo, S K Kim,2017 *Process Biochemistry* **54** 188-194
5. H C Mai, T T Le, T T Diep, T H N Le, D T Nguyen, and L G Bach 2018 *Asian Journal of Chemsistry* **30** 293-297.
6. T T Hien, N P T Nhan, D T Nguyen, V T T Ho, L G Bach 2018 *Solid State Phenomena* **279** 217-221.

7. A R C Braga, D C Murador, L M S Mesquita, V V Rosso 2018 *Journal of Food Composition and Analysis* **68** 31-40
8. A Cassidy,2018 *Molecular Aspects of Medicine* **61** 76-82
9. P Ongkowijoyo, D A Luna Vital 2018 *Food Chemistry* **250** 113-126
10. L Eliasson, L Labrosse, L Ahrné 2017 *Food Science and Technology* **85** 510-516
11. E Agcam, A Akyıldız, V M Balasubramaniam 2017 *Food Chemistry* **237** 461-470
12. M Bonfigli, E Godoy, M A Reinheimer, N J Scenna 2017 *Journal of Food Engineering* **207** 56-72
13. R Romero-Díez, M Matos, L Rodrigues, Maria R Bronze, S Rodríguez-Rojo, MJ Cocero, A A Matias 2019 *Food Chemistry* **272** 258-266
14. T Suppadit, N Sunthorn, P Poungsuk 2011 *African Journal of Biotechnology* **10** 19109-19118
15. Y Tanaka, F Brugliera, S Chandler, *International Journal of Molecular Sciences,* **10**, 2009, 5350-5369
16. J Lee, R W Durst, R E Wrolstad 2005 *Journal of AOAC International* **88** 1269-1278

IOP Conf. Series: Materials Science and Engineering **544** (2019) 012014 doi:10.1088/1757-899X/544/1/012014

A Tough and Efficient Self-Healing Supramolecular Materials modulated with WCNT-NH_2

H R Ma, Y X Li, C Jiang, Y W Yao, C Yao

College of Chemistry and Molecular Engineering, Nanjing Tech University, Nanjing 211816, P. R. China
E-mail: yaochengnjut@126.com.

Abstract. Self-healing polymers crosslinked by solely reversible bonds are intrinsically weaker than common covalently crosslinked networks. In this work, a self-healing materials with excellent mechanical properties was synthesized by a new design strategy, wich is to introduce covalent into reversible networks to improve its mechanical properties. First, polyamidoamine dendrimer oligomers (PAMAM-Oligomers) was synthesized. Then amine functionalized carbon nanotubes (WCNT-NH_2) reacted with PAMAM-Oligomers to form a polyamidoamine carbon nanotubes composite materials (PAMAM-WCNT-NH_2). The synthesized materials behave as rubber and exhibit spontaneous self-healing behaviour at room temperature without external stimuli. The mechanical and self-healing properties are closely related to the content of WCNT-NH2. For 0.5wt% WCNT-NH_2 in the materials, the strain of the material is 330% and tensile strength is 0.6 MPa. Importantly it can improve the self-healing efficiency to 98% of its original tensile strength in 2h.

1. Introduction

Self-healing polymeric materials can revert to their original state with full or partial recovery of mechanical strength after damage. As such, they hold great promise to extend the lifetime of polymeric products in many fields[1]; examples include aerospace, automotive, civil, and medical engineering. Unlike classical polymer networks that are crosslinked by permanent covalent bonds, self-healing polymeric materials are often based on reversible associations, such as hydrogen bonds, metal–ligand coordination, host–guest interactions, ionic interactions, electrostatic interactions, hydrophobic associations, π–π stacking, or polymer entanglements[2]. Such reversible associations can break and reform to enable self-healing ability, but they are nevertheless relatively weak compared to covalent bonds. Thus, the toughness of self-healing polymers does not match that of covalent polymer networks such as natural rubber[3].

Hyperbranched polymer, as one kind of three dimensional macromolecules with globular or dendritic architectures, is widely used in coatings, nanomaterials, drug release and other fields [4, 5]. Compared with common linear molecular structure, hyperbranched structure has some advantages; 1) unique molecules with uniform, repeated branching units, endows it with unique chemical and physical properties. 2) The abundant functional groups in the branched structures allow them to easily form covalent bonds or non-covalent bonds. We designed and fabricate a novel polyamidoamine dendrimer (PAMAM) with a new design strategy, which is to introduce permanent covalent crosslinks into a reversible network to improve its mechanical properties. And hyperbranched structure self-healing marital have few reports.

Here, we report a stiff, self-healing hyperbranched composite materials that contains both covalent and reversible networks. The preparation process is in Figure 1. Firstly, the polyamidoamine (PAMAM) dendrimer (G-0.5PAMAM, G-1.0PAMAM) were synthesized in the step1 and step 2. Then

G-1.0PAMAM and dimer acids form polyamidoamine oligomers (PAMAM-Oligomers) through condensation reaction. As illustrated step 3, the PAMAM-Oligomers contain multifunctional branched with a controlled distribution of shapes and lengths. The oligomers carry motifs that can form reversible bonds (green symbols) along the branched chains of polymers and form covalent bonds (red circles) at the end of branching arms. Then WCNT-NH_2 crosslinks with the end groups of the PAMAM-Oligomers to form a network containing both reversible hydrogen bonds and permanent covalent as described by Step 4. And the formation of reversible associations is illustrated by the change of a pair of empty green symbols to filled ones.

2. Experiment and methods

2.1. Synthesis of G-0.5 PAMAM.

Ethylenediamine (13.3 g, 0.2mol) was dissolved in 20 ml of anhydrous methanol and added drop wise to a solution of methylacrylate (144.0g, 2.4 mmol) in 50 ml of anhydrous methanol under nitrogen in an ice bath. The mixture was kept stirring for 24h at room temperature. The residual methylacrylate and solvent was evaporated under reduced pressure to afford the desired G-0.5.

2.2. Synthesis of G-1.0 PAMAM

G-0.5 PAMAM (25.0 g, 61.8mmol) was dissolved in 30 ml of anhydrous methanol. The solution was stirred under nitrogen for 30 min in an ice bath. Then, diethylenetriamine (75.4g, 714.6 mmol) was dropwise added to the solution and kept stirring for 24 h at room temperature. The residual diethylenetriamine and solvent were evaporated under reduced pressure to afford the desired G-1.0.

2.3. Synthesis of PAMAM-WCNT-NH_2 composite materials

Amine functionalized carbon nanotubes WCNT-NH_2 were synthesized according to the reported procedure[6], then the WCNT-NH_2 was dispersed in anhydrous DMF.

G-1.0 PAMAM (15g, 21.8mmol) was mixed with dimer acid (48.8g, 87.2mmol), and the mixture was stirred at 120 ℃ for 2 h until a homogeneous dispersion was obtained. After a further 4 h stirring at 150℃, the PAMAM-Oligomers was allowed to reach room temperature and dispersed in DMF. 0.5wt%, 1.0wt% or 1.5wt% WCNT-NH_2 was respectively added to the oligomers solution and stirred for 30min. The mixture was heated to 150℃ and kept stirring for 8h while DMF was removed under N_2. Finally, the product was poured into template to obtain a testable sample for test.

2.4. Characterization

Mechanical tensile-stress were performed on a LLOYD LR30k pulls tensile testing machine. Four samples were completely cut perpendicular to the tensile direction and then jointed immediately by contact the fracture surfaces. The healing was lasted for variable times (1h, 2 h, 24h) at room temperature of 25 ℃ The tensile tests were performed at a stretching rate of 5 mm / min at the ambient temperature of 25 ℃. Electron microscopy (TEM) images are collected by using a JEOL JEM-1400 Microscope with an accelerating voltage of 100 kV.

IOP Conf. Series: Materials Science and Engineering **544** (2019) 012014 doi:10.1088/1757-899X/544/1/012014

3. Results and Discussion

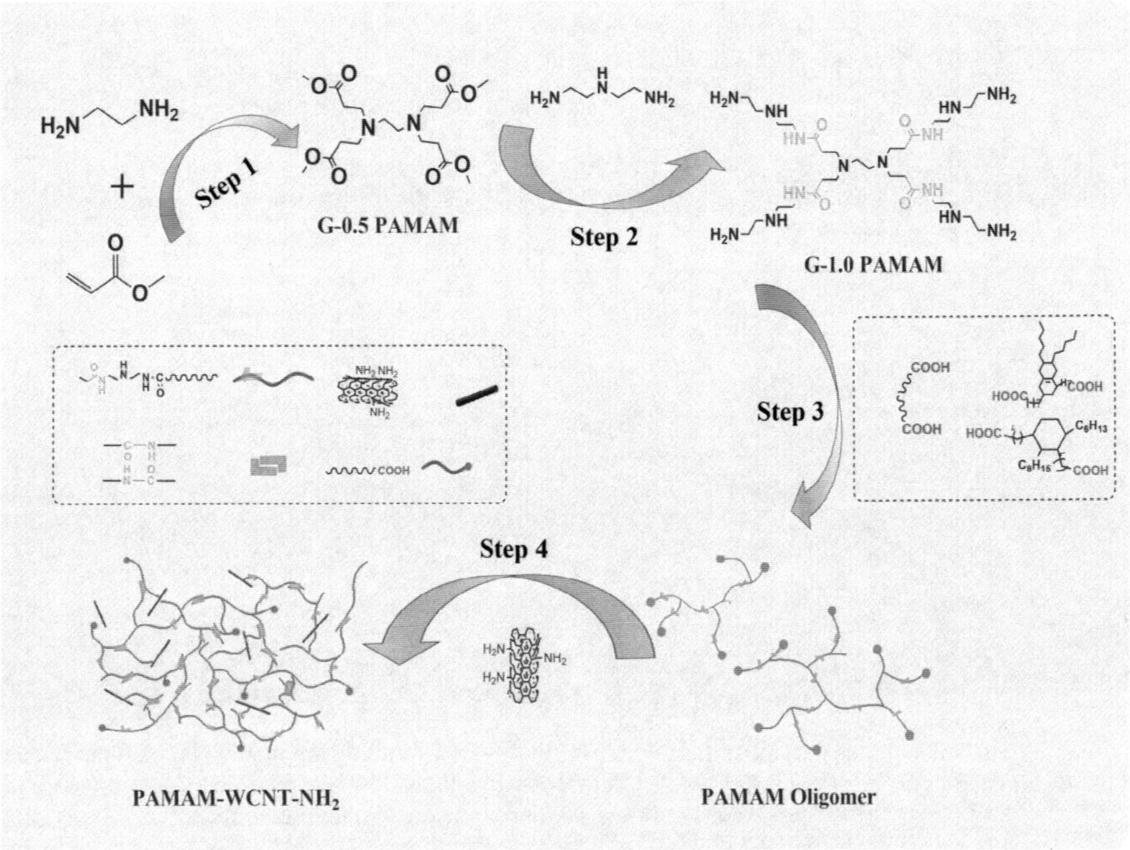

Figure 1 Concept and synthesis of hybrid networks.

3.1. Tensile Properties of the Composites Materials

To investigate the influence of WCNT-NH_2 as a filler on materials, we synthesized three PAMAM-WCNT-NH_2 with different weight percentages of WCNT-NH_2 at 0.5wt%, 1.0wt% and 1.5wt% which we termed PAMAM-0.5wt%WCNT-NH_2, PAMAM-1.0%WCNT-NH_2, PAMAM-1.5wt%WCNT-NH_2. Stress-strain curves are shown in Figure 2a. For PAMAM-1.5wt%WCNT-NH_2, although its tensile strength reaches 2.80 MPa, the elongation at break is only 114%, which is different from a conventional elastomer. However, for PAMAM-1.0wt%WCNT-NH_2, the elongation at break exceeds 275%, the tensile strength is 1.8 MPa. And it behaves as soft rubber. Similarly, for PAMAM-0.5wt%WCNT-NH_2, the elongation at break exceeds 330%, and the tensile strength decreases to 0.60MPa. Tensile stress increases and the strain decreases with the WCNT-NH_2 content increasing. The result means that the hybrid network disturbs the orientation of the polymer chain at high elongation, while increasing WCNT-NH_2 to improve the crosslinking degree of the network. So, the WCNT-NH_2 is the key factor in tuning the mechanical properties of the composite materials. And with only 0.5wt%WCNT-NH_2, the tensile strength of the composite materials is superior to the hydrogen bond self-healing material previously reported[7].

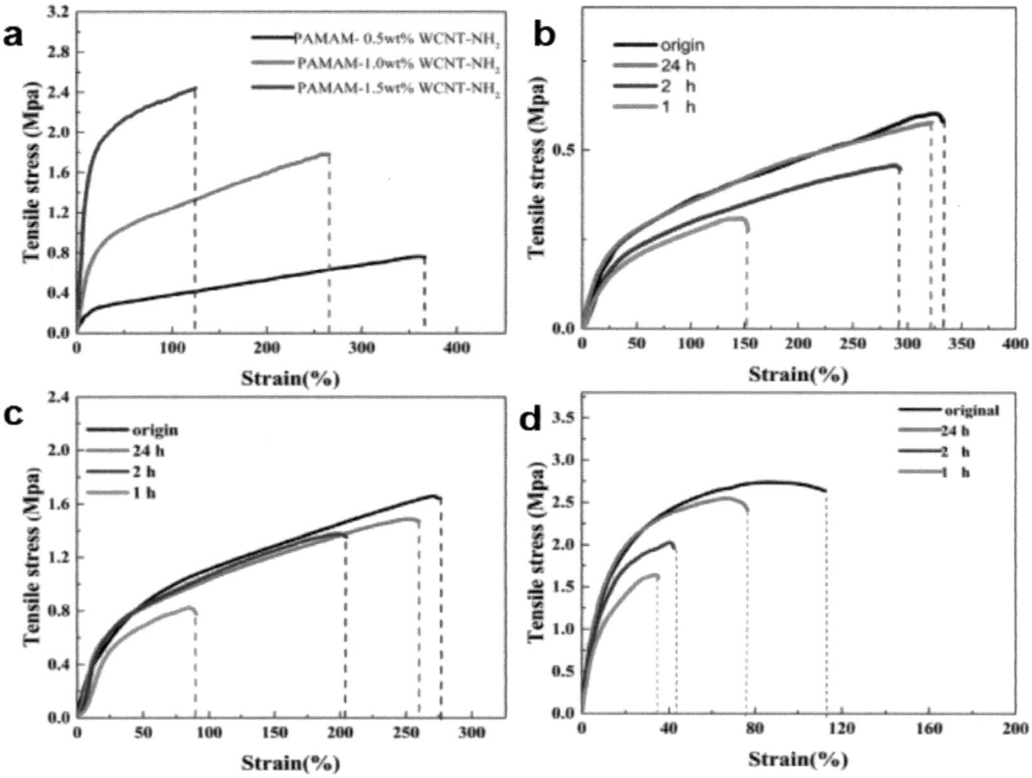

Figure 2. a) Strain-stress curves of the PAMAM-WCNT-NH_2 with different WCNT-NH_2 percentages. Strain-stress curves of the PAMAM-WCNT-NH_2 with WCNT-NH_2 at b) 0.5wt%, c)1.0 wt% and d)1.5wt% upon different healing time at room temperature;

3.2. Self-healing Properties Composites Materials

The hydrogen bond type self-healing mechanism of polymer networks has been reported. Upon a mechanical damage event, the weaker hydrogen bonds are more likely to break as opposed to the stronger covalent bonds. When the ambient temperature is above the glass transition temperature (Tg), the dynamic hydrogen bonds found in the supramolecular components of the fracture area can be re-established as soon as the damaged surface resumes contact[7]. Based on results from the tensile test, we focused on the self-healing ability of the materials. First, we cut the self-healing materials into two separate pieces. Then bring them together with different healing times. As shown in Figure 2b, 2c and 2d, the curves indicate that the PAMAM-0.5wt%WCNT-NH_2, PAMAM-1.0wt%WCNT-NH_2 and PAMAM-1.5wt%WCNT-NH_2 were able to recover 90%, 78% and 45% of their original elongation at break in 2h. Particularly, the healing time of PAMAM-0.5wt%WCNT-NH_2 and PAMAM-1.0wt%WCNT-NH_2 are significantly shorter than other previously described hydrogen bonding self-healing supramolecular materials which generally need more than 48 h heal to 75% of its original strain[8]. As the healing time increased to 24 h, the self-healing efficiency increased to 98.6%, 92.1% and 66%, respectively. Increasing the amount of WCNT-NH_2 lead to more covalent cross-linking and more restricted movement of polymer chains, which may explain the observed trend for increased time for the self-healing process. The self-healing experiment of the PAMAM-1.0wt%WCNT-NH_2 in Figure 3a demonstrates that the broken ends are recombined just by simply bringing them together at room temperature. The SEM images confirm that fracture surface is sufficiently healed with time in the Figure 3b.

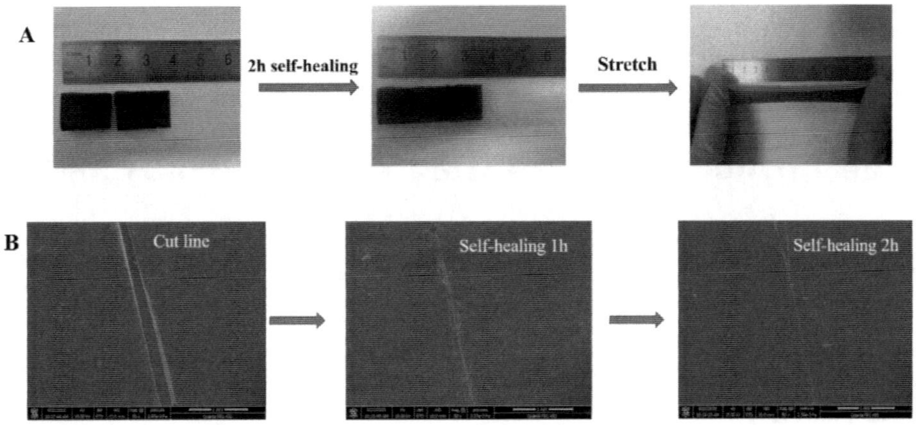

Figure 3. a) Stretching of the PAMAM-1.0wt%WCNT-NH_2 sample after cut and subsequent 2h self-healing at room temperature. The sample can be stretched to 260% while retaining a modulus similar to pristine sample. b) SEM images of the PAMAM-1.0wt%WCNT-NH_2 surface after 1 and 2h.

4. Conclusions

In summary, this research introduces a self-healing materials based on crosslinking hyperbranched structure. Then we described concept of using WCNT-NH_2 as both cross-linkers and fillers to improve the mechanical strength. And by adjusting the content of WCNT-NH_2, the degree of crosslinking and mechanical strength of the materials can be modulated. The final synthesized composite materials are similar to rubber and can spontaneously heal at room temperature in 2 h. In addition, due to the excellent conductivity of WCNT-NH_2, the PAMAM-WCNT-NH_2 is also electrically conductive, making it practical for use in stretchable self-healing conductive elements.

5. References

[1] Roy N, Bruchmann B and Lehn J-M 2015 DYNAMERS: dynamic polymers as self-healing materials *Chemical Society reviews* **44** 3786-807

[2] Wu J, Cai L H and Weitz D A 2017 Tough Self-Healing Elastomers by Molecular Enforced Integration of Covalent and Reversible Networks *Advanced materials* **29**

[3] Liu L, Zhu L and Zhang L 2018 A Solvent-Resistant and Biocompatible Self-Healing Supramolecular Elastomer with Tunable Mechanical Properties *Macromolecular Chemistry and Physics* **219** 1700409

[4] Chávez G, Campos C H, Jiménez V A, Torres C C, Díaz C, Salas G, Guzmán L and Alderete J B 2017 Polyamido amine (PAMAM)-grafted magnetic nanotubes as emerging platforms for the delivery and sustained release of silibinin *Journal of Materials Science* **52** 9269-81

[5] Vivek B, Kumar P and Prasad E 2016 Induction and Tunability of Self-Healing Property of Dendron Based Hydrogel Using Clay Nanocomposite *The journal of physical chemistry. B* **120** 5262-71

[6] Caamaño C, Grady B and Resasco D E 2012 Influence of nanotube characteristics on electrical and thermal properties of MWCNT/polyamide 6,6 composites prepared by melt mixing *Carbon* **50** 3694-707

[7] Wang C, Liu N, Allen R, Tok J B, Wu Y, Zhang F, Chen Y and Bao Z 2013 A rapid and efficient self-healing thermo-reversible elastomer crosslinked with graphene oxide *Advanced materials* **25** 5785-90

[8] Liu L, Pan C, Zhang L and Guo B 2016 A Novel and Non-Cytotoxic Self-Healing Supramolecular Elastomer Synthesized with Small Molecular Biological Acids *Macromolecular rapid communications* **37** 1603-10

Mechanical properties of SiC fibers irradiated by swift heavy ions

L Q Zhang[1, 2], C H Zhang[1] , Q Huang[3], T X Yan[1, 2], Z N Ding[1, 2], X L Zhang[1], Y G Chen[1, 2], B S Ma[1, 2] and J Y Li[1, 2]

1 Institute of Modern Physics, Chinese Academy of Sciences, Lanzhou, China
2 University of Chinese Academy of Sciences, Beijing, China
3 Ningbo Institute of Materials Technology and Engineering, Chinese Academy of Sciences, Ningbo, China
E-mail address: c.h.zhang@impcas.ac.cn, liqingzhang07@impcas.ac.cn

Abstract. Mechanical properties and surface features of SiC fibers irradiated with 246.8-MeV Ar ions at different fluences ($1.5\text{x}10^{14}$, $9.0\text{x}10^{14}$, $1.8\text{x}10^{15}$, and $2.7\text{x}10^{15}$ ions/cm^2) were investigated using a specific single filament tensile test and a field emission scanning electron microscopy (FE-SEM). Results reveal that carbon concentration on the fiber surface decreases while silicon concentration increases, with increasing ion fluence, accompanied by an adsorption of oxygen. Meanwhile, the diameter of fibers first reduces and then increases, with increasing ion fluence. Moreover, at $2.7\text{x}10^{15}$ Ar ions/cm^2 irradiation, some fibers were fractured. As a result, the mean tensile strength and the average elastic modulus of fibers decrease generally with respect to the ion fluence. The degradation mechanisms of mechanical properties of SiC fibers under irradiation were discussed in detail.

1. Introduction

Due to low neutron capture cross-section, low induced radioactivity, low afterheat and quick decay of activity, high thermal stability and excellent high-temperature mechanical properties, continuous silicon carbide fiber reinforced silicon carbide (SiC_f/SiC) composites have been recognized as the promising structure materials of harsh environments, such as aerospace, aviation, fission and future fusion reactors [1-4]. SiC_f/SiC composites are composed of near-stoichiometric SiC fibers, stoichiometric and fully crystalline SiC matrices, and the pyrocarbon (PyC) or BN interphase between the fiber and the matrix [5]. The performance of SiC_f/SiC composites, especially its toughness and high ultimate strength, depends on the effectiveness of the interface component and the high strength of the fiber component [6]. In fission reactors, SiC fibers are inevitably subjected to various fission fragments (heavy ions with up to hundreds of MeV) irradiation, which are produced via chain fission of the uranium. While in fusion reactors, the average energy of recoils produced by neutron irradiation in materials can reach several hundred keV. Most of the energy of fission fragments/recoils is deposited in the materials mainly through electronic processes (excitation and ionization of atoms along the trajective range). Therefore, it is indispensable to acquire a comprehensive knowledge of the mechanical behavior of SiC fibers under energetic-heavy-ion irradiation.

In this work, in order to understand the mechanical stability and the failure mechanisms of SiC fibers induced by energetic heavy ions and forecast the performance of SiC_f/SiC_m compositions in irradiation environments, we investigate changes of surface characteristics and tensile properties of SiC fibers irradiated with 246.8-MeV Ar ions at different fluences, by using a FE-SEM and a specific single filament tensile test.

2. Experiment

The SiC fiber specimens with diameter of about 10 μm were cut into a length of 50 mm and mounted on aluminum sheets with little overlap for irradiation. The irradiation was performed with high-energy Ar ions at a terminal chamber of the sector-focused cyclotron (SFC) at HIRFL (Heavy Ion Research Facility in Lanzhou) located in Institute of Modern Physics. The beam current was monitored with an Al foil assembly placed in the beam line before the chamber, which is composed of an Al foil of 2 μm in thickness to collect electric charge, a Φ18 mm aperture with a bias voltage of -300 V to suppress the secondary electron emission and a Φ15 mm aperture to limit the irradiation area. $^{40}Ar^{16+}$ ions passed through the assembly carry a kinetic energy of 246.8 MeV. The irradiation chamber consists of an energy degrader, a cooling stage (liquid nitrogen) and a heating stage (up to 600°C) of specimens for irradiation (Fig. 1). A description in detail of the irradiation terminal was given in our previous paper [7].

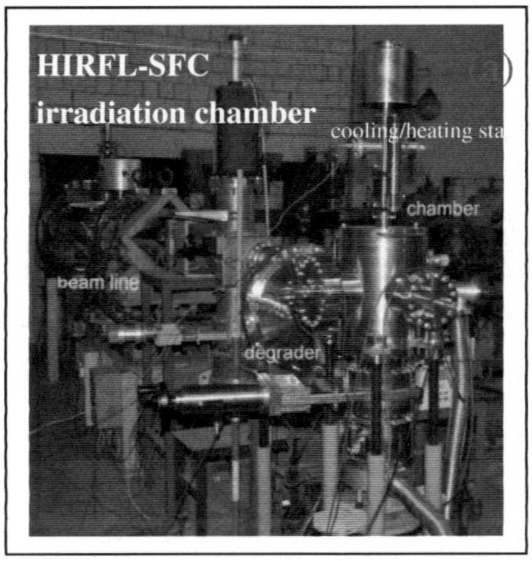

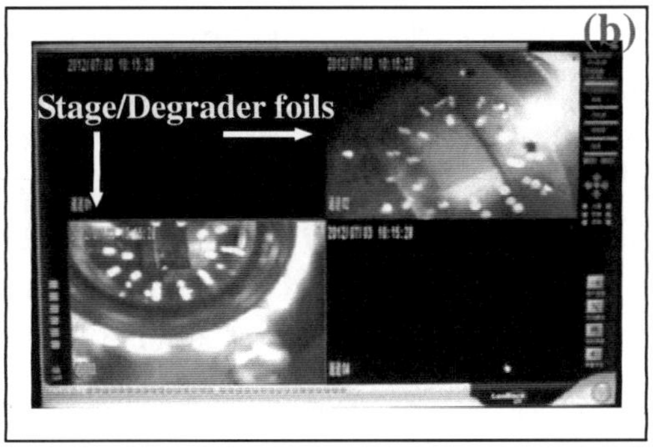

Fig 1. (a) Overview of the HIRFL-SFC irradiation chamber with energy degraders and cooling/heating stages. (b) Screenshot of the heating stage and Al foils of the energy degrader.

In order to increase the electron energy loss of Ar ion in 10-μm SiC fibers, we used the energy degrader, which is composed of 10-Al foils with thickness ranging from 4.899 to 45.009 μm. Fig. 2 shows estimates of the electron energy deposition density (in eV/μm^3) and the displacement damage level (in dpa, induced by nuclear energy loss) of energy-degraded Ar ions in 10-μm SiC fiber specimen, corresponding to a fluence of $1.8x10^{15}$Ar-ions/cm^2. The displacement damage level was estimated by using the "quick calculation" mode of SRIM 2012 code [8]. The displacement threshold energies for C and Si sublattices were 20 and 35 eV, respectively [9]. The 10-Al foils with different thickness in the energy degrader result in a successive increasing electron energy deposition density in the SiC fibers with an average diameter about 10 μm. Ar ions which passed through the two foils with thickness of 40.699 and 45.009 μm could produce a considerable displacement damage via nuclear collision in 10-μm SiC fibers. The irradiation fluence was from 1.5×10^{14} to 2.7×10^{15} Ar ions/cm^2. The temperature of specimens was kept around 40°C during irradiation in a vacuum of 10^{-5} Pa range.

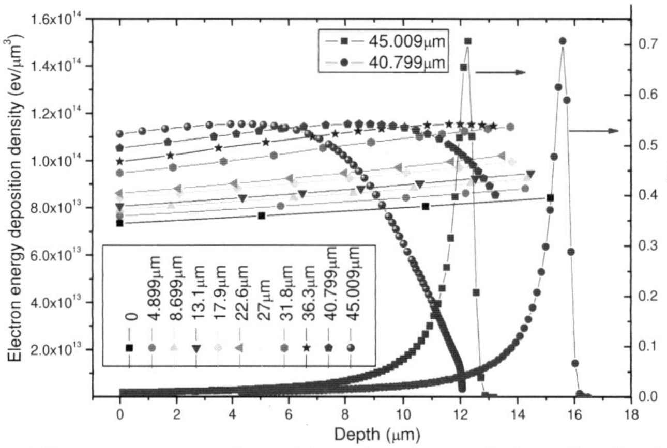

Fig 2. Electron energy deposition density and dpa distribution with respect to depth in SiC fibers under irradiation with 1.8×10^{15} Ar ions /cm^2 through the energy degrader. The horizonal column shows the thickness of 10 Al foils used in the energy degrader. The dpa values corresponding to Ar ions passed the 40.699 and 45.009 μm Al foils were plotted.

After Ar ions irradiation, tensile properties and surface features of SiC fibers were characterized by means of a specific single filament tensile testing machine and a field emission scanning electron microscopy (FE-SEM). The model of the electronic single filament tensile machine is Textechno FAVIMAT$^+$. The two ends of a fiber segment were fixed to two dog-bone shaped grips carefully during the tensile test. The span is 25 mm and the tensile rate is 1 mm/min. The model of the FE-SEM is Nano-SEM 450 with the maximal acceleration voltage of 35 kV. Electron image scanning and energy dispersive spectra (EDS) area scanning with electron energy of 15 keV were carried out at a high-vacuum mode.

3. Results and discussion

3.1. Surface characteristics and diameter analysis

The surface morphology, compositions and the diameter of SiC fibers after Ar ions irradiation were studied by a FE-SEM. The FE-SEM measurement was carried out along the fiber length within a fiber tow. Figs. 3 and 4 show SEM images, energy dispersion spectra and the mean diameter of SiC fibers irradiated with Ar ions at different fluences. For un-irradiated SiC fibers specimen, its average diameter is about 9.7 μm, the surface of fibers is smooth and the atom ratio between silicon and carbon atoms is about 2:3.

After irradiation at a low fluence (9.0×10^{14}Ar ions/cm^2), the average diameter of the SiC fibers decreases to 8.9 μm and the atom ratio between silicon and carbon atoms is above 2:3. That is, the concentration of carbon content reduces at the surface, possibly due to the lower displacement threshold energy of carbon atoms compared with silicon atoms [9]. On the other hand, oxygen atoms also appear at the fibers surface, because the damaged SiC fibers absorb oxygen atoms easily.

After irradiation at a high fluence (2.7×10^{15} Ar ions/cm^2), SiC fibers were fractured and at the fracture site, an enrichment of the silicon and a lack of the carbon were found. Meanwhile, big grains/bubbles appear on the fiber surface and the surface roughness of the SiC fibers also sharply increases. Moreover, the average diameter of fibers increases to about 10 μm.

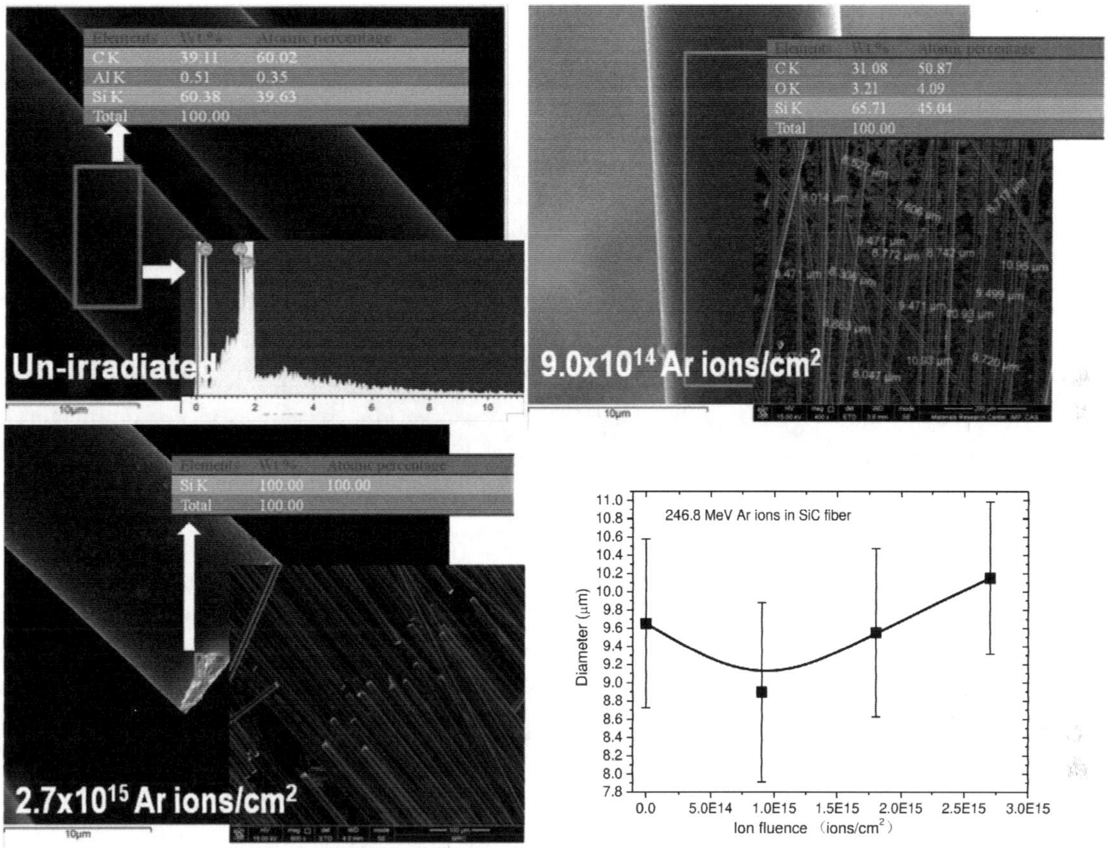

Fig 3. SEM images, EDS and the diameter of SiC fibers irradiated with Ar ions at different fluences.

Fig 4. The average diameter of SiC fibers as a function of the Ar-ion fluence.

3.2. Single filament tensile testing

In order to evaluate the influence of the irradiation on the mechanical stability of fibers, we carried out the single filament tensile test. The single filament tensile test can provide a mean tensile strength and an average elastic modulus. Mean tensile strength and elastic modulus were calculated from the load (F), gauge length (L) and the diameter (d) measurement. The tensile strength (failure strength) was characterized by $4F_{max}/\pi d^2$, where, F_{max} is the load at failure. The average diameter (d) was determined from above FE-SEM views (Figs. 3 and 4). The elastic modulus was characterized by $4LF_{max}/\pi d^2/(L_{max}-L)$, where, L_{max} is the fiber length at failure. The mean failure load is taken as the arithmetic mean for the same irradiation condition. For the same irradiation condition, twenty or more specimens were tested to ensure statistics. The mean diameter is also taken as the arithmetic mean.

Fibers that failed at the grip fiber interface were discarded from the analysis.

Typical single filament stress-strain curves of SiC fibers after they were irradiated with Sn ions at a fluence of 1.8x10^{15} Ar ions/cm^2 were shown in Fig. 5. The scattering of the ultimate tensile strength and the strain is due to the fluctuation of diameters of SiC fibers. All SiC fiber specimens were tested in the same way.

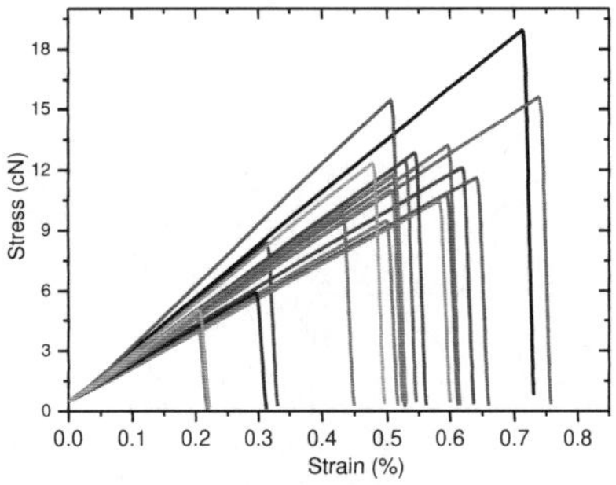

Fig 5. Typical single filament stress-strain curves of SiC fibers after they were irradiated with Ar ions at a fluence of 1.8x10^{15} Ar ions/cm^2.

Figs. 6 and 7 show the mean tensile strength and the elastic modulus of SiC fibers as functions of the ion fluence, after they were irradiated with Ar ions at different fluences. The elastic modulus value of fibers exhibit an obvious decrease as the ion fluence increases up to1.8x10^{15} ions/cm^2. Above this fluence, the elastic modulus slightly increases with the ion fluence (Fig. 7). And the tensile strength always displays a reduced trend with increasing ion fluence (Fig. 6). However, no significant change was observed in tensile-strength values. The mechanical strength of the fibers is determined mainly by the bonds between the particles rather than by the constituent particles themselves [10]. After irradiation, disordered structures were produced because the intergranular amorphous SiC_xO_y phase was decomposed into four fragments (SiC, C, SiO, CO), meanwhile the rearrangement and coalescence of SiC and C phases occurred under the heat effects of electron energy loss of Ar ions [11]. This rearrangement and coarsening of the crystal grains reduce its mechanical retention through the increase in the density of grain boundaries, because bonds between the grains are much weaker than those between atoms in crystalline or amorphous states [10, 12]. Simultaneously, during the decomposition-agglomeration process, the stress between the grains may be produced. The generation of these stresses is also responsible for the failure in fiber mechanical stabilities. On the other hand,

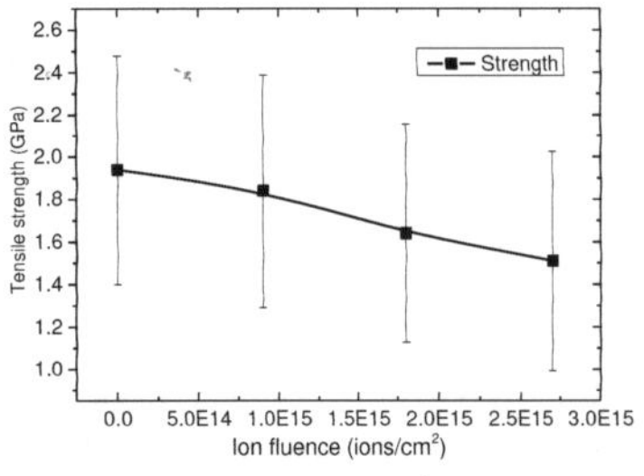

Fig 6. Average tensile strength of SiC fibers versus the irradiation fluence, after SiC fibers were irradiated with Ar ions at different fluences.

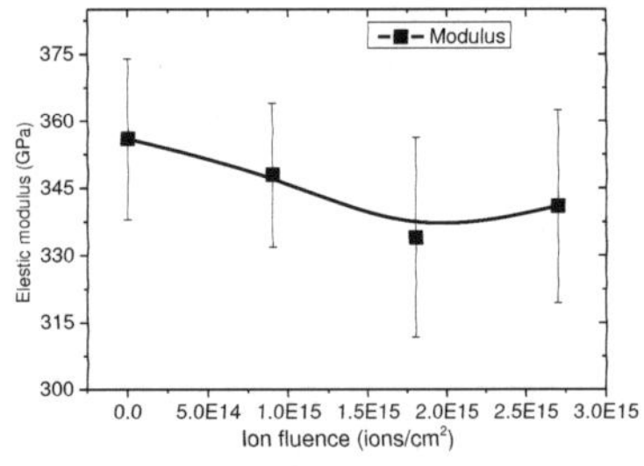

Fig 7. Mean elastic modulus of SiC fibers versus the irradiation fluence, after SiC fibers were irradiated with Ar ions at different fluences.

the formation of surface defects accelerates the attack of oxygen to fibers, resulting in an obvious reduction in fracture toughness (Fig. 3). However, at a fluence of 2.7x10^{15} Ar ions/cm^2 irradiation, the elastic modulus of fibers slightly increases. This is due to the material experiences a recovery from an amorphous phase to crystalline state under the high electronic energy deposition of Ar ions, revealed by TEM analysis in our previous work [13]. Summarily, the strength degradation of SiC fibers induced by irradiation can be ascribed to the decomposition of amorphous phase and the surface damage as well as the generation of stresses.

4. Conclusions

Surface morphology, compositions, and the mechanical properties of SiC fibers irradiated with 246.8-MeV Ar ions at different fluences were investigated using FE-SEM and a special single filament tensile test. Results show the irradiated SiC fibers first radically shrink and then expand, and at the surface, carbon concentration decreases while silicon concentration increases gradually, with increasing ion fluence. At the highest fluence irradiation ($2.7x10^{15}$Ar ions/cm^2), the SiC fibers were fractured. As a consequence, the fiber mechanical stability generally degrades with respect to the ion fluence due to sever stoichiometric imbalance between Si and C atoms and the addition of oxygen. Meanwhile, this degradation in mechanical properties also can be ascribed to that the decomposition and the coalescence of the amorphous phase lead to the grain agglomeration and coarsening, and flaws/stresses generation under the heat effects of electron energy loss of Ar ions.

5. References

[1] Morscher G N, Ojard G and Miller R 2008 *Compos. Sci. Technol.* **68** 3305
[2] Sharafat S, Aoyama A and Morley N 2009 *Fusion. Sci. Technol.* **56** 883
[3] Yvon P and Carré F 2009 *J. Nucl. Mater.* **385** 217
[4] Yu H, Zhou X and Wang H 2010 *Fusion. Eng. Des.* **85** 1693
[5] Yang B, Zhou X and Chai Y 2015 *Ceram. Int.* **41** 7185
[6] Curtin W A 1991 *J. Am. Ceram. Soc.* **74** 2837
[7] Zhang C H, Chen J C, Jang J and Kimura A 2014 *J. Nucl. Mater.* **455** 61
[8] Ziegler J F, Biersack J P and Littmark U 1984 *The stopping and range of ions in solids,* vol. 1 (New York: Pergamon Press)
[9] Weber W J, Gao F and Devanathan R 2004 *Nucl. Instr. Meth. B* **218** 68
[10] Sasaki Y, Nishina Y, Sato M and Okamura K 1987 *J. Mater. Sci.* **22** 443
[11] Shimoo T, Okamura K, Tsukada I and Seguchi T 1999 *J. Mater. Sci.* **34** 5623
[12] Sha J J, Hinoki T and Kohyama A 2007 *J. Mater. Sci.* **42** 5046
[13] Zhang L Q, Zhang C H and Huang Q 2018 *Nucl. Instru. Meth. B* **435** 169

Acknowledgements

This work has been supported by the National Natural Science Foundation of China (Grant Nos. 11675231, 91426304).

IOP Conf. Series: Materials Science and Engineering **544** (2019) 012016 doi:10.1088/1757-899X/544/1/012016

Response surface modeling and optimizing conditions for anthocyanins extraction from *Hibiscussabdariffa* L. (Roselle) grown in Lam Dong, Vietnam

Tri Nhut Pham[1,2], Thuong Nhan Phu Nguyen[1], Lam Tri Duc[1,3], Minh Tien Nguyen[1]Tran Quoc Toan[4], Le Thi Hong Nhan[5], Dai Viet N-Vo[6], Thanh Sang Vo[1], Le Minh Bui[1,*]

[1]NTT Hi-Tech Institute, Nguyen Tat Thanh University, Ho Chi Minh City, Vietnam
[2]Faculty of Chemical Engineering and Food Technology, Nguyen Tat Thanh University, Ho Chi Minh City, Vietnam
[3]Graduate School of Biotechnology, Kyung Hee University, South Korea
[4]Institute of Natural Products Chemistry, Vietnam Academy of Science and Technology, Ha Noi, Vietnam
[5]Department of Chemical Engineering, HCMC University of Technology, VNU-HCM, Ho Chi Minh City, Vietnam
[6]Faculty of Chemical and Natural Resources Engineering, Universiti Malaysia Pahang, Malaysia
*Corresponding author: blminh@ntt.edu.vn, labams2013@gmail.com

Abstract.*Hibiscus sabdariffa* L. (Hs, roselle; Malvaceae) is a common ingredient used in the food and pharmaceutical industries for color and health benefits. Anthocyanins along with protocatechuic acid and quercetin have been recognized as bioactive compounds in *Hibiscussabdariffa* L. aqueous extracts. In this work, Response Surface Methodology was implemented together with Box-Behnken experimental design to investigate the impacts of process variables on anthocyanin extraction yields from *Hibiscus sabdariffa* L. (Roselle) grown in Lam Dong, Vietnam. The effects of three independent variables; namely the extraction temperatures of solvent ethanol (X1); the duration of extraction (X2); solid-liquid ratios (X3) on the total anthocyanins content (Y1) were explored. The optimization model showed that with the solid liquid ratio 7.7:1, extraction time 33.29 min, and temperature 56.9°C, the highest anthocyanin concentration of 186.006 mg/L was obtained. *Hibiscussabdariffa* L. anthocyanin yield detailed significant correlation with high F values, low P values (<0.0001), and desirable determination coefficient (R2 = 0.9770). The resulting quadratic model could be used to predict yields with given conditions. Actual experimental verification and residual testing revealed that the actual anthocyanin yields could be accurately predicted by the model.

1. Introduction

Hibiscus sabdariffa L., commonly known as roselle or red sorrel, is one of the most commonly cultivated plants globally. The plant has tropical origin and belongs to the Malvaceae family. The plant is a common ingredient in traditional medicine and extracts from the plant has been shown to exhibit a wide spectrum of therapeutic effects regarding various diseases such as cardiovascular disease, liver disease, fever and hypertension [1-3]. In addition, due to its colour and unique flavor, *Hibiscus sabdariffa* L. are utilizable in culinary as a cooking ingredient and colorant. A prominent feature of the *Hibiscus sabdariffa* L extract is its high content of antioxidants where anthocyanins are major components. Anthocyanins are water-soluble glycosides and acylglycosides of anthocyanidins, and they are found in the form of polyhydroxylated and or methoxylated heterosides which derive from the flavylium ion or 2-phenylbenzopyrilium in nature. Anthocyanins are compounds of interest because they are suitable replacements for chemical dyes. The compounds is characterized polyphenolic family, commonly existing in the plant kingdom [4-6].

However, a suitable extraction technique is important to maintain the antioxidant activity of the extract from the plant and to obtain the highest yield of anthocyanins. Although parameters including solvent acidity, solid-to-solvent ratio, time or temperature of extraction were individually studied in different studies, a study which include all important parameters is lacking in the literature. The aforementioned issue could be addressed by a technique that accommodates variables in an interactive manner. Optimization, in most cases, is performed with response surface methodology (RSM). The technique is a powerful and widely used technique devised for testing of multiple-process variables and optimizing experimental responses. The advantage of the technique is numerous ranging from simplicity to high efficiency and consistency with comprehensive theories. In addition, it also allows fewer experimental trials in comparison with studies of one variable at a time and is applicable in various fields, including construction field [7-15]. In order to achieve the above purpose, RSM combined with Box-Behnken design was employed in this study to identify the set of conditions which yields the highest anthocyanin yields from Hibiscus sabdariffa L. (Roselle) grown in Lam Dong, Vietnam. Following parameters were optimized: solid-liquid ratios, extraction temperatures and duration of extraction. The established model will serve as reasonable relationship between the independent variables and the response.

2. Material and methods

2.1. Sample preparation

Hibiscus sabdariffa L. dried calyces were used, provided by farmers of Da Lat city, Lam Dong Province, Viet Nam. Dried petals of roselle were ground using a commercial grinder and were weighed to 10g, put in the two neck round bottom flask and was extracted by Ethanol 50° solutions. Ethanol (C_2H_5OH) is purchased from Sigma Aldrich (US). The different particle diameter ground samples were stored in polyethylene vacuum bags at room temperature until extraction was carried out. First of all, each experimental condition will be tested individually to determine the optimum condition. For extraction parameter study, 10g of roselle powder was placed in the two neck round bottom flask as and was extracted by ethanol with concentration at 50. The liquid/solid ratio in this experiment ranges from 2:1 to 10:1 (mL/g) and. The extraction temperature is adjusted from 40 to 80 (^{o}C) and time varies from 15 to 35 (min). Then, centrifugation took place at 4500 rpm for 15 min by high speed centrifuge Model LACE16 (from COLO lab expert). The supernatant was collected and the extract, after being filtered with filter paper, was transferred into plastic bottle to estimate anthocyanin content. The pH scanning of supernatant ranges from λvis_{max} to 700 nm were recognized for anthocyanin content. The total anthocyanin content was determined by using the pH-differential method [16]

2.2. Experimental design using RSM

Following the estimation of the statistical model, analysis of variance (ANOVA) was performed to evaluate the significance of the experimental variables of consideration. These variables include solid/liquid ratio (X_1), extraction temperature (X_2), and time extraction (X_3). The whole optimization process was administered in the Design Expert 7.0 software.Based on the results obtained from the single factor investigation, 3-level factors (Table 1) were produced and attempted in actual

experiments to generate data (Table 2) for estimating the second order polynomial model, which is described as follows:
$Y=\beta_0+\sum_{i=1}^{3}\beta_i x_i+\sum_{i=1}^{3}\beta_{ii}x_i^2+\sum_{i=1}^{2}\sum_{i=j+1}^{3}\beta_{ij}x_ix_j$,where, Y is a response variable, β_0 , β_i , β_{ii} and β_{ij} are the regression coefficients of the intercept, linear, quadratic and interaction terms, respectively. x_i and x_j are coded values of independent variables affecting the response of Y.

Table 1. Coded and levels of the three independent variables for full factorial design of RSM model

Variables	Coded factor	Level		
		Low (-1)	Medium (0)	High (+1)
Solid-liquid ratios	X_1 (mL/g)	7 : 1	8 : 1	9 : 1
Extraction temperatures	X_2 (°C)	60	65	70
Duration of extraction	X_3 (min)	35	40	45

3. Result and discussion

3.1 Process optimisation and statistical analysis of experimental design

The experiments were designed according to the design method of complex CCD center. Experimental results (20 experiments) and predicted values generated by Design-Expert 11 are shown in Table 2. To be specific, variables of interest include concentration of ethanol 50%, liquid/solid ratio ranged from 4:1 to 6:1 (mL/g), temperature ranged from 60 to 70° C and reaction time ranged from 35-45 (min). No.1~14 were the factorial experiments, and No.15~20 were the central experiments. The resulting anthocyanin content ranges from 163.86 mg/L to 186.77 mg/L. The results indicated that all factors showed statistical significance, solid/liquid ratio at the $p \le 0.05$ level and both temperature and time at a ≤ 0.01 level. Additionally, there was a high statisticalsignificance ($p \le 0.01$) in the interaction between solid/liquid, temperature and time; This occurs when one of the independent variables exerts an effect on another of the independent factors of the experimental design. The results implied that the anthocyanin content would be severely altered as survey parameters change (ratio, temperature, time).

The difference between actual values and predicted values was not substantial indicating that the results of the experimental experiments have high accuracy. As indicated by Table 3, determination coefficient $R^2 = 0.9770$ shows a high goodness-of-fit of the predicted values to the real data. Overall, the model is significant as demonstrated by the F-value of 47.18. The probability of the model to obtain such value owing to noise is minimal (0.79%). All model terms are also statistically significant as indicated by low p-values. Insignificance of the lack of fit shows that the model fits well with the data, thus no further specification of the model is required. After estimating model coefficients, the following model is obtained:

$Y = 186.6 - 2.82X_1 - 1.31X_2 + 1.36X_3 - 1.37X_1X_2 - 2.57X_2X_3 - 2.16X_1X_3 - 5.01X_1^2 - 3.00X_2^2 - 3.89X_3^2$(3)

The predictive model calculation (second order polynomial equation model) leads to the optimal theoretical calculation of extraction in order to achieve the maximum value of 187.006 mg/L anthocyanin content, using 50% EtOH, at 56.9 °C, and 7.7:1 (mL/g) solid/liquid ratio.

Table 2. The matrix of observed and predicted values for RSM model

No	Independent factors			Y (mg/L)		No	Independent factors			Y (mg/L)	
	X_1	X_2	X_3	Actual	Predicted		X_1	X_2	X_3	Actual	Predicted
1	7	55	25	173.35	171.37	11	8	51.59	30	180.13	180.32
2	9	55	25	173.47	173.62	12	8	68.41	30	176.15	175.91
3	7	65	25	176.19	175.80	13	8	60	21.59	171.34	173.31
4	9	65	25	173.69	172.56	14	8	60	38.41	179.87	177.8
5	7	55	35	182.44	183.55	15	8	60	30	186.01	186.60
6	9	55	35	175.13	175.51	16	8	60	30	186.73	186.60
7	7	65	35	179.52	179.36	17	8	60	30	186.66	186.60

8	9	65	35	163.86	165.83	18	8	60	30	186.77	186.60
9	6.32	60	30	176.34	177.17	19	8	60	30	186.71	186.60
10	9.68	60	30	168.51	167.68	20	8	60	30	186.74	186.60

Table 3 ANOVA data for removal models

Source	Sum of Squares	Degree of freedom	Mean Square	F-value	Prob. > F	Comment	
Model	862.56	9	95.84	47.18	< 0.0001	significant	SD = 1.43
A-A	108.64	1	108.64	53.48	< 0.0001		Mean = 178.48
B-B	23.50	1	23.50	11.57	0.0068		CV(%) = 0.7986
C-C	25.32	1	25.32	12.46	0.0054		R^2 = 0.9770
AB	15.04	1	15.04	7.40	0.0215		AP = 20.60%
AC	52.99	1	52.99	26.09	0.0005		
BC	37.20	1	37.20	18.31	0.0016		
A^2	361.94	1	361.94	178.16	< 0.0001		
B^2	129.60	1	129.60	63.79	< 0.0001		
C^2	217.76	1	217.76	107.19	< 0.0001		
Residual	20.32	10	2.03				
Lack of Fit	19.89	5	3.98	46.37	0.1503	not significant	
Pure Error	0.4289	5	0.0858				

3.2 Analysis of response surface

The coinciding relationship of various variables and the response could be described by the surface plot. Based on the model estimated by the software, three response surface plots of the anthocyanin content, corresponding to different combinations of the solid-liquid ratios (X_1), extraction temperatures (X_2) and duration of extraction (X_3) were plotted as in Fig.1. The contour plots represented impact intensity of the interaction variables on the content. According to the model results, all three interaction terms representing combinations of three factors are significant (P<0.05). From the response surface analysis, optimum process of anthocyanin extraction are identified with the following parameters: temperature 56.9 (^{o}C), time 33.29 (min), and liquid-solid ratio 7.7:1 (mL/g). The maximum predicted anthocyanin content was 187.006 mg/L.

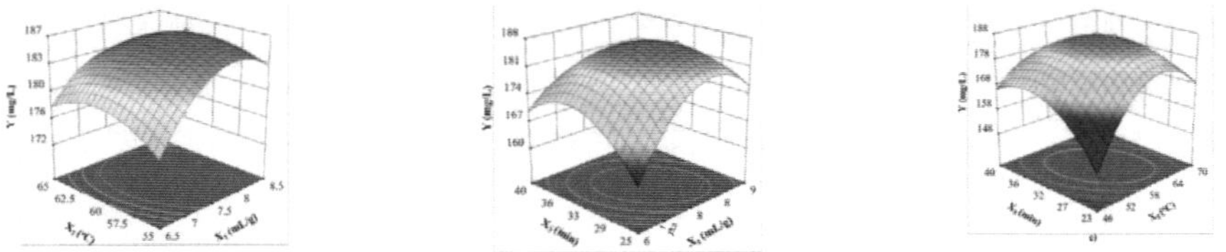

Figure1. 3D model about the influence and interaction of Y with (a) the solid-liquid ratios (X_1) and extraction temperatures (X_2); (b) the solid-liquid ratios (X_1) and duration of extraction (X_3); (c) extraction temperatures (X_2) and duration of extraction (X_3)

3.3. Testing the Model Precision

The yield of anthocyanin extraction could be predicted using the above model. To validate the model, residuals of 20 runs and yields of oil were plotted in the Figure 2. A normal plot of residuals was generated (Figure 2) and it can be observed that most data points are grouped reasonably near to, and

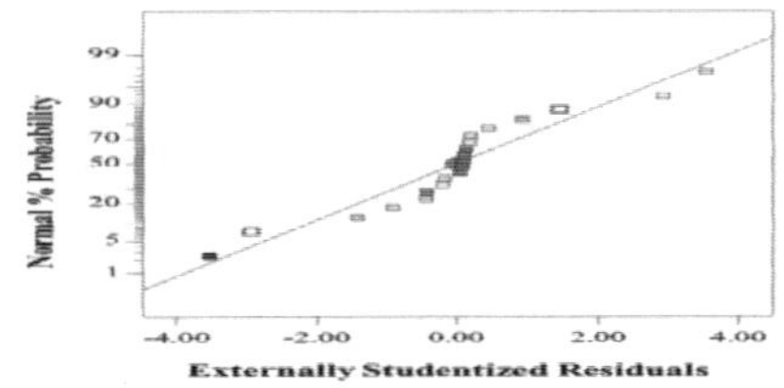

on either side of, the straight 45 ° line.

Figure 2. Normal plot of residuals of anthocyanin content

4. Conclusion

In this work, a RSM procedure was applied to study the effect of three variables (solid-liquid ratios, extraction temperatures, duration of extraction) on anthocyanin extraction yield from Vietnamese Hibiscus sabdariffa L. (*Roselle*). The 20experiments wasplanned by CCD to generate the data for estimation of the model. The best combination of factors yielding the highest anthocyanin content (187.006 mg/L) was as follows: 33.29 (min), 56.9 (^{o}C) temperature and 7.7:1 (mL/g) solid/liquid ratio. The ANOVA results indicated that the solid-liquid ratios (X_1), extraction temperatures (X_2) and duration of extraction (X_3) had significant effects on the responses. The obtained results suggested that RSM is a reasonable tool for optimization not only for extraction processes, but also for other applications involving multiple experimental conditions.

Acknowledgments

This work was supported by grants from Nguyen Tat Thanh University, Ho Chi Minh City, Viet Nam.

Reference

1. G Riaz, R Chopra 2018 *Biomedicine & Pharmacotherapy***102** 575-586
2. I DaCostaRocha, B Bonnlaender, H Sievers, I Pischel, M Heinrich, L 2014*Food Chemistry***165** 424-443
3. Z Aziz, SY Wong, NJ Chong 2013*Journal of Ethnopharmacology***150**442-450
4. ARC Braga, D Murador, VV Rosso 2018*Journal of Food Composition and Analysis***68** 31-40
5. P Ongkowijoyo, DA Luna-Vital, E. Mejia 2018*Food Chemistry***250** 113-126
6. L Xie, H Su, C Sun, X Zheng, W Chen 2018*Trends in Food Science & Technology***72** 13-24
7. THTran, HHHmNguyen, DC Nguyen, TQ Nguyen, H Tan, LTH Nhan, DH Nguyen, LD Tran, ST Do, and TD Nguyen 2018 *Processes***6** 206
8. TT Hien, NPTNhan, DT Nguyen, VTT Ho, LGBach 2018*Solid State Phenomena***279** 217-221.
9. NPNhan, T Hien, LTH Nhan, PNQ Anh, LT Huy, NTC Trinh, DTNguyen, L.G. Bach 2018*Solid State Phenomena***279** 235-239
10. TV Tran, TP Bui, T Nguyen, NTH Le, LG Bach 2017*Adsorption Science & Technology*, vol. **35**, 2017, pp. 72–85
11. VT Tran, DT Nguyen, VTT Ho, PQH Hoang, PQ Bui, LGBach2017*Journal of materials and Environmental Sciences* **8** 426-437
12. TV Tran, QTP Bui, TD Nguyen, VTT Ho, LG Bach, 2017 *Water Science & Technology*, **75**2047-2055.
13. TV Thuan, VTT Ho, ND Trinh, NT Thuong, BTP Quynh, LG Bach 2016 *Carbon – Science and Technology***8**63-69
14. T Van Thuan, BTP Quynh, TD Nguyen, LG Bach,2017*Surfaces and Interfaces,***6**209-217
15. LG Bach, T Tran, TD Nguyen, TV Pham, ST Do*2018 Research on Chemical Intermediates***44**1661-1687
16. J Lee, RW Durst, RE Wrolstad *2005Journal of AOAC International***88**1269-1278

Physical and mechanical properties of Si_3N_4/Mo laminated composite

Wei Chen[1], Zhaoxun Wang[1] and Xin Jia[1]
1. College of Mechanical and Electrical Engineering, Shaanxi University of Science & Technology, Xi'an 710021, P.R. China
E-mail: chenweijd@sust.edu.cn

Abstract. In this paper, a kind of new laminated composite about Si_3N_4/Mo was designed and the physical and mechanical properties of the layered composite were investigated. The results show that layered structure in Si_3N_4/Mo laminated composite promoted an improvement in fracture toughness, but the new kind of laminated composite did not show good properties in other aspects.

1. Introduction

As one of the most important structural materials, silicon nitride (Si_3N_4)-based ceramic composites had been widely used in bearings, piston pumps, engine parts and high-speed cutting tools. However, this kind of composites showed low fracture toughness [1, 2]. In our recent studies, hexagonal boron nitride (hBN) as an additive to Si_3N_4 was shown to cause deterioration of mechanical performance [3-6]. Therefore, a method to effectively toughen Si_3N_4 based composites was greatly needed.

Biological materials, such as nacre, demonstrated a strong fracture resistance. It was found that the "brick and mortar" nanolaminated microstructure of nacre was responsible for the significant improvement of fracture resistance [7]. Inspired by the prominent mechanical property of biological materials, artificial ceramics were therefore designed with a biomimetic laminated structure to enhance the fracture toughness [8, 9]. Zan et al. [10] found the strength of Si_3N_4/BN increased as the number of layers increased. Wang et al. [11] studied the fabrication and mechanical properties of SiC/ZrB_2 laminated ceramic composite, demonstrating a significant improvement of fracture toughness of the composite, as a result of the crack deflection along the interfacial layer as well as the presence of residual stresses.

In this paper, pure molybdenum was sandwiched between Si_3N_4 to form laminated Si_3N_4/Mo composites, and physical and mechanical performances of the laminated composite were investigated.

2. Materials and methods

High purity Si_3N_4 powders (99.99%) with a mean particle size of 0.3 μm were provided by Hefei Aijia

New Materials, China. Fig. 1 shows the X-ray diffraction (XRD) results of the silicon nitride powder. Al_2O_3 and Y_2O_3 powders (purity > 99.5%, a mean particle size of 1 μm, Hefei Aijia New Materials, China) were employed as sinter aids. Mo powders (purity ≥ 99.0%, Sinopharm Chemical Reagent Co., Ltd, China) were used as weak layer material. The powders of Si_3N_4, Al_2O_3 and Y_2O_3 were mixed by ball-milling (KQM-X4Z/B, Xianyang, China) at a speed of 100 rpm for 5 h (zirconia balls, ball to powder ratio = 2:1). After drying, the mixed powder and Mo powder were stacked sequentially in a steel die with the number of ceramic and metal molybdenum layers being 8 and 7, respectively. Afterwards, the billets of the laminated composites were pressed at 35 MPa for 10 minutes. Hot-pressing at 30 MPa and 1800 ℃ for 30 minutes in the nitrogen atmosphere was employed to finally densify the Si_3N_4/Mo laminated composite (Φ 44 mm × 6 mm disc). The composites were cut by diamond saw blade into a dimension of 5 mm × 6 mm × 30 mm for mechanical testing.

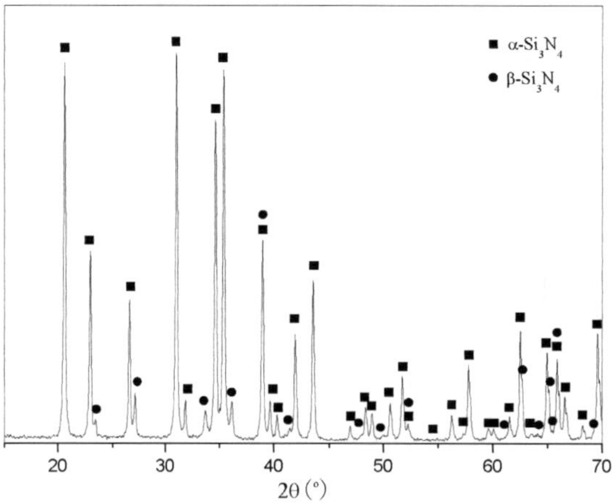

Fig.1 X-ray diffraction spectrum of Si_3N_4 powder

The relative densities of the specimen were estimated according to the Archimedes method. Vickers indentation were used to characterize the hardness of the composites with a load of 1 kg for 15 s [13]. Hardness can be calculated using the following equation:

$$H_V = 1.8544 \times (P/d^2) \tag{1}$$

where P is the indentation load (N) and d is average length of the two diagonals of the indentation (μm).

Fracture toughness based on the nature of the observed cracks was estimated by the following expression [14]:

$$K_{IC} = 0.203 \times (a/c)^{-3/2} \cdot \sqrt{a} \cdot HV \tag{2}$$

where a is the half length of the indentation diagonal and c is the radial half-crack length.

Bending strengths of the composites were tested in a three-point bending fixture (span of 20 mm) with a crosshead speed of 0.5 mm/min at ambient temperature.

3. Results and discussion

IOP Conf. Series: Materials Science and Engineering **544** (2019) 012017 doi:10.1088/1757-899X/544/1/012017

3.1 Phase composition

Fig. 2 shows the XRD pattern of the Si_3N_4/Mo laminated composite. The result shows that a complete transformation from α to β-Si_3N_4 occurred in the final composite. Besides, molybdenum oxide can be detected, indicating that some molybdenum had been oxidized during cutting or fabrication processes. In addition, the existing of silicon-molybdenum phase suggested that there were some reactions between silicon nitride and molybdenum during sintering.

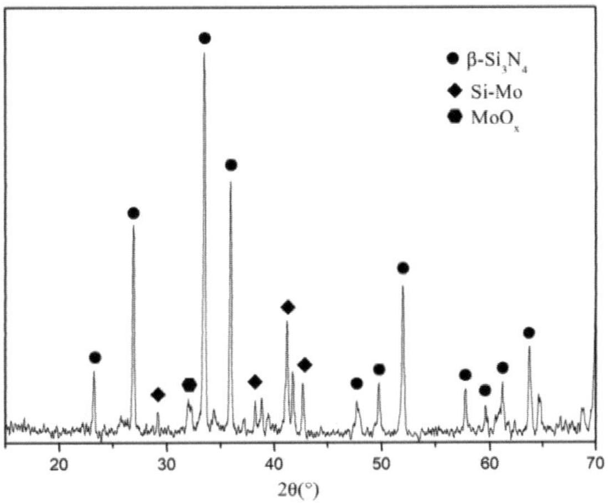

Fig.2 X-ray diffraction spectrum of Si_3N_4/Mo laminated composite

3.2 Physical and mechanical properties of the laminated composite

The porosity of the Si_3N_4/Mo laminated composite was measured to be 14.20%, which was significantly higher than the pure Si_3N_4 ceramic. The large porosity difference could be attributed to the poor binding between silicon nitride and molybdenum. The bending strength (275 MPa) was also affected by the poor interfacial binding. Obviously, it was shown that the Vicker's hardness of the laminated composite (16.1 GPa) was slightly lower than the pure Si_3N_4. Importantly, the layered structure was observed to have a toughening effect: the fracture toughness of the Si_3N_4/Mo laminated composite was shown to have a value of 10.7 MPa•$m^{1/2}$, which was slightly higher than the pure Si_3N_4 ceramic (Table 1).

Table 1. Physical and mechanical properties of laminated composite and silicon nitride

Samples	Laminated composite	Si_3N_4
Density (g/cm^3)	3.23	3.31
Porosity (%)	14.20	0.84
Bending strength (MPa)	16.1	19.9
Vicker's hardness (GPa)	10.7	8.01
Fracture toughness (MPa•$m^{1/2}$)	275	812

4. Conclusion

In this paper, the biomimetic laminated Si_3N_4/Mo composite was fabricated by hot-pressing, and its physical and mechanical properties were investigated. The results show that, benefit from the laminated microstructure, the fracture toughness of the laminated composite was slightly higher than the pure silicon nitride. However, the density and bending strength were remarkably lower than pure Si_3N_4 due to the poor interfacial bonding.

Reference:

[1] Guo S Q 2009 Densification of ZrB2-based composites and their mechanical and physical properties: a review J. Eur. Ceram. Soc. **29** 995-1011

[2] Fahrenholtz W G, Hilmas G E, Talmy I G and Zaykoski J A 2007 Refractory diborides of zirconium and hafnium J. Am. Ceram. Soc. **90** 1347-64

[3] Chen W, Zhang D and Ai X 2017 Effect of load on the friction and wear characteristics of Si_3N_4- hBN ceramic composites sliding against steels *Ceram. Int.* **43** 4379-89

[4] Chen W, Zhang D, Lv Z L and Li H Q 2017 Self-lubricating mechanisms via the in situ formed tribo-film of sintered ceramics with hBN addition in a high humidity environment *Int. J. Refract. Met. H.* **66** 163-73

[5] Zhang D, Chen W and Ai X 2017 Effect of the humidity on the friction and wear characteristics of Si_3N_4-hBN composite ceramic *P. I. Mech. Eng.* J **231** 1517-26

[6] Chen W, Gao Y M, Chen L and Li H Q 2013 Influence of sliding speed on the tribological characteristics of Si_3N_4-hBN ceramic materials *Tribol. T.* **56** 1035-45

[7] Smith B L, Schaffer T E, Viani M, Thompson J B, Frederick N A, kindt J H, Belcher A M, Stucky G, Morse D E and Hansma P K 1999 Molecular mechanistic origin of the toughness of natural adhesives, fibres and composites *Nature* **399** 761-3

[8] Clegg W J, Kendall K, Alford N M, Button T W and Birchall J D 1990 A simple way to make tough ceramics *Nature* **347** 455-7

[9] Huang Y and Wang C 2008 *Multiphase composite ceramics with high performance* (Beijing: Tsinghua University Press)

[10] Zan Q F, Zhang C, Huang Y, Zhao S K and Li C W 2004 Effect of geometrical factors on the mechanical properties of Si_3N_4/hBN multilayer ceramics *Ceram. Int.* **30** 441-6

[11] Wang H L, Fan B B, Lun F, Chen D L, Lu H X, Xu H X, Wang C and Zhang R 2012 The fabrication and mechanical properties of SiC/ZrB_2 laminated ceramic composite prepared by spark plasma sintering *Ceram. Int.* **38** 5015-22

[12] Shi D K and Jin Z H 1998 *Mechanical properties of materials* (Xi'an: Xi'an Jiaotong University Press) p 38

[13] Evans A G, Charles E A 1976 Fracture toughness determinations by indentation *J. Am. Ceram. Soc.* **59** 371-2

Computer simulation, analysis of force and temperature-speed parameters of the process of combined machining of Al-Mg-Sc alloys

S B Sidelnikov[1], O V Yakivyuk[1], V N Baranov[2], I L Konstantinov[1], I N Dovzhenko[1], E S Lopatina[3], D S Voroshilov[1], A P Samchuk[1] and V A Frolov[1]

1 Department of Metal forming, School of Non-Ferrous Metals and Material Science «Siberian federal university», Krasnoyarsk, 660025, Russia
2 Department of Common metallurgy, School of Non-Ferrous Metals and Material Science «Siberian federal university», Krasnoyarsk, 660025, Russia
3 Department of Metallurgy and Heat Treatment of Metals named after V.S. Biront, School of Non-Ferrous Metals and Material Science «Siberian federal university», Krasnoyarsk, 660025, Russia

Email: sibdrug@mail.ru

Abstract. The results of studies of the combined processing of alloys of the Al-Mg-Sc system with known rheological properties under various temperature-rate and deformation conditions are presented. A computer model of the combined machining process was created and the DEFORM 3D complex was used to analyze the metal flow, the distribution of the strain rate and the temperature of the metal over the cross section of the workpiece when receiving round bars of various diameters. The results of the research were used to develop the production technology of welding wire from alloys of the Al-Mg-Sc system.

Keywords: aluminum alloys, scandium, combined machining, rolling-extrusion, temperature, strain rate, drawing ratio, energy-power parameters, welding wire, longish semi-finished products.

1. Introduction

The modeling of metal forming processes has great scientific and applied importance, since the construction of a working model contributes to a better assessment and planning of the results of scientific activity. Preliminary modeling of the metal processing process allows reducing labor and energy costs in comparison with traditional methods of physical modeling. One of such programs is the DEFORM 3D complex of finite element modeling, which allows analyzing and calculating the parameters of combined machining processes to obtain longish deformed semi-finished products [1-3] and establish the patterns of the distribution of metal flow rates, moments and forces on the rolls, the force acting on the die, the temperature distribution in the deformation zone.

2. Experimental procedures

Studies have been performed for one of the alloys of the Al-Mg system doped with scandium. As it is known, scandium is an effective modifier of the cast structure of aluminum alloys. [4-11] and allows to obtain non-dendritic metal structure during the implementation of the method of continuous casting [12]. In this regard, magnalia, economically doped with scandium, are considered the most promising alloys capable of providing a complex of high mechanical and performance properties.

IOP Conf. Series: Materials Science and Engineering **544** (2019) 012018 doi:10.1088/1757-899X/544/1/012018

To create a model used data of combined machining unit CRE-200 [3] and the parameters characterizing the shape of the gauge, the die, the workpiece and the press products are given in Table 1 and Figure 1. For the studied alloy of the Al-Mg-Sc system, the rheological properties were determined (Figure 1), tooling simulation for the combined rolling-extrusion process of metal in the SolidWorks package has been performed. Then, the resulting models and data on the material properties of the workpiece were imported into DEFORM -3D.

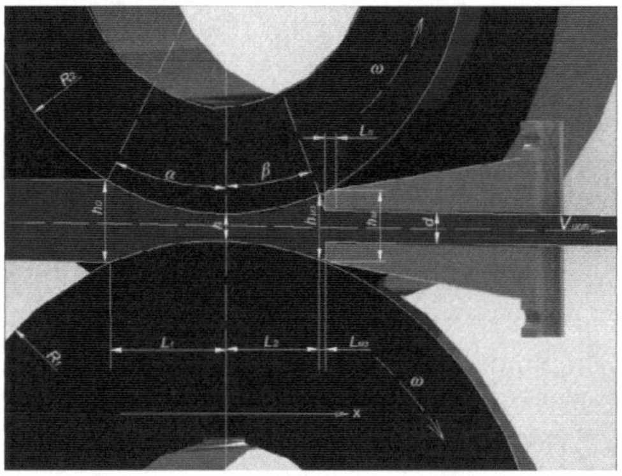

Figure 1. Geometric model of the combined machining process

Table 1. The parameters of the combined machining

Characteristic	Designation	Value
The diameter of the roll with a protrusion, mm	$2R_1$	214
The diameter of the roller with the groove, mm	$2R_2$	164
The average diameter of the rolls, mm	$2R$	189
Minimum height of caliber, mm	h	7
The width of the caliber, mm	b	15
Initial height of the workpiece, mm	h_0	14
Initial width of the workpiece, mm	b_0	14
The diameter of the gage hole of die, mm	d	9
Maximum allowable rolling force, kN	$P_{R.max}$	1000
Maximum allowable extruding force, kN	$P_{D.max}$	300

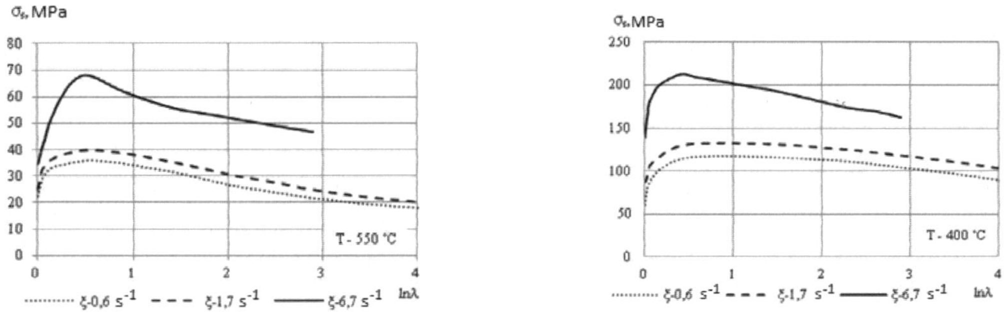

Figure 2. Deformation resistance dependence σ_S of investigated alloy from logarithmic degree of deformation lnλ at different temperatures T and strain rates ξ

When modeling the process used the following boundary conditions:

- workpiece material is isotropic;
- material of rolls and die is rigidly plastic;
- degree of deformation during rolling $\varepsilon = 50$ %;
- Zibel friction indicator when the metal moves along the die ψ_D=0,3;

- Ziebel friction index on the rolls and the calibrating belt of the die ψ_R=0,9;
- the number of finite elements into which the workpiece is divided, 32 thousand pcs.

In order to verify the possibility of implementing the process of combined processing of the investigated alloy, associated with restrictions on energy-power parameters, we calculated the forces acting on the die P_D and on the rolls P_R. It was believed that in the process of combined processing, the temperature of the metal of the workpiece coming into rolling may be equal to T = 450 °C when implementing the process of combined rolling-extrusion [3], either in the implementation of twin roll casting-extruding [2] can reach 550 °C, as this melt of the metal crystallizes in the rolls before the deformation.

3. Results and its discussion

Calculations of power parameters are given in table. 2. As follows from the table, the obtained values do not exceed the allowable values of the forces acting on the die and rolls (table 1). Analysis of the results also shows that with increasing metal temperature at the entrance to the deformation zone, the values of the forces on the rolls and the die decrease, which is associated with a decrease in the resistance of the metal to deformation. An increase in drawing ratio and strain rate leads to hardening of the metal and an increase in the force values, both on the rolls and on the die.

Table 2. Energy-power machining parameters of the investigated alloy on the unit CRE-200

Parameters		T = 450 °C		T = 550 °C	
		μ = 3.7	μ = 11.8	μ = 3.7	μ = 11.8
ξ_1 = 0.78 s^{-1}	P_D, kN	124	237	121	242
	P_R, kN	432	796	372	657
ξ_2 = 1.57 s^{-1}	P_D, kN	108	216	101	241
	P_R, kN	330	807	243	707

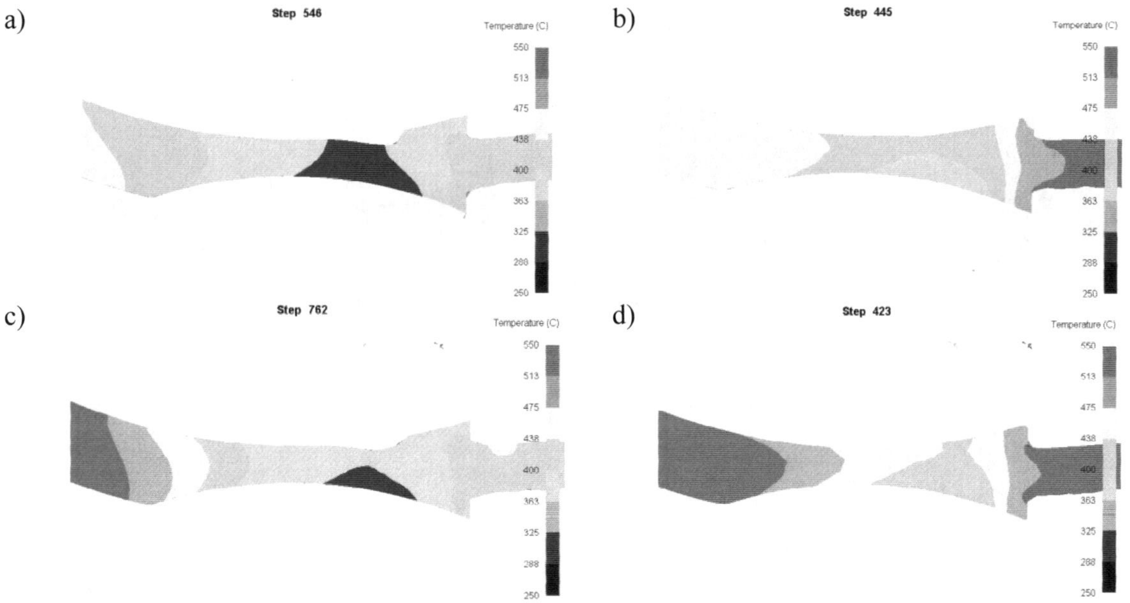

Figure 3. The temperature change in the deformation zone for the studied alloy at drawing ratio μ = 3.7: a, b – T = 450 °C; c, d – T = 550 °C; a, c – ξ_1 = 0.78 s^{-1}; b, d – ξ_2 = 1.57 s^{-1}

The variable factors in the simulation were the temperature of the workpiece (T = 450 °C and 550 °C), rolling strain rate (ξ = 0.78 s^{-1} and 1.57 s^{-1}) and the drawing ratio during extruding to obtain bars with a diameter of 9 and 5 mm (μ = 3.7 and 11.8). By changing the values of the selected parameters, we simulated various technological modes of combined machining.

Figure 3 shows the nature of the forming metal in the process of combined machining and the temperature distribution in the deformation zone.

Earlier studies of the formation of metal in the machining of combined methods on the unit CRE-200 [1-3] suggest that the deformation zone consists of zones of capture and rolling of metal, repressing and extrusion (pressing). At the time of capture of the workpiece by the rolls, it bends toward the roll with a larger diameter (with a protrusion), and after passing the rolling zone, it begins to bend toward the roll with a smaller diameter (with a groove), bounding it. Next, the workpiece reaches the die and begins the process of repressing the metal in the working part of the caliber formed by the rolls and the die. After full filling of the gauge in front of the die, the process of metal outflow begins with the formation of a press product according to the shape and dimensions of the corresponding calibrating hole of die. This sequence of combined machining is confirmed by the simulation data (figure 3, 4). The peculiarity of the change in the investigated alloy is that when drawing ratio μ = 3.7 and strain ratio ξ_1 =0.78 s^{-1} the steady extruding process takes place without completely filling the deformation zone from the side of the roll with a protrusion (figure 3, a, c).

As a result, there is no cooling of the workpiece from the contactless area and the temperature of the workpiece from the side of the roll with a protrusion is higher than from the side of the roll with a groove.

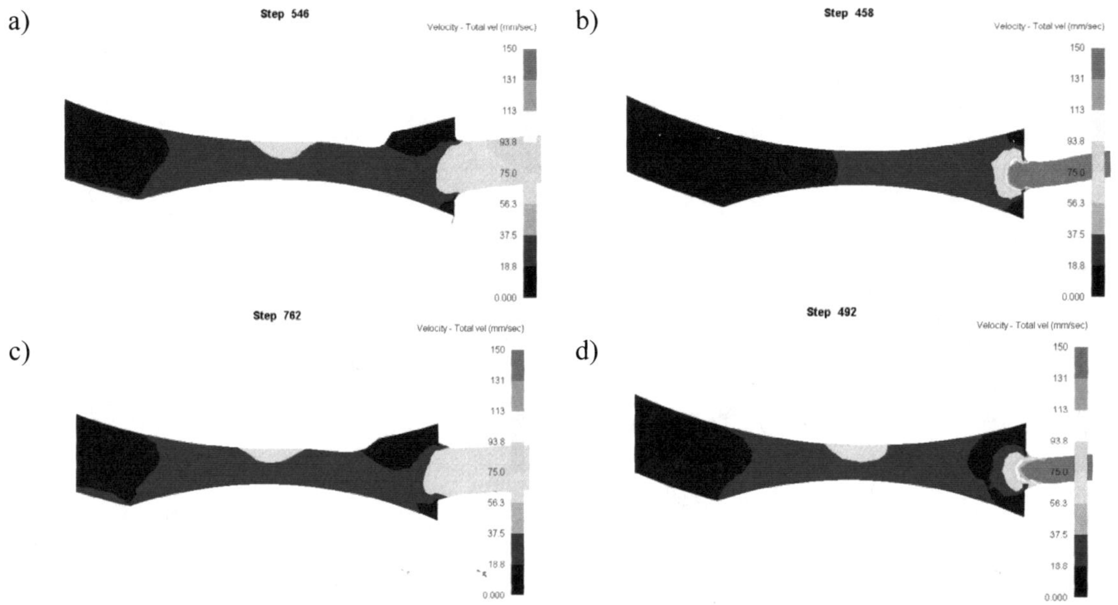

Figure 4. Distribution of strain rate over the cross section of the workpiece for the investigated alloy: a, b – T = 450 °C; c, d – T = 550 °C; a, c – μ = 3.7; b, d – μ = 11.8

An increase in drawing ratio (figure 4) or strain rate leads to a decrease in the contactless zone or to its absence, which is typical for alloys with high strain resistance. The workpiece is cooled unevenly, and a more intense heat sink is carried out in the lower part of the workpiece from the side of the roll with a groove. Throughout the entire length of the deformation zone, the inner layers of the workpiece have a higher temperature.

Increased drawing ratio is characterized by complete contact of the metal with the roll and a more uniform temperature distribution on the metal surface. In the "dead" zones of the metal in front of the die, there is a slight decrease in temperature (figure 3, c, d), and the bar at the exit from the gauge of the die is heated evenly.

The analysis of the data obtained showed that with an increase in the strain rate, the temperature of the semi-finished product at the exit from the die increases, which is associated with an increase in the strain resistance and the amount of heat released. At the same time, the heat transfer time between the workpiece metal and the tool is reduced, thereby increasing the temperature of the workpiece in the rolling and repressing zone. An increase in the drawing ratio from 3.7 to 11.8 leads to an increase in the strain rate in the extruding zone (figure 4).

These patterns of change in the parameters of combined machining are confirmed by experimental data [1] for the preparation of welding wire from alloys Al-Mg-Sc system.

4. Conclusion

Thus, as a result of the research it was confirmed that the process of combined machining of Al-Mg-Sc system alloys with known rheological properties under various temperature-speed and deformation conditions is realizable, since the limitations on the energy load of the equipment are fulfilled. With the help of modeling in the DEFORM 3D package, the laws of the distribution of metal temperature and strain rate over the cross section of the workpiece with combined rolling-pressing of round bars with a diameter of 5 – 9 mm were established.

Acknowledgements

The manuscript was prepared using the results of the work carried out during the project 03.G25.31.0265 “Development of economically alloyed high-strength Al-Sc alloys for use in road transport and navigation” in the framework of the Program for the implementation of complex projects for the creation of high-tech production, approved by the Government of the Russian Federation on April 9, 2010 №218.

References

[1] Vladimir Baranov, Sergey Sidelnikov, Evgeniy Zenkin, Olga Yakivyuk 2018 *Materials Science Forum* **918** 54–62

[2] Sidelnikov S.B., Lopatina E.S., Dovzhenko N.N. 2015 *Features of structure formation and properties of a metal during high-speed crystallization-deformation and modification of aluminum alloys: a collective monograph* (Krasnoyarsk: SibFU) p 179

[3] Sidelnikov S.B., Dovzhenko N.N., Zagirov N.N. 2005 *Combined and complex methods of processing non-ferrous metals and alloys* (M.: MAKS Press) p 344

[4] S. Malopheyev, V. Kulitskiy, R. Kaibyshev 2017 *Journal of Alloys and Compounds* **698** 957–966

[5] Pedro Henrique R. Pereiraa, Ying Chun Wang, Yi Huang, Terence G. Langdon 2017 *Materials Science & Engineering* **A685** 367–376

[6] M. Li, Q. Pan, Y. Shi, X. Sun, H. Xiang 2017 *Materials Science & Engineering* **A687** 298–305

[7] W. Kang, H.Y. Li, S.X. Zhao, Y. Han, C.L. Yang, G. Ma 2017 *Journal of Alloys and Compounds* **704** 683–692

[8] Anthony De Luca, David C. Dunand, David N. Seidman 2016 *Acta Materialia* **119** 35–42

[9] Zhemchuzhnikova D., Kaibyshev R. 2014 *Advanced Materials Research* **922** 862–867

[10] Zhemchuzhnikova D., Kaibyshev R. 2014 *Materials Science Forum* **794-796** 331–336

[11] D. Zhemchuzhnikova, A. Mogucheva, R. Kaibyshev 2013 *Materials Science & Engineering A* **565** 132–141

[12] Dobatkin V.I., Eskin G.I. 1991 *Non-ferrous metals* **12** 64–67

Equivalent moisture diffusion coefficient of fiber reinforced composites

Libao Zhu, Yongqing Li

College of Naval architecture and Ocean Engineering, Naval University of Engineering, Wuhan, China
E-mail: lpzhulibao@163.com

Abstract. Total diffusion time and the total moisture absorption amount were proposed as the measurement standards of equivalent coefficients for composites moisture diffusion with impermeable fibers. Representative volume element (RVE) models and 2D transient finite element models of unidirectional fiber reinforced composites were established and the accuracy of the RVE model for calculating the equivalent diffusivity was verified. The influences of fiber volume fraction, fiber arrangement and interfaces on the equivalent moisture diffusivity were investigated using the RVE models. The results show that the equivalent moisture diffusivity decreases with the increase of fiber volume fraction. The equivalent moisture diffusivity of random fiber distribution is smaller than that of the periodic fibers distribution, and this difference can be neglected when the fiber volume fraction is relative small. The interfaces with higher diffusivity than matrix will apparently increase the composites diffusivity.

1. Introduction

Fiber reinforced resin composites are increasingly used in aerospace, transportation, civil and marine applications owing to their special properties including high specific strength, high specific stiffness and excellent physical and chemical properties. Composite materials may be exposed to environmental loads during service, resulting in weakened performance. Among them, moisture is one of the important factors that cause the decreasing of composites properties. Studying the water diffusion properties of composite materials is the primary task before assessing their long-termbehavior.

Theoretical models and the steady-state finite element models have been proposed to predict the equivalent diffusivity of unidirectional composites. Shen [2] and Halpin [3] drew an analogy between the diffusion process and the heat conduction, giving an analytical expression of equivalent diffusivity for the unidirectional composites. Whitcomb [4] established a steady-state finite element diffusion model and studied the effects of fiber volume fractions. Based on the Halpin model, Gueribiz [5] proposed a self-consistent theoretical model and predicted the effects of fiber volume components, interfaces and voids on the equivalent diffusivity. Rocha [6] established a multi-scale FE diffusion model for unidirectional composite. However, moisture diffusing into composites is a transient process and thus the accuracy of steady state FE model may be questionable [7]. Y. Joliff[8] and M. Meng[9] studied the influence of interfaces on the equivalent diffusivity using 2D transient FE model.

The previous steady-state FE models [4, 5] contain a flux-isolated boundary condition which is inconsistent with the actual situation. Meanwhile, the equivalent diffusivity calculated by the steady-state model is smaller than that calculated by the transient FE model at the same fiber fraction [4-5,9]. In this paper, we propose the total diffusion time and the total moisture absorption amount as the measurement standards of equivalent coefficients for composites with impermeable fibers. A RVE FE model for diffusion was established and verified by the transient model and was then used to analysis the effects of fiber volume fraction, fiber arrangement and interfaces on the composites diffusivity.

2. Numerical models

2.1. Concept of equivalence

A Fickien kinetic can be used to an infinite plane sheet in diffusion [10]:

$$\frac{M_t}{M_\infty} = 1 - \frac{8}{\pi^2}\sum_{n=0}^{\infty}\frac{1}{(2n+1)^2}\exp\left[-\frac{Dt}{l^2}\pi^2(2n+1)^2\right] \tag{1}$$

where, M_t is the water mass fraction which diffuses at the time t, M_∞ is the saturated water mass fraction which diffuses after infinite time, D is the diffusivity and l the plate thickness.The moisture absorption curve will be linear part at initial time and the diffusivity can be expressed as Eq. (2):

$$D = \pi\left(\frac{l}{4M_\infty} \cdot \frac{M_1 - M_2}{\sqrt{t_1} - \sqrt{t_2}}\right)^2 \tag{2}$$

The relationship between the saturation water concentration and the water mass fraction for pure resin can be expressed by Eq. (3):

$$C_m = \frac{\rho_m M_m}{18\times10^{-3}} \tag{3}$$

where, C_m is the saturation water concentration and ρ_m is the density of pure resin.

Assume that the diffusivity and solubility of the polymeric composites' matrix are D_m and C_m, respectively. The existence of impermeable fibers change the diffusion path of water, thus the diffusivity and the amount of absorbed water decreased. As shown in Figure 1, in this paper, we extend the concept of "diffusion equivalence" for composite materials to including both the diffusivity and the solubility: the moisture absorption curves of the original composites model and of the effective model should be coincident. Under this requirement, the equivalent diffusivity D_{eff} can be calculated by Eq. (3), and the equivalent solubility C_{eff} can be expressed by Eq. (4):

$$C_{eff} = C_m V_m \tag{4}$$

in which, V_m is the matrix percentage of the composites.

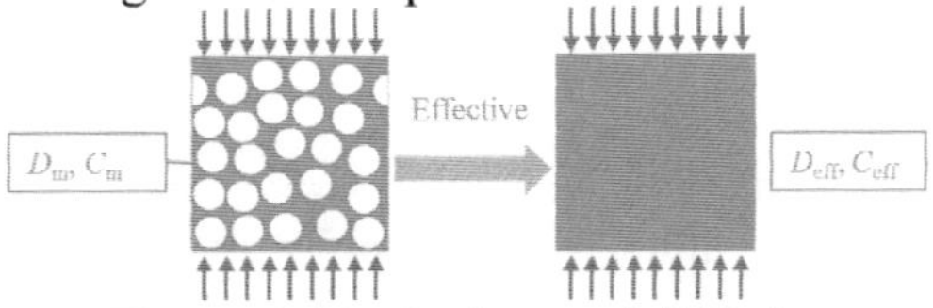

Figure 1. Equivalence of diffusion

2.2. 2D transient model

Commercial FE software Abaqus™ was implemented to simulating the water diffusion process. Transient diffusion FE model was firstly established to calculate the composites equivalent coefficient. As shown in Figure 2 (a), the 2D FE model has length of 990μm and width of 90μm. 275 fibers with diameter of 15μm are randomly distributed to the model, occupying 54.5% of the volume fraction. The fibers are not modeled realistically because of its impermeability and this simplification has been verified [1][8]. Linear quadrilateral element DC2D4 was used in the model. The converged analysis has been conducted and element size of 1μm will be used in the following simulations. Diffusivity of 0.741μm²/s, and solubility of 3.94% are input to the model, while the solubility was calculated as 2845mol/m³ according to Eq. (3). The unit normalized concentration was applied at the external edges of the model as depicted in Figure 3, implying a saturation state during the analysis.

The equivalent diffusivity D_{eff} was calculated by Eq. (2) according to the moisture absorption curve of the above transient diffusion FE model. Under the requirement of ensuring the same total moisture of the original model and the effective model, the equivalent solubility can be calculated to be 1294.5 mol/m³ according to Eq. (4). Meanwhile, in order to validate our concept of equivalence, an equivalent model was built with the same material properties and boundary conditions as the composite model, as shown in Figure 2(b).

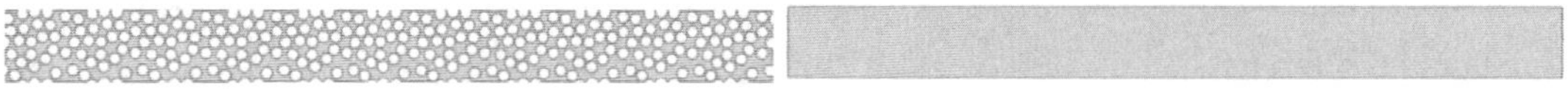

Figure 2. 2D transient FE model (a) composite model; (b) equivalent model

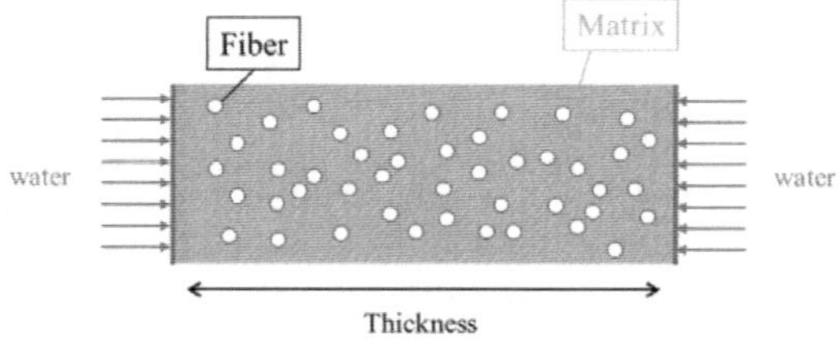

Figure 3. Boundary conditions of the 2D composite model

2.3. RVE model

3D RVE models of pure resin and composites were established and the steady state diffusion analysis was applied to calculate the equivalent diffusivity. As shown in Figure 4, fiber arrangements of hexagon, square and random distributions were taken into account, which all follow the periodic pattern. The material properties and geometry of the matrix and the fiber are as described in Section 2.2. Referring to the previous research results, the RVE model contains 25 fibers to guarantee the result convergence. Linear quadrilateral element (DC2D4) was chosen and the mesh element size was set as 1μm. The mesh nodes on the opposite faces should be located at the opposite position in order to simplify the applying of periodic boundary conditions (PBCs). An Python script was used to apply the PBCs. Unit concentration gradient was applied to all the three directions, and a zero normalized concentration was adopted at a corner to reach a unique solution for all models. Five replicates were conducted for each random condition.

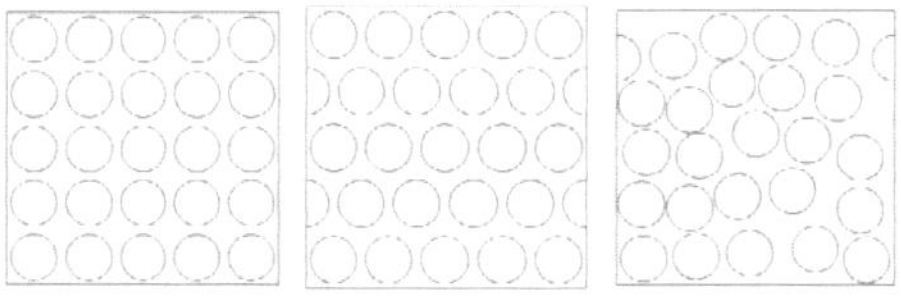

Figure 4. Fiber distribution of (a)Square; (b)Hexagon; (c)Random

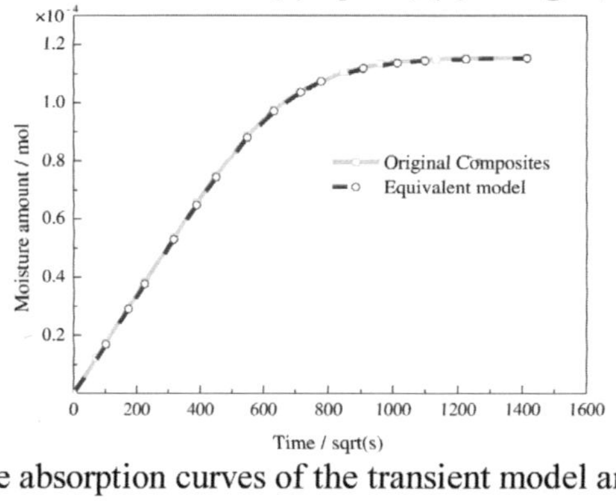

Figure 5. Moisture absorption curves of the transient model and effective model

3. Results and discussion

3.1. Validation of the equivalent concept

Figure 5 shows the moisture absorption curves of the 2D original composites model and the equivalent model. The diffusivity D_{eff} is 0.407μm²/s calculated by Eq. (2). The D_{eff} and C_{eff} was input to the equivalent model and the resulting moisture absorption curve is coincident with that of the original

IOP Conf. Series: Materials Science and Engineering **544** (2019) 012019 doi:10.1088/1757-899X/544/1/012019

composites model, which implies that the previously proposed calculation method for equivalent coefficients is correct.

3.2. Validation of the RVE model

3D RVE steady state models and 2D transient models were established respectively for the pure resin and the composites with 54.5% fiber fraction. The 3D RVE models are 90μm×90μm×90μm cubes and the 2D transient models are 90μm×990μm squares. The composites random fiber distribution of the 2D transient model was achieved by replicating that of the 3D RVE model nine times.

Table 1 shows the equivalent diffusivity of the pure resin and the composites calculated by the 3D RVE steady state model and the 2D transient model, respectively. For the pure resin, the RVE model result is equal to the input value while the 2D model result is a little smaller than it. For composites, the RVE result is about 56.8% of the input value while the 2D model result is still a smaller than the RVE result. The two model results are basically the same and the RVE model may be more accurate than the 2D transient model. Figure 6 shows the moisture absorption curves of the RVE results and the 2D models for both the pure resin and the composites, in which the RVE results were calculated by the 2D equivalent model using the RVE equivalent diffusivity. The curves coincide well for both the resin and the composites, respectively.

Table 1 Effective moisture diffusion coefficient

Materials	Input	Pure resin RVE model	Pure resin 2D model	Random RVE model	Random 2D model
Diffusivity / $\mu m^2 \cdot s^{-1}$	0.741	0.741	0.734	0.417	0.407

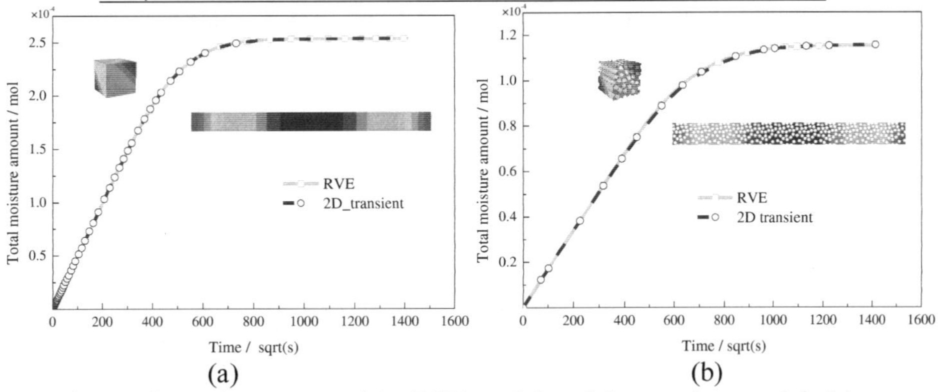

Figure 6. Moisture absorption curves of the RVE model and the transient model: (a) pure resin (b) composite

3.3. Fiber fraction and arrangements

After verifying of the accuracy, the RVE model was used to analysis the influence of fiber fraction and fiber arrangements on the diffusivity of composites. Figure 7 (a) shows the relationship between the fiber fraction and the resulting relative diffusivity (D_{eff} / D_m). The relative diffusivity decreases along with the increasing fiber fraction. While the "Analytical" results [2, 3] and the "Whotcomb" results [7] are quite similar, the present results are much larger and the ratio increases with increasing fiber fraction. The reason of this phenomenon may be the different boundary conditions between these models. Figure 7(b) shows the RVE results of the hexagon, square and random fiber distributions. The equivalent diffusivities are quite coincident when fiber fraction are smaller than 20%. However, when fiber fraction is larger than 20%, the difference between the three models become more evident and the hexagon fiber distribution seems to be more conducive for moisture diffusing. The periodic fiber distributions, namely hexagon and square are special cases of random fiber distributions. From the above analysis we can conclude that the periodic fiber distributions may be proper substitutes for the random one when the fiber fraction is smaller than 20%, while random fiber distribution may be more accurate with over 20% fiber fraction.

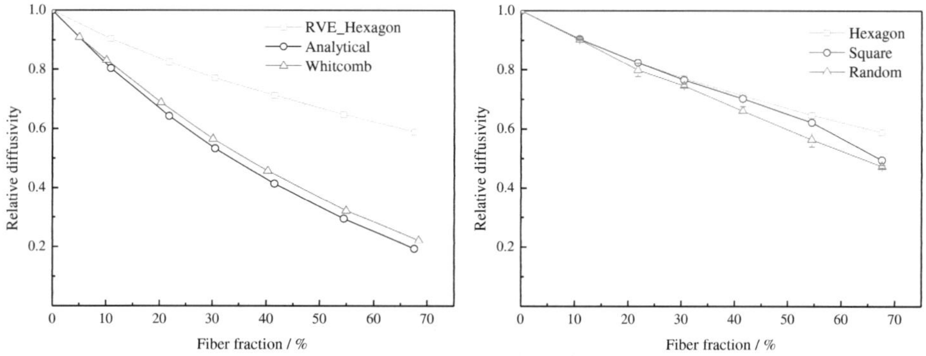

(a) (b)

Figure 7. Relative diffusivity of the composite

3.4. Interface

Researchers have found that interface layer, with a thickness of 1-4μm, exists between the fiber and matrix and its diffusivity is larger than the composite matrix. In this section, an interface-embedded RVE model was established to figure out its influence on the composites diffusivity. Only square fiber distribution was modeled for simplifying.

The accuracy of the interface RVE model was firstly validated by the 2D transient model. For the composites with 54.5% fiber fraction, 1μm interface thickness and 1.556$\mu m^2 \cdot s^{-1}$ interface diffusivity, its moisture absorption curves of the RVE model and the 2D transient model were shown in Figure 8 (a). The two curves coincide well and this proves a good accuracy of the interface RVE model.

The relative equivalent diffusivities under different interface diffusivity ratio and different fiber fractions are shown in Figure 8 (b), in which the ratios of 2, 5, 10, and 20 of the interface diffusivity to the resin diffusivity were taken into account. Generally, the relative equivalent diffusivities increase along with the interface diffusivity ratio. However, there may exist a critical value for interface diffusivity ration to get unit relative diffusivity. When interface diffusivity ration is smaller than the critical value, the relative diffusivity decreases along with increasing fiber volume fraction. While when interface diffusivity ration is larger than the critical value, the relative diffusivity increases along with increasing fiber volume fraction. This may because that the fiber volume fraction, which decreases the composites diffusivity, dominates the influence on the composites relative diffusivity when interface diffusivity is small. While the interface diffusivity ratio, which increases the composites diffusivity, will dominate the influence with larger interface diffusivity.

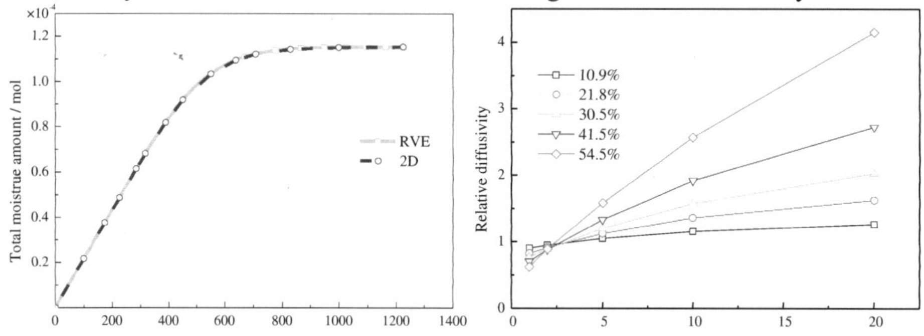

(a) (b)

Figure 8. Effects of interface on the relative effective moisture diffusion coefficient

4. Conclusion

In this paper, total diffusion time and total moisture absorption amount were proposed as the measurement standards of equivalent coefficients for composites moisture diffusion. A steady-state RVE FE model with periodic boundary conditions was established and was validated by the 2D transient model. Based on the analysis we conclude that:

(1) The proposed standards can accurately describe the moisture absorption process for composite materials.

(2) The difference of diffusivity between the random fiber distribution and the periodic fiber distribution increases with the increasing fiber volume fraction; the hexagon fiber distribution has the largest diffusivity among the three ones.

(3) There may exist a critical value for interface diffusivity ration to get unit relative diffusivity. The opposite influences of fiber fraction and interface diffusivity ratio may be the reasons.

References

[1] Bond D A. Moisture Diffusion in a Fiber-reinforced Composite: Part I - Non-Fickian Transport and the Effect of Fiber Spatial Distribution[J]. Journal of Composite Materials, 2005, 39(23):2113-2141.

[2] Shen C H, Springer G S. Moisture Absorption and Desorption of Composite Materials:[J]. Journal of Composite Materials, 1976, 10(1):2-20.

[3] Halpin J C. Primer on composite materials : analysis[M]. Washington Univ, 1984.

[4] Whitcomb J, Tang X. Micromechanics of Moisture Diffusion in Composites with Impermeable Fibers[J]. Journal of Composite Materials, 2002, 36(9):1093-1101.

[5] Gueribiz D, Rahmani M, Jacquemin F, et al. Homogenization of Moisture Diffusing Behavior of Composite Materials with Impermeable or Permeable Fibers -- Application to Porous Composite Materials[J]. Journal of Composite Materials, 2009, 43(12):1391-1408.

[6] Rocha I B C M, Raijmaekers S, Meer F P V D, et al. Combined experimental/numerical investigation of directional moisture diffusion in glass/epoxy composites[J]. Composites Science & Technology, 2017, 151.

[7] Vaddadi P, Nakamura T, Singh R P, et al. Inverse analysis for transient moisture diffusion through fiber-reinforced composites[J]. Acta Materialia, 2003, 51(1): 177-193.

[8] Joliff Y, Belec L, Chailan J F. Modified water diffusion kinetics in an unidirectional glass/fiber composite due to the interphase area: Experimental, analytical and numerical approach[J]. Composite Structures, 2013, 97(2):296-303.

[9] M. Meng Meng M, Rizvi M J, Le H R, et al. Multi-scale modelling of moisture diffusion coupled with stress distribution in CFRP laminated composites[J]. Composite Structures, 2016, 138:295-304.

[10] Crank J. The Mathematics of Diffusion[M]. Clarendon Press, 1956.

IOP Conf. Series: Materials Science and Engineering **544** (2019) 012020 doi:10.1088/1757-899X/544/1/012020

Simulation of fabric drying based on heat and mass transfer theory in porous media

Shuangqing Wang*, Huimin Chen, Huile Zhang and Xiaoli Yue*
Mechanical Engineering Department, Donghua University, Shanghai, China
E-mail: sq_wangsq@163.com
E-mail: xlyue@dhu.edu.cn

Abstract. Nowdays, the process of fabric drying inside the drying equipment is critical in practical production, this process affect the quality of dried fabric and energy utilization rate directly. There is a problem that the distribution of the percentage of water in fabric is uneven in drying process in practical production. In order to provide a method to solve this problem, this paper develop an finite element model(FE) on the platform of FLUENT by combining heat and mass transfer theory in porous media and transport theory of gas species organically, On the base of Mathis DHe steaming machine, the FE model is verified by carrying out an experiment.After that, this numerical model is used to study the process of fabric drying and the distribution of the percentage of water. The results indicate that the change of the weight of water in the fabric is nearly linear relationship with time and the percentage of water is low in the margin of fabric in the process of drying. In the later analysis, there have enough confidence that this finite element model can be used to analyse the structure and technological parameters of drying device.

1.Introduction

In open-width printing and dyeing process of fabric, the process of drying has a great influence on fabric productions[1]. Nowdays, there are still some faults in our country's drying technology. In practical producing process, the energy utilization efficiency is low and the difference of the percentage of water in different regions of the fabric is distinct[2]. The percentage of water is low in edge of the fabric, this will cause water moving from interior to edge. As a result, the distribution of water is not uneven in fabric and there will have colour difference . In order to increase uniformity of the distribution of the percentage of water in the process of fabric dyeing. Polat and Douglass[3] studied how the velocity of wind of the nozzles and the distance between the rectangular nozzles affect the the distribution of the percentage of water respectively. The result shows that when the rate of the distance between nozzles and the width of nozzle is 5 while the rate of the distance between nozzles and the distance between nozzles and fabric is 1, heat transfer coefficient is high relatively and the distribution of heat transfer is very even. Min Ming Gu[4] used numerical simulation method to investigated the airflow distribution inside the drying device, and he suggested a method of improving big size eddy and designed an air channel backflow structure which can guide the exchanged gas, these above researches improved airflow distribution and the stability of temperature on the surface of fabric. To reduce the colour difference in dried fabric, Pautea Sarafle[5] carried out two experiments, first, reduced the percentage of water; second, added migration inhibitor in fabric before drying process. The result shows both two methods can reduce the colour difference by making the distribution of water more even in the process of fabric drying. Many scholars has studied the distribution of the percentage of water in fabric in the process of drying, however, most of them just consider the effect of airflow and temperature distribution or just consider the initial situation of moist fabric.

In practical process of fabric drying, evaporated water will change the airflow and temperature distribution inside the equipment, and changed airflow and temperature distribution will also affect the process of water evaporation process in fabric; thus, it is key to consider the interaction between the evaporated water from moist fabric and the conditions inside the drying device including the airflow and the distribution of temperature .

This manuscript develop a numerical simulation model through combining the heat and mass transfer theory in porous media and species transport theory organically. In this paper, the model is used to predict the change of the weight of water with time in the fabric and the distribution of the percentage of water in the process of drying.

2.Build mathematic model

2.1.Control equation

This paper consider the flowing of water and gas. Thus, standard k-ε turbulence model is chosen. Relative to other turbulence models, this model can reach the accuracy requirement and has smaller computation.

(1) Mass conservation equation

$$\frac{\partial u_j}{\partial x_j} = 0$$

(2) Momentum conversation equation

$$\frac{\partial u_i}{\partial t} + u_j \frac{\partial u_i}{\partial x_j} = -\frac{1}{\rho}\frac{\partial p}{\partial x_i} + v\frac{\partial^2 u}{\partial x_j^2} + \frac{\partial}{\partial x_j}\left[v_r \left(\frac{\partial u_i}{\partial x_j} + \frac{\partial u_j}{\partial x_i} \right) \right] - g_i$$

(3) Turbulence momentum conversation equation (k)

$$\frac{\partial (pk)}{\partial t} + \frac{\partial_i}{\partial x_i}(\rho u_i k) = \frac{\partial_i}{\partial x_j}[(\mu + \frac{\mu_t}{\sigma_k}) + \frac{\partial k}{\partial x_j}] + G - \rho\varepsilon$$

(4) Turbulence momentum conversation equation (ε):

$$\frac{\partial(\rho\varepsilon)}{\partial t} + \frac{\partial}{\partial x_i}\left(\rho u_j \varepsilon\right) = \frac{\partial}{\partial x_i}\left[\left(\mu + \frac{\mu_t}{\sigma_E}\right) + \frac{\partial \varepsilon}{\partial x_i}\right] + C_{1\sigma}\frac{\varepsilon}{k}G - C_{2\sigma}\rho\frac{\varepsilon^2}{k}$$

2.2.Mixture multiphases model

Both Eular model and Mixture model can compute flow and heat transfer between gas and water. Eular model not only need more compute but also can not obtain a converged solution easily. Relative to Eular model, Mixture model can reach all requirements and needs less compute and can obtain a converged solution easily. This manuscript use the following equation to describe the percentage of volume of different phases.

$$\frac{\partial}{\partial t}\left(\alpha_p \rho_p\right) + \nabla\cdot\left(\alpha_p \rho_p v_m\right) = -\nabla\cdot\left(\alpha_p \rho_p v_{dr,p}\right)$$

where α_p, v_m, $v_{dr,p}$ represents the percentage of volume of phase p, mass average velocity, the drift velocity of phase p , respectively.

2.3.Species transport model

Both Eular model and Species transport model can compute flow and heat transfer between different gases. Relative to Eular model, Species transport model can calculate the interaction between air and water vapour rapidly and accurately, including the change of the percentage of the volume of different phases and temperature transfer of different phases.

$$\frac{\partial}{\partial t}\left(\rho Y_i\right) + \nabla\cdot\left(\rho \mathbf{v} Y_i\right) = -\nabla \mathbf{J}_i + R_i + S_i$$

IOP Conf. Series: Materials Science and Engineering **544** (2019) 012020 doi:10.1088/1757-899X/544/1/012020

where Y_i represents the percentage of mass of phase i, R_i is the production rate of chemical reaction, S_i denotes the extra production caused by diffusion item and source item.

3.Physical model and boundary conditions

3.1.Physical model

On the base of Mathis DHe steaming machine(as shown in figure 1 a)), this paper build the geometry (as shown in figure1 b)), the sizes of the geometry is the same with the real sizes of this streaming machine. Then, to reduce the computational amount and improve efficiency, the two-dimensional FE model (as shown in figure1c)) is built by the plane cut from the geometry(as shown in figure 1 b)). It is remarkable that in order to ensure the accuracy around the fabric, the grid around the fabric is refined(as shown in figure 1 c)) and the partial enlarged figure is shown in figure 1 d).

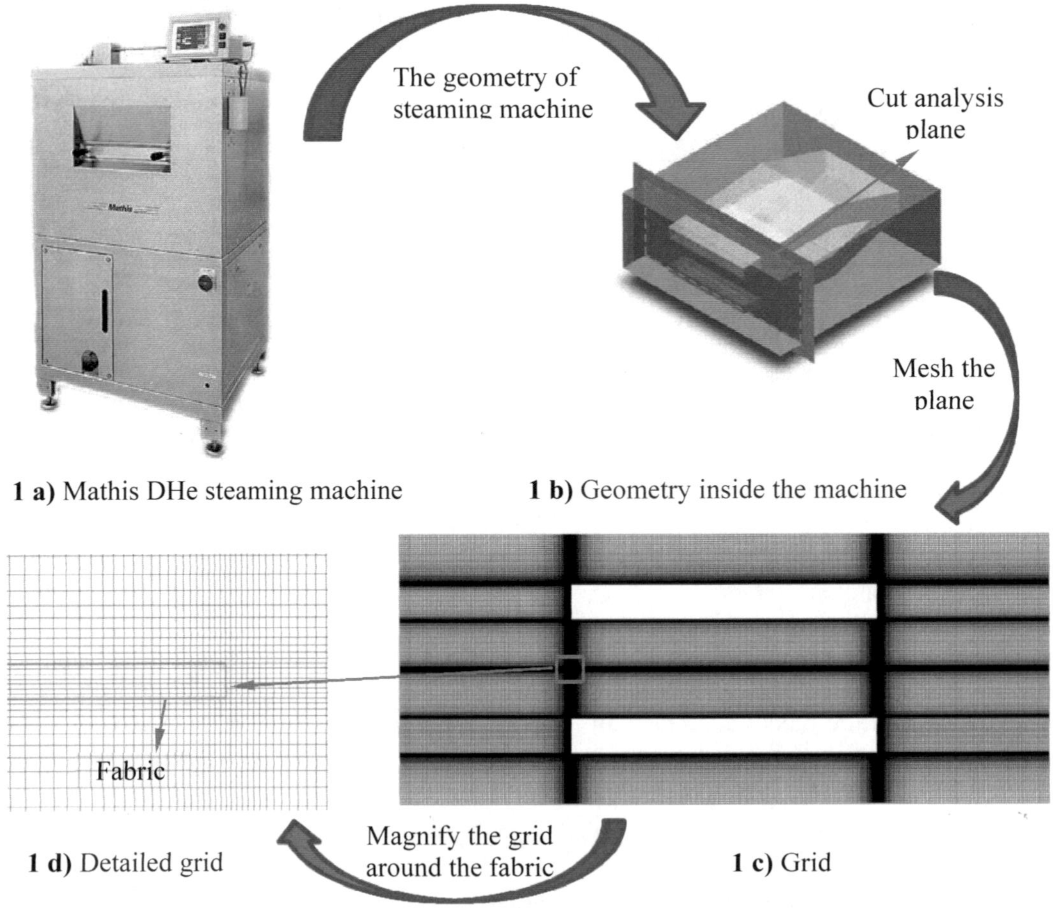

Figure 1 flowchart of building geometry and grid

3.2.Boundary conditions

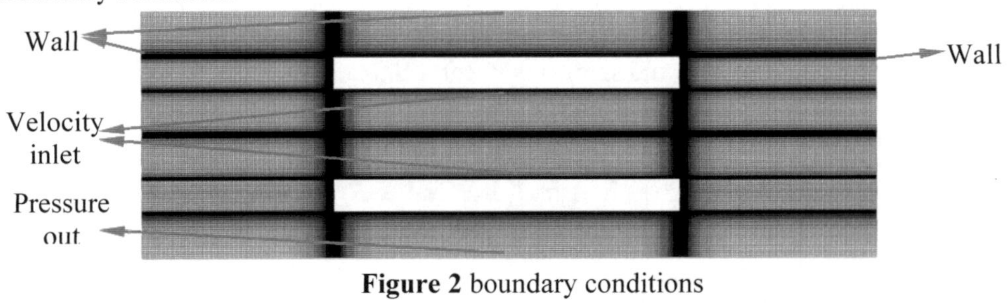

Figure 2 boundary conditions

Figure 2 shows the grid in the cut plane. The boundary condition of the wall is 0 heat flux and the velocity of wind in inlet is 1.6 m/s and the temperature is 130°C. The boundary condition of the outlet is pressure outlet and the pressure is -0.1013MPa. The initial temperture inside the device is 130°C and the initial temperature of the moist fabric is 20°C. Table 1 shows the parameters of the fabric, these parameters affect the process of fabric drying directly.

Table 1 fabric parameters

Yarn density （kg/m^3）	Yarn thermalconductivity λ_t(W/m ·K)	Yarn specific heat capacity c_t （J /(kg·K)）
265	0.5	300

4.Experiment

4.1 The process of experiment

In order to verify the results of numerical computation, this paper carry out an experiment by Mathis DHe steaming machine. The boundary conditions and initial condition are the same with numerical model. The type of fabric of this experiment is weft-knitted fabric, the thickness is 0.67mm and the width and the length is 330×330mm^2. First, measure the initial weight of the fabric; second, soak the fabric entirely and then roll the soaked fabric by the roller; third, rapidly fix the rolled fabric by needle board and put it into the machine. Passing the fixed time, the fabric is taken from the machine and measure the weight again. From the change of the weight of the fabric, the change of the weight of water in the fabric with time will be known. Table 2 shows the equipment used in the process of the experiment.

Table 2 Experiment equipment

Equipment	Type	Manufacturer
Mathis steaming machine	DHe	Shanghai cenxin company
Electric scale	Series YP	Shanghai youkecompany
Electric evenness padder	Model P-BU	Chinese Taiwan Rapid company

4.2.Compare the results between experiment and numerical computation

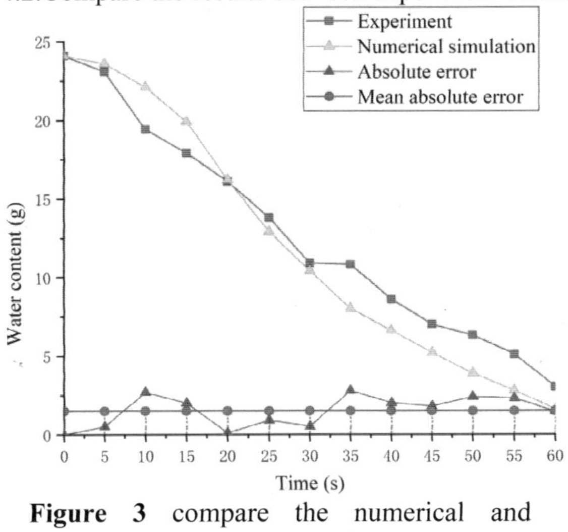

Figure 3 compare the numerical and experiment results

The experiment's result and numerical result are put in one figure to be compared clearly (as shown in figure c). As time goes on, the weight of water in fabric is smaller. Figure 3 shows the regularity of the reduction of water, within 30s, the velocity of the reduction of water is a little speedy in numerical simulation result, after 30s it becomes a little slow, it is closed to positive correlation between the weight of water and time. The error is small between the numerical simulation's result and experiment's result, the biggest absolute value of absolute error is 2.6g in 35s, and the average of absolute value of absolute error is 1.98g. These shows the numerical simulation method used in this paper can predict the process of fabric drying accurately.

5.The analysis of difference of the percentage of water in weft direction in fabric

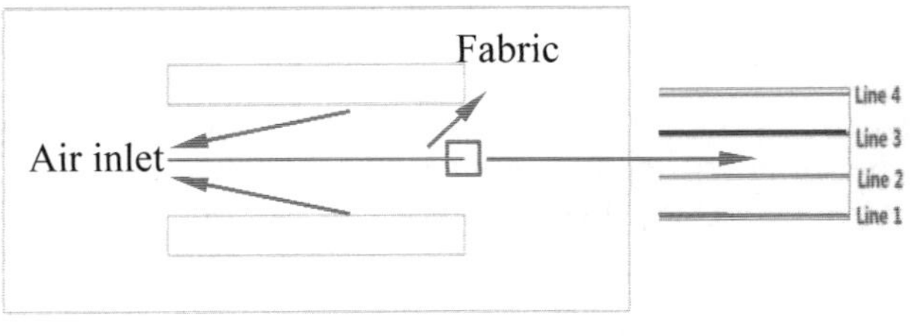

Figure 4 the location of the lines along the thickness

This manuscript also analyse the distribution of the percentage of water in the fabric in four lines through numerical simulation result. The four lines are taken along the thickness of the fabric, it is Line1, Line2, Line3, Line4 from bottom to top(as shown in figure 4) .

Figure 5 shows the difference of distribution of the percentage of water in weft direction of the fabric in different times, relative to the interior of the fabric, the percentage of water in the margin is low. And the difference is distinct. As time goes on, the region of low percentage of water in the fabric increase, and the difference between the margin and the interior of the fabric is more distinct.

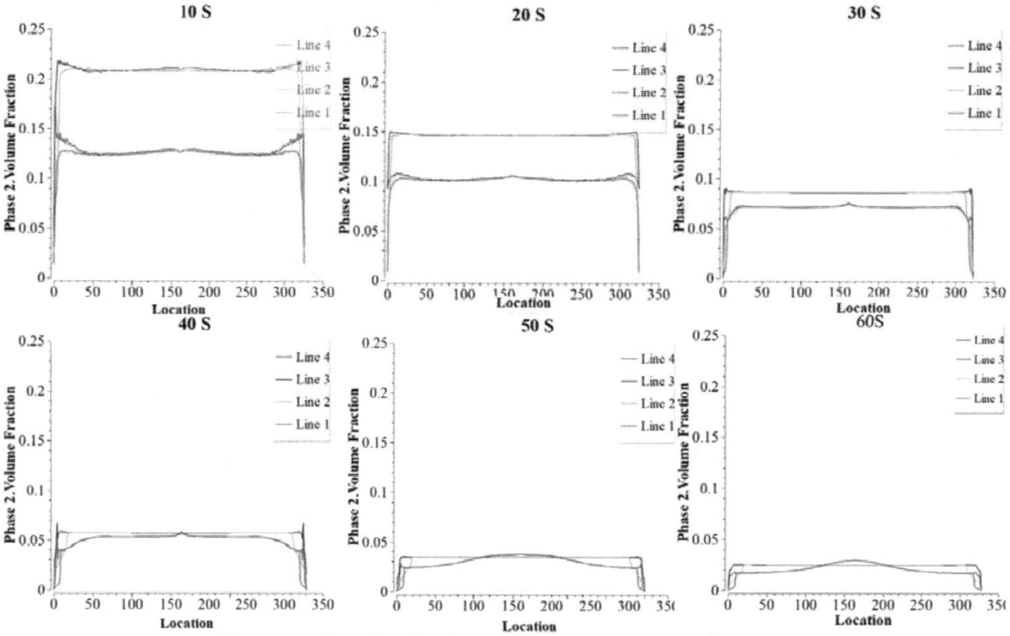

Figure 5 the distribution of the percentage of water

6.Conclusion

On the base of FLUENT, this paper developed fabric drying model by combining standard k-ε turbulence model and Species transport model and mixture multiphase model organically. On the base of Mathis DHe steaming machine , the FE model is verified by experiment. It is worth noting that all the boundary conditions and initial conditions in experiment are identical with the conditions in numerical model; after that, this numerical model is used to study the process of fabric drying inside the drying equipment and the distribution of the percentage of water. The results reveal that the change of the weight of water in the fabric is nearly positive correlation with time and the distribution of the percentage of water is uneven in the fabric in the process of drying, this tendency is conformable with the problem found in the practical production. In later analysis, this numerical model can be used to analyse the structure and technological parameters of drying device, it can improve the distribution of the percentage of water in the fabric and make it more even in the process of drying. Finally this research will provide a method to solve the problem.

7.References:

[1] CAYA, TARAKCIOGLU I , HEPBASLI A.Exergetic analysis of textile convective drying with stencers by subsystem models : part 1-Exergetic modeling and evaluation [J]. Drying Technology,2010,28(12):1359-1367.

[2] SCHMIDT M,CHLOSSER U, BAHNERS T. CFD simulation of a drawing dryer [J]. International Textile Guide, 2009, 37(9): 42-44.

[3] Polat S, Douglas W J M. Heat transfer under multiple slot jets impinging on a permeable moving surface , AICHE J ,1990,36(9) : 1370-1378.

[4] Min Ming Gu. The Research on Modeling and Optimization of the Stener Chamber [D]. Zhe Jiang Zhejiang Sci-Tech University.2018.

[5] Pautea Sarafle, Jie Min. Inhibitory effect of vacuum dehydration on disperse dyes swimming before drying [J]. foreign textile technology: chemical fiber. Dyeing and finishing, 1993(5):22-26.

IOP Conf. Series: Materials Science and Engineering **544** (2019) 012021 doi:10.1088/1757-899X/544/1/012021

Buckling and Postbuckling Analyses of Structure using Absolute Nodal Coordinate Formulation

Abdur Rahman Shaukat[1], Tengfei Wang[2] and Peng Lan[3,4]

[1,2,3] School of Mechatronics Engineering, Harbin Institute of Technology, Harbin, China

E-mail: [1] arshaukat@gmail.com, [2] tfwangsme@hit.edu.cn, [3] lan_p@sina.com

Abstract. In this paper, the three dimensional higher order beam element based on the absolute nodal coordinate formulation (ANCF) is used to study both the classical buckling and the nonlinear postbuckling problem. The analyses are performed using the Newton-Raphson method and the arc length method. The Newton-Raphson method is used for the Euler buckling whereas the Crisfield's arc length method is applied to track the equilibrium path of the William's Toggle. The solutions agree well with the analytical one or that given by the commercial finite element software ANSYS. Hence, the validity of the analyses is demonstrated.

1. Introduction

The absolute nodal coordinate formulation (ANCF) was originally proposed by Shabana in 1996 [1] for the dynamic analysis of flexible multibody system, especially for problems in which the components exhibit large deformation and rotation. The kinematic description based on the global coordinate system endows ANCF with the advantages of constant mass matrix, zero Coriolis and centrifugal force in the system equation of motion. Such valuable properties promoted fast development of the ANCF finite element (FE). Correspondingly, extensive applications of ANCF in different fields can be seen in the last two decades [2]. In the literature, most of the implementations are limited to the dynamic analysis. Few papers focus on the buckling and postbuckling analyses. As a nonlinear formulation, the ANCF is suited for the large deformation analysis. Additionally, it allows for the development of the non-incremental solution procedure. The implementation of nonlinear general continuum mechanics theory [3] has made it a general formulation for the structural analysis. All the illustrated advantages indicate that the nonlinear structural analysis including the buckling and postbuckling problem can be a potential application direction of the ANCF. It is the objective of this paper to show the validity of the ANCF application in the buckling and postbuckling analyses. In this paper, the application of ANCF element in the study of classical buckling by the means of Newton-Raphson (NR) method and nonlinear postbuckling response using the arc length method (ALM) are demonstrated. The remaining paper is organized as follows: In section 2, the adopted three dimensional ANCF higher order beam element is introduced briefly. In section 3, the Euler buckling of a column is shown. In section 4, the postbuckling of the William's Toggle is presented. At last, the obtained conclusions are provided in section 5.

[4] Author to whom any correspondence should be addressed

2. Three dimensional ANCF higher order beam element

It is reported that the initially proposed fully parameterized ANCF beam elements [2] suffer from locking. Since the unbalanced order of displacement interpolation in axial and lateral directions is the main reason behind the locking phenomenon. An efficient way to alleviate the locking problem is to increase the interpolation order in the lateral directions. Correspondingly, a higher order beam element with 42 nodal coordinates (HOBE42) was proposed by Shen et al. [4]. Alleviating the locking effect, the HOBE42 is expected to correctly implement the nonlinear structural analysis. The element is introduced briefly herein for the sake of completeness.

The HOBE42 adopts quadratic interpolation in transverse directions whereas cubic polynomial in longitudinal direction. Besides all the gradient vectors, it also uses the curvature vectors as the nodal coordinates [5]. The position vector is represented as below:

$$\mathbf{r} = \mathbf{S}\mathbf{e} \tag{1}$$

Where, the vector of the element coordinate is $\mathbf{e}=\left[(\mathbf{e}^i)^{\mathrm{T}}\ (\mathbf{e}^j)^{\mathrm{T}}\right]^{\mathrm{T}}$ and the nodal coordinate vector is:

$$\mathbf{e}^i = \left[\mathbf{r}^{i\mathrm{T}}\ \ \mathbf{r}_x^{i\mathrm{T}}\ \ \mathbf{r}_y^{i\mathrm{T}}\ \ \mathbf{r}_z^{i\mathrm{T}}\ \ \mathbf{r}_{yz}^{i\mathrm{T}}\ \ \mathbf{r}_{yy}^{i\mathrm{T}}\ \ \mathbf{r}_{zz}^{i\mathrm{T}}\right]^{\mathrm{T}} \tag{2}$$

Where, $\mathbf{r}_x^i = \partial\mathbf{r}^i/\partial x$, $\mathbf{r}_y^i = \partial\mathbf{r}^i/\partial y$, $\mathbf{r}_z^i = \partial\mathbf{r}^i/\partial z$, $\mathbf{r}_{yz}^i = \partial^2\mathbf{r}^i/\partial y\partial z$, $\mathbf{r}_{yy}^i = \partial^2\mathbf{r}^i/\partial y^2$ and $\mathbf{r}_{zz}^i = \partial^2\mathbf{r}^i/\partial z^2$ are gradient and curvature vectors at node i. The shape function matrix is:

$$\mathbf{S} = \left[s_1\mathbf{I}\ \ s_2\mathbf{I}\ \ \ldots\ \ s_{14}\mathbf{I}\right] \tag{3}$$

Where, $\mathbf{I}$ is the 3×3 identity matrix, s_i $i=1,2,\ldots,14$ are shape functions shown as below:

$$\left.\begin{aligned}
&s_1 = 1-3\xi^2+2\xi^3,\ s_2 = l\left(\xi-2\xi^2+\xi^3\right),\ s_3 = l(\eta-\xi\eta)\\
&s_4 = l(\zeta-\xi\zeta),\ s_5 = l^2(\eta\zeta-\xi\eta\zeta),\ s_6 = \frac{l^2}{2}\left(\eta^2-\xi\eta^2\right)\\
&s_7 = \frac{l^2}{2}\left(\zeta^2-\xi\zeta^2\right),\ s_8 = 3\xi^2-2\xi^3,\ s_9 = l\left(-\xi^2+\xi^3\right)\\
&s_{10} = l\xi\eta,\ s_{11} = l\xi\zeta,\ s_{12} = l^2\xi\eta\zeta,\ s_{13} = \frac{l^2}{2}\left(\xi\eta^2\right),\ s_{14} = \frac{l^2}{2}\left(\xi\zeta^2\right)
\end{aligned}\right\} \tag{4}$$

Where, the dimensionless coordinates are defined as $\xi=x/l$, $\eta=y/l$ and $\zeta=z/l$, l is the length of the element.

3. Euler Buckling of a column

Under the buckling phenomenon of a column, a very slight increment in the axial load leads to significant change in the lateral deflection. In other words, the small increase in the axial force will cause the loss of the lateral stiffness of the column making the structure unstable. Based on this property, one can predict the buckling load numerically. The classical Euler buckling problem is used in this section to show the validity of the method.

A beam with one end clamped and the other end free is adopted. The dimension of the square cross section is $A=0.01\times0.01\mathrm{m}^2$, the length $L=1\mathrm{m}$ and the corresponding area moment of inertia is $I=8.333\times10^{-10}\mathrm{m}^4$. The Young's modulus is $E=2\times10^{11}\mathrm{N/m}^2$ and the Poisson's ratio is $v=0.3$. The beam is meshed using 10 HOBE42 elements.

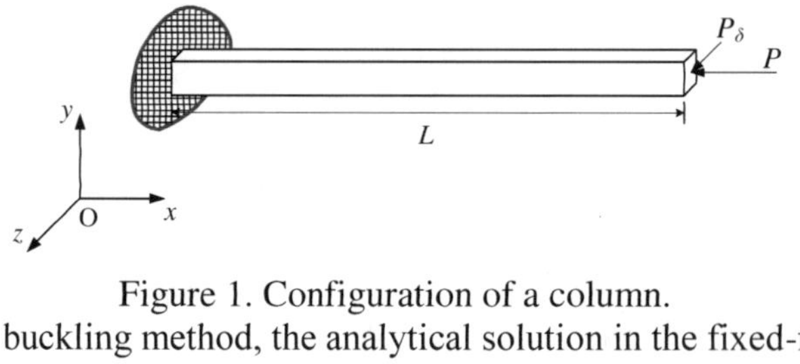

Figure 1. Configuration of a column.

According to the Euler's buckling method, the analytical solution in the fixed-free case is:

$$P_E = \frac{\pi^2 EI}{(2L)^2} = 411.3\text{N} \tag{5}$$

Based on the above discussion, the Euler's buckling load can be obtained numerically. The axial compressive force P is applied at the free end and is gradually increased in discrete load steps. The small constant lateral perturbation force $P_\delta = 0.01\text{N}$ is applied at the free end to model the imperfection in the system. The nonlinear static analysis using full Newton-Raphson iteration is performed at each load step to obtain the corresponding lateral deflection.

Then, the full loading equilibrium path, which is the change of the dimensionless axial load P / P_E with respect to the ratio of lateral deflection δ / δ_o, is plotted as shown in figure 2, where δ_o is the lateral deflection at $P = 0$. Due to the P-δ effect [6], the lateral deflection at the free end keeps on increasing with the increment of the axial load instead of remaining constant. At the beginning, as the axial compressive force increases gradually, the lateral deflection increases slowly. On the contrary, as the axial load approaches the Euler's buckling load, the change in lateral deflection becomes asymptotic which symbolizes loss of the lateral stiffness. The corresponding axial load can be identified as the critical buckling load P_{cr} predicted numerically.

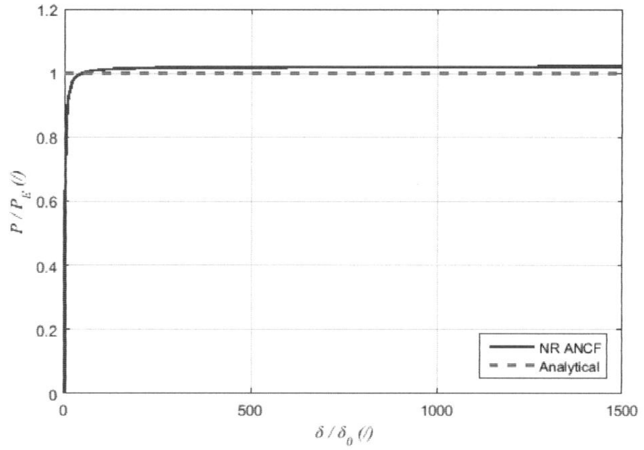

Figure 2. Buckling of a column.

It can be seen in the figure 2 that the result obtained is in good agreement with the analytical result, i.e. $P_{cr} / P_E = 1$. It can be concluded that the ANCF can successfully predict the Euler's buckling load.

4. Postbuckling of a William's Toggle

In the field of nonlinear finite element method, the Newton-Raphson method is widely used to follow the equilibrium path as shown in section 3. But the NR method is not capable to follow the equilibrium path after the buckling phenomenon occurs. To circumvent this problem, certain load control and displacement control techniques were proposed but they are not very general [7]. Wempner [8] introduced the idea of using the arc-length control to track the nonlinear equilibrium path. Later, Riks [9] formulated the general equations which can be used to predict the complete equilibrium path. However, the solution techniques adopted by those equations destroy the banded nature of the tangential stiffness matrix. Hence, its implementation in the nonlinear finite element code is not straight forward. To solve the aforementioned problems, Crisfield [10] proposed a modification of the arc length method which utilizes the tangential stiffness matrix directly from the nonlinear FEM theory and exhibits better convergence properties as well. This made the arc length method compatible with the existing nonlinear FE code.

The Crisfield's ALM can be readily used in combination with the ANCF theory to study the nonlinear postbuckling response. In this section, the William's toggle problem is studied and the dimensions adopted in [11] are used. As shown in figure 3(a), two bars of length $L = 2.5\text{m}$ are connected by pin joint at the middle. The height of the middle pin joint is $d = 0.025\text{m}$ from the base line and both the other ends are pinned. The dimensions of the bar cross section are $b = h = 0.01\text{m}$, Young's modulus

IOP Conf. Series: Materials Science and Engineering **544** (2019) 012021 doi:10.1088/1757-899X/544/1/012021

$E = 5\times10^{11}\,\text{N/m}^2$ and Poisson's ratio is $v = 0.3$. Both bars have the same material and geometric properties. Hence, the principle of symmetry is applied. As shown in figure 3(b), only the left bar is analyzed and meshed using 20 HOBE42 elements. The vertical downward force is gradually applied.

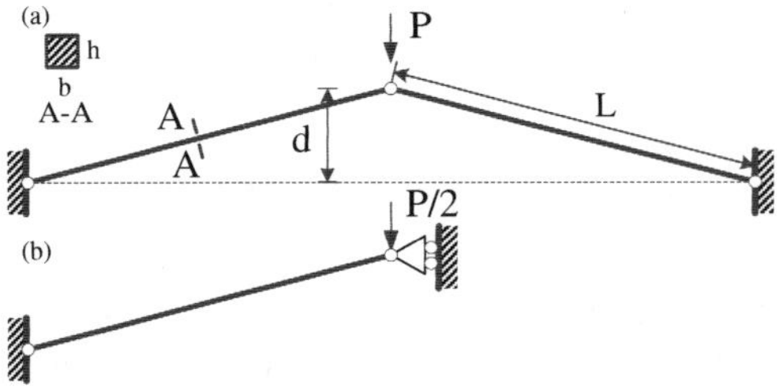

Figure 3. William's Toggle. (a) The whole structure. (b) The simplified half structure.

The equilibrium path obtained by the arc length method, the Newton-Raphson method based on the ANCF HOBE42 and the three dimensional Timoshenko beam element BEAM188 using the commercial FE software ANSYS are shown in figure 4. The converged result obtained by using 100 BEAM188 elements with nonlinear geometric option on in ANSYS is used as the reference. It can be found that the three solutions agree well. While the ALM based on ANCF is able to follow the full path, the Newton-Raphson method fails to follow after the limit point, instead it jumps past the postbuckle snap through region of the path.

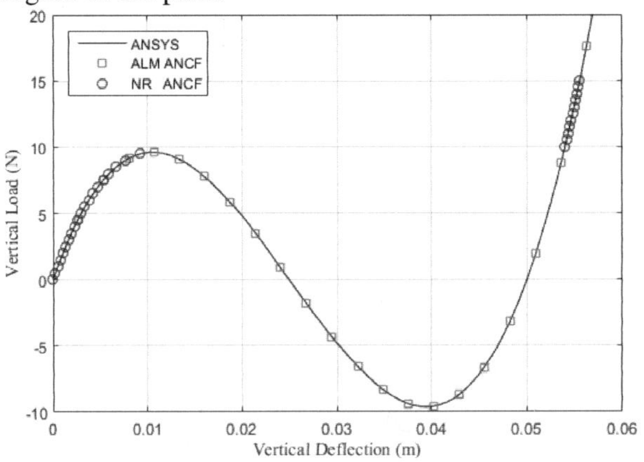

Figure 4. Postbuckling of a William's Toggle (Pin-Pin).

In the second case, the William's Toggle is considered with both ends clamped. The equilibrium path obtained is different from that of the previous case but still follows the snap through pattern. Both results are compared with the plot obtained from the commercial finite element software ANSYS. The comparison is shown in figure 5.

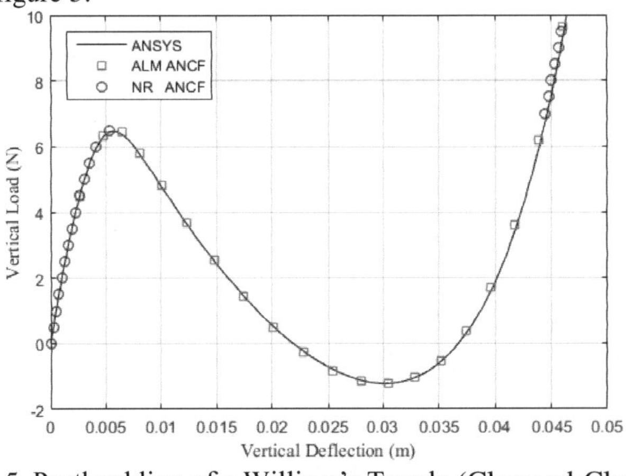

Figure 5. Postbuckling of a William's Toggle (Clamped-Clamped).

IOP Conf. Series: Materials Science and Engineering **544** (2019) 012021 doi:10.1088/1757-899X/544/1/012021

Same conclusion can be drawn in the clamped-clamped case. Thus, the validity of the ANCF application in the postbuckling analysis is verified.

5. Conclusion

The absolute nodal coordinate formulation based on the nonlinear continuum mechanics can not only be used in the dynamic analysis but can also perform well in the structural analysis. In this paper, it is utilized to predict the Euler buckling load of a column. The natural characteristic of ANCF makes it suitable for the nonlinear structural analysis. Therefore, combined with the Crisfield's arc length method, the ANCF is applied to obtain the nonlinear postbuckling equilibrium path of the William's Toggle. Hence, it is shown that the ANCF can successfully be applied in the study of buckling and nonlinear postbuckling analyses.

References

[1] A A Shabana 1997 Definition of the slopes and the finite element absolute nodal coordinate formulation *Multibody Syst. Dyn.* **1** 339–348

[2] J Gerstmayr, S Hiroyuki, and M Aki 2013 Review on the absolute nodal coordinate formulation for large deformation analysis of multibody systems *J. Comput. Nonlinear Dyn.* **8**

[3] A A Shabana 2012 *Computational continuum mechanics* (NewYork: Cambridge University Press)

[4] Z Shen, P Li, C Liu, and G Hu 2014 A finite element beam model including cross-section distortion in the absolute nodal coordinate formulation *Nonlinear Dyn.* **77** 1019–1033

[5] G Orzechowski and A A Shabana 2016 Analysis of warping deformation modes using higher order ANCF beam element *J. Sound Vib.* **363** 428–445

[6] S P Timoshenko and J M Gere 1963 *Theory of Elastic Stability* (Singapore: McGraw Hill)

[7] B A. Memon and X Su 2004 Arc-length technique for nonlinear finite element analysis *J. Zhejiang Univ. Sci.* **5** 618–628

[8] G A Wempner 1971 Discrete approximations related to nonlinear theories of solids *Int. J. Solids Struct.* **7** 1581–1599

[9] E Riks 1978 An incremental approach to the solution of snapping and buckling problems *Int. J. Solids Struct.* **15** 529–551

[10] M A Crisfield 1981 A fast incremental/iterative solution procedure that handles 'Snap-through' *Comput. Struct.* **13** 55–62

[11] M A Crisfield 1991 *Non-linear Finite Element Analysis of Solids and Structures* (West Sussex: John Wiley & Sons)

IOP Conf. Series: Materials Science and Engineering **544** (2019) 012022 doi:10.1088/1757-899X/544/1/012022

Research on Optimization Design of Shipborne Carbon Fiber Reinforced Polymer Frame Structure

H Wang[1,2], RC Yu[1,2], YZ Chen[1,2,*] and L Hua[1,2,*]

1. Hubei Key Laboratory of Advanced Technology for Automotive Components, Wuhan University of Technology, Wuhan 430070, China
2. Hubei Collaborative Innovation Center for Automotive Components Technology, Wuhan 430070, China
E-mail: yzchen@whut.edu.cn; hualin@whut.edu.cn

Abstract. Display consoles are the core frame structure of the command and control system on the military combat unit, which can be called the "brain" of the equipment. With the great development of new research on materials and applications, using Carbon Fiber Reinforced Polymer (CFRP) instead of aluminum alloys to fabricate the display console frame structure has become an international hot point. However, whether the components obtained by CFRP can meet the requirements of complex working conditions, and whether the structural design is reasonable, are still existing problems. This study analyzed the stress distribution of the bottom frame structure of CFRP display console under impact and vibration conditions. The frame structure was optimized based on the stress distribution and topology. The results show that the maximum stress value of the frame structure under complex conditions was 16.97MPa, the maximum stress after optimization was 22.94MPa. The reduction of weight of this complex structure was 13%. The prototypes have passed the actual working condition test. This study shows that the route method based on impact stress to optimize the wall thickness of CFRP structures is feasible and effective.

1 Introduction

The display consoles carry the command and control system of combat unit. These components are highly integrated with the detection, monitoring and controlling of signals, so as to realize the functions of alert detection, precise guidance and electronic countermeasures. Therefore, it is essential to ensure the reliability of these frame structures in complex service conditions such as explosion or strong vibration.

The display console products used in China is the second-generation fabricated by aluminum alloys[1]. However, the third-generation display console using Carbon Fiber Reinforced Polymer (CFRP)has been widely applied in western counties. In order to achieve military products with lightweight and high-strength, it is very urgent to develop shipboard display consoles using CFRP. The CAE analysis method has been widely used for the sake of improving forming efficiency and strength, reducing costs. With the help of ANSYS software, some modal analysis on an airborne CFRP display console was performed [2]-[3], the structure was optimized according to the obtained modes and natural frequencies. Yu Yang[4] used ANSYS software to establish a finite element model of a shipborne display console, performed modal analysis on the display console. The overall structural strength of

* Corresponding author: YZ Chen(yzchen@whut.edu.cn) and L Hua(hualin@whut.edu.cn)

the display console according to the loading condition was evaluated. YAN Hong[5] combined the fault-tolerant hierarchical sequence method to optimize the platform under an airborne display console to achieve a weight loss of 1.3 kg.

However, there is more attention should be paid to the reliability of shipborne equipment when it was subjected to explosions such as mines and torpedoes. At the same time, in order to meet relevant requirements, it is necessary to optimize the structure of CFRP components to achieve weight reduction. In this paper, the stress analysis of the CFRP frame structure under impact condition was analyzed. The wall thickness was optimized based on the impact stress distribution. Then, vibration simulation was carried out basing on optimization results. The final specimen manufactured according to the optimization results passed the test successfully.

2. Research methods and solutions

2.1 Research route and optimization method

This study was carried out according to the following method. Firstly, the finite element model of the CFRP frame structure was established and the original analysis was carried out. Then, the structure optimization was performed basing on the stress distribution under impact condition. Thirdly, whether the structural strength of the optimization result satisfies the requirements of complex working conditions such as vibration and impact was analyzed.

The analysis of the CFRP frame structure finite element model was carried out by OptiStruct[6]. After determining the constraints of the optimization model, according to the optimization purpose of this paper, the wall thickness of the frame structure was taken as the design variable, the impact stress was used as the constraint, and the minimum mass was used as the objective function to optimize.

2.2 Analysis conditions of impact and vibration

According to the GJB-150.16-1986/GJB-150.18-86 environmental test method for military equipment, the impact and vibration conditions of the frame structure are shown in the table1. The table shows that the vibration test is mainly divided into two frequency ranges including1~16Hz and 16~60Hz. Among them, the displacement value of the component is required to reach 1.0mm in the range of 1~16Hz.The acceleration of the component is required to reach 10m/s2 in the range of 16~60Hz.

Table 1. Vibration test value

Partition	Test parameters		
	Frequency(Hz)	Displacement(mm)	Acceleration(m/s^2)
Main area	1~16	1.0	----
	16~60	----	10

Impact tests are generally tested using a hammer method. Before the test, the hammer is lifted to a certain height and then falls freely. The peak acceleration requirement is 7g, and there are certain requirements for the duration and speed variation, as shown in table 2.

Table 2. Impact test value

Waveform	Test parameters		
	Peak acceleration(g)	Duration(mm)	Speed change(m/s)
Half-sine wave	7	11	(2Ad/π)2.10

2.3 Finite element model

Accurate modeling of the CFRP frame structure before finite element analysis is the basis for obtaining acceptable simulation results. In this paper, HyperMesh was used to simplify the CAD model of the CFRP frame structure and extracted the middle surface. After the necessary repaired, the mesh was accurately divided. In view of the fact that the material of the frame structure was carbon fiber composite material, the layer thickness was set using HyperLaminate including proper layup angle and layup order of carbon fiber. The frame structure of this paper used Japan Toray T700 carbon fiber, adapting orthogonal layering method in the direction of 0° and 90°. The material performance parameters are as follows in table 3:

Table 3. Material Parameters

Material	Density (g/cm3)	Elastic Modulus(Gpa)	Poisson's ratio	Tensile strength (Mpa)	Bending strength(Mpa)	Elongation
T700	1.35	26.935	0.3	555	530	1%
Resin	1.0	0.7	0.45	18	---	70%

In order to obtain more accurate simulation results and analyze the optimization characteristics of composite materials, this study selected the typical framework structure for analysis. Its three-dimensional picture is shown in the figure 1. Its general size was about 700*450*600mm and consisted of 7 parts with weighs about 20.1kg.

There were seven parts in the CFRP frame structure, which were connected by bolts. In order to improve the accuracy of CAE analysis and obtain more accurate simulation results, the method of adding rigid connection bolt was used to simulate the connection state of each component of the frame structure. 56 simulation bolts were added to the key parts. The bolt position is shown in the white dot of the figure 2 below.

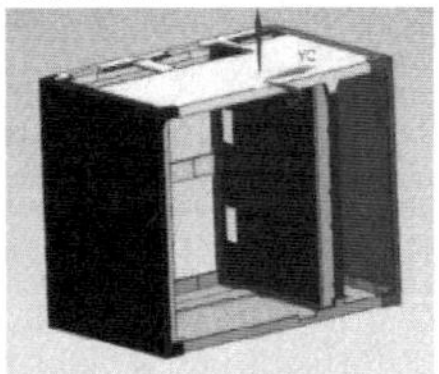

Figure1. Three-dimensional model

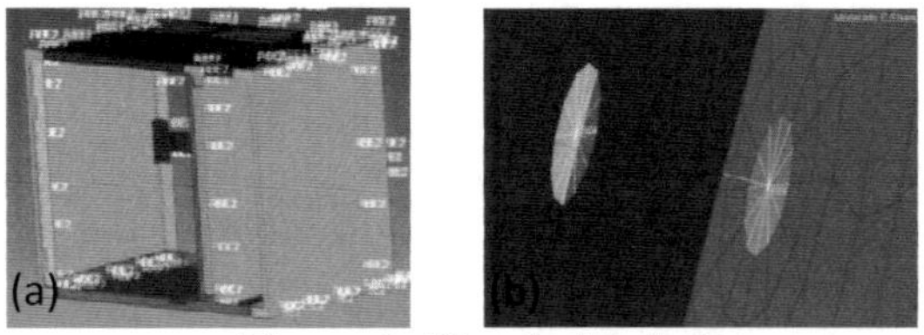

Figure 2. Simulation bolts

In the service process of frame structure, there were extra assembly parts in the components. Therefore, the exact load was added in finite element model according to the true weight of the extra assembly parts. The main weight is shown in the figure 3 and figure 4, which mainly includes the chassis and the wiring harness, and the total weight is 77 kg. According to the connection form of the chassis and the wiring harness board with the frame structure, weightless rigid connecting parts RBE2 was created at each connection position in the frame structure and intersected at a point as shown in figure 5. F1 and F2 were set as the chassis and wiring harness using the concentrated force loading mode.

Figure 3. The chassis

Figure 4. The wiring harness

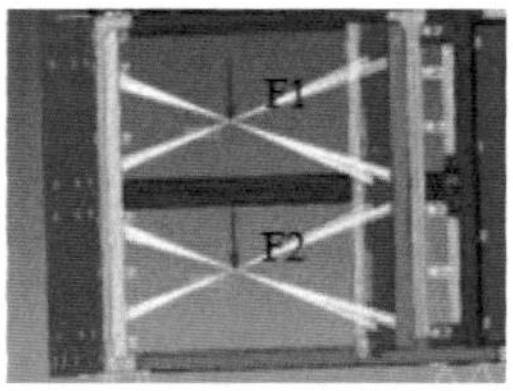

Figure 5. Loading position

3. Results and discussion

3.1 Stress distribution of CFRP frame structure

The stress distribution of the frame structure under the impact condition is shown in the figure 6 below. It can be seen from the figure that the maximum stress of the frame structure was 16.97Mpa, which located at a bolt hole position of the top plate. This means that the structure meet the requirements for the use of CFRP without any damage. At the same time, the overall stress value of the bottom plate is about 16.8Mpa. From the simulation results, it can be found that the stress of the plate 1, the plate 2 and the back plate were actually very small, indicating a larger optimization space for the following analysis.

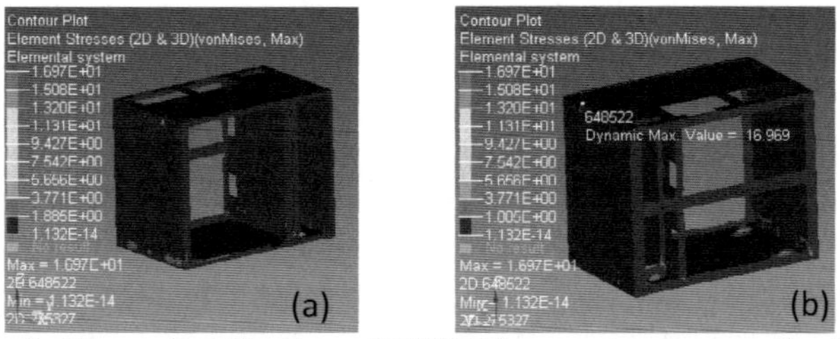

Figure 6. Stress distribution of CFRP frame structure under impact.

3.2 Stress analysis after optimization

Using the optimization method mentioned, the thickness of the frame structure was decreased. The optimized stress distribution is shown in figure 7. It can be seen from the figure that the maximum stress of the frame structure after optimization was 22.94Mpa, which is located at a bolt hole position of the bottom plate and meets the requirements for the use of carbon fiber composite materials without damage. At the same time, the bottom plate is more stressed, and the overall stress value is about 22Mpa. In general, the optimization results meet the strength requirements of vibration and impact conditions.

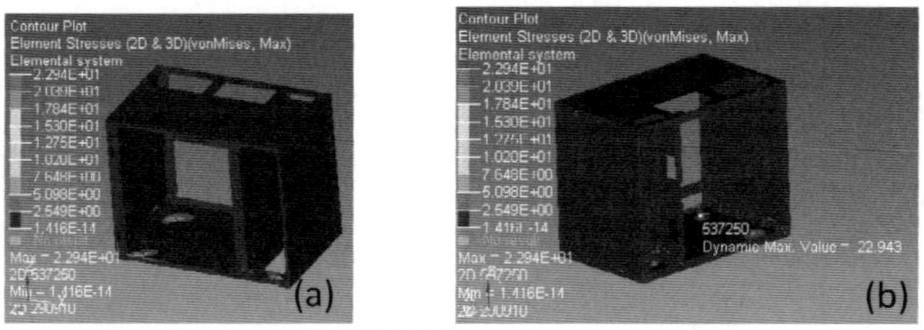

Figure 7. Stress distribution of CFRP frame structure after optimization.

3.3 Comparison of the results

The stress and weight of the CFRP frame structure before and after optimization are shown in the table 4. It can be seen from the table that before optimization, the plate 1, 2 and the back plate had less stress and had a larger optimization space. After optimization, the side plate 1, the side plate 2 and the back plate are respectively reduced by about 0.53 kg,0.32 kg,and 0.09kg, while the maximum stress of other large stress components, were increased compared to before optimization, but they are still much smaller than the allowable stress of T700 carbon fiber composites, and the optimization results meet the requirements.

Table 4. Comparison of the results

optimization		Bottom plate	plate 1	plate 2	plate 3	plate 4	Back plate	Top plate	Total
before	Max stress/Mpa	16.8	13	7.5	15.1	14.2	0.36	16.97	16.97
	weight/kg	4.59	3.72	3.71	2.99	0.90	0.67	3.54	20.12
after	Min stress/Mpa	22.94	16.52	9.9	16.98	20.87	0.51	22.63	22.94
	weight/kg	3.95	3.19	3.39	2.57	0.78	0.58	3.04	17.5

3.4 Specimen and test results

According to the results of CAE analysis, we used RTM technology to manufacture various parts of CFRP frame structure, and carried out the assembly work such as punching, connecting and repairing, and obtained the finished products as figure 8 and figure 9 According to the relevant national test standards, the vibration(figure 10.) and impact(figure 11.) tests of the frame structure were carried out. The results showed that the optimized frame structure has successfully passed the impact and vibration tests, and met the requirements of use. It showed that the method provided in this paper can significantly improve design efficiency, reduce costs and achieve products lightweight.

Figure 8 .Specimen (front) Figure 9.Specimen (back) Figure 10. Vibration test Figure 11. Impact test

4. Conclusion

1) This study used Hypermesh and Optistruct to analyze the stress distribution of carbon fiber composite frame structure under impact and vibration conditions, and the topology of the structural wall thickness was optimized based on the impact stress rather than the modal analysis.
2) In original analysis, the maximum stress of the CFRP frame structure is 16.97Mpa.After optimizing the frame structure, the structure weight loss was 13%, and the maximum stress was 22.94MPa, which meets the requirements of parts. The prototype components passed the actual working condition test.
3) This study shows that the route method based on impact stress to optimize the wall thickness of composite structures is feasible and effective.

5.Acknowledgments

This study is supported by National Key R&D program of China No. 2018YFB1106700, National Natural Science Foundation of China (No. 51775398 and 51805392), the Fundamental research Funds for the Central Universities (WUT: 2018III067GX, 2018III074GX and 2018IVA06).

References:

[1] Chen C and Yang N 2009 Development of modeling and structural design of radar display console *Selected Works of Mechanical and Electronic Science Conference of Electronic Mechanical Engineering Branch of China Electronics Society*

[2] Sun F 2015 Structural dynamics simulation analysis and optimization of a carbon fiber display console *J. Electronic Mechanical Engineering* **31** pp 58-64

[3] Hang W, Sun F and Shen W 2015 Design and experiment of helicopter loaded display console *J. Electromechanical Engineering* **31** pp 39-41,45

[4] Yu Y and Yang W 2011 Structural dynamics simulation analysis of a shipboard display console *J. Electronic Mechanical Engineering* 27 pp 58-61

[5] Yan H 2016 Lightweight design of an airborne display console based on topology optimization *J. Machinery and Electronics* **34** pp 22-25

[6] Hong Q, Zhao K and Zhang P 2012 *OptiStruct & HyperStudy theoretical basis and engineering application* (Beijing: Mechanical Industry Press)

Investigation of metal oxides toward organic dyes decolourisation and environmental remediation

ThianKhoon Tan[1,*], PoiSim Khiew[1], WeeSiong Chiu[2] and ChinHua Chia[3]

[1] Engineering Foundation, Faculty of Engineering, University of Nottingham Malaysia, Jalan Broga, 43500 Semenyih, Selangor Darul Ehsan, Malaysia
[2] Low Dimensional Materials Research Center, Department of Physics, Faculty of Science, Universiti Malaya, 50603 Kuala Lumpur, Malaysia
[3] School of Applied Physics Studies, Faculty of Science & Technology, Universiti Kebangsaan Malaysia, 43600 Bangi, Selangor Darul Ehsan, Malaysia
[*] E-mail: TK.Tan@nottingham.edu.my

Abstract. Enormous interest has been focused on metal oxides in its photocatalytic capabilities toward oxidizing organic dyes and environmental remediation. Zinc oxide (ZnO), titanium dioxide (TiO_2) and iron oxide (Fe_3O_4) have been used as the heterogeneous photocatalysts to photodegrade three organic dyes, which are methylene blue (MB), phenol red (PR) and methyl orange (MO). Series of 3.0 ml dyes samples in aqueous solutions were placed into a UV chamber under UV-C (254 nm) light irradiation. The oxides powder was characterised by XRD, FESEM and TEM analysis, where ZnO nanoparticles was in hexagonal wurtzite crystal structure (SG: $P6_3mc$) with average crystalline size of (91 ± 12) nm and TiO_2 nanoparticles was pure anatase of tetragonal crystal structure (SG: $I4/_amd$) with average crystalline size of (117 ± 16) nm. Fe_3O_4 nanoparticles was in cubic spinel crystal structure with average crystalline size of (62.1 ± 5.7) nm. The photocatalyst loading (0.5 – 3.5wt%), initial dye concentration (MB: 2 – 12 ppm, PR: 8 – 48 ppm, MO: 12 – 32 ppm) and irradiation duration on the decolourisation of dye samples were examined. Based on time requirement, the photocatalytic decolourisation rate increased with increasing metal oxide loading and decreased with the increasing of the initial dye concentration. The superior photocatalyst was ZnO with a loading of 2.5 wt% followed by TiO_2 (3.0 wt%) and least photocatalytic capability was Fe_3O_4 (3.0 wt%). On the other hand, the most photostable dye is MO followed by PR and MB.

1. Introduction

In general, metal oxides can take a variety of structural geometries together with an electronic structure that can exhibit different characteristics, such as metallic, semiconductor and insulator. Therefore, they have emerged as a very important substance and contribute in many areas of physics, chemistry and materials science. Recently, metal oxides have gained enormous interest in heterogeneous photocatalysis due to its emerging applications toward environmental remediation and organic synthesis (Amini and Ashrafi 2016). Many researchers attempted to study photocatalytic activity of different metal oxides like ZnO, CeO_2, WO_3, TiO_2, Cu_2O, MnO_2, Fe_2O_3, etc. (Ray and Pal 2017).

Both ZnO and TiO_2 have been extensively used as photocatalyst which have become a promising environmental remediation technology, water photoelectrolysis and dye-sensitized solar cells, due to their low cost, environmental friendly, nontoxicity, high chemical stability and excellent degradation for organic pollutants (Amini and Ashrafi 2016). On the other hand, little is known about the use of iron oxide as a photocatalyst. There are numerous conditions determining the effectiveness of this magnetic nanoparticles in wastewater treatment system and a number of studies have demonstrated the success of Fe_3O_4 nanoparticles on the removal of heavy metals from contaminated water (Hou, Tian et al. 2015, Kim, Woo et al. 2016). Besides, this magnetic nanoparticle also showed a sustainable treatment process through reusing and still regaining the removal capacity in few treatment cycles (Chiu, Khiew et al. 2010, Salamat, Younesi et al. 2017).

In the present study, the photodegradation of the three type of dyes, namely methylene blue (MB), phenol red (PR) and methyl orange (MO) were investigated and the most optimum amount of loading for the three metal oxides, ZnO, TiO_2 and Fe_3O_4 nanoparticles photocatalysts, were evaluated. These dyes were chosen as the organic compound since they are complex organic substances which contain a significant amount of various functional groups and are well dissolved in water. MB dye is easy to monitor by simple techniques such as UV-visible spectroscopy at its maximum absorption wavelength (Nogueira, Castro et al. 2014), with its four main peaks at 665, 614, 293 and 247 nm (Kuan and Chan 2012). On the other hand, PR is a weak organic acid and is a typical reversible pH-sensitive dye. Its absorption maximum wavelength is at 435 nm. Finally, MO is a typical azo dye and generally used in the textile industry. Its absorption maximum wavelength, λ_{max}, at 466.5 nm (Hakamada, Hirashima et al. 2012).

2. Materials and Methods

2.1. Materials

Zinc oxide (ZnO, R & M Chemicals, U.K.), titanium dioxide (TiO_2, R & M Chemicals, U.K.) and iron oxide (Fe_3O_4, Systerm, ChemPur) purchased are of analytical grades and used without any modification. The dyes include methylene blue ($C_{16}H_{18}N_3SCl{\bullet}3H_2O$; M_w=373.90 g/mol), phenol red ($C_{19}H_{14}O_5S$; M_w=354.38 g/mol), and methyl orange ($C_{14}H_{14}N_3NaO_3S$; M_w=327.33 g/mol were purchased from R & M Chemicals, UK. All the chemical products were used as received without any further purification. The deionised water obtained from Favorit Water Distiller (Favorit Water Still W4L; conductivity of 3-4 μs/cm and resistivity of 0.25-0.30 MΩ·cm) was used to prepare all the aqueous solutions.

2.2. Experimental

All metal oxides nanopowder was characterised by X-ray diffraction (XRD) using Bruker AXS-D8 Advance diffractometer equipped with CuK_α radiation over the scan range 2θ of 10° to 80° at room temperature. The surface and subsurface imaging were conducted by using the field emission scanning electron microscopy (FESEM) from FEI, Quanta 400 SE/SEM, which is also coupled with energy dispersive X-ray analyser (EDX). The morphology of the sample was studied using a LEO 912AB high-resolution transmission electron microscope (TEM) equipped with energy filter. The specific surface area was determined on the basis of BET instrument using Micromeritics ASAP 2020 V3.01 H.

Individual metal oxides powder was premixed with distilled water in beakers and vials to obtain different concentration of the catalysts loading, that is between 0.50 wt% to 3.50 wt%. In addition, the organic dyes, such as MB, PR and MO, were also premixed by adding the weighted dyes powder with distilled water to obtain aqueous solutions of the desired concentration, in the range of 2.00 ppm to 12.0 ppm for MB, 8.00 ppm to 48.0 ppm for PR and 12.0 ppm to 32.0 ppm for MO. After that, 3.00 ml of organic dye solutions were added with 0.20 g (0.20 mL) of premixed photocatalyst of different concentrations. These mixtures were allowed to be in the dark for about 30 minutes and then shifted and placed under UV light irradiation for various durations, with 10 minutes interval, to allow the degradation process to take place. The photocatalytic degradation was carried out in the UV chamber

(Uvitec Cambridge ultraviolet crosslinker, CL-508.G, with 5 of 8 W tube bulbs) equipped with five UV tube each producing light source of 254 nm wavelength, as shown in Figure 1.

After the irradiation, each individual photocatalyst sample was placed into the centrifuge (Kubota Tabletop Centrifuge 2420). A 2 mL aliquot was taken from the solutions to measure the absorption spectra of the dyes. The first aliquot was withdrawn from the solutions right after the dark absorption/desorption equilibrium period to determine the first absorbance spectrum (A_o) of the dye, which represents the initial concentration (C_o) of the dye. The absorption spectra of the dyes were evaluated using a UV-Visible Spectrophotometer (Varian, Cary 50) detector with scan rate of 120 nm/min for wavelength range from 200 nm to 800 nm. The degradation was studied by observing at the changes of the absorption maximum of the organic dye used.

2.3. Decolourisation analysis

The absorbance calibration curve of the dye solution was used as according to Beer-Lambert law (Kelsall, Hamley et al. 2006). Therefore percentage degradation was calculated from the initial and final absorption of the UV-vis experiments as follows:

$$Degradation\ (\%) = \frac{C_O - C}{C_O} \times 100 = \frac{A_O - A}{A_O} \times 100 \quad (1)$$

where C_o and C are the initial and post-irradiation concentration of the organic dye, respectively, whereas A_o and A are the initial and post-irradiation absorbance of the dye, respectively, as measured by the UV-vis spectrophotometer.

3. Results and Discussion

3.1. XRD analysis

Figure 2 shows the XRD characterization for ZnO, TiO_2 and Fe_3O_4 nanoparticles. First diffraction pattern shows all the peaks are consistent with the database in JCPDS file (#79-0208), which are the standard peaks of the ZnO planes with a hexagonal wurtzite crystal structure (SG: $P6_3mc$) (Saravanan, Mansoob Khan et al. 2015). The diffraction pattern also indicates high purity of the nanopowder as there was no other impurity peaks are observed. For the TiO_2 nanoparticles, the 2θ values match the characteristic pattern of pure anatase phase of TiO_2 tetragonal structure (SG: $I4/_amd$), with no other crystalline byproducts, in agreement with the literatures JCPDS file #21-1272 (Adachi, Jiu et al. 2008, Liu, Chu et al. 2015).

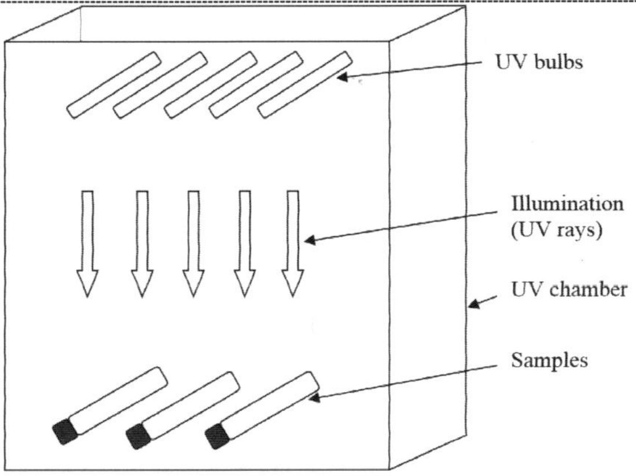

Figure 1. Schematic diagram of the experimental set up.

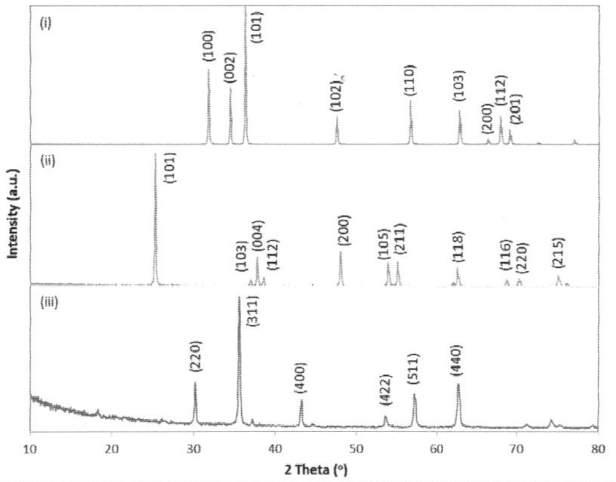

Figure 2. X-ray diffraction pattern of (i) ZnO, (ii) TiO_2, and (iii) Fe_3O_4 nanoparticle photocatalysts.

The XRD diffraction peaks for Fe_3O_4 nanoparticles conform to the Fe_3O_4 crystal of cubic spinel structure, consistent with literature JCPDS file #19-0629 (Wu, Shen et al. 2011). Its purity, could be supported by the EDX analysis, which was discussed in section 3.5. On top of that, the average crystallite size was evaluated, which are (91 ± 12) nm, (117 ± 16) nm, and (62.1 ± 5.7) nm, for ZnO, TiO_2 and Fe_3O_4 oxides, respectively.

3.2. FESEM analysis

Figures 3, 4 and 5 show the FESEM photos for morphologies of ZnO, TiO_2 and Fe_3O_4 nanoparticles. All the three oxides show no aggregation with a well disperse nanoparticle. They are fairly homogeneous in size for TiO_2 and Fe_3O_4, except for ZnO nanoparticles with a slightly wider range of size distribution.

Figure 3. FESEM picture of ZnO.

Figure 4. FESEM picture of TiO_2.

3.3. TEM analysis

The ZnO nanoparticles were irregular in shape with spherical to rod-like, as well as particle in blocks. It shows a wide particle size distribution with an average particle size of (93.2 ± 4.6) nm, which was in agreement with the crystallite size obtained from XRD. The TiO_2 metal oxide particles were generally spherical with few rod-like shape and fairly monodispersed. The particle size distribution was quite narrow as compared to the other two nanoparticles. The TEM image for the Fe_3O_4 nanoparticles was uniformly dispersed throughout with the particles were globular in shape and quite evenly distributed in size, where its average particle size is (64.0 ± 4.6) nm.

3.4. BET analysis

The BET measurement for the metal oxides nanoparticles was given in Table 1. The results clearly show that ZnO nanoparticles were having smaller surface area and pore volume if compared to the TiO_2 photocatalyst, which might impact its photodegradation capability.

Table 1. BET data for ZnO, TiO_2 and Fe_3O_4 nanoparticles.

Measurement Properties	ZnO	TiO_2	Fe_3O_4
BET surface area (m^2/g)	2.406	6.795	12.458
Pore volume (cm^3/g)	0.009720	0.04519	0.04367
Average pore size (nm)	16.156	26.606	14.023

3.5. Energy Dispersive X-ray Spectroscopy (EDX) analysis

IOP Conf. Series: Materials Science and Engineering **544** (2019) 012023 doi:10.1088/1757-899X/544/1/012023

The atomic fraction of each materials elements, ZnO, TiO_2 and Fe_3O_4, were in close agreement with their chemical formulae, with the values (1:1.12), (1:2.96) and (1:1.52), respectively. The presence of each element for each respective spectrums strengthen the arguments on the purities of these nanoparticles (Hafez, Lan et al. 2010). Together with the XRD analysis, it conforms that the samples are pure, single phase and nanocrystalline in nature.

3.6. The effects of photocatalysts loading on photocatalytic decolourisation

There have been many studies being focused on the effects of photocatalyst loading on the photocatalytic degradation of organic dyes in wastewaters (Sugiyana, Handajani et al. 2014, Regulska, Brus et al. 2016). The loading of the metal oxides photocatalysts was varied from 0.50 to 3.50 wt% in increment of 0.50 wt% with the sample dye used was 12.0 ppm MB.

3.6.1. ZnO photocatalyst. First-order kinetics study could easily convey the information of rate of the organic substance being degraded (Xie, Wang et al. 2011). The first-order kinetics results for the ZnO photocatalyst can be observed in Figure 6. It is obvious that the degradation rate was lowest for solution without catalyst, at 0.0080 min^{-1}. The degradation rate immediately improved with the addition of the ZnO photocatalyst, as the catalyst provides active radicals as well as interaction sites on its surface (Xu, Zhang et al. 2011). It increased until 2.5 wt% (0.2221 min^{-1}) of the catalyst loading, but dropped slightly with 3.0 wt% (0.2180 min^{-1}) and further decreased to 0.1856 min^{-1} for 3.5 wt% loading. With the increased of the photocatalyst loading, the number of active sites increases. Further increases beyond the optimum level will block the photon light from reaching the other photocatalyst (Velmurugan and Swaminathan 2011).

Figure 5. FESEM picture of Fe_3O_4.

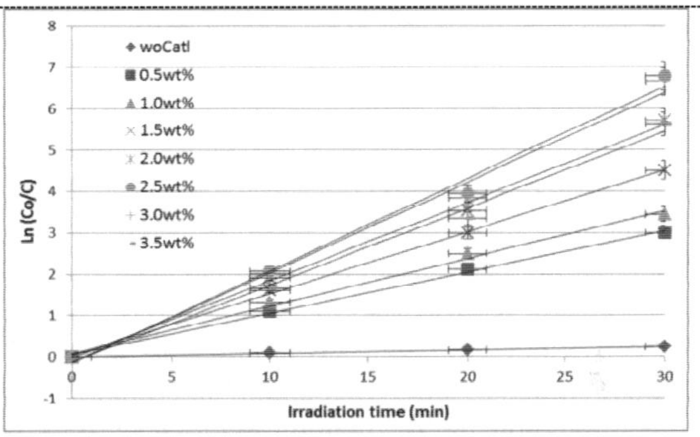

Figure 6. First-order kinetics of MB with ZnO photocatalyst.

3.6.2. TiO_2 photocatalyst. For the TiO_2 photocatalyst, as seen in Figure 7, similar pattern of degradation rate was observed. The degradation rate was lowest (0.0051 min^{-1}) when no catalyst was added and the degradation rates increased significantly (0.0487 min^{-1} for 0.50 wt% loading) when the TiO_2 photocatalyst was present in the solutions. The optimum loading for TiO_2 photocatalyst was achieved at 3.0 wt% (0.1012 min^{-1}) with a slight dropped in photodegradation rate when the loading was increased to 3.5 wt% (0.0976 min^{-1}).

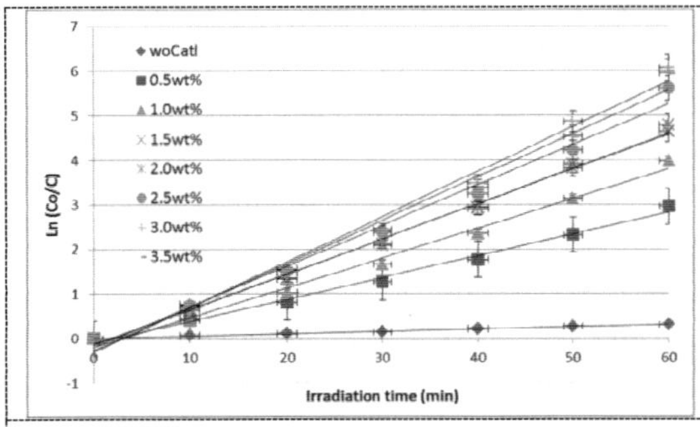

Figure 7. First-order kinetics of MB with TiO_2 photocatalyst.

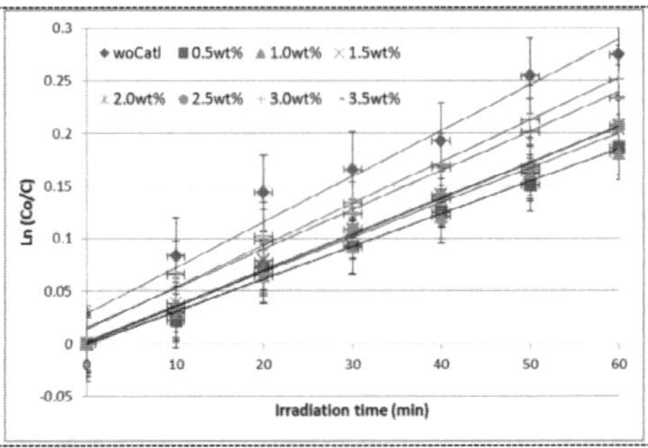

Figure 8. First-order kinetics of MB with Fe_3O_4 photocatalyst.

3.6.3. Fe_3O_4 photocatalyst. This metal oxide did not show any photocatalytic capability if compared to the other two photocatalysts, as shown in Figure 8. It is clearly seen that the highest degradation rate was without photocatalyst (0.0043 min^{-1}), where the highest degradation was only (0.0040 min^{-1}) for 3.0 wt% of Fe_3O_4 loading. It is very apparent that Fe_3O_4 nanoparticles do not exhibit any photocatalytic capability that able to enhance the photodegradation process. This is in agreement with a report by Singh and his co-researchers on Fe_3O_4 which found that Fe_3O_4 nanoparticles has insignificant photocatalytic capability. Nonetheless, Fe_3O_4 nanoparticles was generally used as a composite material with other photocatalyst materials, such as ZnO, TiO_2 etc., due to its magnetic capabilities (Singh, Barick et al. 2013).

3.7. The effects of initial dye concentration on decolourisation

In investigating this effect, the samples of dyes used were methylene blue (MB), phenol red (PR) and methyl orange (MO). The photocatalyst loading for ZnO, TiO_2 and Fe_3O_4 were 2.5, 3.0 and 3.0 wt% respectively.

3.7.1. Methylene blue. The concentration of the MB was ranging from 2.00 to 12.0 ppm. The kinetic constant are highest for the lowest dye concentration, especially for the outstanding photocatalyst of the ZnO nanoparticles which shows the steepest slopes. Although the amount of loading for ZnO is slightly less than TiO_2 and Fe_3O_4 nanoparticles, it might be due to the ability of the ZnO nanoparticles to absorb more light quanta than TiO_2 (Gupta, Saurava et al. 2015). The kinetic constant values of ZnO, TiO_2 and Fe_3O_4 photocatalysts are recorded as 1.2160 (R^2 = 1.0000), 0.2582 (R^2 = 0.9929) and 0.0237 min^{-1} (R^2 = 0.9963), respectively for 2.00 ppm MB.

3.7.2. Phenol red. The result shows that the degradation rates decreased with the increases of the PR dye concentration, especially for the ZnO and TiO_2 photocatalysts. However, when the concentration was higher, the degradation rate gradually decreases due to the competition between the contaminant with the hydroxyl ions to be absorbed on the active site of the photocatalyst surface. On the other hand, the degradation rate for the Fe_3O_4 photocatalyst was stagnant for all the concentrations.

3.7.3. Methyl orange. Both ZnO and TiO_2 photocatalysts show very obvious changes of the degradation rates, started with highest degradation rate for the lowest concentration (12.0 ppm) with 0.0868 min^{-1}, R^2 = 0.9427 for ZnO and 0.0103 min^{-1}, R^2 = 0.9925 for TiO_2. Subsequently, with the increased of the MO concentration, it leads to the distortion of degradation efficiency. The first-order kinetics for the Fe_3O_4 photocatalyst shows an inconsistent degradation rate with even a negative value for one of the concentration. This was due to the very small changes on the measurements across the irradiation duration; indicating no degradation took place, as MO is an azo and very stable dye (Hakamada, Hirashima et al. 2012).

4. Conclusion

From this study, all the oxide nanoparticles used were of standard crystal structures and well dispersed. There were no impurities observed during the characterization of the nanoparticles. By using MB as a model dye, optimum photocatalyst loadings for the three photocatalysts were obtained. It was found that the optimum loading for ZnO was 2.5wt%, whereas for TiO_2 and Fe_3O_4 was 3.0wt%. MB was completely decomposed between 10 to 20 minutes of irradiation duration in the presence of ZnO photocatalyst, whereas for TiO_2 photocatalyst, MB was decomposed in between 20 to 30 minutes of irradiation. Nonetheless, Fe_3O_4 photocatalyst was just able to decomposed 72.8% of the MB in 60 minutes of irradiation. For the decomposition of PR, ZnO and TiO_2 photocatalysts were able to degrade PR solutions within 30 and 50 minutes of irradiation, respectively. However, Fe_3O_4 showed only very mild decomposition even after 60 minutes of irradiation. In the photodecomposition of MO, it was completely decomposed by ZnO in 50 minutes of irradiation, whereas TiO_2 decomposed 45.9% of MO in 60 minutes of irradiation. On the other hand, no trace of decomposition could be detected for the 60 minutes irradiation with the presence of Fe_3O_4 nanoparticles. It can be concluded that ZnO is the superior photocatalyst followed by TiO_2, whereas Fe_3O_4 showed very mild photocatalytic capability. In term of photostability of the dyes, it showed that MO is the most stable dye followed by PR and MB. This result was consistent for all the dyes whether they are irradiated under UV light irradiation with or without the presence of photocatalysts.

Acknowledgments

We are greatly acknowledged the financial support obtained from MOSTI, Government of Malaysia, with the grant No. 03-02-12-SF0019.

5. References

[1] Adachi, M, Jiu, J, Isoda, S, Mori, Y, and Uchida, F 2008 *Nanotechnol. Sci. Appl.* **1** 1-7
[2] Amini, M, and Ashrafi, M 2016 *Nano. Chem. Res.* **1**(1) 79-86
[3] Chiu, W S et al. 2010 *Chemical Engineering Journal* **158**(2) 345-352
[4] Gupta, A, Saurava, J R, and Bhattacharya, S 2015 *RSC Adv.* **5** 71472-71481
[5] Hafez, H, Lan, Z, Li, Q, and Wu, J 2010 *Nanotechnol Sci Appl.* **3** 45–51
[6] Hakamada, M, Hirashima, F, and Mabuchi, M 2012 *Catalysis Science & Technology* **2**(9) 1814
[7] Hou, X, Tian, Y, Zhang, X, Dou, S, Pan, L, Wang, W, Li Y, and Zhao, J 2015 *Journal of Alloys and Compounds, 638*, 214-220
[8] Kelsall, R, Hamley, I, and Geoghegan, M 2006 *Nanoscale Science and Technology* West Sussex, England: John Wiley and Sons Ltd
[9] Kim, S -E, Woo, J -Y, Kang, S -Y, Min, B K, Lee, J K, and Lee, S -W 2016 *Journal of Industrial and Engineering Chemistry* **43** 142–149
[10] Kuan, W -H, and Chan, Y -C 2012 *Journal of Hazardous Materials* **239-240** 152-159
[11] Liu, S, Chu, J, Guo, X, Ge, J, and Wu, H 2015 *Colloids and Surfaces A: Physicochemical and Engineering Aspects* **484** 434-440
[12] Nogueira, A E, Castro, I A, Giroto, A S, and Magriotis, Z M 2014 *Journal of Catalysts* **2014** 712067
[13] Ray, C and Pal, T 2017 *J. Mater. Chem. A* **5** 9465
[14] Regulska, E, Brus, D M, Rodziewicz, P, Sawicka, S, and Karpinska, J 2016 *Catalysis Today* **266** 72-81 doi:10.1016/j.cattod.2015.08.010
[15] Salamat, S, Younesi, H and Bahramifar, N 2017 *RSC Adv.* **7** 19391
[16] Saravanan, R, Mansoob Khan, M, Gupta, V K, Mosquera, E, Gracia, F, Narayanan, V, and Stephen, A 2015 *Journal of Colloid and Interface Science* **452** 126–133
[17] Singh, S, Barick, K C, and Bahadur, D 2013 *Journal of Materials Chemistry A* **1**(10) 3325
[18] Sugiyana, D, Handajani, M, Kardena, E, and Notodarmojo, S 2014 *Journal of JSCE* **2** 69-76
[19] Velmurugan, R, and Swaminathan, M 2011 *Solar Energy Materials and Solar Cells* **95**(3) 942-950 doi:10.1016/j.solmat.2010.11.029
[20] Wu, C, Shen, L, Zhang, Y -C, and Huang, Q 2011 *Materials Letters* **65**(12) 1794-1796
[21] Xie, J, Wang, H, Duan, M, and Zhang, L 2011 *Applied Surface Science* **257**(15) 6358-6363
[22] Xu, T, Zhang, L, Cheng, H, and Zhu, Y 2011 *Applied Catalysis B: Environmental* **101**(3-4) 382-387 doi:10.1016/j.apcatb.2010.10.007

Effect of friction history of workpiece on friction behaviour in metal forming

Z G Wang[1] and Y Kozakai[2]

1 Professor, Department of Mechanical Engineering, Gifu University, Gifu, Japan
2 Master Course Student, Gifu University, Gifu, Japan
E-mail: zgwang@gifu-u.ac.jp

Abstract. This paper describes the influence of friction histories of the specimen surface on frictional behaviour measured by a side-stretching type testing method under lubricated conditions with various loading and unloading patterns of the average contact pressure. The lubricant with lower viscosity has no lubricating effect and the influence of friction histories on frictional behaviour is hardly found. The lubricating effect of the lubricant with higher viscosity is recognized under higher pressures and the specimens with different friction histories give a little higher frictional stress. It is concluded that the influence of the friction history of the specimen surface on the friction law in metal forming is negligibly small.

1. Introduction

Tribological phenomena in metal forming have been studied because they greatly affect the forming load [1], product surface qualities [2-4], tool life [5], environmental impact [6] etc. Recently, friction modelling is attracting the most attention again since the numerical simulation is used widely in metal forming industries. In the numerical simulation, two friction laws are generally used. One is Coulomb's law in which the frictional shear stress is proportional to the normal stress and the friction coefficient is used for the simulation of processes under low pressure. Another one is frictional shear stress law in which the frictional shear stress is dependent on the flow stress of the bulk material and the friction shear factor is used for the simulation of processes under high pressure. The basic problem to be addressed is that the limit of the application of Coulomb friction is not clear [7]. This problem becomes particularly severe in the simulation of the combined processes, such as plate forging [8-11], in which the contact pressure changes over a wide range.

In a previous paper, the friction law for materials without work hardening under dry condition was established based on experimental results [12]. In the friction law, Coulomb friction holds up to a critical pressure and then the frictional stress becomes constant for higher pressures. The critical pressure is the flattening pressure of the surface asperities without the deformation of the bulk material and is given by the slip line field method. The friction law can be integrated in the numerical simulation codes easily. The friction law is also applicable to materials with work hardening and pre-strain [13].

In the oil-lubricated forming processes, the lubrication mode is determined by the amount of lubricant carried into the frictional interface and the distribution of lubricant at the interface [14]. Mizuno [14] showed the amount of lubricant carried into the interface is proportional to the entraining velocity of the lubricant at the inlet of the interface. This was modified by taking the thermal effects due to viscous shear at the inlet and the effect of surface roughness into account [15,16]. The friction coefficient under hydrodynamic lubrication is proportional to a product of the viscosity of lubricant

under high pressure and the relative sliding velocity and is in inverse proportion to the lubricant film thickness [17]. The relative sliding velocity especially gives a large impact on the existing state of lubricant at the interface [18]. A rolling-type testing method was developed to investigate the effects on friction behaviour of the entraining velocity of lubricant and the relative sliding velocity independently [19]. Experimental results using mild steel and stainless steel sheets with various surface textures showed that effects on friction behaviour are quite different according to the surface texture of the tested sheet [20, 21]. These researches suggested that the friction coefficient under the hydrodynamic lubrication mode and micro plasto-hydrodynamic lubrication mode [22] is determined by the performance of the lubricant film associated with the change of viscosity of the lubricant.
However, the pressure dependence of the friction coefficient under the hydrostatic lubrication mode is still not to be understood well [23]. Azushima proposed a lubricating model based on their observation results on the forming interface and suggested that surface roughening at the real contact area occurs and the lubricant in closed pools flows into the real contact area when the average contact pressure is close to the flow stress of the specimen material [24].
In the previous paper, the pressure dependence of the friction coefficient was considered based on the pressure of lubricant trapped in closed pools and a new friction law under the oil-lubricated condition was proposed based on the friction law under dry condition [25].
In this paper, the effect of friction history on friction behaviour is discussed. This issue is important since the products manufactured by metal forming processes are generally made by multi stages and the nominal contact pressure and surface texture of the workpiece changes significantly during the forming processes.

2. Experimental methods and conditions

Figure 1 shows the principle of the side stretching type tribometer used in the present paper [12]. A specimen is fastened at one end, and pulled at the other end by a constant load. The compression tool is moved upwards to press the specimen with a predetermined average contact pressure and then the friction tool is slid on the specimen surface. The sliding speed is 0.1 mm/s to avoid the temperature rise during testing. The sliding distance is 5 mm. During testing, the compression load and the frictional force are measured to calculate the friction coefficient.

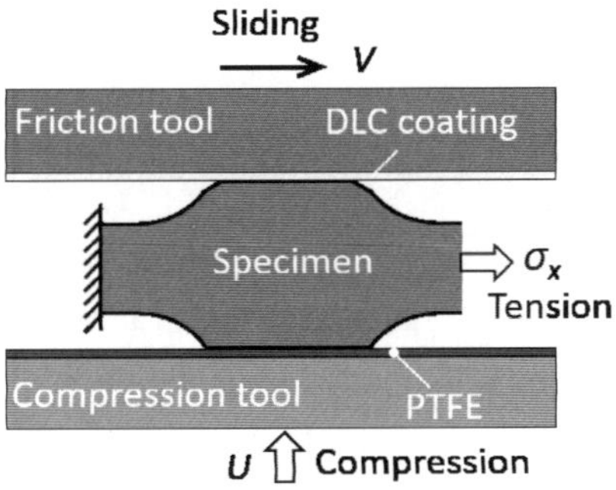

(a) Principle of tribometer

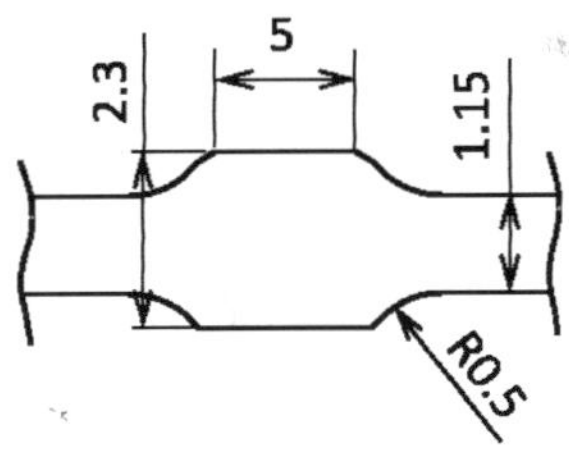

(b) Specimen dimensions (mm)

Figure 1. Schematic illustration of testing method [12].

The specimen is a cold rolled steel sheet SPCC (by JIS). The flow curve is $Y = 548(\varepsilon + 0.005)^{0.21}$ MPa measured by a tension test. The initial flow stress Y_0 is 180 MPa and the flow stress used in the dimensionless average pressure $Y_{\varepsilon=0.08}$ is 326 MPa [13]. The maximum surface roughness is 10.2 μmR_z. Three paraffinic mineral oils with different viscosity are used: P460 (viscosity at 313K: 460×10^{-6} m^2/s), P100 (viscosity at 313K: 100×10^{-6} m^2/s) and P10 (viscosity at 313K: 10×10^{-6} m^2/s).

3. Experimental results

Figure 2 shows an example of the friction history during testing. As shown in figure 2(a), the average pressure $p_a/ Y_{\varepsilon=0.08}$ is set to 1.0 first and the friction test is carried out. Then the average pressure is suddenly decreased to $p_a/Y_{\varepsilon=0.08} = 0.8$ and the friction test is continued. By this method, the frictional behaviour during decreasing average pressure is measured and shown in figure 2(b). This friction history appears well when the workpiece surface slides along the tool surface. As shown in figure 2(b), the average frictional stress changes during sliding even when the average pressure is kept constant and thus the maximum τ_{max} and minimum τ_{min} of the average frictional stress during sliding of 5 mm are used to express the frictional behaviour.

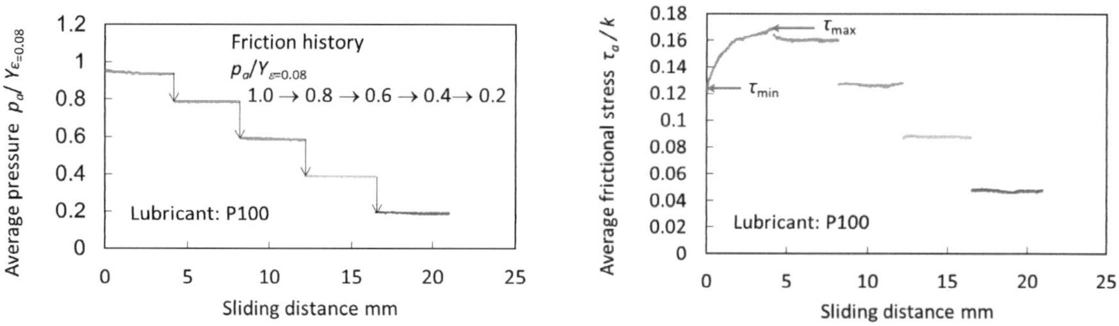

(a) Applied average pressure (b) Average frictional stress

Figure 2. Changes of average frictional stress during testing.

Figures 3-5 show the effects of friction history on frictional behaviour for different lubricants. The friction test with a constant compression load is the standard testing method generally used. The friction test with a friction history of ($p_a/Y_{\varepsilon=0.08}$ = 1.0 → 0 → 0.8 → 0) means that the average pressure $p_a/Y_{\varepsilon=0.08}$ is set to 1.0 first and the friction test is carried out. Then the average pressure is decreased to zero and the specimen surface is separated from the friction tool surface. After unloading, the specimen is compressed again with the average pressure $p_a/Y_{\varepsilon=0.08} = 0.8$ and the friction test is continued. This friction history simulates the sliding interrupting between forming stages.

As shown in figure 5, in the case of the lubricant P10, all of experimental results are on the line of dry condition. This means that the lubricant with low viscosity has no lubricating ability and no effects of friction history on frictional behaviour can be found.

As shown in figures 3-4, in the case of the lubricant P460 and P100, the average frictional stress measured in the constant pressure test leaves the straight line of dry condition when the average pressure $p_a/Y_{\varepsilon=0.08}$ is higher than 0.4 for

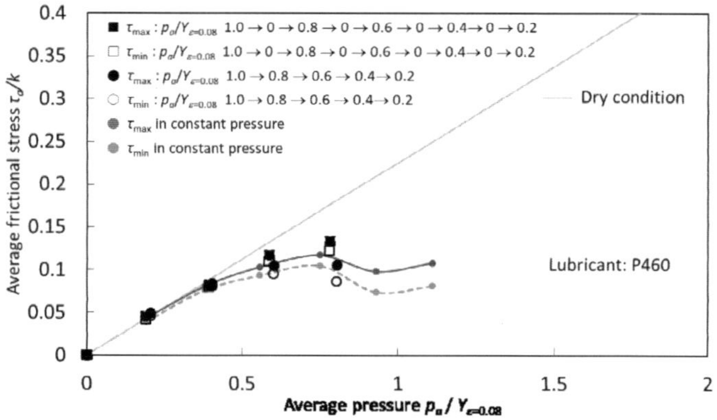

Figure 3. Effects of friction history on frictional behaviour for lubricant P460.

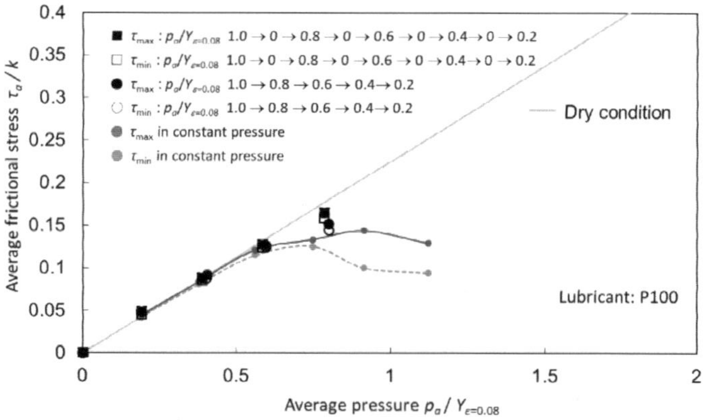

Figure 4. Effects of friction history on frictional behaviour for lubricant P100.

the lubricant P460, 0.6 for the lubricant P100, respectively. The lubricating effect of the lubricant with high viscosity under higher pressures is due to the contribution of lubricant in the closed pools in supporting the nominal contact pressure [25]. Although the specimens with different friction histories show a little higher frictional stress, the effect of friction history on frictional behaviour can be ignored. This means that the friction law proposed for the virgin specimen surface [25] can be applied to the specimen surface with friction histories.

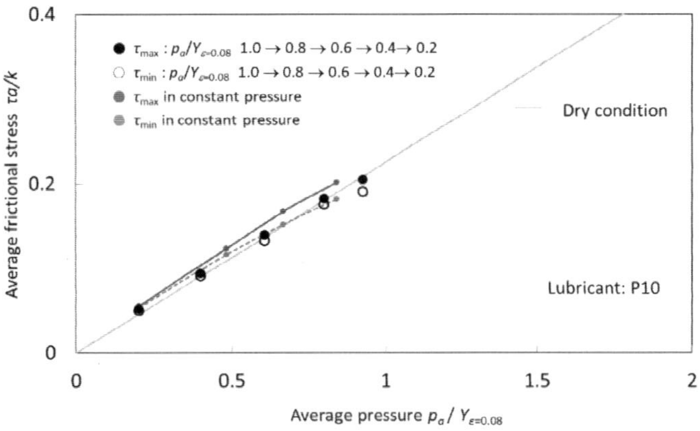

Figure 5. Effects of friction history on frictional behaviour for lubricant P10.

4. Conclusion

Friction tests under oil-lubricated conditions with various loading and unloading patterns of normal load are carried out by a side stretching type tribometer. It is concluded that the influence of the friction history of the workpiece surface on the friction law is negligibly small.

5. References

[1] Kawai N, Dohda K, Saito M, Hayashi N and Wang Z G 1992 *Journal of Engineering for Industry* **114**(2) *175*
[2] Kawai N, Dohda K, Wang Z G and Akiyama K 1990 *Journal of Materials Processing Technology* **22**(2) 123
[3] Wang Z G, Kondo K and Mori T 1995 *Journal of Engineering for Industry* **117**(2) 259
[4] Wang Z G, Dohda K and Jeong Y 2001 *Journal of Materials Processing Technology* **113**(1-3) 705
[5] Soejima T, Yang M, Wang Z G and Yoshikawa Y 2013 *Key Engineering Materials* **535-536** 334
[6] Wang Z G 2004 *Journal of Materials Processing Technology* **151**(1-3) 223
[7] Bay N and Wanheim T 1976 *Wear* **38** 201
[8] Merklein M, Allwood J M, Behrens B A, Brosius A, Hagenah H, Kuzman K, Mori K, Tekkaya A E and Weckenmann A 2012 *CIRP Annals* **61** 725
[9] Wang Z G, Yoshikawa Y and Osakada K 2013 *CIRP Annals* **62** 291
[10] Wang Z G and Yoshikawa Y 2014 *Procedia Engineering* **81** 389
[11] Wang Z G, Hirasawa K, Yoshikawa Y and Osakada K 2016 *CIRP Annals* **65** 293
[12] Wang Z G, Yoshikawa Y, Suzuki T and Osakada K 2014 *CIRP Annals* **63** 285
[13] Wang Z G and Suzuki T 2018 *Procedia Manufacturing* **15** 475
[14] Mizuno T 1966 *J. JSTP* **7** 383 (in Japanese)
[15] Wilson W R D and Mahdavian S M 1974 *Journal of Lubrication Technology* **95** 572
[16] Sutcliffe M P F and Johnson K L 1990 *Proc. Inst. Mech. Eng.* **204** 249.
[17] Kudo H, Tsubouchi M, Takada H and Okamura K 1982 *CIRP Annals* **31** 175
[18] Wang Z G, Kondo K and Mori T 1995 *Journal of Engineering for Industry* **117**(3) 351
[19] Dohda K and Wang Z G 1995 *Journal of Tribology* **117**(3) 529
[20] Dohda K and Wang Z G 1998 *Journal of Tribology* **120**(4) 724
[21] Wang Z G, Dohda K and Haruyama Y 2006 *Wear* **260**(3) 249
[22] Mizuno T and Hasegawa K 1982 *Journal of Lubrication Technology* **104** 23
[23] Azushima A 1995 *CIRP Annals* **44** 209
[24] Azushima A, Miyamoto J and Kudo H 1998 *CIRP Annals* **47** 479
[25] Wang Z G, Dong W Z and Osakada K 2018 *CIRP Annals* **67** 257

IOP Conf. Series: Materials Science and Engineering **544** (2019) 012025 doi:10.1088/1757-899X/544/1/012025

Essential oil from fresh and dried Rosemary cultivated in Lam Dong province, Vietnam

Trieu Tuan Anh[1*], Lam Thi Thu Ngan[1], Tri Duc Lam[2]

[1]Faculty of Chemical Engineering and Food Technology, Nguyen Tat Thanh University, Ho Chi Minh City, Vietnam
[2]NTT Hi-Tech Institute, Nguyen Tat Thanh University, Ho Chi Minh City, Vietnam
*Email: anhtt@ntt.edu.vn; labams2013@gmail.com

Abstract. Rosemary essential oil has been widely used in folk medicine for treatment of anxiety, memory deficit and cancer thanks to its high antioxidant activity and antibacterial properties. Composition of rosemary essential oil largely depends on the geographical position of the cultivated plant and conditions of the extraction process. In this study, fresh and dried rosemary leaves were used for extraction of essential oil using hydrodistillation method. The extraction performance, composition and antioxidant activity of the essential oil were measured. The oil yield for dried leaves (1.2 ml/g) is significantly lower than that for fresh leaves (3.16 ml/g). However, the difference on chemical profile and antioxidation activity of the two oil samples was indistinguishable. The most remarkable finding was the presence of Levoverbenone at very high concentration in rosemary essential oil, accounting for 10.87% and 12.12% of the oil sample extracted from fresh and dried leaves respectively.

1.Introduction

Nowadays, traditional medicines make use of natural products (such as plants, animals, microorganisms, and marine organisms), which are of great importance [1-4]. Rosemary(R. officinalis) is the plant that is cultivated mainly in Mediterranean region. Rosemary essential oil was also found to have the capability to relieve the symptoms caused by respiratory disorders, to stimulate hair growth, to reduce stress and mental alertness, and to treat Rheumatoid disease[5-7]. The benefits and application of rosemary essential oil are diverse [8-10]

The essential oil could be extracted from both flower and leaf organs of the plant, oils extracted from leaves often display higher quality. The chemical composition of the essential oil largely depends on the extraction conditions. Methods for extraction of rosemary essential oil range from mechanical pressing to hydrodistillation distillation and microwave assisted distillation[11,12].Among such methods, microwave-assisted distillation,albeit not economically feasible for large-scale production, was found to give higher extraction yield [13,14]. In this research, the essential oil from rosemary was obtained by conventional hydrodistillation to find out the optimal parameters with regards to storage duration, and drying temperature. In addition oil samples obtained from fresh and dried Rosemary leaves were compared in terms of extraction performance and antioxidant activity. The chemical profiles of essential oil from Rosemary cultivated in Lam Dong province were also bestudied to compare with other research.

2. Material and methods

2.1. Sample preparation

Rosemary plants were harvestedfrom The Seed Garden located Lam Dong Province, Vietnam. It was cleaned and split into 3 parts. In order to investigate the impacts of different storage conditions on the compositions of distilled oils, the material samples were subjected to different storage temperatures such as fresh, refrigerator (4°C), different drying temperatures (40°C to 65°Cand sample to be dried until constant weight) and at room temperature (30°C). The oils analysis of all storage treatments performed daily.

2.2. Extraction of rosemary essential oil by Hydrodistilltion method

Before extracting, the moisture of each sample moisture was measured by Electronic moisture driers Ohaus MB25.100g sample wastransferred into the 1L fit round bottom flask containing 500mL of water. Depending on the requirements of the experiments, parameters may vary. To be specific, water/material ratio varies from 1:2 to 1:5, distillation time varies from 1 hour to 4 hours and material sample alternates from fresh todry.The collected essential oil was anhydrated with Na_2SO_4 and stored in small glass bottles sealed with teflonsealed caps and covered with foil to prevent oxidation.Bottles are stored in dry and cool condition with the absence of direct sunlight and extraction yield (Y):

$$Y(ml/100g)=\frac{\text{Vessential oil}}{m(1-\alpha)}x\ 100$$

Where V (mL) is the volume of oil, m (g) is the mass of water and α (%) is the moisture content of the material.

2.3. DPPH (1,1-diphenyl-2-picrylhydrazyl) radical scavenging activity

Two samplesof essential oils extracted from fresh and dry leaves were analyzed for radical scavenging activity at Academy of Science and Technology,Ho Chi Minh City, Vietnam.50 µL of DPPH (OD 520 nm = 0.0403 ± 0.013) was introduced into 150µL of sample solution. The mixturewas then mixed at room temperature in the dark for 30 min till stable state. The optical measurement of the mixture was performed byUV/VIS - PowerWave HT Microplate Spectrophotometer at 517 ηm. Blank sample, but 150 µL solution replaced EtOH 99.7%. Standard sample: Vitamin C 400ppm.

$$\%\ DPPH = \frac{A_b - A_s}{A_b}\ x100$$

where:Ab isthe optical density in the blank sample; As is the optical density of sample solution %DPPH ispercentage of free radical DPPH.

2.4. Gas chromatography – mass spectroscopy (GC-MS)

The gas chromatography-mass spectroscopy (GC-MS) was employed to analyze the chemical composition of the essential oil samples. GC-456 SQ with SCION performance RESTEK Rxi-5ms (30 m x 0.25 mm (for example), 0.25 µm df), bring the gas Helium constant flow rate: 1 mL/min. Injector temperature is 250° C the rate of Division: 30

3.Result and discussion

3.1. Effects of storage duration, distillation time, material/water ratio and drying temperature on extraction yield

The distillation yield of rosemary essential oil was examined with respect to three variables including storage duration, distillation time, material to water ratio, and drying temperature. From Figure 1,it is observed that the preservation time of raw materials shows an inverse relationship with the essential oil production. This is possibly due to the low temperature which increases moisture, and in turn, incites decomposition of the plant. In addition, prolonged storage duration also creates favorable environments for microorganisms to flourish, causing oil degradation in terms of quality and composition. Regarding influence of extraction, figure 2 showed that the oil yield was proportional with increased extraction time, peaking at 3 4.33% at 3 hours. This indicates that almost essential oil in the materials has been fully extracted after 3 hours. Therefore, to save energy and time, the appropriate time period is 3 hours.

Figure 3 expresses the influence of material/water ratio on production Rosemary essential oil. Accordingly, the highest yield was achieved at 1:5 ratio. Under effect of heating, the penetration of water through cell membrane was enhanced, rupturing oil sacs and releasing the essential oils out in the form of evaporation. On the contrary, insufficient water impairs the viscosity of the membrane, reducing the permeability of water vapor and quantity of essential oils. The drying temperature is one of the important factors in the process of distillation of rosemary oil as it affects the color, smell and yield of essential oil. The experimental results obtained from Figure 4 shows that, at 40-45°C,the materials gave the highest yield of essential oil. Minimum oil yield is reached at temperature of 65°C Therefore, the suitable temperature range is with 40-55°C.

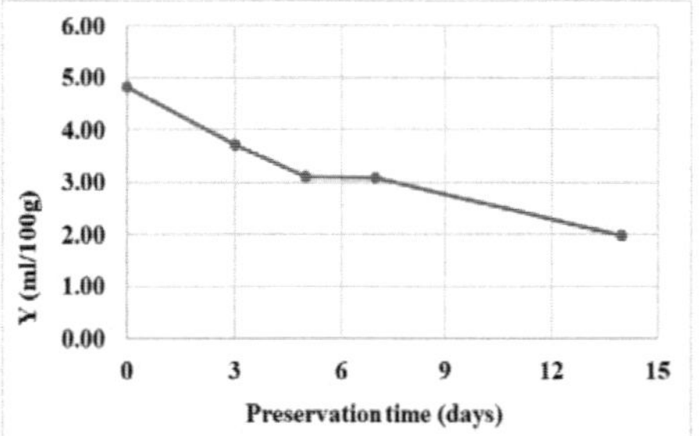

Figure 1. Effect of storage durationextraction yield from fresh leaves

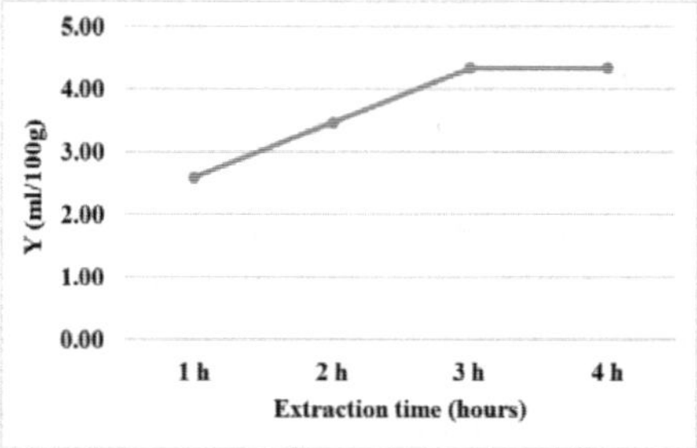

Figure 2. Effect of distillation time on extraction yield

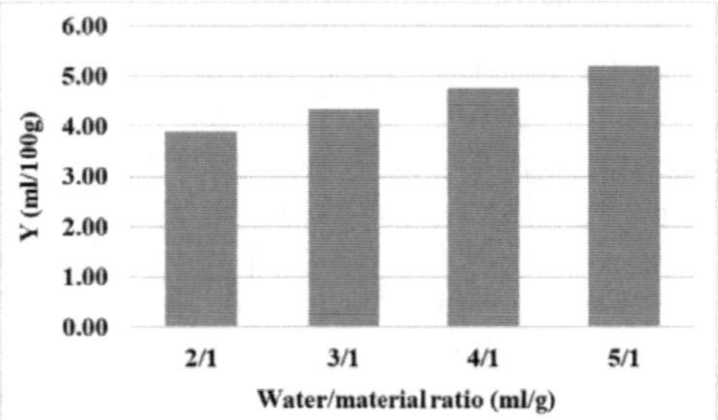

Figure 3. Effect of material/water ratio on extraction yield

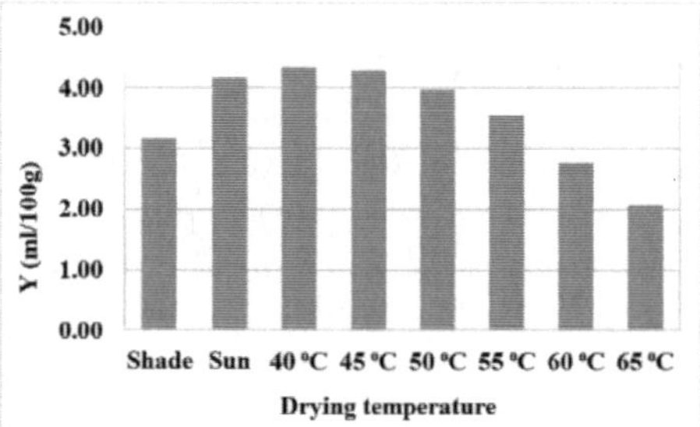

Figure 4. Effect of drying temperatue on extraction yield

3.2. The extraction yield, antioxidant activity and GC-MS of fresh and dried leaves

Two samples, fresh leaves and dried leaves (at 45°C), were used in this analysis. After pre-treatment, the sample was distilled in 500 ml of water in 4 hours.Figure 5 shows the oil yield of fresh material (4.818%), dried material (4.334%). In addition, the two samples also exhibited similar antioxidant activities where antioxidant activity of the fresh sample is recorded at %DPPH 49.6% and that for dried sample was % DPPH 52.73%.The essential oil of fresh rosemary leaves obtained using hydro-distillation was analyzed by gas chromatography-mass spectroscopy (GC-MS). 20 components were identified in Table 1. The major components were 1R-α-Pinene (26.252%), followed by Eucalyptol (Cineole) (14.490%), Levoverbenone(12.121%), Geraniol(6.361%), Bornyl acetate (4.938%), and Camphol(4.235%). For dry leaves, table 2 shows 17 components, in which the major components were 1R-α-Pinene (23.001%), Eucalyptol (Cineole) (11.858%), Levoverbenone(10.866%), Geraniol (4.024%), Camphor (3.177%), and Bornyl acetate (3.046%). From the results, we found that the essential oil extracted from rosemary leaves from Lam Dong province is characterized by the abundance of L-Verbenone, which has been-proved to be a health beneficial and active substanceused in the treatment of respiratory diseases. In comparison to results of other studies [15, 16] where L-Verbenone content was either nonexistent or very low, it is suggested that the oil is suitable to support the treatment of allergic rhinitis [17] and that the substance found in the fresh material is higher than that found in the dried raw material. Regarding other components, it is implied that trans-Verbenol (0.795%), Levomenthol (2.622%), and L-Borneol (1.013%) were lost in the drying process.

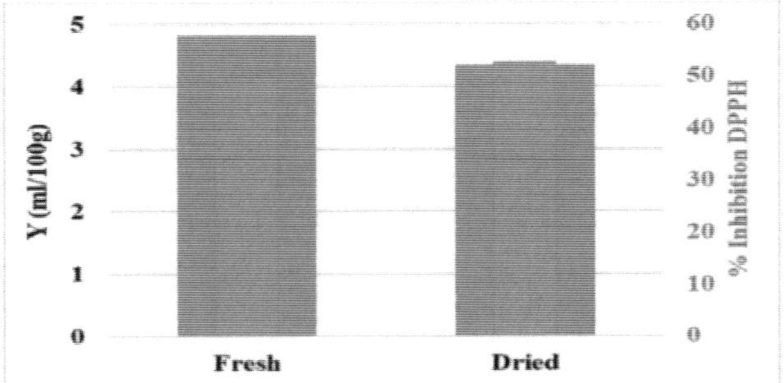

Figure 5. Changing in extraction yield and antioxidant actitity of fresh and dried leaves

Table 1. Chemical compound of Rosmarinus offcinalis essential oils obtained using fresh ingredients.

S/N	Rt	Compounds	Concentration (%)	Mass	Homology mass spectrometry
1	5.090	1R-α-Pinene	26.252	136	947
2	5.349	Camphene	3.189	136	951
3	5.832	β-Pinene	2.159	136	933
4	5.976	β-Myrcene	1.140	136	938
5	6.781	D-Limonene	3.016	136	934
6	6.851	Eucalyptol (Cineole)	14.490	154	954
7	7.392	γ-Terpinene	1.400	136	944
8	8.054	Terpinolene	1.129	136	941
9	8.228	Linalool	3.045	154	929
10	9.377	trans-Verbenol	0.795	152	937
11	9.415	(-)-Camphor	2.627	152	937
12	9.913	Camphol	4.235	154	931
13	10.030	Levomenthol	2.622	156	957
14	10.175	Terpinen-4-ol	1.641	154	914
15	10.482	Terpineol	2.665	154	928
16	10.759	L-Borneol	1.013	154	797
17	10.982	Levoverbenone	12.121	150	959
18	11.973	Geraniol	6.361	154	941
19	12.874	Bornyl acetate	4.938	196	934
20	16.327	Caryophyllene	2.616	204	942

Table 2 Chemical compound of Rosmarinus offcinalis essential oils obtained using material is dried

S/N	Rt	Compounds	Concentration (%)	Mass	Homology mass spectrometry
1	5.087	1R-α-Pinene	23.001	136	946
2	5.346	Camphene	2.521	136	948
3	5.829	β-Pinene	1.440	136	938
4	5.973	β-Myrcene	0.803	136	937
5	6.778	D-Limonene	2.178	136	937
6	6.847	Eucalyptol (Cineole)	11.858	154	953
7	7.389	γ-Terpinene	1.056	136	942
8	8.051	Terpinolene	0.736	136	945
9	8.224	Linalool	1.988	154	935
10	9.412	(-)-Camphor	2.517	152	938
11	9.909	Camphol	3.177	154	937
12	10.171	Terpinen-4-ol	0.814	154	914
13	10.478	Terpineol	1.834	154	935
14	10.978	Levoverbenone	10.866	150	961
15	11.966	Geraniol	4.024	154	936
16	12.870	Bornyl acetate	3.046	196	938
17	16.324	Caryophyllene	1.000	204	939

4. Conclusion

We attempted the extraction and compared composition and antioxidant activity of the essential oils extracted from dried and fresh rosemary leaves, harvested in Lam Dong province, Vietnam.It is found that some components in the essential oil of Rosemary were lost due to the drying process affect the material. In addition, the oil sample from dried leaves exhibited higher antioxidant activity compared to the sample from fresh leaves. Storage time and drying temperature were found to be inversely proportional to the yield of the essential oil. We also found high concentration of Levoverbenone in the oil, accounting for 10.87% and 12.12% of the oils extracted from fresh and dry leaves respectively. This remarkable result suggested that Rosemary planted at Lam Dong Provine may act a good source for alternative medicine for respiratory diseases and also support treatment methodsfor allergic rhinitis.
Acknowledgments: This work was supported by grants from Nguyen Tat Thanh University, Ho Chi Minh City, Viet Nam.

References

[1] Vo T-S, Ngo D-H, Bach L G, Ngo D-N and Kim S-K 2017 *Process Biochemistry* **54** 188–94
[2] Mai H C, Le T T T, Diep T T, Le T H N, Nguyen D T and Bach L G 2018 *Asian Journal of Chemistry* **30** 293–7
[3] Hien T T, Nhan N P T, Trinh N D, Ho V T T and Bach L G 2018 *Solid State Phenomena* **279** 217–21
[4] Nhan N P T, Hien T T, Nhan L T H, Anh P N Q, Huy L T, Nguyen T C T, Nguyen D T and Bach L G 2018 *Solid State Phenomena* **279** 235–9
[5] Oluwatuyi M, Kaatz G and Gibbons S 2004 *Phytochemistry* **65** 3249–54
[6] Zhang A, Sun H and Wang X 2013 *European Journal of Medicinal Chemistry* **63** 570–7
[7] Rahila M P, Surendra Nath B, Laxmana Naik N, Pushpadass H A, Manjunatha M and Franklin M E E 2018 *Journal of Food Processing and Preservation* **42** e13443
[8] Farshchi H K, Azizi M, Jaafari M R, Nemati S H and Fotovat A 2018 *Biocatalysis and Agricultural Biotechnology* **16** 54–62
[9] Bousbia N, Abert Vian M, Ferhat M A, Petitcolas E, Meklati B Y and Chemat F 2009 *Food Chemistry* **114** 355–62
[10] Ngo S N T, Williams D B and Head R J 2011 *Critical Reviews in Food Science and Nutrition* **51** 946–54
[11] Uysal B, Sozmen F, Aktas O, Oksal B S and Kose E O 2011 *International Journal of Food Science & Technology* **46** 1455–61
[12] Bousbia N, Abert Vian M, Ferhat M A, Petitcolas E, Meklati B Y and Chemat F 2009 *Food Chemistry* **114** 355–62
[13] Tran T, Nguyen H, Nguyen D, Nguyen T, Tan H, Nhan L, Nguyen D, Tran L, Do S and Nguyen T 2018 *Processes* **6** 206
[14] Hien T T, Nhan N P T, Trinh N D, Ho V T T and Bach L G 2018 *Solid State Phenomena* **279** 217–21
[15] Hcini K, Sotomayor J A, Jordan M J and Bouzid S 2013 *Asian Journal of Chemistry* **25** 2601–3
[16] Lo Presti M, Ragusa S, Trozzi A, Dugo P, Visinoni F, Fazio A, Dugo G and Mondello L 2005 *Journal of Separation Science* **28** 273–80
[17] Jiang Y, Wu N, Fu Y-J, Wang W, Luo M, Zhao C-J, Zu Y-G and Liu X-L 2011 *Environmental Toxicology and Pharmacology* **32** 63–8

IOP Conf. Series: Materials Science and Engineering **544** (2019) 012026 doi:10.1088/1757-899X/544/1/012026

Effect of extracting process on saponin content and antioxidant activity of Gleditschia fera (Lour.) Merr dried fruit extract

Trieu Tuan Anh[1*], Nguyen Thi Thanh Huyen[1], Tri Duc Lam[2]

[1]Faculty of Chemical Engineering and Food Technology, Nguyen Tat Thanh University, Ho Chi Minh City, Vietnam
[2]NTT Hi-Tech Institute, Nguyen Tat Thanh University, Ho Chi Minh City, Vietnam
*Email: anhtt@ntt.edu.vn; labams2013@gmail.com

Abstract. Gleditschia fera (Lour.) Merr (GfL) was a traditional herbal used for treatment of diabetes mellitus and for arresting and/or slowing down hair loss. In this study, powder of GfL dried fruit was immersed in the pure water to extract saponin and other components atdifferent material/ solvent ratios, extraction time (hours) and extraction temperature (°C). Saponin content of the final product (extract) was evaluated by Oleanolic acid-H_2SO_4 assay. The antioxidants activity of the extract was determined by DPPH method. High extraction yield (74g/100 g) was obtained at the material/ solvent ratio 1:15 (w/v), 50°C and 4 hours extraction time. Under these optimized conditions, the highest saponin content of was 37.06 mg/ g. The extract can harvest 45.30 % of DPPH at 10000 μg/ml concentration

1. Introduction

Saponins consist of a large group with similarly structured compounds containing a steroid or triterpenoid aglycone (sapogenin) linked to one or more oligosaccharide moieties by glycosidic linkage. The substancescomprise both polar (sugar) and nonpolar (steroid or triterpene) and are characterized by their hemolytic activity and foaming properties [1,2]. Other studies showed that saponins isolated from Gleditsia members exhibited anti-ulcerogenic, anti-inflammatory and anti-allergic activities [3,4]. Another study showed that triterpene saponins displayed significant anti-HIV activity, demonstrating the antiproliferative activity of Gleditsia sinensis fruit extract (GSE) on various solid tumor and leukaemia cell lines as well as primary cultured bone marrow cells isolated from patients with acute and chronic myelogenous leukaemia [5,6].

Plants have been for long considered an important source of bioactive compounds useful forthe safety of herbal medicines [7-10]. Gleditschia fera (Lour.) Merr (GfL) is a plant specices in the Gleditsia genus, belonging to the family of Caesalpiniaceae. The plant is also known as Mimosa fera Lour and could grow up to 6 to 8m in height. Genus of Gleditsia (GoG) is consisted ofdeciduous trees distributed mainly in tropical regions such as Asia, Africa,America. The fruit is used to produce soaps, shampoo, and medicinal drugs . The beneficial effects of the fruits in traditional medicine, including treatment of kidney stones and diabetes mellitus, have been mainly attributed anti-inflammation.Saponin is the main constituent of Gleditsia fruits[11-14]

Extraction of GlF can be carried out by several methods. The maceration extraction is a solid-liquid procedure in which bioactive compoundscontained in the plant material are extracted by submerging the material in a solvent, usually ethanol due to its environment friendliness, for anextended period [15,16]. However, the extraction period is often long and this could be remedied by the assistance of amechanical shaker. Other more commonly used methods are reflux with ethanol solvent, soxhlet extraction, ultrasound-assisted extraction, accelerated solvent extraction, and microwave-assisted extraction[17-19]. However, few studies have been attempted to systematically investigate parameters for the extract of GfL. Therefore, the purpose of this research is to find the optimal conditions (time, material/solvent ratio, extraction temperature) which gives the highest saponin content and antioxidation activity. The result of this study could act as a precursor for the further research in establishment of saponin production in larger scale.

2. Material and methods

2.1. Sample preparation

Fruits of G. fera (Lour.) Merr was purchased from Thanh Binh Herbal Tea Company, Ho Chi Minh city. Chemicals included H_2SO_4, DPPH, EtOH, Vanillin, MeOH, and Vitamin C were purchased from Sigma-Aldrich.

2.2. Extraction Procedure

Raw material is ground into particles having a size 0.4÷ 0.8 mm, preserved by ziplock at room temperature. Moisture content of raw material and products are measured by Ohaus Moisture Analyzer MB – 45 for each sample.10g ± 0.2 g of material power was added into 100ml EtOH 70 % by heating under reflux about 45min. The resultingsolution was filtered. After evaporation, saponins were qualitatively measured by methods such as Foaming, Fontan – Kaudel , Liebermann – Burchard, and Acid phosphoric $_{cc}$ reaction.The presence of flavonoid was evaluated by methods including Sodium hydroside, Reduction – oxidation, and Acetat lead (Appendix II).

The power of the G. fera (Lour.) Merr (5g ± 0.2) was introduced into beaker with distilled water.The material:solvent ratio varies from 1:05 to 1: 25.The temperature varies from 30°C to 80°C, $\Delta T = 10°C$, the time (2h to 8h, $\Delta t = 2h$). Beaker is placed stably on the water bath (Daihan General Purpose Water Bath Digital Fuzzy Control System). Extracts of GfL are filtered by a vacuum filtration system and are condensed. The extract was weighted for calculation yield.

Total of extraction yield was calculated based on the absolute dry mass weight by the following formula:

$$\% Y = \frac{m_{dry\ extract}}{m_{material}} = \frac{m_{extract} \times (1 - a_{extract})}{m_{material} \times (1 - a_{material})}$$

Where : $m_{extract}$: mass of extract (g); $m_{material}$:mass of material (g); $a_{extract}$: moisture of extract (%); $a_{material}$: moisture of material (%)

2.3. Saponin content's

1.00 g of extract was dissolved in 100.00 mL EtOH 80 % (C_s = 10 000 µg/ mL). A solution of vanillin, acid sulphuric 72 % and sample solution were cooled in ice water. After that, 100 µL of vanillin and 1000 µL of acid sulphuric72% were added into 100 µL of sample solution and shaken in an ice bath. The solution was allowed to stabilize at 60 ^{0}C in water bath in 10 mins before measuring the optical absorption. The absorptiometry is used in 538 ηm by UV/VIS – 1800 Shimadzu Spectrometer.Blank sample: 100 µL solution sample was replaced EtOH 80%.

Standard sample : oleanolic acid in EtOH 80 %.

Saponin content:

$$S\left(\frac{mg}{g}\right) = C_{sample} * \frac{V}{m} * 1000$$

Where : C_{sample} : saponin concentration in samples (µg / ml); m: mass the material (g); V: the volume 80 % ethanol soluble material (ml)

2.4. Antioxidant activity by DPPH method

The antioxidant activity of extracts was evaluated basing on the modified stable 1, 1-diphenyl-2-picrylhydrazyl (DPPH)-free radical activity. First, 0.1 g of sample was dissolved in 10.00 mL of MeOH 80 % (C_s = 10.000 µg/ mL). Following that, 780 µL of DPPH (O_D 517 nm = 0.7 ± 0.02) was introduced into 520 µL of sample solution. The sample solution was mixed stably at room temperature in the dark for 30 min. The optical measurement of the mixture was performed by UV/VIS – 1800 Shimadzu Spectrometer at 517 ηm.

Blank sample: but 520 μL solution in MeOH 80 %.
Standard sample: Vitamin C (0.1g ± 0.01) was dissolved into MeOH 80 % in volume flask 100 mL, in the dark (C = 100 μL / mL). Percentage inhibition (% I) of DPPH activity was calculated using the formula given below:

$$\% I = \frac{A_b - A_s}{A_b} x\, 100$$

Where : A_b: Optical density of the blank and A_s: Optical density of the sample

3. Results and discussion

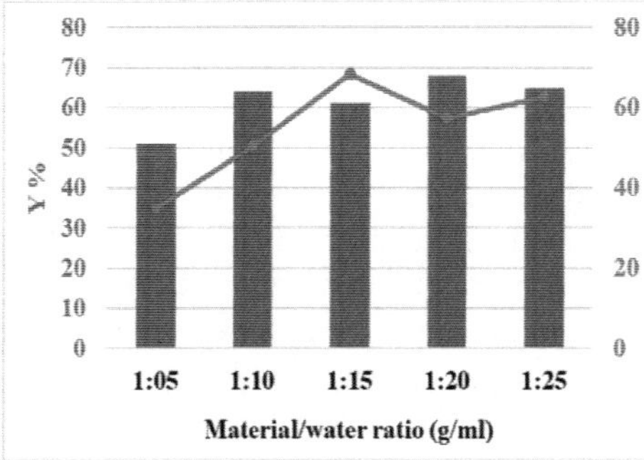

Figure 1. Effect of material / solvent ratio on the extraction yield and content of saponin

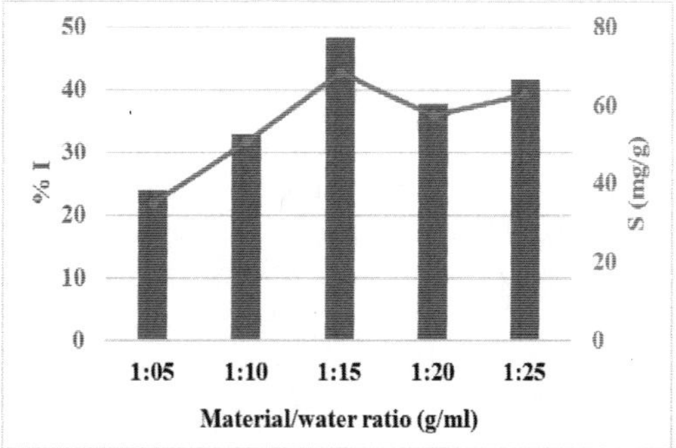

Figure 2. Effect of material/ solvent ratio on content saponin and antioxidant activity

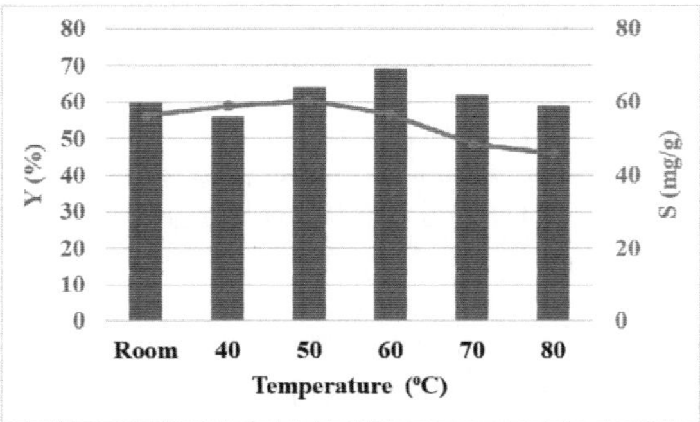

Figure 3. Effect of temperature on extraction yield and saponin content

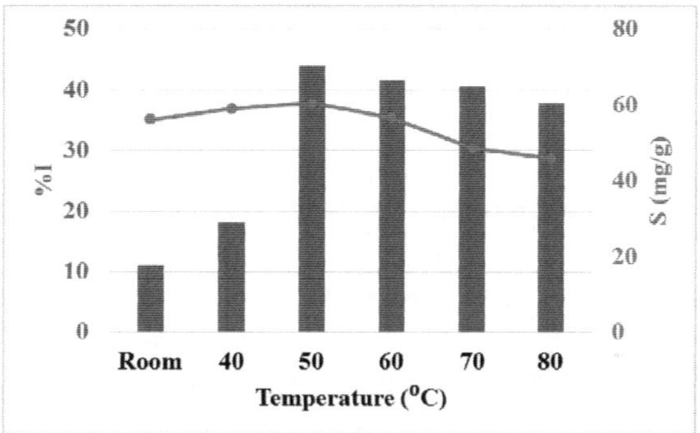

Figure 4. Effect of temperature on the saponin content and percentage inhibition

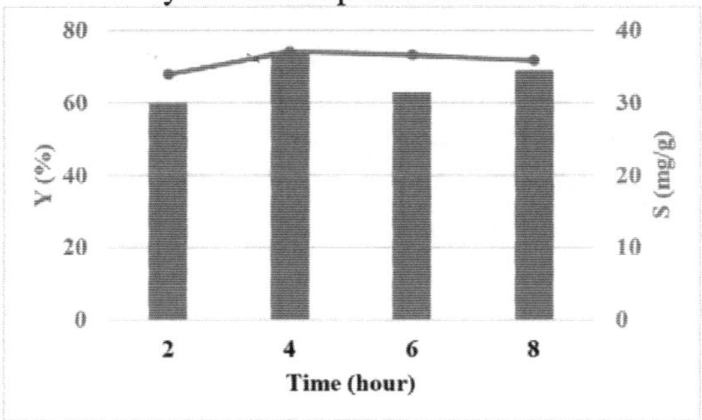

Figure 5. Effect of extract time on the extraction yield and saponin content

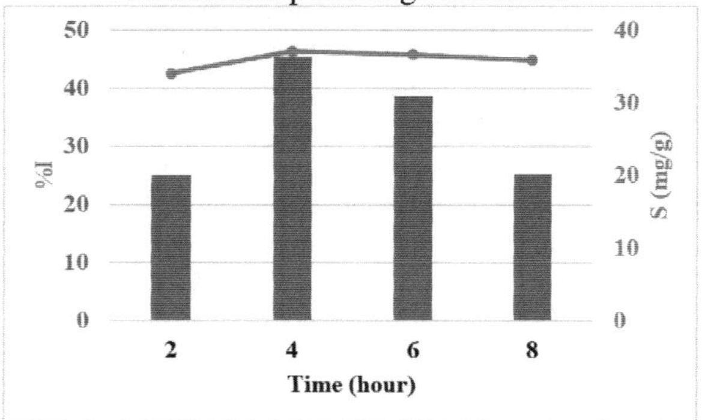

Figure 6. Effect of extract time on the saponin content and percentage inhibition

Figure 1 shows the effect of the material/solvent ratio ranging between 1:05 and 1:25 on saponin yield. The result showed the highest extraction yield was obtained at the material /solvent of 1:20 (73g/100 g). However, the saponin content is highest at ratio of 1:15 (68.39 mg/g). From 1: 20 to 1: 25, saponin content reached the equilibrium phase. Since the saponin content could only be fully extracted from the cell with sufficient solvent, further introduction of solvent promotes extraction yield and contentof saponin.Figure 2 describes the effect of material/solvent ratio on saponin content and percentage inhibition. The highest percent inhibition was obtained at material /solvent of 1:15 (48.40 %) and decreased gradually until the material / solvent ratio reached 1:20 where a mild increase took place. With the results of the current study, the proportion of material/extraction solvent of 1:15 (w / v) was selected to carry out further research.

Figure 3 displays the variations of extraction yield and saponin content as a function of temperature at various temperature. Visually, the extraction yield increases as temperature increases.The maximum yield occurred at about 60°C (at 69g/100g). The content of saponin increased from 56.16 mg/g to 60.41 mg/g when temperature increases from 30°C to 50°C. However, when the temperature grows from 60°C to 80°C, the saponin content go down. That is because the heat was making the extraction easier by destroying the cell membrane by the denature membrane. This result agrees with Cacace et.al (2003b) who mentioned that increasing the temperature reduces viscosity and in turn increases diffusion rate, creating conditions conducive to the transfer of compounds from raw material to solvents [20]. Several studies have confirmed the effect of temperature on the extraction capability of biological compound from plant materials [21, 22].The saponin content and percentage inhibition were showed in Figure 4. When the temperature grew from 30°C to 50°C , the inhibition activity increased from 10.97%to 43.92%. However, as the temperature continues to increases, the percentage inhibition decrease afterwards.The variation of extraction yield and content as a function of temparature is illustrated in Figure 3. The corresponding percent inhibition is presented in Figure 4. From these results, the temperature of 50 °C was chosen as the optimum extraction temperature.

The effect of time ratio on extraction yield and the saponin content is presented in Figure 5. The result indicates that the highest extraction yield and saponin content wasobtained at 4 hours (74g/100g and 37.06 mg/g).The may be explained high diffusion of saponin content from the cell. On the other hand, increasing time of extraction exceeding 4h causes saponin inside and outside the cell to reach the equilibrium phase. This in turn causes oxidization and hydrolysis, reducing the amount of saponins obtained. The biologically active compounds are sensitive to high temperature and longer extraction time shoud lead to the breakdown of compounds having biological activity. Figure 6 describes the variations saponin content and percent inhibition as a function of time at various exposure periods. The inhibitory activity increases as time increases, but peak inhibition occurred at about 4 hours (at 45.30 %). Thus, extraction time of 4 hours was selected as the optimum point due to the practical and economicconsiderations.

4. Conclusion

This research has identified the appropriate optimal parameters to maximize the yield and inhibition activity of the GIF extract. The optimal conditions were found at material/ solvent ratio 1:15 (w/v), operating temperature of 50°C extraction time of 4 hours. We also found high extraction yield (74g/ 100g), saponin content (37.06 mg/ g extract). In addition 10.000 μg/mL extract can inhibit 45.30% of DPPH. These optimum conditions can be useful for further optimization of the extraction of Gleditschia fera (Lour.) Merr. Further studies are needed to examine effects of other conditions such as solvent (EtOH, MeOH, etc.) and methods (microwave, ultrasound, pressurized liquid extraction (PLE), etc.) on extraction of Gleditschia fera (Lour.) Merr.

Acknowledgments: This work was supported by grants from Nguyen Tat Thanh University, Ho Chi Minh City, Viet Nam.

References

[1] Singh D and Chaudhuri P K 2018 *Integrative Medicine Research* **7** 33–43
[2] Guo N, Tong T, Ren N, Tu Y and Li B 2018 *Phytochemistry* **149** 42–55
[3] Sidana J, Singh B and Sharma O P 2016 *Phytochemistry* **130** 22–46
[4] Saleh D O, Kassem I and Melek F R 2015 *Pharmaceutical Biology* 1–5
[5] Chui C, Tang J, Lau F, Teo I, Yau M, Wong R, Cheng G, Ho S, Leung T, Hui K, Wong M, Fatima S, Cheng C, Cheung F, Tan W, Chow L, Guo D and Chan A 2004 *International Journal of Molecular Medicine*

IOP Conf. Series: Materials Science and Engineering **544** (2019) 012026 doi:10.1088/1757-899X/544/1/012026

[6] Singh B and Kaur A 2018 *LWT* **87** 93–101
[7] Phan A N Q, Bach L G, Nguyen T D and Le N T H 2019 *Journal of Nanoscience and Nanotechnology* **19** 974–8
[8] Hien T T, Nhan N P T, Trinh N D, Ho V T T and Bach L G 2018 *Solid State Phenomena* **279** 217–21
[9] Nhan N P T, Hien T T, Nhan L T H, Anh P N Q, Huy L T, Nguyen T C T, Nguyen D T and Bach L G 2018 *Solid State Phenomena* **279** 235–9
[10] Tran T, Nguyen H, Nguyen D, Nguyen T, Tan H, Nhan L, Nguyen D, Tran L, Do S and Nguyen T 2018 *Processes* **6** 206
[11] Tran T, Nguyen H, Nguyen D, Nguyen T, Tan H, Nhan L, Nguyen D, Tran L, Do S and Nguyen T 2018 *Processes* **6** 206
[12] Mohammed R S, Abou Zeid A H, El Hawary S S, Sleem A A and Ashour W E 2014 *Saudi Journal of Biological Sciences* **21** 547–53
[13] Melek F R, Kassem I A A, Miyase T and Fayad W 2014 *Phytochemistry* **100** 110–9
[14] Yamazaki K and Takakura K 2011 *Journal of Asia-Pacific Entomology* **14** 179–81
[15] Jian H-L, Lin X-J, Zhang W-M, Sun D-F and Jiang J-X 2013 *International Journal of Biological Macromolecules* **62** 596–602
[16] Lian X-Y and Zhang Z 2013 *Journal of Pharmaceutical and Biomedical Analysis* **75** 41–6
[17] Hien T T, Nhan N P T, Trinh N D, Ho V T T and Bach L G 2018 *Solid State Phenomena* **279** 217–21
[18] Tran T, Nguyen H, Nguyen D, Nguyen T, Tan H, Nhan L, Nguyen D, Tran L, Do S and Nguyen T 2018 *Processes* **6** 206
[19] Wen C, Zhang J, Zhang H, Dzah C S, Zandile M, Duan Y, Ma H and Luo X 2018 *Ultrasonics Sonochemistry* **48** 538–49
[20] Marić M, Grassino A N, Zhu Z, Barba F J, Brnčić M and Rimac Brnčić S 2018 *Trends in Food Science & Technology* **76** 28–37
[21] Cacace J E and Mazza G 2003 *Journal of Food Science* **68** 240–8
[22] Santoyo S, Rodríguez-Meizoso I, Cifuentes A, Jaime L, García-Blairsy Reina G, Señorans F J and Ibáñez E 2009 *LWT - Food Science and Technology* **42** 1213–8
[23] Zaibunnisa A H, Norashikin S, Mamot S and Osman H 2009 *LWT - Food Science and Technology* **42** 233–8

IOP Conf. Series: Materials Science and Engineering **544** (2019) 012027 doi:10.1088/1757-899X/544/1/012027

Effect of extraction process on extraction yeild, total polyphenol content and antioxidant activity of Jasminum Subtriplinerve

Trieu Tuan Anh[1*], Luu Tu Hao[1], Tri Duc Lam[2]

[1]Faculty of Chemical Engineering and Food Technology, Nguyen Tat Thanh University, Ho Chi Minh City, Vietnam
[2]NTT Hi-Tech Institute, Nguyen Tat Thanh University, Ho Chi Minh City, Vietnam
*Email: anhtt@ntt.edu.vn; labams2013@gmail.com

Abstract. Jasminum Subtriplinerve Blume (Oleaceae) is herbal plants widely a tea for weight loss, stimulates milk glands. Extract product it easy used yet had little research. This study was carried out to evaluate the effect of extracting conditions such as water/material ratio, extraction temperature and extracting time on extracting yield, polyphenol content and antioxidant activity of J.subtriplinerve extract. The total phenolic compounds were determined according to the Folin–Ciocalteu method. Extraction yield (0.22g/100 g) was obtained at water/material ratio 15:1 (ml/g), 40°C and 4 hours extraction time. Under these optimized conditions, the polyphenol content of was 2640.4 μg/1g extract. And the extract can harvest 46.11 % of DPPH (30 μg/ mL) at 10 000 μg/ml concentration

1. Introduction

The development of natural biomaterials is not regarded as a new area of science, but has existed for centuries. The use of natural products as a biomaterial is currently undergoing a renaissance in the biomedical field [1-5]. Phenolic compounds are important components of the color, flavor, and aroma of fresh fruits, vegetables, and their products. Phenolic compounds, in addition to antioxidative roles, anticarcinogenic, anti-inflammatory, or antimicrobial activities[6,7]. Phenolics are widespread constituents of plant foods (fruits, vegetables, cereals, olive, legumes, chocolate, etc.) and beverages (tea, coffee, beer, wine, etc.), and partially responsible for the overall organoleptic properties of plant foods [8].

Jasminum Subtriplinerve Blume (Oleaceae) is herbal plants is the most used for the purpose as weight loss, stimulates milk glands for postpartum women and so on.. It thrives well under wet and warm climatic conditions. Oleaceae istwigs spreading, leaves opposite oval-lanceolate, white flowers, fragrant, calyx campanulate, berry the oval or spherical[9]. Jasminum Subtriplinerve in many chemical components such as alkaloids, glycosides, flavonoids, terpenes, and plastic [10]. The main components identified in Oleaceae oxidation mainly represented by linalool monoterpenes (14.8% - 44.2%), α-terpineol (15.5%), geraniol (19.4 %), cis-linalool oxide (8.8%), benzopyrone coumarin (48.9%) [11,12]. Ethyl acetate extract of the aerial parts of Jasminum subtriplinerve Blume (Oleaceae), 6'-O'-menthiafoloylverbascoside, rutin, isoverbascoside, isooleoverbascoside, apiosylverbascoside, astragalin, isoquercitrin, and verbascoside were isolated. Chemical investigation of Chardonnay grape pomace has resulted in the isolation of 17 polyphenols. Therefore, the purpose of this research is to find the optimal conditions (solvent,temperature, time…) on the polyphenol content and to evaluate its antioxidation activity.The result of this study could be a hot trend in the organic markets and may provide the useful parameters for further research.

2. Material and methods

2.1. Sample preparation

NaOH 10%, $FeCl_3$ 5%, Cyanidin reagents, $Pb(CH_3COO)_2$ 10%.Jasminum Subtriplinerve were purchased from Herbal Tea Thanh Binh Company, Ho Chi Minh City. The raw material was grinded into particles having a size 0.2 ± 0.5 mm , preserved by ziplock and room temperature. The moisture content of raw material and products are measured by Ohaus Moisture Analyzer MB-45 for each sample.

2.2. Extraction procedure

The power of the J.subtriplinerve (10 g ± 0.2) was soaked into beaker with distilled water the proportion of material : solvent respectively surveyed: the ratio (from 1:5 to 1: 20), the temperature (room temperature to 80 ^{0}C, Δ T = 10 ^{0}C) , the time (2 h to 8 h , Δ t = 2 h). Beaker is placed in the stable on the water bath (Daihan General Purpose Water Bath Digital Fuzzy Control System) is adjusted. Extracts of J.subtriplinerve are filtered by a vacuum filtration system and are condensed by bain marie. The extract was weighted for calculation yield.

Total of efficiency process were calculated based on absolute dried mass weight

$$Y(\%) = \frac{m_{driedextract}}{m_{material}} * 100 = \frac{m_{extract} \times (1 - a_{extract})}{m_{material} \times (1 - a_{material})} * 100$$

Where:Y-Total of extraction yield, $m_{extract}$ - mass of extract(g), $m_{material}$ - mass of material (g), $a_{extract}$ - moisturre of extract (%), $a_{material}$ - moisture of material (%).

2.3. Polyphenol content's computation

0.1g of sample add 20 ml of distilled water for 15 minutes decanting ultrasound services, repeated 4 times and add water to 100 ml (flask). Diluted 5 times and filtered. Take 1 ml of gallic acid solution (10, 20, 30, 40, 50 µg / mL) into 5 ml of 10% Folin-Ciocalteu reagent and react for 3-8 minutes; After that, add 4mL of Na_2CO_3 solution. After 1 h reaction at room temperature, the absorbance was determined by a spectrometer at 765 nm. The experiment was repeated three times. The OD value was recorded and a linear calibration plot was used to determine the total phenolic content in the extracts. The extracts were carried out similarly to standard phenolic[13]

Formula the total polyphenol content: $P = \frac{f(A)}{m*(1-X)}$,

where: P total polyphenol content (µg/g),X - extract moisture (%),m - mass extraction of quantified (g),f(A) - gallic acid concentrations calculated from standard line equation (µg). Standard calibration curve of gallic acid to determine polyphenol content. Figure 2 shows the construct phenolic datum line with the standard gallic acid in the concentration (10 - 50) ÷ 0.3 (µg /ml). Results of a linear regression equation Y = 0.1091X + 0.1355; R^2 = 0.9933.

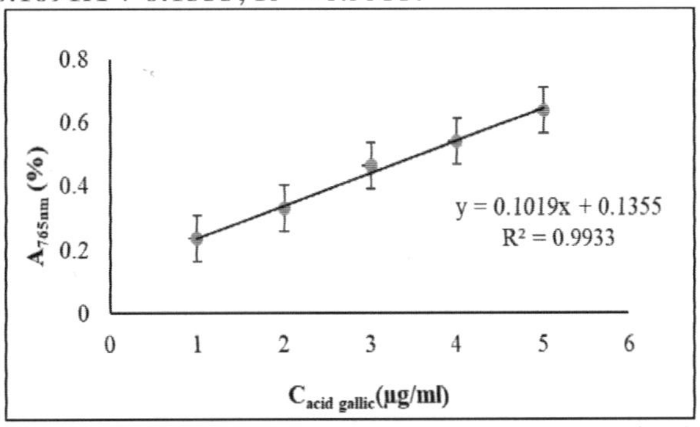

Figure 1. Results survey of gallic acid standard solution

2.4. Antioxidant activity by DPPH method

The antioxidant activity of extracts and the standard was assessed on the basis of the radical scavenging effect of the stable 1, 1-diphenyl-2-picrylhydrazyl (DPPH)-free radical activity by modified method.0.1g sample dissolved in 10.00 mL MeOH 80 % (C = 10.000 µg/ mL) . 780 µL DPPH (OD 517 nm = 0.7 ÷ 0.02) into 520 µL solution sample. The sample solution with preconcentration , and the mixed fermentation at room temperature in the dark within 30 min. The optical measurement of the mixture a UV-Vis spectrophotometer, at 517 ηm.Blank sample, but 520 µL solution replaced MeOH 80 %.Standard sample: Vitamin C (0.1g ÷0.01) was dissolved MeOH 80% into volume flask 100 mL, in the dark (C = 100 µL / mL). Percent inhibition [14]:$\%I = \frac{A_b - A_s}{A_b} x\ 100$, where: A_bOptical density blank sample,A_s: Optical density sample, %I : Precent free radical DPPH. Standard calibration curve of vitamin C. Figure 2 shows construct phenolic datum line with the standard gallic acid in the concentration (40000 - 10000) ÷ 0.3 (µg /ml). Results of a linear regression equation Y = 0.0045X – 1.0984; R^2 = 0.9857

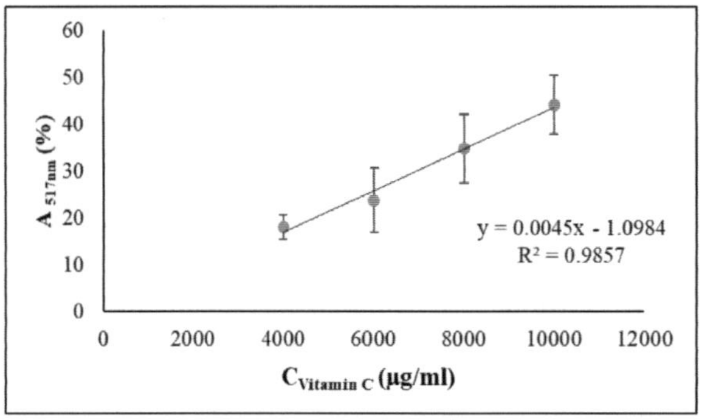

Figure 2. Results survey of vitamin C standard solution

3. Result and discussion

3.1. Effect material/ solvent ratio

Figure 3 describes the effect of material/water ratio on extraction yield and the polyphenol content. The results showedthe highest polypenol yield was obtained at solid/solvent 1:15 (18g/100g),polyphenol content (276.65µg/g). High solvent ratios can promote a higher solubility concentration gradient, resulting in faster diffusion rates that allow for better solid solvent extraction [15]. However, the total polyphenol content will not continue to increase once the balance has been reached [16].Figure 4 shows the effect of material/solvent ratio on content polyphenol percent of inhibition.The results showed at solid/solvent 1:15 (65,42%) percent of inhibition the highest and decreases gradually until it comes the material/solvent ratio 1:15. Both polyphenol content and inhibition activity are proportional.

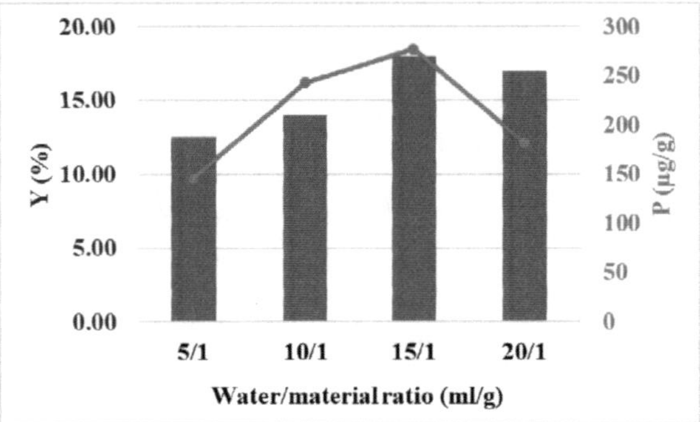

Figure 3 Effect of material / water ratio on the extraction yield and content

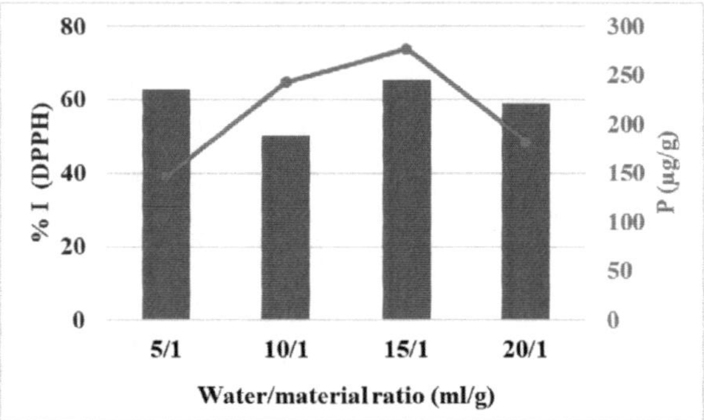

Figure 4 Effect of material / water ratio on the content and percent of inhibition

3.2. Effect of extract time

Figure 5 indicates results showed the highest polyphenol yield was obtained at 4h (0.23g/100g) when increasing the extraction time from 2 to 4h this content increases rapidly and 4h extraction time was highest. As the extraction time increases, the content of the material in the diffused material from the cell is increased [17]. However, when the extraction time of polyphenol compounds reached equilibrium, the polyphenol content increased slowly. In addition, these compounds can be oxidized by elements from the environment (temperature, light, oxygen)[18].

Figure 6 point the effect of time on content polyphenol and the percent inhibition was decreased after reaching a maximum value at 6h. Prolonged extraction would increase the chance for occurrence of oxidation of phenolic compounds thus, the DPPH free radical scavenging ability was decreased after extraction time of 6h [18]. However, 4 hours extracting process could be more convenient and energetically less expensive. That is why extraction at 4h was selected, even though leading to lower polyphenol content.

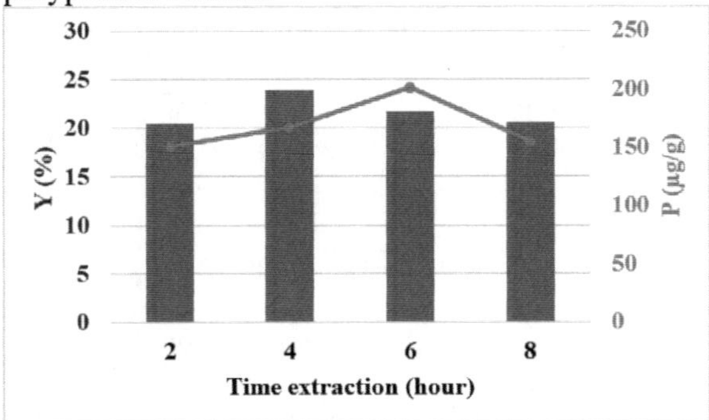

Figure 5 Effect of time on the extraction yield and content.

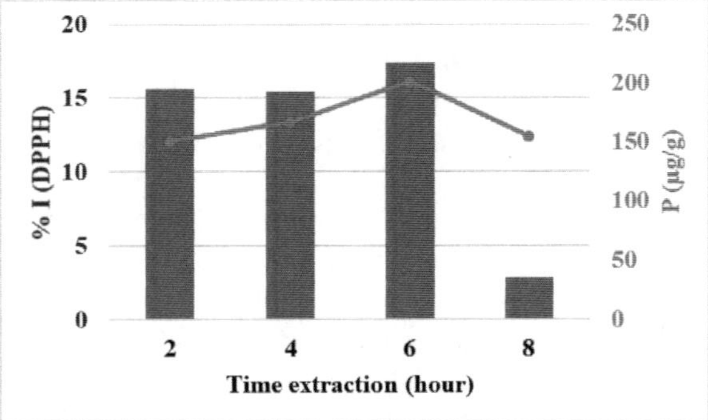

Figure 6 Effect on time to the extraction content and percent radical scavey.

3.3. Effect of temperature

Figure 7 nominate the effect of tempertature on extraction yield and the polyphenol content. The results showed the yields were not different when the extraction temperature above 40°C. However, the polyphenol content is a highest at 80°C (892 µg/g), it could be explainedat higher temperature the higher solubility. Not only that viscosity, surface tension also decrease at high temperature lead to higher velocity, that contributes to increasing the rate of extracting process[15,19].

Figure 8 shows the effect of temperature on content polyphenol and the percent inhibition. The results showed 80°C (65.28%) percent radical scavey the highest. As the temperature increases, the degradation of plant cell components increases, thereby increasing the permeability of the cell membrane (Wang et al., 1998)[20]. However, high temperatures denature the compounds matter. If the extract at 40°C it could be more convenient and energetically less expensive, shorter times, cost lowly. That is why extraction at 40°C was selected, even though leading to yields lowly.

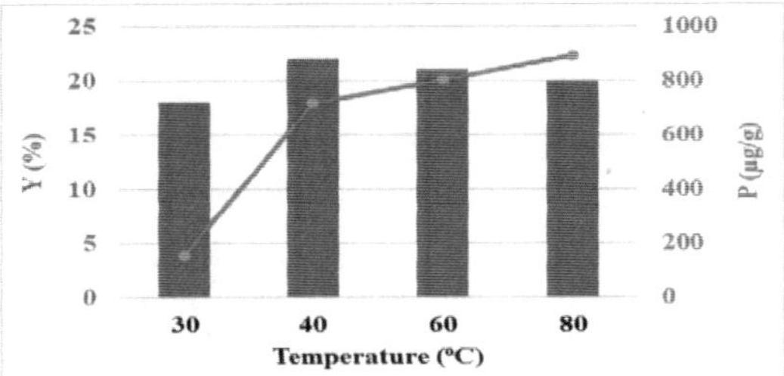

Figure 7. Effect of temperature on the extraction yield and content.

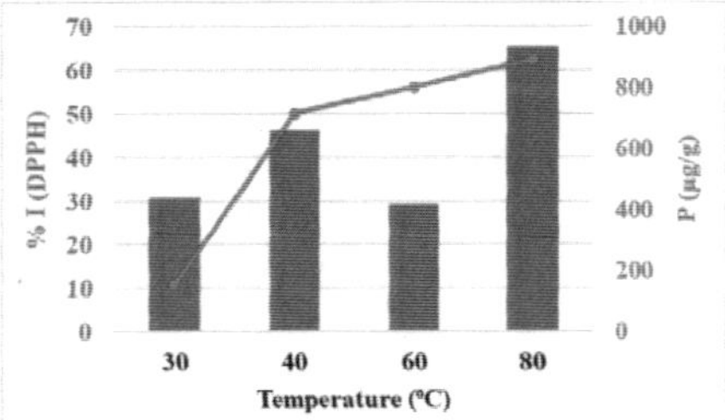

Figure 8. Effect on tempertature to the extraction content and percent inhibition

4. Conclusion

This research has identified the appropriate extraction conditions to obtain extract and saponins content from gums of J.sub evaluate its percent inhibition active. The optimal conditions were found ratio 1:15, temperature 40°C extraction time 4 hours. these optimum conditions can be useful for further optimization of the extraction of J. Subtriplinerve. Further studies are needed to examine the effect conditions such as different solvent (EtOH, MeOH, etc.) and methods (microwave, ultrasound, pressure liquid extraction (PLE), etc.) of Jasminum Subtriplinerve.

Acknowledgments: This work was supported by grants from Nguyen Tat Thanh University, Ho Chi Minh City, Viet Nam.

Reference

[1] Vo T-S, Ngo D-H, Bach L G, Ngo D-N and Kim S-K 2017 *Process Biochemistry* **54** 188–94
[2] Mai H C, Le T T T, Diep T T, Le T H N, Nguyen D T and Bach L G 2018 *Asian Journal of Chemistry* **30** 293–7
[3] Hien T T, Nhan N P T, Trinh N D, Ho V T T and Bach L G 2018 *Solid State Phenomena* **279** 217–21
[4] Nhan N P T, Hien T T, Nhan L T H, Anh P N Q, Huy L T, Nguyen T C T, Nguyen D T and Bach L G 2018 *Solid State Phenomena* **279** 235–9
[5] Tran T, Nguyen H, Nguyen D, Nguyen T, Tan H, Nhan L, Nguyen D, Tran L, Do S and Nguyen T 2018 *Processes* **6** 206
[6] Eberhardt M V, Lee C Y and Liu R H 2000 *Nature* **405** 903–4
[7] Lee J Y, Chang E J, Kim H J, Park J H and Choi S W 2002 *Archives of Pharmacal Research* **25** 313–9
[8] Dai J and Mumper R J 2010 *Molecules* **15** 7313–52
[9] Hue Ngan D, Hoai H T C, Mai Huong L, Hansen P E and Vang O 2008 *Natural Product Research* **22** 942–9
[10] Hue Ngan D, Hoai H T C, Mai Huong L, Hansen P E and Vang O 2008 *Natural Product Research* **22** 942–9
[11] Kraus W, Ngoc L H, Conrad J, Klaiber I, Reeb S and Vogler B 2002 *Phytochemistry Reviews* **1** 409–11
[12] Dai D N, Thang T D, Ogunwande I A and Lawal O A 2016 *Natural Product Research* **30** 860–4
[13] Nyalambisa M, Oyemitan I A, Matewu R, Oyedeji O O, Oluwafemi O S, Songca S P, Nkeh-Chungag B N and Oyedeji A O 2017 *Saudi Pharmaceutical Journal* **25** 381–6
[14] Gajula D, Verghese M, Boateng J, Walker L T, Shackelfor L, Mentreddy S R and Cedric S 2009 *International Journal of Cancer Research* **5** 130–43
[15] Alkhatib R, Joha S, Cheok M, Roumy V, Idziorek T, Preudhomme C, Quesnel B, Sahpaz S, Bailleul F and Hennebelle T 2010 *Planta Medica* **76** 86–7
[16] Al-Farsi M A and Lee C Y 2008 *Food Chemistry* **108** 977–85
[17] Sarawong C, Schoenlechner R, Sekiguchi K, Berghofer E and Ng P K W 2014 *Food Chemistry* **143** 33–9
[18] Cracolice M S and Peters E I 2011 (Belmont, CA: Brooks/Cole Cengage Learning)
[19] Naczk M and Shahidi F 2004 *Journal of Chromatography A* **1054** 95–111
[20] Wang J, Sun B, Cao Y, Tian Y and Li X 2008 *Food Chemistry* **106** 804–10
810, 2008.

IOP Conf. Series: Materials Science and Engineering **544** (2019) 012028 doi:10.1088/1757-899X/544/1/012028

Response surface methodology optimization for extraction of natural anthocyanins from Vietnamese *Carissa carandas* L. fruit

Tri Nhut Pham[1,2], Thi Ha Xuan Huynh[1,2], Thien Hien Tran[1], Dai-Viet N. Vo[3], Tran Quoc Toan[4], Tri Duc Lam[1,5], Xuan Tien Le[6], Hieu Vu-Quang[1,*]

[1]NTT Hi-Tech Institute, Nguyen Tat Thanh University, Ho Chi Minh City, Vietnam
[2]Faculty of Chemical Engineering and Food Technology, Nguyen Tat Thanh University, Ho Chi Minh City, Vietnam
[3]Faculty of Chemical and Natural Resources Engineering, Universiti Malaysia Pahang, Malaysia
[4]Institute of Natural Products Chemistry, Vietnam Academy of Science and Technology, Ha Noi, Vietnam
[5]Graduate School of Biotechnology, Kyung Hee University, South Korea
[6]Department of Chemical Engineering, HCMC University of Technology, VNU-HCM, Ho Chi Minh City, Vietnam
Email: vqhieu@ntt.edu.vn, labams2013@gmail.com

Abstract. Anthocyanins are naturally occurring compounds that are responsible for a wide variety of colors in many plants, fruits and vegetables. In this study, the extraction of natural anthocyanins from Vietnamese *Carissa carandas* L. beverage was optimized using response surface methodology (RSM). We applied a Box–Behnken design consisting of three levels and three factors. Examined factors are extraction temperature (ranging from 40 to 60°C), liquid to solid ratio (ranging from 2:1 to 4:1), extraction time (ranging from 30 to 60 min). Using 60% ethanol as solvent for the process, we determined the maximum yields of anthocyanin was 273.786 mg/L. This yield corresponds to extraction conditions of 3:1 (v/w) liquid to solid ratio, temperature of 48.10 °C with a 44.08 min extraction time. The experimental results also fit well with the proposed response model of anthocyanin yield (R^2 = 0.9992). Therefore, this study suggested optimization of different extraction methods for the defatted fruit parts.

1. Introduction

Carissa carandas L., categorized as herb in Apocynaceae family, is an underutilized fruit plant which thrives well throughout tropical and sub-tropical climate [1-3]. They are also edible and rich in minerals such as iron, calcium, magnesium and phosphorus. After being dried, the fruit is shrunk and changed to dark brown colour. Naturally, fruits owe their ability to dye fabrics and threads to the pigment that they contain. Such pigment could be anthocyanins, which are a very large group of red-blue plant pigments, present in the *Carissa carandas* L. fruit [4-7]. The uses of natural bio-pigments, such as anthocyanins from fruit and food products, are an excellent alternative to synthetic colors. Anthocyanins are also non-toxic and the source from which they are extracted are renewable.
Extraction of anthocyanin was conventionally carried out with solvents, such as ethanol, methanol, water, acetone, or mixtures, and often with of small acid addition to prevent nonacylated compounds

from deteriorating [8,9]. To optimize such parameters and produce an efficient experimental procedure, Response surface methodology (RSM) is commonly used. The method whose core is a nonlinear model and is widely applied in chemistry, biology, and agriculture for condition optimization of the systems. RSM is superior to "one to one factor" method, owing to its ability to allow for simultaneous and mutual interaction of process variables [10-20].

Considering the aforementioned background, this study aimed to optimize the parameters for anthocyanin extraction and maximize the anthocyanin yield from Vietnamese *Carissa carandas* L. Beverage by employing RSM with three-variable, three-level Box-Behnken design (BBD). The results were used to evaluate the influence of temperature, liquid-solid ratio, and time factor on the extracted yields. The extract obtained from this extraction method could be further isolated and purified for a certain anthocyanin or can be used as food colorants or in parapharmaceutical products.

2. Experimental Section

Sample Preparation and Chemicals

Fresh *Carissa carandas* L. fruit were obtained in mature stage in July 2018 from Tien Giang province, Viet Nam. The harvested *Carissa carandas* L. fruit were washed and kept in a non-hygroscopic bag. Lastly, 10g *Carissa carandas* L. fruit was cut into 1-2 mm portions. The extraction of anthocyanins was carried out with ethanol (C_2H_5OH), purchased from Sigma Aldrich (US).

Extraction of Anthocyanins

Various amounts of 60% ethanol containing 0.1 (v/v) hydrochloric acid (pH~1.0) were mixed with Freeze-dried powder samples (1.0 g) [21] to produce mixtures with liquid-solid ratios ranging from 2:1 to 4:1 (v/w). The mixtures were then placed in thermostatic water bath with temperatures and durations ranging from 40-60 (°C) and 30–60 (min), respectively, followed by centrifugation at 4000 rpm for 10 min at 25 °C. The resulted supernatant was filtered through filter paper for total anthocyanin evaluation. The pH of supernatant was scanned from 400 nm to 700 nm. Spectrophotometric pH-differential method was used to determine the total anthocyanin content [22].

Experimental Design for Optimization

RSM, in conjunction with a Box-Behnken design, was employed using Design-Expert software (Version 11.0) to optimize the parameters of anthocyanin extraction from the fruit. The independent variables were liquid to solid ratio, extraction time, and extraction temperature and were termed as X_1, X_2 and X_3, respectively. The response variable (Y) was the anthocyanin yield. Details of the independent factors and their corresponding ranges are described in the table 1. Experimental data were fitted to the following second-order model (Equation 2):

$$Y = \beta_0 + \beta_1 X_1 + \beta_2 X_2 + \beta_3 X_3 + \beta_{11} {X_1}^2 + \beta_{22} {X_2}^2 + \beta_{33} {X_3}^2 + \beta_{12} X_1 X_2 + \beta_{13} X_1 X_3 + \beta_{23} X_2 X_3 \quad (2)$$

where Y, β_0, and β_n are predicted response, intercept and coefficients of model terms, respectively. The statistical significance of coefficient estimates and statistical indicators of the model was calculated by analysis of variance (ANOVA).

Table 1. Levels and Independent factors of the natural anthocyanin extraction process from Vietnam *Carissa carandas* L. fruit.

Levels	Independent factors		
	X_1 (v/w)	X_2 (Min)	X_3 (°C)
-α	1.32:1	33.18	19.77
-1	2:1	30	40
0	3:1	45	50
+1	4:1	60	60
+α	4.68:1	66.82	70.23

Statistical Analysis

Actual experiments results, derived from triplicate trials, and were summarized as means ± SD. Accrued data were analyzed with the SPSS 14.0 package (SPSS Inc., Chicago, IL, USA, 2005). Significance of a term is recognized if its p-value is lower than 0.05

3. Results and Discussion

The 20-experiment Box-Behnken design was detailed in the Table 2. The yields of anthocyanins were described in both experimental and predicted values. The following equation showed fitted model in terms of response variable and actual factors: $Y = 273.65 - 0.3116X_1 - 0.1712X_2 - 1.36X_3 + 0.1420X_1X_2 - 1.67X_1X_3 + 0.8953X_2X_3 - 3.15X_1^2 - 2.79X_2^2 - 3.72X_3^2$ (3)

Table 2. Box-Behnken design arrangement and response.

No	Independent factors			Y (mg/L)		No	Independent factors			Y (mg/L)	
	X_1	X_2	X_3	Actual	Predicted		X_1	X_2	X_3	Actual	Predicted
1	2:1	30	40	265.32	265.20	11	5:1	19.77	50	265.91	266.04
2	4:1	30	40	267.68	267.64	12	5:1	70.23	50	265.39	265.46
3	2:1	60	40	262.82	262.78	13	5:1	45	33.18	265.31	265.42
4	4:1	60	40	265.87	265.79	14	5:1	45	66.82	260.73	260.83
5	2:1	30	60	264.09	264.03	15	5:1	45	50	273.93	273.65
6	4:1	30	60	259.87	259.77	16	5:1	45	50	273.83	273.65
7	2:1	60	60	265.29	265.19	17	5:1	45	50	273.73	273.65
8	4:1	60	60	261.52	261.50	18	5:1	45	50	273.56	273.65
9	1.32:1	45	50	265.14	265.26	19	5:1	45	50	273.33	273.65
10	4.68:1	45	50	264.14	264.22	20	5:1	45	50	273.57	273.65

Analysis of variance (ANOVA) for the resulted equation was shown in the Table 3. The model F-value of 1397.98 implied that the signal to noise ratio was adequately measured. Low p-value of F-value also suggested that "model F-Value" of this large could occur due to noise with the chance of lower than 0.01% [23]. In addition, the coefficient of determination (R^2) was higher than 0.99 and *Coefficient of Variation %* was lower than 6.98%, showing reasonable fitness and reproducibility of the model. As reported by the ANOVA results, X_1, X_2, X_3, X_1X_3, X_2X_3, X_1^2, X_2^2 and X_3^2 were significant model terms. The F-value of lack-of-fit statistic of 0.49 rejected the null hypothesis that the data was unfitted to the proposed model, and subsequently, affirmed the validity of the produced quadratic model. Regarding R^2, Le Man et al. and Chauhan and Gupta proposed that R^2 of any model is acceptable when it is greater than 0.75 [24, 25], which is also in line with the ANOVA results of the model.. Therefore, except for X_2X_3, remaining linear and non-linear terms of the model are showed to be significant predictors of anthocyanin yield. Hence, the relevance of independent variables on the TAC is ranked as follows: $X_1 > X_2 > X_3$.

Figure 1 showed surface and contour plots demonstrating the effects of two variables on the response function (anthocyanin yield). In each plot, one variables is fixed while other variables vary, as reflected by two bottom horizontal axes. In figure 2A, given any value of X_1, anthocyanin yield (Y) mounted up proportionally with X_2 to the peak value, and started to diminish after that. Similar relationship could also be observed for X_1 when X_2 was fixed. Figure 2B illustrated relationship between yield, liquid-solid ratio and temperature. The plot indicated that, compared to that of temperature the variations of solid-liquid ratio were more influential on Y The effect of temperature (X_3) and time (X_2) is shown in Figure 1C. Visually, when X_3 exceeds 45 °C, a slight increase of X_2 could lead to a considerable decline of yield. Therefore, from the slope of the surface, one may conclude that temperature and liquid-solid ratio are two key factors in achieving high extraction efficiency of anthocyanins. To obtain the optimal values of the factors, the resulted equation was evaluated by the software. It is revealed that the optimal conditions of the extracting process from *Carissa carandas* L. fruit were 3:1 (v/w) liquid to solid ratio, temperature 48.1 °C with 44.08 min of extraction time, corresponding to Y = 273.786 mg/L. A set of triplicate experiment was conducted following the calculated optimum conditions. These confirmation experiments showed the results of 273.784 ±0.45 mg/g. This concurs with the predicted optimal value, suggesting that the actual data is well fitted and the extraction process is optimized.

Table 3. Analysis of Variance (ANOVA) for the regression equation

Source	Sum of Squares	Degree of freedom	Mean Square	F-value	Prob. > F	Comment
Model	437.60	9	48.62	1397.98	< 0.0001	significant

X_1	1.33	1	1.33	38.14	0.0001	
X_2	0.4001	1	0.4001	11.50	0.0069	
X_3	25.42	1	25.42	730.81	< 0.0001	
X_1X_2	0.1613	1	0.1613	4.64	0.0567	
X_1X_3	22.43	1	22.43	644.95	< 0.0001	
X_2X_3	6.41	1	6.41	184.35	< 0.0001	
X_1^2	143.04	1	143.04	4112.56	< 0.0001	
X_2^2	112.39	1	112.39	3231.46	< 0.0001	
X_3^2	199.68	1	199.68	5741.10	< 0.0001	
Residual	0.3478	10	0.0348			
Lack of Fit	0.1140	5	0.0228	0.4878	0.7752	not significant
Pure Error	0.2338	5	0.0468			
Std. Dev.= 0.1865		**Mean = 267.05**	**C.V. % = 0.0698**			**R^2 = 0.9992**
Adjusted R^2 = 0.9985		**Predicted R^2 = 0.9972**	**Adeq. Precision = 105.2679**			

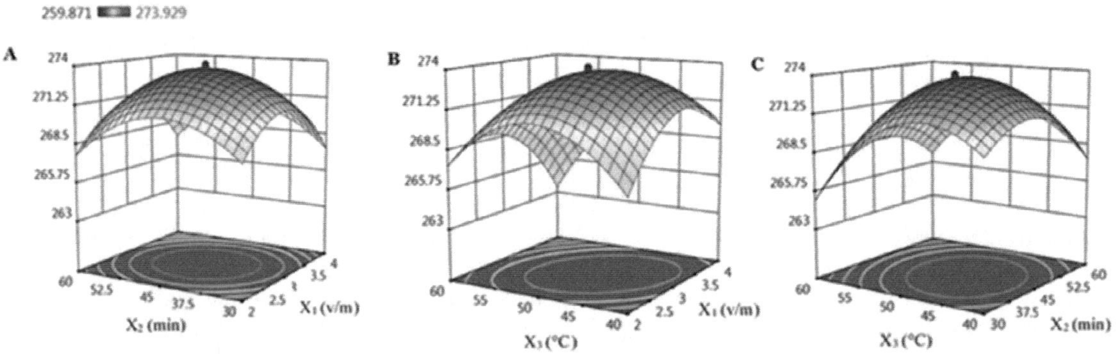

Figure 1. Response surface plots of the natural anthocyanin yielded (Y) from the Vietnamese *Carissa carandas* L. fruit extract, corresponding to pairwise comparisons of three factors.

Conclusions

We successfully attempted RSM with an experimental design to optimize the extraction process of natural anthocyanins from the Vietnamese *Carissa carandas* L. fruit. We showed that the highest anthocyanin yield was 273.786 mg/L, corresponding to optimal combination of independent variables of temperature at 48.1°C, time at 44.08 min, and liquid-solid ratio at 3:1. Of the three examined factors, temperature was more dominant when it comes to influencing the efficiency of anthocyanin extraction. The resulted optimum condition could be applied in methanolic extraction of anthocyanin pigments, facilitating the production of natural colourants in textile, solar cell, and food industry.

1. **Acknowledgments:** This work was supported by grants from Nguyen Tat Thanh University, Ho Chi Minh City, Vietnam

References

[1] Kittur Rani Channamma College of horticulture, Arabhavi, University of Horticultural Sciences, Bagalkot-587 103 and Patil U S 2017 *International Journal of Pure & Applied Bioscience* **5** 374–8
[2] Kingcha Y, Tosukhowong A, Zendo T, Roytrakul S, Luxananil P, Chareonpornsook K, Valyasevi R, Sonomoto K and Visessanguan W 2012 *Food Control* **25** 190–6
[3] Abeyrathne E D N S, Lee H Y and Ahn D U 2013 *Poultry Science* **92** 3292–9
[4] Sudjaroen Y 2017 *Journal of Pharmaceutical Negative Results* **8** 31
[5] Sarkar R, Kundu A, Banerjee K and Saha S 2018 *LWT* **93** 673–8

[6] Ali J and Mueen Ahmed K 2011 *International Journal of Pharmaceutical Investigation* **1** 1
[7] Pewlong W, Sajjabut S, Eamsiri J and Chookaew S 2014 *Chiang Mai University Journal of Natural Sciences* **13**
[8] Khazaei K M, Jafari S M, Ghorbani M, Kakhki A H and Sarfarazi M 2016 *Food Analytical Methods* **9** 1993–2001
[9] Guldiken B, Boyacioglu D and Capanoglu E 2016 *Food Analytical Methods* **9** 1876–86
[10] Van Tran T, Bui Q T P, Nguyen T D, Le N T H and Bach L G 2017 *Adsorption Science & Technology* **35** 72–85
[11] V T Tran, D T Nguyen, V T T Ho, P Q H Hoang, P Q Bui, L G Bach, *Journal of materials and Environmental Sciences* **8** 2017 426-437
[12] Van Tran T, Bui Q T P, Nguyen T D, Ho V T T and Bach L G 2017 *Water Science and Technology* **75** 2047–55
[13] Nhan N P T, Hien T T, Nhan L T H, Anh P N Q, Huy L T, Nguyen T C T, Nguyen D T and Bach L G 2018 *Solid State Phenomena* **279** 235–9
[14] Tran T, Nguyen H, Nguyen D, Nguyen T, Tan H, Nhan L, Nguyen D, Tran L, Do S and Nguyen T 2018 *Processes* **6** 206
[15] Bach L G, Van Tran T, Nguyen T D, Van Pham T and Do S T 2018 *Research on Chemical Intermediates* **44** 1661–87
[16] Van Thuan T, Quynh B T P, Nguyen T D, Ho V T T and Bach L G 2017 *Surfaces and Interfaces* **6** 209–17
[17] Hien T T, Nhan N P T, Trinh N D, Ho V T T and Bach L G 2018 *Solid State Phenomena* **279** 217–21
[18] Liu G-L, Guo H-H and Sun Y-M 2012 *International Journal of Molecular Sciences* **13** 6292–302
[19] Zou T-B, Wang M, Gan R-Y and Ling W-H 2011 *International Journal of Molecular Sciences* **12** 3006–17
[20] Fan G, Han Y, Gu Z and Chen D 2008 *LWT - Food Science and Technology* **41** 155–60
[21] Guo H, Ling W, Wang Q, Liu C, Hu Y, Xia M, Feng X and Xia X 2007 *Plant Foods for Human Nutrition* **62** 1–6
[22] Bi S, Tang W, Gan N, Ye L, Liu J, Chen G, Dai L, Cao M and Chen Y 2000 *Analytical Letters* **33** 677–89
[23] Yang Z and Zhai W 2010 *Innovative Food Science & Emerging Technologies* **11** 470–6
[24] Le Man H, Behera S K and Park H S 2010 *International Journal of Environmental Science & Technology* **7** 157–64
[25] Chauhan B and Gupta R 2004 *Process Biochemistry* **39** 2115–22

Chemical Constituents of the Aerial Parts of Scoparia dulcis and Anti-cancer, Anti-inflammatory Activities

Nguyen Ngoc Thanh Tin[1], Nguyen Dang Thanh Truc[1], Hoang Thi Thu Hang[1,4], Pham Thi Nhat Trinh[2,4], Tri Duc Lam[3] and Le Tien Dung[5,6,*]

[1]Nguyen Thong High School, Vinh Long Province, Vietnam

[2]Faculty of Natural Sciences, Tien Giang University, Tien Giang Province, Vietnam

[3]NTT Hi-Tech Institute, Nguyen Tat Thanh University, Ho Chi Minh City, Vietnam

[4]Faculty of Chemical Engineering and Food Technology, Nguyen Tat Thanh University, Ho Chi Minh City, Vietnam

[5]Institute of Applied Materials Science, Vietnam Academy of Science and Technology, Ho Chi Minh City, Vietnam;

[6]Graduate University of Science and Technology, Vietnam Academy of Science and Technology, Ha Noi, Vietnam;

*Email: inpcdung@yahoo.com, labasm2013@gmail.com

Abstract. In Vietnam, the medicinal plant *Scoparia dulcis* is used for the treatment of bronchitis, gastric disorders, antidiabetes, hypertension, hepatitis etc. In this paper, experiments was designed to evaluate in vitro anticancer and anti-inflammatory activities and isolate phytochemicals from methanol extracts of *Scoparia dulcis* whole plants. The chemical investigation of methanol fraction of Scoparia dulcis led to the isolation of benzoxazinone (1), phenylethanoid (2), flavone (3), and lignan (4) glycosides. The bioactivity results indicated that crude ethanol and ethyl acetate extracts had potent cytotoxic activity toward the HepG2 cancer cell with IC50 of 47.03 μg/mL and 36.04 μg/mL, respectively. Interestingly, crude ethanol and ethyl acetate extracts had the NO inhibitory activity, with inhibition of 56.9 % or 74.7 % at 30 μg/mL and 97.3% or 82.9 % at 100 μg/mL, respectively. In addition, the hexane extract at concentrations of 100 μg/ml demonstrated anti-inflammatory activity through the inhibition of nitric oxide production of 76.7%.

1. Introduction

Fruits, herbs and aromatic plants are important sources of phenolic compounds such as phenolic acids, flavonoids and anthocyanins, which have attracted considerable attention for their application in natural biomaterials [1-4]. *Scoparia dulcis*, (family: Scrophulariaceae), a perennial herb which is commonly found in tropical and subtropical regions around the world. It has been used to treat bronchitis, gastric disorders, antidiabetes, hypertension, hepatitis in Vietnamese folk medicine [5]. The results of previously phytochemical investigation indicated that this plant contained flavonoids, diterpenoids, triterpenoids, sterol, benzoxazinoids etc [6-10]. Morever, *Scoparia dulcis* possessed antioxidant, anti inflamatory, analgesic, anti cancer, antidiabetic activities [11-15]. As parts of our continuing studies on phytoconstituents from Vietnam medicinal plants [16-19], we report the anti-cancer and anti-inflammatory activities and the separation and characterization of four glycoside compounds of methanol fraction.

2. Experimental

2.1. General experimental procedures

Specific rotation were measured on digital polarimeter (Kruss, Hamburg, German). NMR spectra were recorded on a Bruker AM500 FTNMR spectrometer (Bruker, Karlsruhe, Germany), Institute of Chemistry (VAST). ESI-MS were performed on a MicroOTOF-Q mass spectrometer (Bruker, Karlsruhe, Germany), University of Science (National University, HoChiMinh city, Vietnam). TLC was performed on silica gel 60 F254 (Merck, Darmstadt, Germany) and RP-18 F254S plates (Merck, Darmstadt, Germany). The zones were detected by H_2SO_4/EtOH. Column chromatography was performed on silica gel (Merck, Darmstadt, Germany), ODS (Merck, Darmstadt, Germany) and Sephadex LH- 20 (GE Healthcare Bio-Sciences AB, Uppsala, Sweden).

2.2. Plant materials

Whole plants of *Scoparia dulcis* were collected in Vinhlong province, Vietnam in June, 2018. A voucher specimen (015842) has been deposited in Institute of Applied Materials Science (VAST) by Dr. Luu Hong Truong.

2.3. Extraction and isolation

The aerial parts of *S.dulcis* (4.5 kg) was extracted at room temperature with ethanol 96% (5x10 L) for 4 weeks. The crude extract was concentrated to dryness in rotary evaporator at 60^0 C to yield a residue (280 g). The residue was chromatographed on silica gel column to give n-hexane, ethyl acetate and methanol soluble fractions. The methanol fraction (126.7 g) was fractionated by column chromatography on silica gel (EtOAc: MeOH:H_2O=25:1:1 to 5:1:1) to yield three fractions (SDM1–3). Fraction SDM2 was chromatographed eluting with $CHCl_3$: MeOH (30:1 to 8:1) to obtain eight subfractions (SDM2.1-8). Subfraction SDM2.7 was chrommatographed on a silica gel column eluting with $CHCl_3$:MeOH (8:1 to 5:1) to obtain six subfractions (SDM2.7.1-6). Fraction SDM2.7.5 was rechromatographed on a silica gel column eluting by $CHCl_3$: MeOH (8:1) and then followed by sephadex LH20 column (MeOH) to give **1** (218 mg). Fraction SDM2.7.6 was passed over Sephadex LH 20 column (MeOH) to give three fractions. Subfraction SDM2.7.6.2 was repeatedly purified through a ODS column (MeOH:H_2O = 1:1) to give **2** (18.2 mg). Fraction SDM2.7.3 was subjected to a ODS comlumn (MeOH:H_2O=1:3) as eluent to give seven subfractions (SDM2.7.3.1-7). Subfraction SDM2.7.3.1 was purified through sephadex LH20 (MeOH), then subjected to a silica gel column eluting with $CHCl_3$: MeOH (6:1) to obtain **3** (8.0 mg). Using a Sephadex LH-20 column (MeOH) to purify subfraction SDM2.7.3.4 and then followed by a ODS column eluting MeOH:H_2O (1:2) to obtain **4** (30 mg).

2.4. Cytotoxicity assay

Human hepatocarcinoma HepG2 cells were maintained in Dulbecco's modified Eagle's medium containing 10 % fetal bovine serum, 100 units/mL penicillin, and 10 µg/mL streptomycin at 37 °C and 5 % CO_2. Cell viability was assessed after the incubation with extracts at a concentration from 6 - 100 µg/mL for 24h. The plates were read immediately at 540 nm on a microplate reader. All the experiments were performed three times and the mean absorbance values were calculated.

2.5. Anti-inflammatory asay

RAW 264.7 macrophages were pretreated for 3 h with extracts, and treated for 24 h with LPS. The nitrite concentration in the medium was measured as an indicator of NO production by the Griess reaction. The inhibitory effect on the NO production was estimated by calculating the concentration ratio of nitrite produced in the cells treated with/without extracts.

IOP Conf. Series: Materials Science and Engineering **544** (2019) 012029 doi:10.1088/1757-899X/544/1/012029

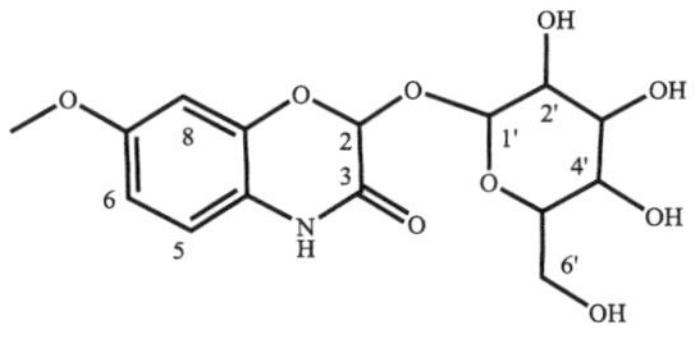

1

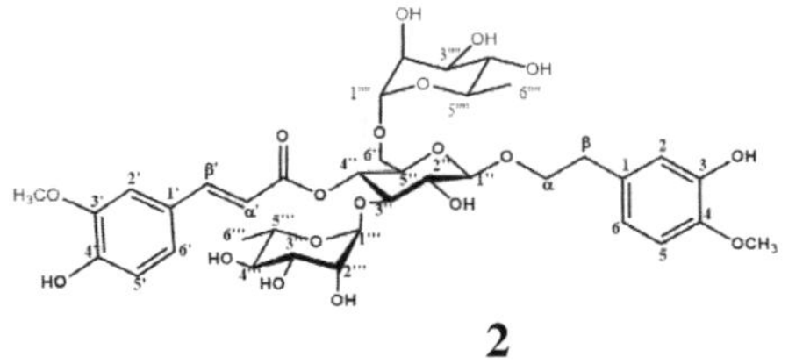

2

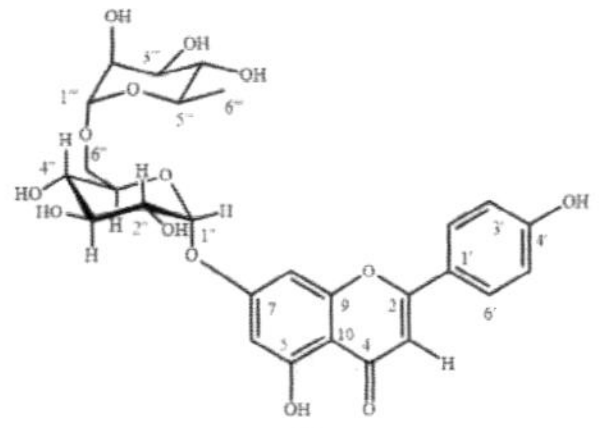

3

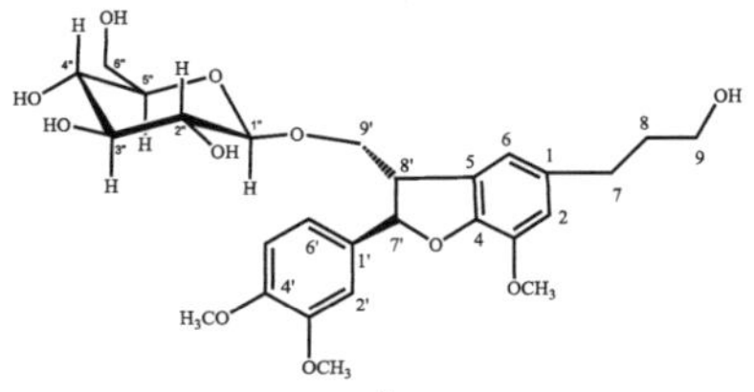

4

Fig.1. Isolated compounds (1-4)

3. Result and Discussion

The ^{1}H NMR spectrum of **1** indicated the presence of a 1,2,4-trisubstituted benzene; one methoxyl group; one acetal proton; one anomeric proton as well as oxymethine proton of a glucose varied from 3.23 to 3.90 ppm. The correlation observed between H-2 δ_H 5.74 with C-3 δ_C 160.7 and C-8a δ_C 142.9 in HMBC of **1** indicated that **1** possess 1,4-benzoxazin-2-hydroxy-3-one skeleton with C-7 bearing methoxy.[16] The glucose moiety was located at C-2 of benzoxazinone framework due to the HMBC correlations of H-2 with C-1'. On the basis of the above evidence, compound **1** was indicated as 2-hydroxy-7-methoxy-1,4(2H)-benzoxazin-3-one 2-O-β-D-glucopyranoside [20,21].

The NMR spectrum of **2** exhibited the resonance of 37 carbon signals including two benzen rings, one double bond, an ethyloxyl moeity, one carbonyl carbon along with three sugar moieties consisting of one glucose and two rhamnose. The existence of feruloyl and 3-hydroxy-4-methoxy-phenylethanol moieties in **2** structure were proved due to correlations in HMBC spectrum. HMBC correlations between H-4'' with carbonyl carbon δ_C 167.9 indicated that feruloyl moiety attached to glucose at C-4''. This was confirmed through lowfield shift of H-4'' at δ_H 5.03 (t, 9.5Hz, H-4''). Similarly, the attachment of 3-hydroxy-4-methoxy-phenylethanol to C-1'' was established due to the cross-peak between H_2-α with C-1'' of glucose moiety. The interglycosidic linkage between glucose C-6'' and rhamnose C-1'''' was determined through correlation between H-1'''' with C-6'' in the HMBC experiment. The anomeric proton δ_H 5.21 correlated with glucose C-3'' in HMBC spectrum indicated this rhamnose to be connected at C-3'' of glucose. Thus, compound **2** was established as Ferruginoside C [22]. Compounds **3** and **4** were determined as isorhoifolin and 3,4'-O-dimethylcedrusin 9'-O-glucopyranoside, respectively (Fig.1) [23,24].

The bioactivity results (Table 1 and 2) indicated that crude ethanol extract and ethyl acetate fraction had potent cytotoxic activity toward the HepG2 cancer cell with IC_{50} of 47.03 μg/mL and 36.04 μg/mL, respectively. Interestingly, crude ethanol and ethyl acetate extracts had the NO inhibitory activity, with inhibition of 56.9 % or 74.7 % at 30 μg/mL and 97.3% or 82.9 %at 100 μg/mL, respectively. In addition, the hexane extract at concentrations of 100 μg/ml demonstrated anti-inflammatory activity through the inhibition of nitric oxide production of 76.7%.

Table 1. Cytotoxicity of extracts against HepG2 cell

Samples	**IC_{50} (μg/mL)**
Hexan fraction	≥100
Ethyl acetat fraction	36.04
Methanol fraction	≥100
Paclitaxel	2.9

Table 2. Inhibition values of extracts on LPS-stimulated NO production in RAW 264.7 macrophages

Samples	Concentration	Inhibition NO production (%)	Cell viability (%)
Crude extract	30 µg/mL	56.9±0.84	68.3±1.0
	100 µg/mL	97.3±0.4	38.4±1.0
Hexan fraction	30 µg/mL	27.4±0.7	103.8±2.1
	100 µg/mL	76.7±0.4	95.8±0.9
Ethyl acetat fraction	30 µg/mL	74.7±0.61	58.5±0.5
	100 µg/mL	82.9±0.2	33.4±1.5
Methanol fraction	30 µg/mL	14.8±0.5	109.4±1.3
	100 µg/mL	47.3±0.8	81.2±1.6
Cardamonin	0.3 µM	17.1±0.8	85.8±0.1
	3.0 µM	96.6±0.6	85.8±0.1

4. Conclusion

From methanol fraction of *Scoparia dulcis*, 2-hydroxy-7-methoxy-1,4(2H)-benzoxazin-3-one 2-O-β-D-glucopyranoside (**1**), ferruginoside C (**2**), isorhoifolin (**3**) and 3,4'-O-dimethylcedrusin 9'-O-glucopyranoside (**4**) were isolated. Crude ethanol and ethyl acetate fractions showed cytotoxic activity toward the HepG2 cancer cell with IC_{50} of 47.03 µg/mL and 36.04 µg/mL, respectively. In addition, crude ethanol and ethyl acetate extracts at 30 and 100 µg/mL inhibited NO production. The hexane extract at concentration of 100 µg/ml showed the inhibition of nitric oxide production of 76.7%.

Acknowledgments: This paper was partially supported by the Nguyen Tat Thanh University, Ho Chi Minh City, Vietnam. We thank Mr. Dang Vu Luong at the Institute of Chemistry for their aid in obtaining the NMR spectra.

References

1. T H Tran, H H H Nguyen, D C Nguyen, T Q Nguyen, H Tan, L T H Nhan, D H Nguyen, L D Tran, S T Do, and T D Nguyen 2018 *Processes* **6** 206
2. T S Vo, D H Ngo, L G Bach, D N Ngo, S K Kim, 2017 *Process Biochemistry* **54** 188–194
3. A N Q Phan, L G Bach, T D Nguyen, N T H Le 2019 *Journal of Nanoscience and Nanotechnology* **19** 974-978
4. T T Hien, N P T Nhan, D T Nguyen, V T T Ho, L G Bach 2018 *Solid State Phenomena* 279 217-221.
5. Loi DT, 2004 *Medicinal plants and remedy of Vietnam* 871-872.
6. Chen, C M and Chen M T 1976 *Phytochemistry* **15** 1997– 1999.
7. Mahato, S B, Das, M C and Sahu, N P 1981 *Phytochemistry* **20** 171–173.
8. Kawasaki M, Hayashi 1988 *Phytochemistry* 27 3709-3711.
9. Ahsan, M, Haque, M R 2012 *Phytochemistry Letters* **5** 609–612.
10. Wu, W H, Chen, TY 2012 *Phytochemistry* **83** 110–115.
11. Latha, M, Pari, L 2009 *Natural Products Research* **23** 1528–1540
12. Beh, J E, Latip 2010 *Journal of Ethnopharmacology* **129** 23–33.
13. Beh, J E, Khoo, L T 2013 *Journal of Ethnopharmacology* **150** 339–352.

14. Zulfiker, A H, Siddiqua, M, 2011 *International Journal of Pharmacy and Pharmaceutical Sciences* 3 198–203.
15. Tsai, J C, Peng, W H 2011 *The American Journal of Chinese Medicine* **39** 943–956.
16. Dat, B T Phat, N T 2015 *Phytochem Lett* **11** 102-105.
17. Tri, D M, Minh, P N 2016 *Phytochem Lett* **18** 10-13.
18. Trinh, P T N, Tri, M D, 2016 *Nat Prod Res* **30(7)** 761-767.
19. Trinh, P T N, Luan, N Q 2017 *Nat Prod Res* **31(15)** 1733-1738.
20. Tsuneatsu, N, Hideaki O 1985 *Phytochemistry* **24(12)** 2959-2962.
21. Sasai, H, Ishida M, 2009 *Biosci Biotechnol Biochem* **73(6)** 1333-8.
22. Calis, I, Akbay, P 1999 *Pharmazie* **54** 926-930.
23. Adam, K, Maria, L, 2012 *Food Chemistry* **133** 1373–1382.
24. Fumiko, Tatsuo, Y 1986 *Chemical & Pharmaceutical Bulletin* **34(10)** 4340-4345.

Effect of Heat treatment on microstructure and properties of Cu-Ni-Co-Si alloy

Lijun Peng [1*, 2], Jimiao Ma [1], Xingyu Liu [2], Feng Liu [1], Guojie Huang [2], Songbai Hong [1], Haofeng Xie [2], Zhen Yang [2]
1 NingBo XingYe Sheng Tai Group Co. Ltd, Provincial enterprise technology center, Zhejiang, NingBo, 315336, China
2 State Key laboratory of Nonferrous Metals and process, General Research Institute for Nonferrous Metals Engineering Institute, Beijing, 101400, China
E-mail: penglj@grinm.com

Abstract. In the paper, the effect of different solution and aging treatment on microstructure and properties of Cu-Ni-Co-Si alloy is investigated in detail. The results show that the recrystallization, grain growth and second phase solubilization are observed during the continuous solution treatment on line of Cu-Ni-Co-Si alloy. When the Cu-Ni-Co-Si alloy is solutioned at 970 ℃ and speed of annealing is 10 m/min, aged at 500 ℃ for 8 h, 40 % cold roll and aged at 325 ℃ for 12 h, it has the excellent comprehensive properties, which can meet the needs of high-end lead frame and connectors. The Disk-shaped precipitates have been confirmed as $(Ni, Co)_2Si$ phase. The crystal orientation relationship between matrix and precipitates is$(\bar{2}\bar{2}1)_p \parallel (200)_{Cu}$, $[102]_p \parallel [001]_{Cu}$.

1. Introduction

Age-hardening Cu-Ni-Si system alloys have high strength, high elasticity, excellent electrical conductivity and anti-stress relaxation, owing to the precipitation of the fine and uniform distribution of Ni-Si particles in the Cu-Ni-Si alloy during the aging process [1-5]. Therefore, there are many attractive candidates for connectors used in cars, intergrated circuit lead frame, which can be widely used in many fields of aerospace, automobile and electronic communication [6-9].

The effect of precipitation behavior and alloying element on microstructure and properties on Cu-Ni-Si alloy had been systematic invesigated by many researchers at home and abroad [10-21].The industrialization production of a typical Cu-Ni-Si alloy, C70250, had been realized in many contries, such as China, Japan, USA and so on. A new Cu-Ni-Co-Si alloy, C70350 alloy, had been developed by Olin company in USA, when some Co elements were in place of Ni in the Cu-Ni-Si alloy. Compared to C70250 alloy, The C70350 alloy has he excellent comprehensive properties, higher strong and

IOP Conf. Series: Materials Science and Engineering **544** (2019) 012030 doi:10.1088/1757-899X/544/1/012030

electrical conductivity, which can meet the needs of high-end elastic components.
In the paper, the effect of different heat treatments, solution and aging treatment, on the microstructure and properties of Cu-Ni-Co-Si alloy can be investigated in detail during the industrial production, which can provide theoretical foundation and operation instruction for the alloy.

2. Experimental

The chemical composition of cold rolled Cu-Ni-Co-Si alloy with average of 0.5 mm thickness was shown in **Table.1**. The cold rolled alloy was solutioned at different temperature and speed on the EBNER Vertical bright continuous withdrawal furnace. The different solution treatments were shown in **Table.2**. Then the specimens aged at 500 ℃ for 8 h. The aged strips were subsequent cold rolled at 30 % and 40 % and aged at 325 ℃ for 12 h. TEM samples were fabricated by double jet electropolishing techniques using a mixed solution (nitric acid: methanol =1:4) at about -30 ℃. The TEM samples are observed using a JEM 2100F transmission electron microcope with 200 kV. The hardness of these specimens was measured on an HXD-1000 testing machine. The tensile strength for aged alloys were tested on an MTS-810 testing machine for three measurements. The electrical conductivity for aged alloys were tested using a 7501 Eddy current conductometer with three measurements.

Table 1. The chemical composition of Cu-Ni-Co-Si alloy (wt.%)

Element	Cu	Ni	Co	Si
Conent	Bal.	1.4	1.13	0.53

Table 2. The different solution treatments process

process	parameters
I	T=930 ℃, v=8 m/min
II	T=970 ℃， v=8 m/min
III	T=970 ℃ ，v=10 m/min

3 Results and Discussion

The microstructure of Cu-Ni-Co-Si alloy after different continuous solution treatment on line is shown in **Figure.1**. The recrystallization, grain growth and second phase solubilization can be observed when the cold-rolled Cu-Ni-Co-Si alloy are solution treatment on line. The solution temperature and speed of anneal are higher, those phase transformation changes faster. When the speed of annealing and solution temperature are 8 m/min and 930 ℃, respectively, the recrystallization and second phase solubilization can be obviously found. The average grain size is about 15 μm and some grains size are about 50 μm, as shown in **Figure.1(a)**. When the solution temperature increases to 970 ℃, the average grain size are 15～30 μm and few of second phase can be observed, as shown in **Figure.1**(b). From the **Figure.1(c)**, the average grain size is about 15 μm and the grain uniformity is greatly improved when the solution temperature and speed of annealing are 970 ℃ and 10 m/min, respectively.

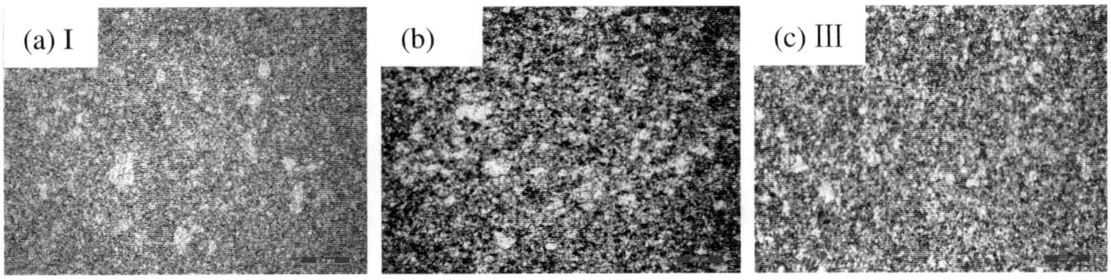

Figure.1 Microstructure of Cu-Ni-Co-Si alloy after different continuous solution treatment on line (Cooling fan speed is 1900rpm)

Physical properties of Cu-Ni-Co-Si alloy after different solution treatment on line is shown in **Figure.2**. It can be known in **Figure.2** that the tensile and yield strength increase by 51 MPa and 73 MPa, respectively, when the solution temperature is increased from 930 ℃ to 970 ℃. The elongation and electrical conductivity show little change. Because the crystallization, grain growth and second phase solubilization can be found during the solution treatment process. The process of crystallization and grain growth can promote the increasing of electrical conductivity and decrease the strength of alloy. But the second phase solubilization has the opposite effect. Therefore, the physical properties depend on the combined effort of these transformations. Based on these results, solution temperature is higher, the average grain size is larger and second phase is less, which can be increase the strength of alloy. Therefore, the intensity of cooling is increased and grain size uniformity is improved when the speed of annealing increase to 10 m/min, which decrease the strength and electrical conductivity, and increase elongation of Cu-Ni-Co-Si alloy.

Physical properties of Cu-Ni-Co-Si alloy after solution treatment and aged at 500 ℃ for 8 h is shown in **Figure.3**. From the **Figure.3**, the physical properties of solutioned alloy aged at 500 ℃ for 8 h show a great improve, and The comprehensive properties of process II and III are optimal.

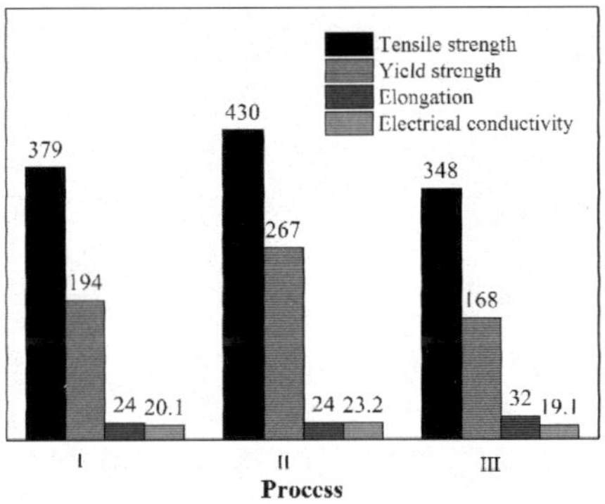

Figure.2 Physical properties of Cu-Ni-Co-Si alloy after different continuous solution treatment (Cooling fan speed is 1900 rpm)

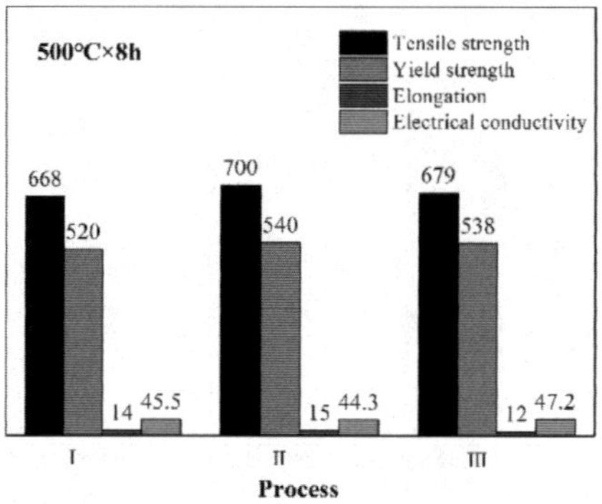

Figure.3 Physical properties of Cu-Ni-Co-Si alloy after solution treatment and aged at 500 ℃ for 8 h (Cooling fan speed is 1900 rpm)

The hardness curve of Cu-Ni-Co-Si alloy after solution treatment, first aged, cold rolled, and second aged are shown in **Figure.4**. The variation trend of hardness curve can be divided to three stages from the **Figure.4(b)**. At the initial aging stage, it shows a sharp increase in hardness when the aging temperature is 22-325 ℃. The second stage happens at about 325 ℃, the hardness increase gradually until it goes to the peak value (Hv 260). At the last aging stage, the hardness begins to decrease because of overaging. It can be known in **Table.3** that the tensile and yield strength, elognation, electrical conductivity and elastic modulus can be obtained, 839 MPa, 810 MPa, 2 %, 48.5 %IACS and 126 GPa, whicn can meet the needs of high-end lead frame.

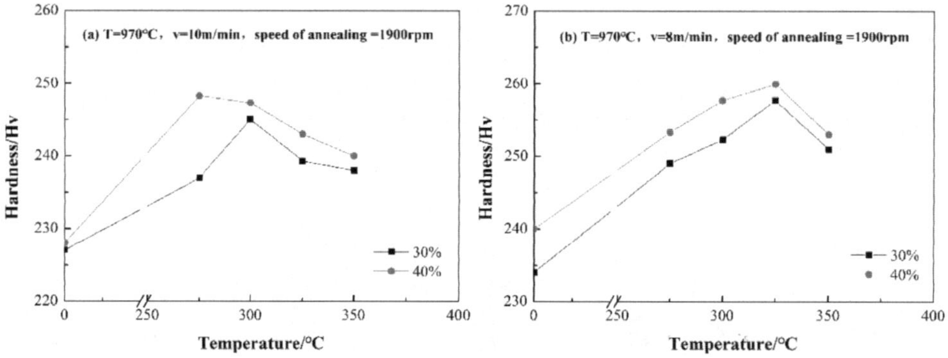

Figure.4 The hardness curve of different state Cu-Ni-Co-Si alloy aged at different temperature for 12 h

Microstructure and select electron area diffraction pattern image of Cu-Ni-Co-Si alloy after solutioned, aged at 500 ℃ for 8 h, 40 % cold-rolled, and aged at 325 ℃ for 12 h is shown in **Figure.5**. Lots of dislocation and precipitates with average grain size, 5 nm, are found in the **Figure.5(a)**. The increase in hardness is ascribed to the interaction between dislocation and precipitates. According to the SEAD in **Figure.5(b)**, these precipitates have been confirmed as $(Ni, Co)_2Si$ phase. The crystal orientation relationship between matrix and precipitates is $(\bar{2}\bar{2}1)_p \parallel (200)_{Cu}, [102]_p \parallel [001]_{Cu}$.

Table.3 Physical properties of different state Cu-Ni-Co-Si alloy

process	Deformation	Age treatment	Tensile strength /MPa	Yield strength /MPa	Elongation /%	Electrical conductivity %IACS	elasticity modulus /GPa
Process II +aged at 500 ℃ for 8 h	30 %	325 ℃×12 h	823	793	2	45.5	125
	40 %		839	810	2	48.5	126

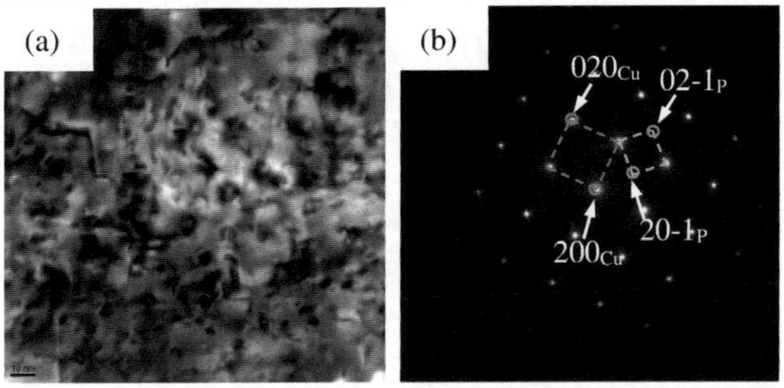

Fig.5 Microstructure and select electron area diffraction pattern image of Cu-Ni-Co-Si alloy after solutioned, aged at 500 ℃ for 8 h, 40 % cold-rolled, and aged at 325 ℃ for 12 h

4. Conclusion

The recrystallization, grain growth and second phase solubilization are observed during the continuous solution treatment on line of Cu-Ni-Co-Si alloy. When the Cu-Ni-Co-Si alloy is solutioned at 970℃ and speed of annealing is 10m/min, aged at 500 ℃ for 8 h, 40 % cold roll and aged at 325 ℃ for 12 h, it has the excellent comprehensive properties, i.e., tensile and yield strength, elognation, electrical conductivity and elastic modulus is 839 MPa, 810 MPa, 2 %, 48.5 %IACS and 126 GPa, respectively. The Disk-shaped precipitates have been confirmed as $(Ni, Co)_2Si$ phase. The crystal orientation relationship between matrix and precipitates is $(\bar{2}\bar{2}1)_p \parallel (200)_{Cu}, [102]_p \parallel [001]_{Cu}$.

Acknowledgement

Our work is supported by National Key R& D Program of China (**No.2016YFB0301300**) and the Natural Science Foundation of China (**No. 51601017**)

5. Reference

[1] Cheng J Y, Tang B B, Yu F X, Shen B 2014 *J. Alloys. Comp.* **614** 189.
[2] Lei Q, Li Z, Zhu A Y, Qiu W T, Liang S Q 2011 *Mater Character.* **62** 904.
[3] Jia Y L, Wang M P, Chen C, Dong Q Y, Wang S, Li Z 2013 *J. Alloys. Comp.* **557** 147.
[4] Monzen R, Watanable C 2008 *Mater. Sci. Eng. A.* **483-484** 117.
[5] Chenna Krishna S, Srinath J, Jha K, Sharma S C, Koshy M 2013 *J. Mater. Eng. Perform* **22** 2115.
[6] Lei Q, Li Z, Dai C, Wang J, Chen X, Xie J M, Yang W W, Chen D L 2013 *Mater. Sci. Eng. A.* **572** 65.
[7] Larbi E H, Azzeddine H, Baudin T, Mathon M H, Brisset F, Helbert A L, Kawasaki M, Bradai D, Langdon T 2015 *J. Alloys. Comp.* **638** 88.
[8] Semboshi S, Sato S, Iwase A, Takasugi T 2016 *Mater Character.* **115** 39.
[9] Wang W, Kang H J, Chen Z N, Chen Z J, Zou C L, Li R G, Yin G M, Wang T M 2016 *Mater. Sci. Eng. A.* **673** 378.
[10] Hu T, Chen J H, Liu J Z, Liu Z R, Wu C L 2013 *Acta Mater* **61** 1210.
[11] Eungyeong L, Seungzeon H, Kwangjun E, Kim S 2011 Metals. Mater. Inter. **17** 569.
[12] Suzuki S, Shibutani N, Mimura K, Isshiki M, Waseda Y 2006 *J. Alloys. Comp.* **417** 116.
[13] Seungwon L, Hirotaka M, Xavier S, Zenji H 2014 *Mater Character.* **90** 62.
[14] Azzeddine H, Mehdi B, Hennet L, Thiaudiee D, Alili B, Kawasaki M, Bradai, Langdon T G 2014 *Mater. Sci. Eng. A.* **587** 288.
[15] Abdel Y K, Faycal H L, Megumi K, Baudin T, Bradai D, Langdon T G 2013 *Mater. Sci. Eng. A.* **576** 149.
[16] Atapek S H, Pantelakis S G, Polat S, Chamos A N, Aktas G 2016 *Theor. App. Fra. Mech.* **83** 60.
[17] Lei Q, Xiao Z, Hu W P, Derby B J, Li Z 2017 *Mater. Sci. Eng. A.***697** 37.
[18] Chenna Krishna S, Srinath J, Jha Abhay K, Pant B, Sharma S C, George K M 2013 *J. Mater. Sci.* **22** 1357.
[19] Rzawski Z, Stobrawa J 1993 *Mater. Sci. Tech.* **9** 142.
[20] Kaneko H, Eguchi T 2012 *Mater Trans.* **53** 1847.
[21] Lei Q, Li Z, Wang J, Xie J M, Chen X, Li S, Gao Y, Li L 2013 *Mater. Des.* **51** 1104.

Effects of electrochemical chloride extraction on the bonding properties of corroded reinforced concrete by the anode of magnesium phosphate cement bonding carbon fiber-reinforced plastics (CFRP)

T Y Hao[1], H Lin[2,3], Y Li[2] and C Zhao[2]
(1 Central Research Institute of Building and Construction Co., Ltd, 100088, China)
(2 The Key Laboratory of Urban Security and Disaster Engineering, MOE. Beijing University of Technology, 100124, China)
E-mail:lin52hui90@126.com

Abstract. This paper studied the effects of electrochemical chloride extraction (ECE) on the bonding properties of corroded reinforced concrete with the anode of Magnesium phosphate cement (MPC) bond CFRP. The results indicated that ECE can efficiently remove the chloride ions in the concrete. The interfacial bonding properties of steel-concrete decreased with the increase of chloride removal current density. Proper corrosion ratio of steel bars can reduce the bonding strength loss caused by ECE treatment.

1. Introduction

The most commonly form of deterioration of reinforced concrete is corrosion of reinforcement. The greatest threat for the corrosion of reinforcement is corrosion due to chloride. The passivation film of the steel bar can be destroyed by chloride ions, resulting in the corrosion of steel bar. The expansive corrosion products leads the concrete structure cracking and peeling, thereby reducing its service life [1]. The conventional repair methods for the corroded reinforced concrete structure is to determine the corroding areas and then replace the chloride-contaminated concrete with the repair mortar [2]. However, after reinforcement, the residual chloride ions continue to cause the corrosion of steel bar and the reinforcement effects is weakened.

ECE is a newly developed nondestructive method for preventing the corrosion of steel bar in concrete [3]. During the ECE process, an electric field is applied between the steel bar (the cathode) and an external electrode (the anode) placed in an alkaline electrolyte around the surface of concrete.

[3] Corresponding author: H. Lin, E-mail:lin52hui90@126.com

Under the driving force of electric field, the negatively charged chloride ions migrates from the steel bar to the exterior electrode through the pores of concrete. Meanwhile, the cation migrates to the cathode, as shown in figure 1.

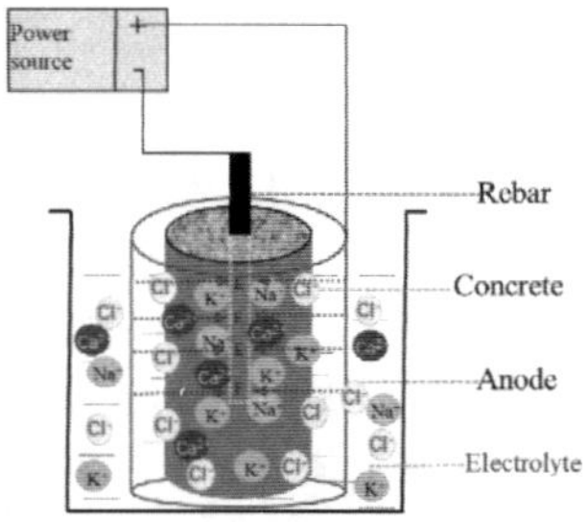

Figure 1 Schematic diagram of ECE

Currently, the commonly used ECE anode materials are mainly titanium mesh, wire mesh and cement-based conductive materials [4, 5]. In addition, for the corroded concrete structures, the method is ECE treatment first and then reinforcement, which is very intricate. Therefore, the integration method of chloride removal and strengthening was proposed. However, to achieve this method, the selection of the anode material is particularly important.

Magnesium phosphate cement (MPC) is a new inorganic cementitious material with excellent bonding strength, high early compressive strength, favorable volume stability and good corrosion resistance in low concentration alkaline solution [6, 7], which has been widely used in the field of repairment and reinforcement of building structures. CFRP is a good corrosion resistance, high specific strength and conductive material. Previous studies has confirmed the feasibility of using CFRP as anode for ECE treatment [8]. Therefore, MPC bonding CFRP (MPC-CFRP) is a potential composite anode material for ECE treatment.

This paper adopted MPC-CFRP as an ECE anode and studied the effect of ECE on the bonding properties of reinforced concrete with different corrosion degree after 28 days treatment in the $Ca(OH)_2$ electrolyte with the pH value 10 at three different current densities of 1.0 A/m^2, 2.0 A/m^2 and 3.0 A/m^2.

2. Experimental details

2.1 Raw materials

The MgO for the preparation of MPC directly calcined from magnesite at 1600℃, the used cement was P.O42.5, and their chemical compositions and physical properties are shown in table 1 and table 2, respectively. The coarse aggregates was continuous grading crushed stone with the particle size of 5~25mm. The fine aggregate was the siliceous sea sand with the chloride ions content of 0.113%. The superplasticizer was polycarboxylate superplasticizer with the water reduction rate of 31%. Tap water was used in concrete specimen and the deionized water was utilized to prepare the electrolytic solution. The involved chemical reagents, such as KH_2PO_4, $Na_2B_4O_7\cdot 10H_2O$, NaCl and $Ca(OH)_2$ were all chemically pure and the purity was larger than 99%. The round steel bar was plain carbon steel (Q235) with a diameter of 10mm. The morphology of the used CFRP is shown in figure 2. The thickness and surface density of CFRP are 0.19mm and 300g/m^2, respectively. Its ultimate tensile strength, elastic modulus and fracture strain are 4200 MPa, 210GPa and 0.02, respectively.

Table 1 Chemical composition and physical properties of calcined MgO

Sample	MgO (%)	CaO (%)	SiO_2 (%)	Al_2O_3 (%)	Fe_2O_3 (%)	Density (g/cm^3)	Bulk density(g/cm^3)	Specific surface area(cm^2/g)
MgO	91.7	1.4	1.6	4.0	1.3	3.46	1.67	805.9

Table 2 Chemical composition and physical properties of cement

Chemical composition (wt%)							Physical properties	
Loss	SiO_2	Fe_2O_3	Al_2O_3	CaO	MgO	SO_3	Specific gravity(g/cm^3)	Specific surface area(m^2/kg)
2.12	22.9	3.8	5.7	63.5	2.2	1.9	3.16	354

Figure 2 The morphology of CFRP

2.2. Mix proportion design

2.2.1. MPC. The existing indicated that the properties of MPC with the molar ratio of KH_2PO_4/MgO (P/M), borax (B) and water binder ratio of W/C were 1/4~1/5, 4~8% and 0.14~0.16 were relatively good [9]. The proportions and physical properties of MPC in this paper are shown in table 3.

Table 3 Materials proportions and physical properties of MPC

P/M	B	W/C	Compressive strength(MPa)				Porosity (%)	Bond strength with C30 concrete at 28days
			3h	1d	7d	28d		
1/4.5	5%	0.14	>20	>30	>40	>60	15~18	Broken in Concrete matrix

2.2.2. Concrete. The water-cement (W/C) ratio of the concrete is 0.57 and the materials proportions are presented in table 4. The chloride ions in sea sand, corresponds to 0.23% by weight of cement.

Table 4 Proportions and properties of concrete

Cement (kg/m^3)	Aggregate (kg/m^3)	Sand (kg/m^3)	W/C ratio	Water Reduce (kg/m^3)
325	1228	662	0.57	0.65

2.3. Laboratory studies

2.3.1. Preparation of concrete samples. The concrete was cast in a PVC mold with the size Ø100×200mm. A length of 250mm steel bar with the diameter of 10mm was centrally embedded in each sample and the embedded depth was 150mm. After demolding, the samples were cured 28 days at room temperature. Then, accelerated corrosion of the samples was conducted by the method of electrochemical technology, as shown in figure 3. The degrees of corrosion were set at 5%, 15% and

IOP Conf. Series: Materials Science and Engineering **544** (2019) 012031 doi:10.1088/1757-899X/544/1/012031

30% using Faraday's law. The corroded samples are shown in figure 4. Subsequently, MPC paste was spread evenly on the CFRP. Then the concrete sample was put on the MPC paste and rolled gently until completely wrapped, as shown in figure 5. After curing at room temperature for 7 days, all the samples were used to ECE treatment.

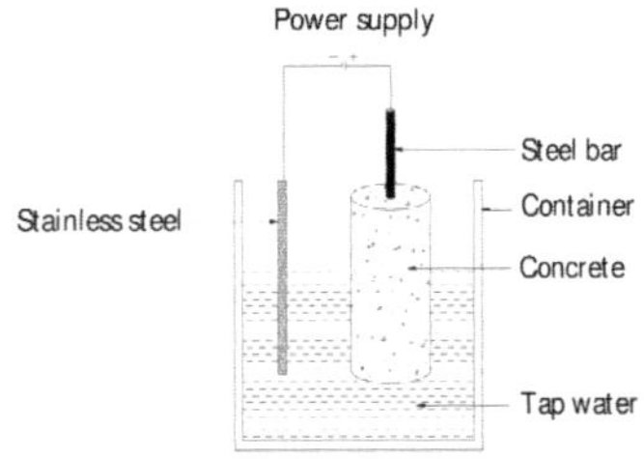

Figure 3 Electrochemical accelerated corrosion

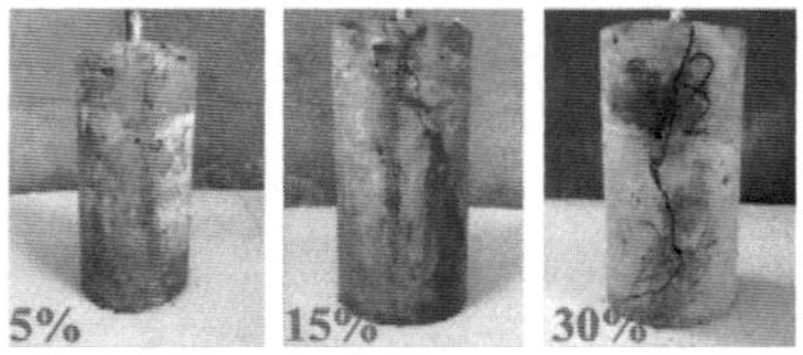

Figure 4 The corroded concrete samples

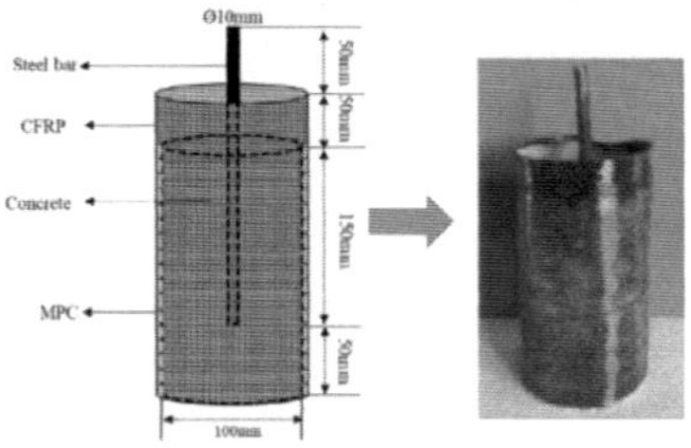

Figure 5 Concrete sample

2.3.2 ECE treatment. The ECE device is shown in figure 6. Steel bar is the cathode, MPC-CFRP is the anode. To prevent the chlorine overflow from the $Ca(OH)_2$ electrolyte, the electrolyte is renewed every two days and the chloride ions concentration re-confirmed. In order to eliminate the influence of temperature change, the temperature of electrolyte was kept at 25°C by a heating rod. In addition, to prevent the evaporation of electrolyte, cling film covered the opening of the container. The current density (*i*) was applied by a DC regulated power supply.

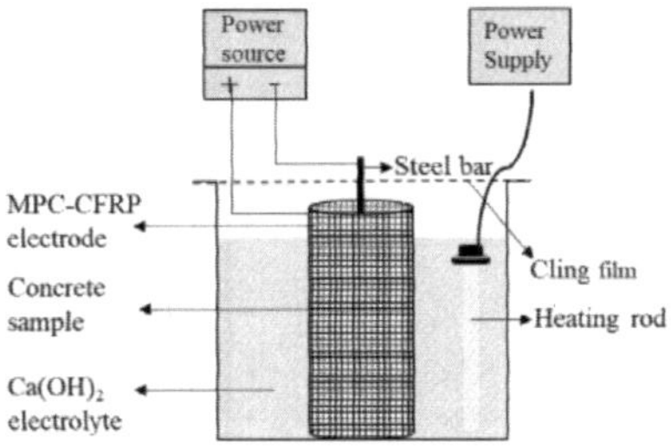

Figure 6 Schematic diagram of ECE device

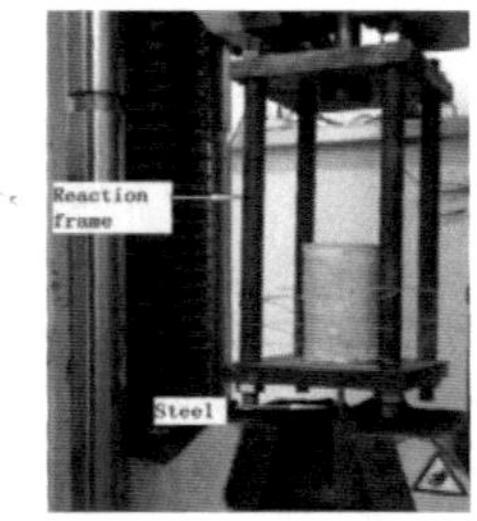

Figure 7 Experimental setup for pull-out tests

2.3.3 Pull-out test of steel bar. Interfacial bond properties were tested by pull-out test. Universal testing machine and reaction frame were used to induce tension on both ends of the column with the loading speed of 0.05MPa/min, as shown in figure 7.

3. Results and discussion

3.1 Chloride removal efficiency

After ECE treatment, the cumulative chloride removal efficiency of reinforced concrete calculated by equation (1) & (2) and the results are presented in Fig. 8.

$$M = V_S \times M_L \times W \tag{1}$$

$$Cl^{-}\% = \frac{M_{Cl^{-}} \times \sum C_i \times V_i}{M} \times 100\% \tag{2}$$

Where, M is the initial chloride ions content of the reinforced concrete. V_s is the volume of each sample. M_L is the cement content ratio of each sample. W is the mass ratio of chloride ions and cement. M_{Cl} is the molar mass of chloride ions. C_i is the chloride ions concentration of electrolyte at two-day intervals. V_i is the volume of the renewed electrolyte.

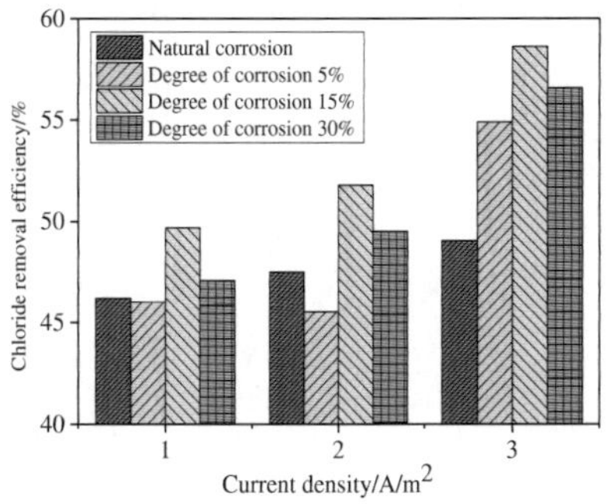

Figure 8 Chloride removal efficiency

Figure 8 indicated that the chloride removal efficiency of concrete with different corrosion degree increased with the increase of current density. The chloride removal efficiency of concrete with the corrosion rate of 15% was relatively good at all current densities. When the current density was 3A/m^2, the chloride removal efficiency was up to 58.62%.

3.2 Pull-out test of steel bar

After ECE treatment, the pull-out test of steel bar in different corrosion degree concrete is shown in figure 9. The bond strength loss of steel-concrete was calculated using equation (3) and the results are shown in table 5.

$$\delta_{loss} = \frac{f_{max} - f_0}{f_{max}} \times 100\% \tag{3}$$

Where, δ_{loss} is the bond strength loss. f_{max} is the maximum bond load of the natural corrosion. f_0 is the maximum bond load of the concrete with the degree of corrosion of 5%, 15% and 30%.

IOP Conf. Series: Materials Science and Engineering **544** (2019) 012031 doi:10.1088/1757-899X/544/1/012031

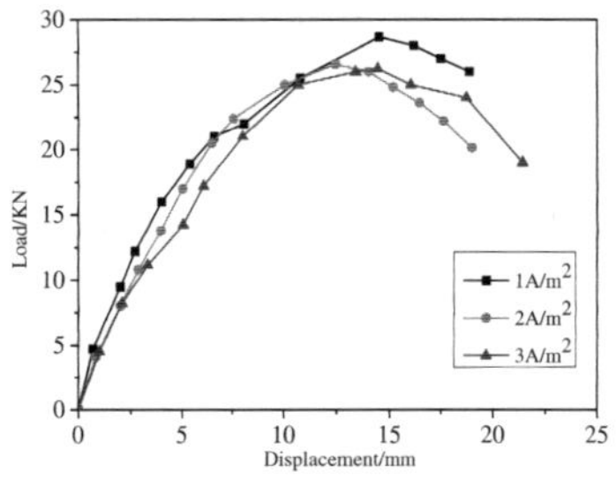

(a)

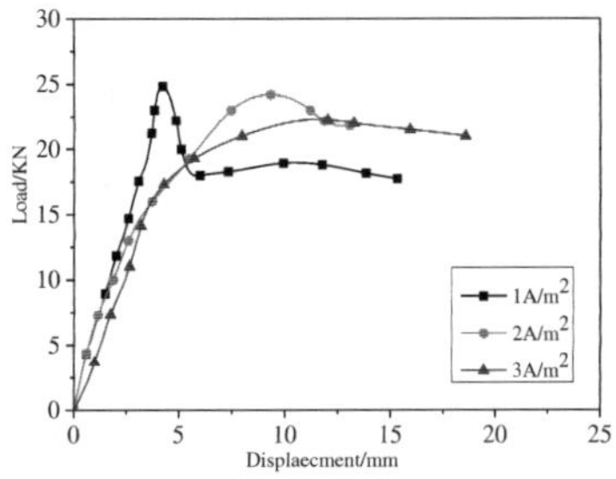

(b)

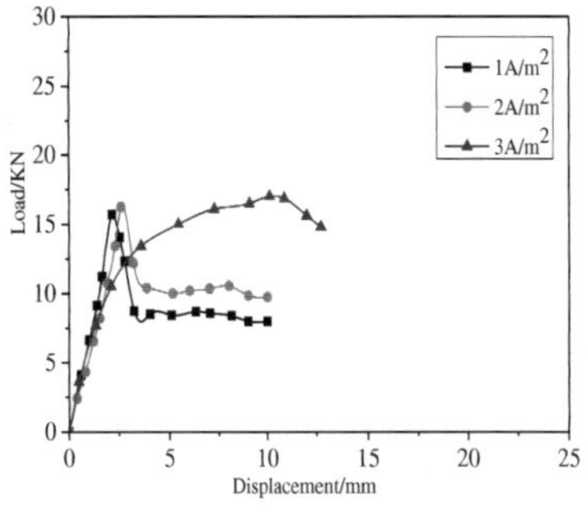

(c)

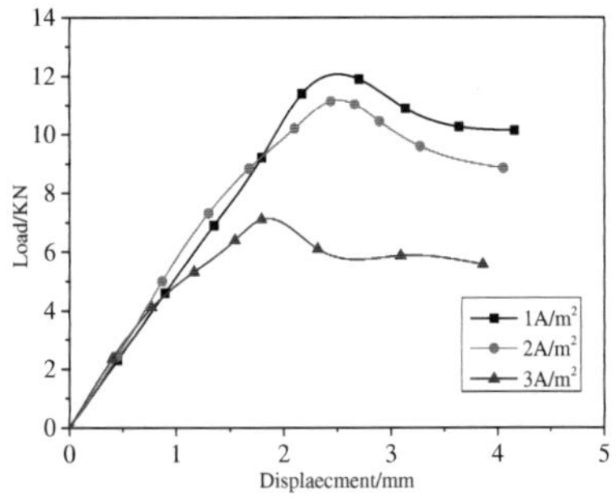

(d)

Figure 9 Load-displacement curves of pull-out test of steel bar:((a) Natural corrosion, (b) The corrosion degree of 5%, (c) The corrosion degree of 15%, (d) The corrosion degree of 30%.)

Table 5 The maximum bond load and bond strength loss of steel-concrete

Corrosion degree	*i* (A/m^2)	*fmax* (KN)	δ_{loss} (%)
Nature corrosion	0	30.2	—
	1	28.6	5.2
	2	26.5	12.1
	3	26.2	13.3
5%	0	25.6	—
	1	24.8	1.6
	2	24.2	5.5
	3	22.2	13.1
15%	0	17.5	—
	1	15.6	10.4
	2	16.2	7.1
	3	17.0	2.8
30%	0	12.9	—
	1	11.9	7.7
	2	11.1	13.9
	3	7.1	44.8

The pull-out curves and bond strength loss of the concrete with different corrosion degree are presented in figure 9 and table 5. The results indicated that the maximum bond load of steel-concrete without ECE treatment decreased with the increase of corrosion. After ECE treated, the bonding strength losses of the samples increased significantly with the increase of current density. When the current density was 3A/m^2, the bond strength losses of the concrete with natural corrosion, 5% corrosion degree and 30% corrosion degree were 13.31%, 13.11% and 44.8%, respectively. However, when the corrosion ratio of the sample was 15%, the bonding strength loss decreased with the increase of current density. The bond strength loss was decreased from 10.4% to 2.8% when the current density increased from 1A/m^2 to 3A/m^2. The results indicated that proper corrosion degree of 15% could reduce the interfacial bonding strength loss caused by ECE treatment.

4. Conclusion

(1) ECE treatment can efficiency remove the chloride ions in the reinforced concrete and achieve the wide application of sea sand by proper ECE parameters.

(2) ECE treatment has negative effects on the interfacial bond properties of steel-concrete. Proper corrosion degree of 15% can reduce the bonding strength loss caused by ECE treatment.

Acknowledgement

The author would like to acknowledge the financial support provided by Open Foundation of National accredited Enterprise Technology Center of China Jingye Engineering Technology Co., Ltd (No. JZA2017Kj02) and National Natural Science Foundation of China (No.51678011)

References

[1] Shi X M， Xie N， Fortune K and Gong J 2012 *J. Constr. Build. Mater*, **30** 125-38.

[2] Orellan J C, Escadeillas G and Arliguie G 2004 *J. Cem. Concr. Res*, **34(2)** 227-34.

[3] Cañón A, Garcés P, Climent M A, Carmona J and Zornoza E 2013 *J.* Corr. Sci, **77** 128-34.

[4] Saraswathy V, Lee HS, Karthick S and Kwon SJ 2018 *J. Constr. Build. Mater*, **158** 549-62.

[5] Jin Z Q, Hou D S, Zhao T J 2018 *J. Constr. Build. Mater*, **173** 149-59.

[6] You C, Qian J S, Qin J H and Wang H T 2015 *J. Cem. Concr. Res*, **78** 179-89.

[7] Bayuseno A P and Schmahl W W 2018 *J. J. Crys. Grow*, **498** 336-45.

[8] Sun H F, Wei L L, Zhu M C and Zhu J H 2016 *J. Constr. Build. Mater*, **112** 538-46.

[9] Xu B W, Ma H Y and Li Z J 2015 **J.** *Cem. Concr. Res*, **68** 1-9.

IOP Conf. Series: Materials Science and Engineering **544** (2019) 012032 doi:10.1088/1757-899X/544/1/012032

A finite element analysis on compressive properties of ECC with PVA fibers

C Y Jin[1], Y Q Li[2,3], J C Zhu[2], Y Li[2].
1. College of Applied Sciences, Beijing University of Technology, 100124, China
2. The Key Laboratory of Urban Security and Disaster Engineering, MOE. Beijing University of Technology, 100124, China
E-mail:yaqiangli@163.com

Abstract. In order to analyze the compressive performance of ECC with PVA by finite element method, in this paper the three-dimensional distribution model of PVA fiber was established by compiling the random sequence program of Matlab considering the random distribution of PVA in matrix ECC. On this basis, the finite element model of ECC was established by ABAQUS software to study the stress and deformation properties of ECC with different PVA fiber content under axial compression.

1. Introduction
Engineered Cementitious Composites (ECC) with strain hardening characteristics were developed by using cement-based cementitious materials, fine sand and PVA fibers[1, 2]. Under tensile and bending loads, multi-cracks was appeared and the maximum crack width could be controlled within 0.1 mm, which could effectively improve the structural resistance, ductility, toughness and energy absorption capacity[3, 4]. In recent years, the researches of ECC materials were mainly focused on macroscopic tensile properties[5-7]. The research of mesoscopic numerical simulation was relatively few. In this paper, ECC was simplified as cement mortar matrix and fiber two-phase composite material, and the mesoscopic numerical model was established based on Matlab software. Then the finite element model was established by ABAQUS, and the stress and deformation performance of ECC with different PVA fiber content under axial compression was studied.

2. Establishment of randomly distributed PVA fiber model
Based on the required dosage, the randomly distributed of PVA fibers in the ECC matrix can be realized by compiling the Matlab random sequence program. The length and diameter of the PVA fiber

[3] Corresponding author: Y Q Li, E-mail:yaqiangli@163.com

were 12mm and 4μm, respectively. The ECC specimen was adopted cube with 13mm in side length. The fiber numbers can be calculated by Equation 1. The results were shown in table 1.

$$N = \frac{4 \times V \times V_f}{100 \times \pi \times d^2 \times L} \quad (1)$$

Where V was the volume of specimen (mm³), V_f was the volume fraction of PVA fiber (%), L was the length of the fiber (mm), D was the diameter of the fiber (mm).

Table 1. The number of fiber in different volume content

V_f (%)	0.5	1.0	1.5	2
Number	766	1533	2299	3066

The coordinates of fibers produced in Matlab were imported into Autocad. In addition, the model was stored in the IGES format which can be recognized by Abaqus. Finally, the IGS file was imported into Abaqus to generate the finite element model. The ECC finite element model with 1.5% fiber content was shown in figure1.

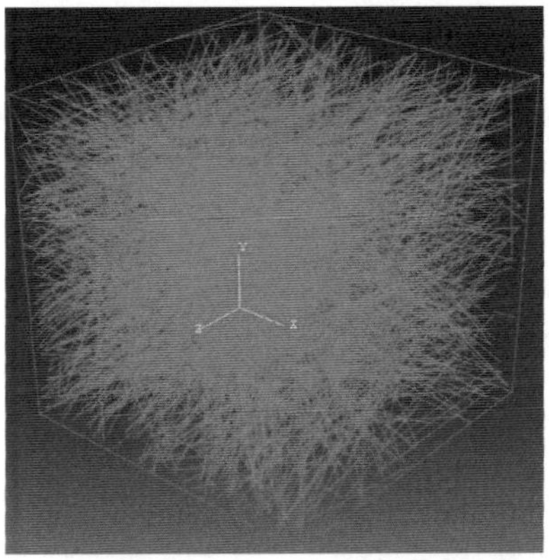

Figure 1. ECC finite element model with 1.5% fiber content

3. Constitutive model of concrete and PVA fiber

The plastic damage model of concrete can well describe the damage change law of cement-based materials. Therefore, the elastic-plastic damage model was adopted in the constitutive model of mortar in this paper. The standard of GB50010-2010 provided a classical plastic damage constitutive relation for concrete, which was widely used in cement-based materials. The stress-strain curves of cement mortar under uniaxial compression can be determined according to equation 2.

$$\sigma = (1 - d_c)E_c\varepsilon \quad (2)$$

$$d_c = \begin{cases} 1 - \frac{\rho_c n}{n-1+x^n} & x \le 1 \\ 1 - \frac{\rho_c}{\alpha_c(x-1)^2+x} & x > 1 \end{cases} \quad (3)$$

$$\rho_c = \frac{f_{cr}}{E_c \varepsilon_{cr}} \quad (4)$$

$$n = \frac{E_c \varepsilon_{cr}}{E_c \varepsilon_{cr} - f_{cr}} \quad (5)$$

$$\alpha_c = 0.157{f_{cr}}^{0.785} - 0.905 \quad (6)$$

$$x = \frac{\varepsilon}{\varepsilon_{cr}} \quad (7)$$

Compared to concrete, mortar has better ductility[9]. The peak strain can be calculated by equation 8.

$$\varepsilon_{cr} = 4.75 \times (700 + 172\sqrt{f_{cr}}) \times 10^{-6} \quad (8)$$

Where, d_c was the damage evolution parameter under uniaxial compression; $\alpha_c$was the descending section parameter of stress-strain curve; f_{cr} was the uniaxial compressive strength value; ε_{cr} was the strain corresponding to peak compressive stress.

The stress-strain curves of cement mortar under uniaxial tension can be determined according to equation 9.

$$\sigma = (1 - d_t)E_c\varepsilon \quad (9)$$

$$d_t = \begin{cases} 1 - \rho_t(1.2 - 0.2x^5 & x \le 1 \\ 1 - \frac{\rho_t}{\alpha_t(x-1)^{1.7} + x} & x > 1 \end{cases} \quad (10)$$

$$\rho_t = \frac{f_{tr}}{E_c \varepsilon_{tr}} \quad (11)$$

$$\alpha_t = 0.312{f_{tr}}^2 \quad (12)$$

$$x = \frac{\varepsilon}{\varepsilon_{tr}} \quad (13)$$

The peak strain can be determined by equation 14.

$$\varepsilon_{tr} = f_{tr}^{0.54} \times 65 \times 10^{-6} \quad (14)$$

Where, d_t was the damage evolution parameter under uniaxial tension; α_t was the descending section parameter of stress-strain curve; f_{tr} was the uniaxial tension strength value; $\varepsilon_{tr}$was the the strain corresponding to peak tension stress.

The mechanical parameters of mortar matrix were shown in table 2.

Table 2. Mechanical parameters of mortar

Mechanical parameters	Elastic modulus/GPa	Compressive strength/Mpa	Tensile strength/Mpa
Mortar	16.2	41.2	2.4

Elastic-plastic model was used for the constitutive relation of PVA fiber. In this study, the elastic modulus of the PVA fiber was 42.8 GPa, and the yield strength was 1600 Mpa.

4. Finite element analysis result

Damage nephograms of mortar with 1.5% fiber content under different steps were shown in figure 2.

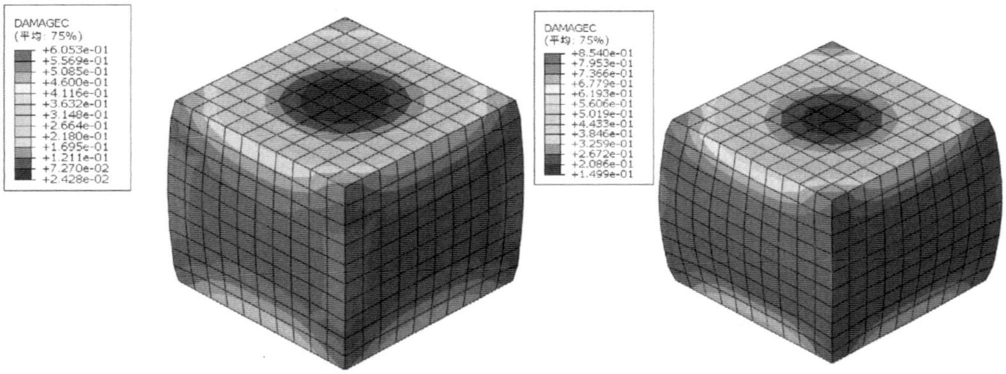

(a) 15 steps (b) 45 steps

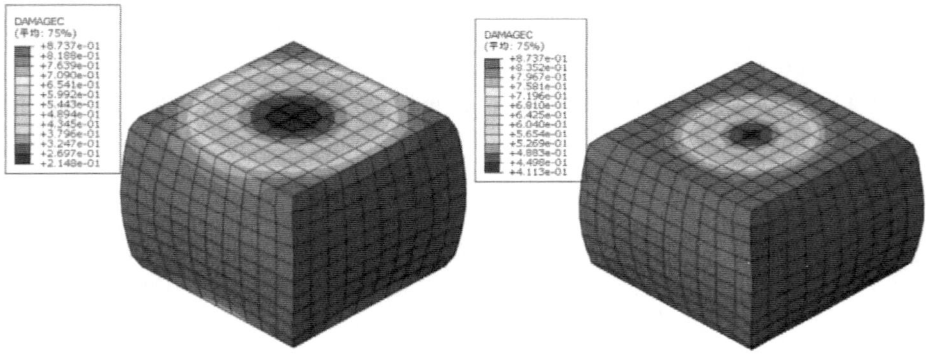

(c) 60 steps (d) 80 steps

Figure 2. Damage nephograms of mortar under different steps

From figure 2, it can be seen that with the increase of steps, the damage of mortar increased gradually. The stress-strain curve was shown in figure 3.

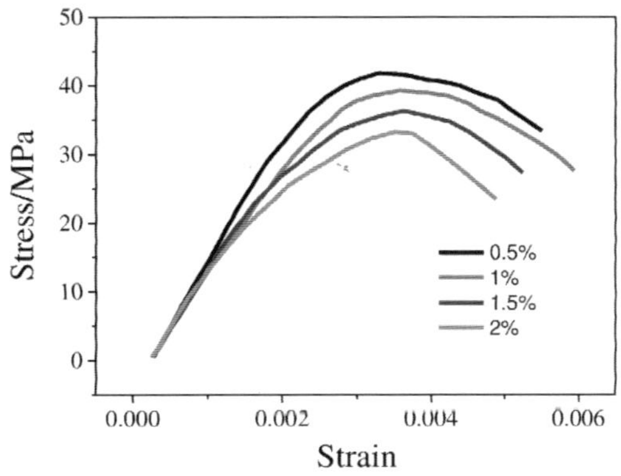

Figure 3. Stress-strain curve

From figure 3 , it can be seen that with the increase of fiber content, the peak load of samples decreased gradually. With the increase of fiber content, the strain corresponding to the peak stress of the samples increased gradually, which indicated that PVA fiber improved the deformability of mortar matrix.

5. Conclusions

(1) By compiling the random sequence program of MATLAB, the mesoscopic models of ECC material with different PVA fiber contents were established according to the content of PVA fibers. Using ABAQUS finite element software, the compression performance of ECC can be simulated using ABAQUS finite element software.

(2) According to stress-strain curve, with the increase of fiber content, the peak load of samples decreased gradually. With the increase of fiber content, the strain corresponding to the peak stress of the samples increased gradually, which indicated that PVA fiber improved the deformability of mortar matrix.

Acknowledge

The authors would like to acknowledge the financial support provided by National Key R&D Program of China-Key materials and preparation technology of high crack resistant ready-mixed concrete(2017YFB0310100), National Natural Science Foundation of China (51678011); Beijing Natural Science Foundation (8162005).

References

[1] Mohamed Maalej I, Victor C. Li Z M A 1994. *J. mater. Civil Eng*, **6**:513-528.
[2] VICTOR C. LI D K 1995. Mater. Struct, **28**: 586-595.
[3] Li V C K T 1998. Materials in Civil Engineering,**10**(2): 66-69.
[4] Patricia Kim Nelson V C L 2002. *J. mater. Civil Eng*, **14**: 384-391.
[5] Zhang Jun, Ju Xianchun, Guo Zili 2009. J. Build. Mater, **12**(6), 706-710.
[6] Pang Chaoming, LEUNG C K Y, Sun Wei. J 2009. Chin. Ceram. Soc, **37**(12), 2071-2077.
[7] Li Yan, Liu Zejun, Liang Xingwen 2013. Eng. Mech, **30**(1), 322-330.
[8] Kaushik H B, Rai D C, Jain S K 2007. *J. mater. Civil Eng*,**19**(9):728-739.

The thermal conductivity of the variband crystals of the Bi_x-Sb_{1-x} system in the semiconductor region of the composition at the temperature of about 100 K

Bochegov V.I., Melnikov N.A.
Kurgan State University, 640000, Kurgan, Russia
E-mail: max_vib@mail.ru

The previously observed experimental difference in the thermal conductivity of gradient-inhomogeneous (variband) single-crystal samples of the Bi_x-Sb_{1-x} system in the case of parallel and antiparallel directions of the temperature gradients and composition in the semimetallic range of composition values (0.03 <x <0.10) under similar temperature conditions was also experimentally confirmed in other samples of this material having a composition in the interval of the semiconductor region ($0.11 < x < 0.19$). A theoretical model of the possible mechanism of this effect for the semiconductor region of the composition is proposed.

It is customary to refer to semiconductor materials with a variable in volume or in the direction of the forbidden band as variband. The author offers a broader interpretation of the concept of a variband material, namely: it is a material in which the position of the actual energy extrema of charge carriers on the energy scale changes with at least one spatial coordinate in it.

An experimentally observed dependence of the thermal conductivity κ on the mutual direction of the temperature gradient $\vec{\nabla} T$ and the concentration gradient $\vec{\nabla} x$ of one of the components in a single-crystal sample of the Bi_x-Sb_{1-x} binary system was presented in [1]. The composition of this sample changed monotonically with the longitudinal coordinate l (measured along the heat flux) in the range of $x \approx 0.97$ at $l = 0$ to $x \approx 0.90$ at $l = 15$ mm. The bulk of this interval of values of x corresponds to a semimetallic state, which is characterized by the presence of a non-direct overlapping of electronic and hole extrema in the space of the wave vector.

The same effect was experimentally confirmed in another variband sample of the same length l = 15 mm with a composition change from $x \approx 0.885$ at one end to $x \approx 0.805$ at the other. This composition interval corresponds to the semiconductor state of the material. The composition of the Bi-Sb system in all figures and in the tables of this work is estimated using the percentage of antimony in the system $C=(1-x) \cdot 100\%$, not bismuth. Evaluation of the composition with the help of parameter C is more common for researchers involved in this system. In this case, in the investigated sample, $C(x=0)$ = 11.5at%, and at l=15mm $C(l)$ = 19.5at%. Figure 1 shows a fragment of the composition distribution along the length of a single-crystal ingot, corresponding to the portion of the sample taken from this ingot. The dots indicate the values of the composition measured by the X-ray fluorescence method, and the line indicates spline interpolation

We used the technology of obtaining the variband single crystals of the Bi-Sb system described in [2]. The thermal conductivity of this sample was measured by a direct method according to the procedure of [3] with an error not exceeding 5%, as in the previous case [1].

The energy spectrum of charge carriers at these values of composition is simpler than in the previous case [1] (fewer actual extrema in the valence band and the conduction band) [4]. Fig. 2 shows the change in the band spectrum of the charge carriers of a given material (homogeneous in volume) with a change in the composition of the samples at three different temperatures. The data for the formation of graphs in Fig. 2, previously not published, were kindly provided to us by the authors of [4,5] and partially were obtained by the author of this work personally. In this figure, the semiconductor region of the material composition is distinguished by background fill at three different temperatures.

If we consider the graphs in Fig. 2, at least on a qualitative level, it can be concluded that the band spectrum of a given material depends both on the composition of C and on the temperature T. As we know, this conclusion is not disputed by the scientific community.

Since this sample is mainly semiconductor, in order to elucidate the mechanism of this new effect, the author has found it reasonable to apply the foundations of the theory of thermal

IOP Conf. Series: Materials Science and Engineering **544** (2019) 012033 doi:10.1088/1757-899X/544/1/012033

conductivity of semiconductors. It is generally accepted that the thermal conductivity κ of a semiconductor consists of three components:

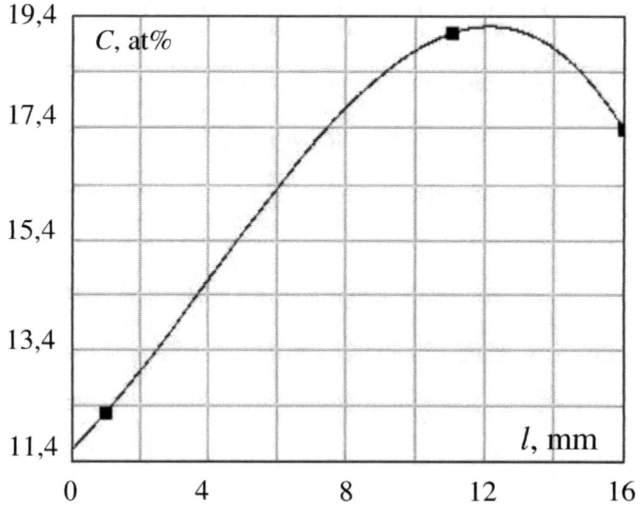

Figure 1. Dependence of the concentration of antimony C on the longitudinal coordinate l in the sample under study.

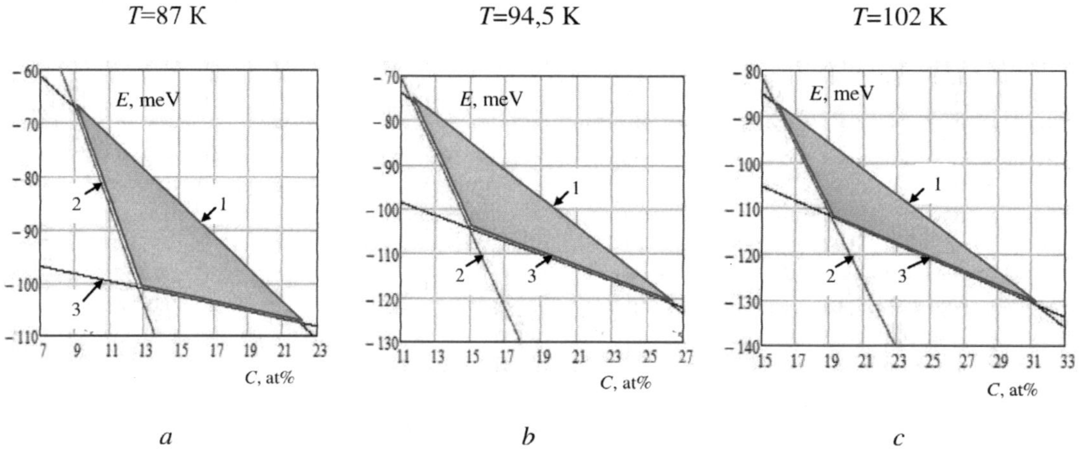

Figure 2. The change in the band spectrum in the Bi_{100-C} - Sb_C system with the change in the parameter C. Number 1 corresponds to the motion line of the extremum L of the conduction band, number 2 corresponds to the T-extremum of the valence band, and number 3 corresponds to the H-extremum of the valence band. The L-extremum of the valence band existing in the system is not shown here, since it is well below the forbidden energy interval and is not relevant.

$$\kappa = \kappa_f + \kappa_e + \kappa_b, \tag{1}$$

where κ_f - phonon, κ_e - electronic, and κ_b - bipolar components of thermal conductivity. In what follows we shall use the notation $\kappa_{el} = \kappa_e + \kappa_b$, so

$$\kappa = \kappa_f + \kappa_{el} \qquad (1a')$$

At the same time, based on the values of the thermal conductivity measured outside the magnetic field and in the transverse magnetic field B up to $B \approx 1200$ mTl, we can separate the phonon thermal conductivity and the sum of the electron and bipolar components of the thermal conductivity (see Table I).

Table 1.

κ, Wt/m·K							
$\vec{\nabla}T \uparrow\uparrow \vec{\nabla}C$, $T_0 = 89{,}4\text{K}$, $T_l = 102{,}5\text{K}$. Heat flux density q=1,907 mWt/mm^2.				$\vec{\nabla}T \uparrow\downarrow \vec{\nabla}C$, $T_0 = 101{,}5\text{K}$, $T_l = 87{,}6\text{K}$. Heat flux density q=1,798 mWt/mm^2.			
Experiment			Calculation	Experiment			Calculation
κ	κ_{fon}	κ_{el}	κ_{el}	κ	κ_{fon}	κ_{el}	κ_{el}
4,526	2,344	2,182	2,181	4,108	2,356	1,752	1,944

From the data of Table I it is seen that the phonon component is practically independent of the mutual direction of the composition and temperature gradient (the difference in the third significant digit). It follows that the mechanism of the observed effect should be sought in the theory of the electronic component of thermal conductivity $\kappa_{el} = \kappa_e + \kappa_b$, the electronic part κ_e, which is determined only by thermal diffusion and the bipolar part κ_b associated with the generation (recombination) of electron-hole pairs in the process of thermal diffusion [6, 8]. According to publications [6, 8], for the total electronic thermal conductivity of a semiconductor the following is valid:

$$\kappa_{el} = T \cdot L \cdot \left\{ \begin{array}{l} \left(\sigma_p + \sigma_n\right) \cdot A + \\ + \left[E_g / kT + (r+2) \cdot (F_{(r+1)}(\mu_p) / F_{(r)}(\mu_p) + F_{(r+1)}(\mu_n) / F_{(r)}(\mu_n) \right]^2 \times \\ \times \sigma_p \cdot \sigma_n / \left(\sigma_p + \sigma_n\right) \end{array} \right\}$$

or

$$\kappa_{el} = T \cdot \left[\left(\sigma_p + \sigma_n\right) \cdot A \cdot L + (\alpha_p - \alpha_n)^2 \cdot \sigma_p \cdot \sigma_n / \left(\sigma_p + \sigma_n\right) \right]. \qquad (2)$$

Here T - the temperature, $L = (k/e)^2$ - the Lorentz number, k - the Boltzmann constant, σ_p and σ_n - the partial conductivity of holes and electrons, respectively, α_p and α_n the partial thermoelectric power of holes and electrons, respectively, E_g - the width of the forbidden band, r - is an indicator of the degree of the dependence of the free path length of charge carriers on energy, $\mu_{n,p}$ - presented charge carrier chemical potential (of the electron or hole), $F_k(\mu_i)$ - the Fermi integral of degree k with respect to the energy of the i-th carrier. The dimensionless coefficient A in the expressions (2) is as follows:

$$A = \left[\frac{r+3}{r+1} \cdot \frac{F_{r+2}(\mu)}{F_r(\mu)} - \frac{(r+2)^2}{(r+1)^2} \cdot \frac{F_{r+1}^2(\mu)}{F_r^2(\mu)} \right] \qquad (3)$$

To calculate the chemical potential of electrons and holes on the basis of the principle of electroneutrality, the value of the ratio of the effective mass in the valence band to the effective mass in the conduction band is necessary. The meaning of this value varies from 5 to 13 in different authors. This is especially true for the range of values of C, in which the value of Eg decreases with increasing C, and where the extremum in the direction of the point H of the Brillouin zone is actual. For this

IOP Conf. Series: Materials Science and Engineering **544** (2019) 012033 doi:10.1088/1757-899X/544/1/012033

extremum, the dispersion law for holes is poorly studied. We used the value $m_p/m_n \approx 7$. The topology of the Brillouin zone of the Bi-Sb system can be found in [7].

Using the experimental data of conductivity $\sigma(T,C)$ and thermopower $\alpha(T,C)$ obtained by the authors of [4,5], as well as personally obtained by the author of this work, the heat conductivity of the material was calculated. The calculation was carried out by numerical methods in the approximation $r = 0$, which corresponds to the preferential scattering of charge carriers on the neutral impurity.

Fig. 3 shows in the polynomial approximation the graph of the dependence on the temperature T and composition C of the value of the forbidden band, referred to the energy of thermal motion, $\eta = E_g(T,C)/k\cdot T$ and Fig. 4, also in the polynomial approximation, shows the graph of the dependence on the temperature T and the longitudinal coordinate l for the calculated thermal conductivity $\kappa_{el}(T,l) = \kappa_{el}(T,C(l))$.

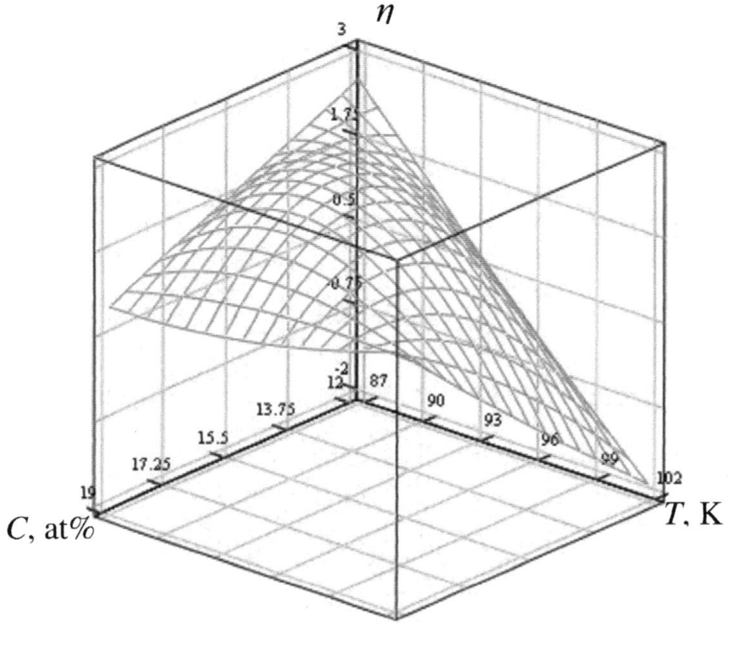

Figure 3. The graph of the dependence of the presented width of the the forbidden band on the temperature T and the amount of antimony C in the $Bi_{100\text{-}C}$ - Sb_C system.

Table 2. The ratio of electron (together with bipolar) thermal conductivity κ_{el} at $\vec{\nabla}T \uparrow\uparrow \vec{\nabla}C$, $T_0 = 89{,}4\text{K}$, $T_l = 102{,}5\text{K}$ to the electronic thermal conductivity (together with bipolar) at $\vec{\nabla}T \uparrow\downarrow \vec{\nabla}C$, $T_0 = 101{,}5\text{K}$, $T_l = 87{,}6\text{K}$.

Experiment	Calculation
1,245	1,123

Using the obtained dependence $\kappa(T,l)$, the temperature is calculated at equidistant fifteen points along the entire length of the sample by numerically solving the differential equation of the heat balance of the type:

$$dT/dl = \kappa(l,T)^{-1} \cdot q\,. \qquad (4)$$

On the basis of the tabulated temperature, the heat conductivity averaged over the length of the sample is calculated for the case $\vec{\nabla}T \uparrow\uparrow \vec{\nabla}C$ and $\vec{\nabla}T \uparrow\downarrow \vec{\nabla}C$ case (see Table I and Table II). The calculated difference in thermal conductivity in these two cases turned out to be noticeably less than that obtained with direct measurement. Such a discrepancy between the measured and the calculated difference in thermal conductivity when the direction of the heat flow changes in very close temperature intervals is, probably, connected with the fact that the smooth polynomial approximation of the function *Eg (T, C)* was used for calculation. In fact, judging by Fig. 2 this function in parameter *C* should consist of a combination of two approximately rectilinear functions with a jump in the value and a sign of the angle of inclination at the point of the largest value of Eg. The underestimation of the result of the calculation could also be affected by ignoring the dependence for the ratio of the effective masses of the m_p/m_n carriers on the composition (coordinates) and the inaccuracy of its absolute value.

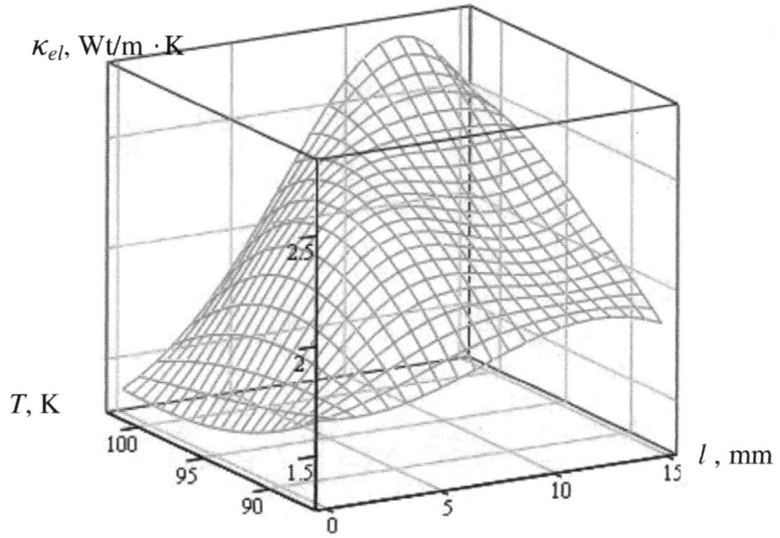

Figure 4. Dependence of the calculated electronic thermal conductivity κ_{el} (together with the bipolar one) on the temperature *T* and on the longitudinal coordinate *l* in the sample under study.

It should be noted that the author also carried out calculations for the case when the dependence of the width of the forbidden band on temperature was ignored, and only the dependence on the composition was preserved. In this case, the calculation gave an almost complete coincidence of the thermal conductivity before and after the change in the direction of the heat flux in a given temperature interval. The difference in this case was the third and even the fourth significant figure. It follows that the experimentally observed effect of the change in the thermal conductivity when the direction of the heat flux changes in the $Bi_x - Sb_{1-x}$ system with a variable volume x is determined by the fact that the energy spectrum of the charge carriers depends both on the composition and on the temperature, and the phonon component of thermal conductivity does not contribute to this effect.

IOP Conf. Series: Materials Science and Engineering **544** (2019) 012033 doi:10.1088/1757-899X/544/1/012033

References

1. V.I. Bochegov, V.M. Grabov. On the thermal conductivity of the gradient-inhomogeneous branches of thermoelements at a difference in the operating temperature. //Semiconductors. July 2017, Vol. 51, Issue 7, P. 874-875.
2. V.I. Bochegov, V.M. Grabov, V.A Komarov, A.S. Parakhin. Segregatsiya pri napravlennoy kristallizatsyi binarnykh sistem tipa Bi–Sb, raschyot i eksperimentalnaya proverka [Segregation with directional crystallization of binary systems of the Bi-Sb type, calculation and experimental verification.] Rasplavy. 2014, No. 3, S. 85–92. [Melts, No. 3, P. 85–92].
3. V.I. Bochegov, V.M. Grabov, A.S. Parakhin. Technical Devices for the Direct Measurement of the Thermal Conductivity of Solids. Measurement Techniques, July, 2014, Vol. 57, No. 4, P. 401-408.
4. V.A. Kulikov. Issledovanie zony provodimosti splavov vismut-surma s kontsentratsyeyi surmy 0-22 at%, legirovannykh tellurom i selenom, po yavleniyam perenosa v intervale temperatury 77-300K.: Dis. … kand. fiz.-mat. nauk. – L.: LGPI, 1970. 132 s. [Investigation of the conduction band of bismuth-antimony alloys with the antimony concentration of 0-22 at% doped with tellurium and selenium, according to transport phenomena in the temperature range of 77-300K: Dis. ... Cand. Sc. in Physics and Mathematics. – L.: LGPI, 1970. 132 p.]
5. T.A. Yakovleva. Issledovanie elektricheskikh i galvanomagnitnykh svoistv splavov vismut-surma s kontsentratsyei surmy do 22 at% v intervale temperatury 77-300K.: Dis. … kand. fiz.-mat. nauk. – L.: LGPI, 1970. 132 s. [Investigation of the conduction band of bismuth-antimony alloys with the antimony concentration of 0-22 at% doped with tellurium and selenium, according to transport phenomena in the temperature range of 77-300K: Dis. ... Cand. Sc. in Physics and Mathematics. – L.: LGPI, 1971. 130 p.]
6. B.I. Davydov, I.M. Shmushkevich. Teoriya elektronnykh poluprovodnikov [Theory of electronic semiconductors]. Uspekhi fizicheskikh nauk. – 1940, T. 24, vyp. 1, s. 21-67. [Successes in physical sciences. – 1940, Vol. 24, no. 1, pp. 21 - 67].
7. L.A. Falkovskii. Fizicheskie svoistva vismuta. [Physical properties of bismuth]. Uspekhi fizicheskikh nauk. – 1968, T. 94, vyp. 1, s. 3-41. [Successes in physical sciences. – 1968, Vol. 94, no. 1, pp. 3 - 41].
8. Shanyu Wang, Jiong Yang, Trevor Toll, Jihui Yang, Wenqing Zhang, and Xinfeng Tangc. Conductivity-limiting bipolar thermal conductivity in semiconductors. Retrieved July 30, 2017, from https://www.ncbi.nlm.nih.gov/pmc/articles/PMC4650813/

IOP Conf. Series: Materials Science and Engineering **544** (2019) 012034 doi:10.1088/1757-899X/544/1/012034

Analysis on End-face Temperature of High-speed Oil-Gas Two-Phase Backflow Pumping Seal

S X Li*, L Chen, H Li, S C Li and Q Z Li

College of Mechanical and Electrical Engineering, Beijing University of Chemical Technology, Beijing, China

E-mail: buctlsx@126.com

Abstract. This paper aims to analyze the end-face temperature in the open state & unopened state, and to verify the feasibility of oil-gas two-phase backflow pumping seal (OG-BPS) in oil-gas two-phase lubrication conditions. The thermal boundary, shear heat, heat distribution, convection heat transfer coefficient and other relevant conditions are applied to the theoretical models establishing. The effect of the seal operation parameters such as the speed, pressure difference and liquid-gas ratio on the end-face temperature is analyzed by calculating the thermo-structural coupling models and conducting experiments. The results show that the end-face temperature in the unopened state is much higher than that in the open state, which can be regarded as a criterion for whether the seal works normally. In the open state, the end-face temperature increases slightly with the increase of the rotation speed and liquid-gas ratio. In the unopened state, the end-face temperature increases sharply with the increase of the rotation speed and the pressure difference. However, with the increases of the liquid-gas ratio, the end-face temperature will decrease rapidly, so the oil-gas two-phase medium will be beneficial to start-stop operation. This paper verifies the feasibility of the OG-BPS, and provides a basis for further study of the effect of end temperature on seal performance.
Key words: oil-gas, end-face temperature, backflow pumping, thermo-structural coupling

1. Introduction

In recent years, as the speed of rotating equipment continues to increase, oil-gas two-phase lubrication technology is widely used in high-speed conditions, because of its high efficiency, good lubrication effect, energy consumption and other advantages [1-3]. However, the sealing problem of oil-gas two-phase lubrication has become a bottleneck restricting its development [4]. Under the oil-gas two-phase condition, the traditional single-medium seal is no longer applicable [5].

Nowadays, a non-contact dynamic pressure seal has been developed for high-speed bearing cavity of aircraft engines under oil-gas two-phase lubrication conditions, such as a high-performance fluid film seal device developed by NASA, which mainly used in aviation high-speed bearing cavity by using hydrodynamic and hydrostatic lubrication principle [6]. Research on single medium seal such as dry gas seal and upstream pumping seal are mainly focused on the influence of dynamic pressure tank groove design & optimization, operating & structure parameters, dynamic state, fluid film thermal behavior and so on [7-9].

However, the theoretical research on end-face temperature in the gas-liquid two-phase mixed medium is relatively less. During the start-stop phase of the seal, the dry friction occurs between the end-faces and the end-face temperature rises sharply, which in turn causes thermal deformation and thermal stress. At the same time, changes in temperature may result in changes in the phase state of the seal medium, and it can easily lead to seal failure [10]. The end-face temperature not only affects the end-face deformation but also characterizes the friction state of the end-face. Monitoring and analyzing the end-face temperature is conductive to study the mechanism of high temperature failure. Therefore, the theoretical and experimental research on the end-face temperature of the OG-BPS is of great significance.

2. Working principle

Figure 1a &1b is the structure and grooved face of OG-BPS. The outer side of the seal is oil and gas, the inside is the atmosphere. When the rotation speed exceeds a certain range, the oil will be disintegrated into small oil droplets under the combined action of high stirring force of the rotating element and its own centrifugal force, and the oil droplets are fully mixed with air to form the oil-gas two-phase medium.

There are micro scale spiral grooves at low-pressure side of rotary ring. When the seal is working, the fluid at high-pressure side enters the end-face from the outside of the seal ring, part of the fluid leaks to the low-pressure side, due to the existence of the spiral groove, the leaking fluid will be passively sucked into the slot and will rotate along with the groove. When the fluid moves to the root of the spiral groove, the flow path narrows and the fluid accumulates here, producing a high-pressure zone and acting as a sealing effect. At the same time, the fluid is pumped back from the low-pressure side to the high-pressure side, which is the effect of the "backflow pumping", thereby achieving "zero leakage".

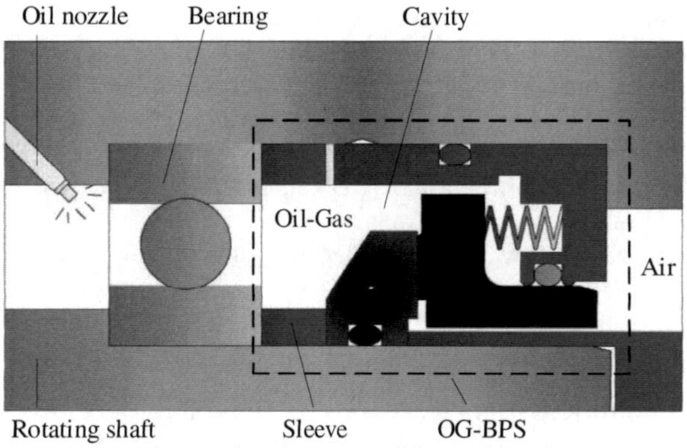

Figure 1a. The structure of OG-BPS.

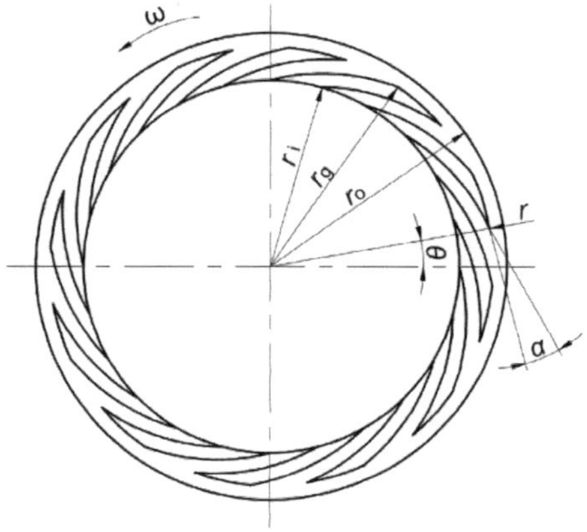

Figure 1b. Grooved face of OG-BPS.

When the rotation speed is lower than the opening speed, the seal rings are in contact with each other, said the state is **unopened state**. When the rotation speed is higher than the opening speed, the seal rings are not in contact with each other, said the state is **open state**. Due to the different heat transfer mechanism of the two states, the following analysis is performed separately.

3. Thermal-solid coupling analysis model

3.1. Physical model

The shape of the groove is a logarithmic spiral, the expression is:

$$r = r_g e^{\theta \tan \alpha} \tag{1}$$

Where r_g is the groove root radius; θ is the rotate-angle; α is the spiral groove angle.

The specific structure size and operating parameters of OG-BPS are shown in Table 1.

Table 1. The structure and operating parameters of OG-BPS.

Parameter	Value
Outside radius r_o/mm	57.5
Groove root radius r_g/mm	55
Inside radius r_i/mm	50
Spiral groove angle α/°	20
Spiral groove depth h_g/μm	5
Land-to-groove ratio γ	0.5
Dam-to-groove ratio δ	0.7
Number of spiral grooves N_g	12
Medium	Air and Mobil jet Oil II
Pressure of oil-gas p_i/MPa	0.15
Pressure of air p_o/MPa	0.05
Operating temperature T/°C	26
Shaft speed n/rpm	8000
Rotary ring material	SiC
Static ring material	C

In view of the centrally symmetric and periodic distribution of the OG-BPS geometric model, a cycle of the seal ring is selected as the calculation area for analysis. The periodic boundary condition is:

$$\begin{cases} \varphi(r,\theta_2,z)=\varphi(r,\theta_1,z) \\ \theta_2=\theta_1+2\pi/N_g \end{cases} \tag{2}$$

The model in the open state is divided into two parts: fluid domains and solid domains.

3.2. Thermal balance analysis

Due to the complicated seal boundaries, the following assumptions are made [11]:

a. Assuming a steady-state temperature field;
b. All the heat is transmitted by the seal rings;
c. Heat flux density on the seal ring end-faces is evenly distributed;
d. The place where the seal elements contact the seal rings is convection heat transfer;
e. The boundary where contacts the O-ring is the adiabatic boundary;
f. The nature of medium and seal rings does not change with temperature.

Figure 2a &2b is the theoretical model of seal rings. In the open state, the viscous shear heat is QV. In the unopened state, the friction heat is QF.O-ring friction heat and the spring vibration heat is QG, which is small and can be neglected. Heat dissipation includes QA1, QA2 (convection heat transfer of surface and oil-gas), QB1, QB2 (convection of static ring and oil-gas, air), QC1, QC2 (the heat rotary ring delivered to the shaft, static ring delivered to the static seat), QD1, QD2 (the heat of the fluid flowing out of the seal end-faces). According to the calculation results of the micro-flow field, the mass flow of the medium at the inlet and outlet is very small, which is on the order of 10-6g·s-1, so it can be regarded as zero leakage [12].

The theoretical thermal equilibrium condition is:

$$Q_V(Q_F)+Q_G=Q_{A1}+Q_{A2}+Q_{B1}+Q_{B2}+Q_{C1}+Q_{C2}+Q_{D1}+Q_{D2}$$

IOP Conf. Series: Materials Science and Engineering **544** (2019) 012034 doi:10.1088/1757-899X/544/1/012034

After simplified:

$$Q_V(Q_F) = Q_{A1} + Q_{A2} + Q_{B1} + Q_{B2} + Q_{C1} + Q_{C2}$$

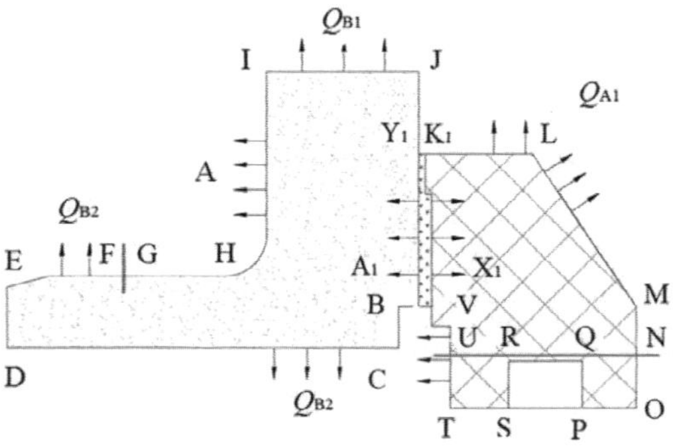

Figure 2a. Theoretical model in open state.

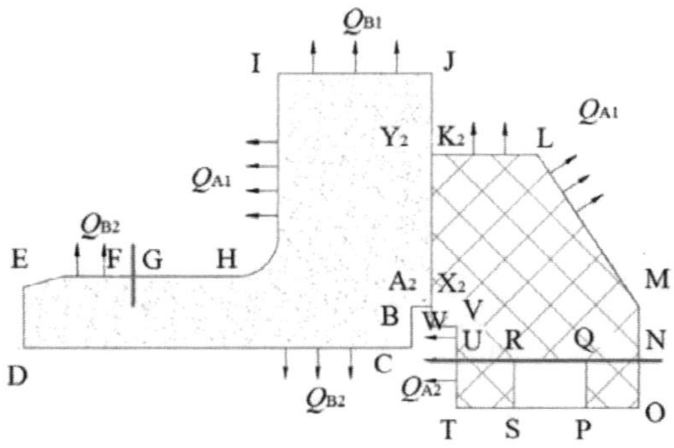

Figure 2b. Theoretical model in unopened state.

3.3. Boundary conditions

According to the structure, the thermal boundary conditions [13] is in Table 2.

Table 2. Thermal boundary of OG-BPS

Thermal boundary conditions	Boundary name
Viscous shear heat generation of end-faces and two-phase fluid	Y_1A_1、K_1X_1
Friction heat between end-faces	Y_2A_2、K_2X_2
Convection heat transfer between rotary ring and medium	KL、LM、MN、NO
Convection heat transfer between rotary ring and air	TS、TU、UV、VW、WX
Static ring and oil-gas two-phase medium	YJ、JI、IH、HG
Convection heat transfer between static ring and air	EF、ED、DC、CB、AB
Contact with the O-ring is considered adiabatic	SR、RQ、QP、FG

The force boundary conditions in the unopened state is in Table 3.

Table 3. Force boundary conditions setting

Force boundary conditions	Boundary name
Oil and gas two-phase medium pressure	GH、HI、IJ、JY、KL、LM、MN、PO、PQ、RQ
Atmospheric pressure	GF、FE、ED、DC、CB、BA、XW、WV、VU、TS
Spring pressure	HI
O-ring pressure	RS
Mixed fluid membrane back pressure	Y_2A_2、K_2X_2
Full constraint	ED、JK
Contact constraint	VM、WL

3.4. Viscous shear heat (Q_V) calculation

The viscous shear heat can be obtained as follows [14]:

$$Q_V = q_V \times A_f \quad (3)$$

Where q_v is the heat flux density; A_f is the area of the seal end-face.

3.5. Heat distribution of the seal rings

The heat distribution on the two seal rings is different, and the thermal steady-state temperature can be obtained as follows [15]:

$$T = q_V (h_l - y) / \lambda \quad (4)$$

Where h_l is the axial thickness of the seal ring; y is the distance from the end-face; λ is the thermal conductivity of the seal ring.

The temperature of the two rings at the end-face is equal, so when $y = 0$, there is:

$$\begin{cases} q_r h_r / \lambda_r = q_s h_s / \lambda_s \\ q_w + q_s = q \end{cases} \quad (5)$$

The heat flux density of the rotary ring & static ring can be obtained as follows:

$$\begin{cases} q_s = q / (1 + h_s \lambda_w / h_w \lambda_s) \\ q_w = q - q_s \end{cases}$$

The heat flux density in unopened state can be obtained as follows [16]:

$$q_{(i)} = 2\pi n r_{(i)} f p_{(i)} \quad (6)$$

Where $q_{(i)}$ is the heat flux density of point i; $p_{(i)}$ is the contact pressure of point i; n is the rotating speed; $r_{(i)}$ is the radius of point i.

3.6. Convection heat transfer coefficient [16]

The convection heat transfer coefficient of rotary ring and oil-gas can be obtained in equation (7). The convection heat transfer coefficient of rotary ring and air can be obtained in equation (8). There is no swirling flow in the flow around the static ring and its surroundings, so the convective heat transfer coefficient is given by equation (9).

$$\alpha_1 = \frac{\lambda}{d_o} \cdot 0.135 \cdot \left[\left(0.5Re_e^2 + Re_f^2 \right) \cdot Pr \right]^{0.33} \quad (7)$$

$$\alpha_2 = \frac{0.21\lambda}{2S_s} \cdot \left(Ta^2 \cdot Pr \right)^{0.25} \quad (8)$$

$$\alpha_3 = \frac{0.023\lambda}{2S_s} \cdot \left(\varepsilon_1 \cdot Re^{0.8} Pr^{0.4} \right) \quad (9)$$

4. Temperature field calculation and analysis

4.1. Temperature field cloud char

The temperature field cloud chart in open state is shown in Figure 3a &3b. The temperature of rotary ring is lower than that of static ring because the former is at oil-gas side and rotates at high speed, so the convective heat transfer of rotary ring is stronger than that of static ring.

IOP Conf. Series: Materials Science and Engineering **544** (2019) 012034 doi:10.1088/1757-899X/544/1/012034

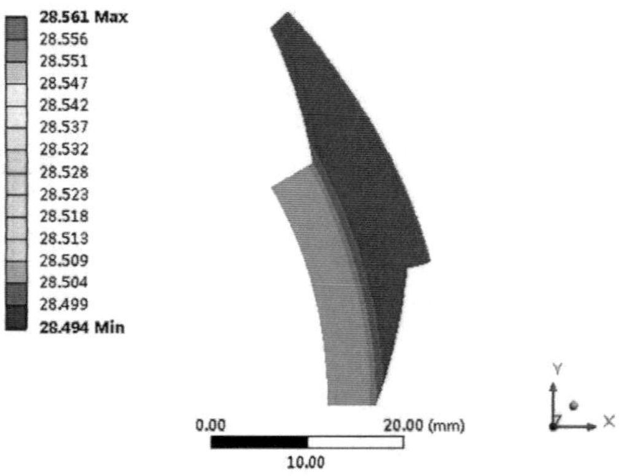

Figure 3a. Rotary ring temperature field.

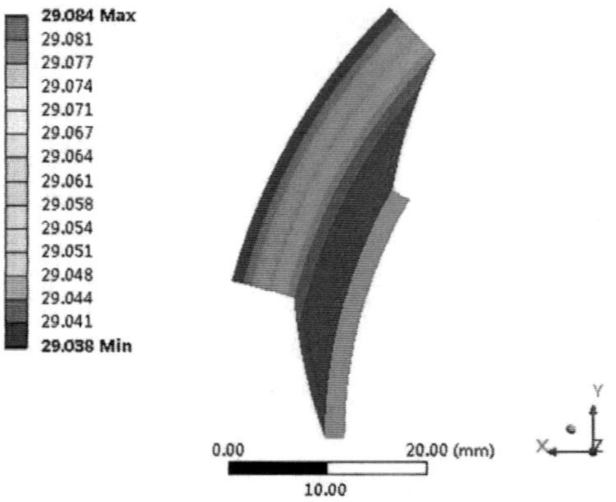

Figure 3b. Static ring temperature field.

The temperature field cloud chart in unopened state is shown in Figure 4a &4b. The heat flux density will be larger with the contact pressure or speed increases; therefore, the highest temperature point is at the outside radius side where the seal rings contact. The static ring (graphite ring) has a larger thermal conductivity, so its temperature gradient is greater than that of the rotary ring.

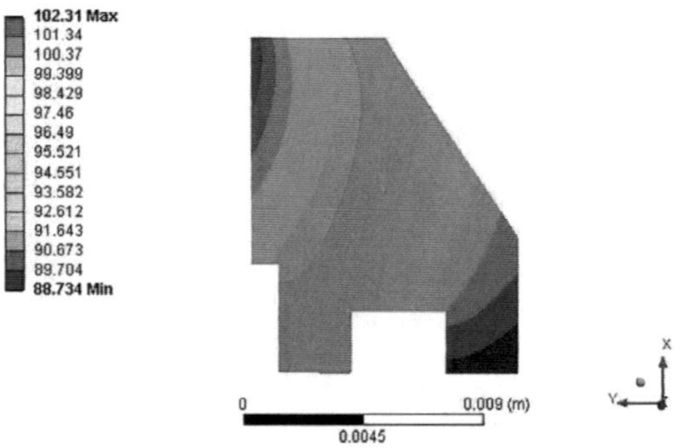

Figure 4a. Rotary ring temperature field.

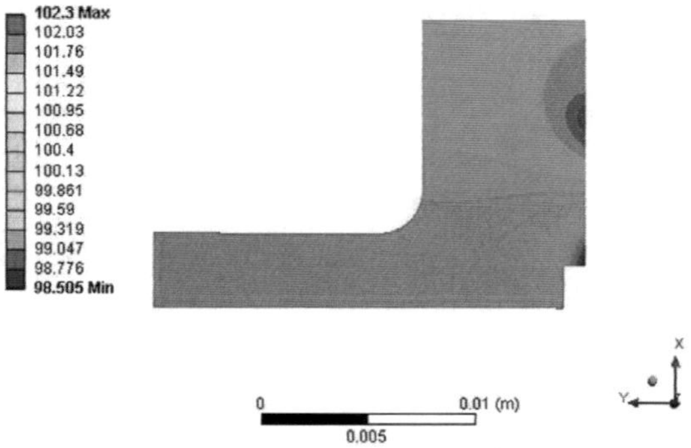

Figure 4b. Static ring temperature field.

4.2. Effect of operating parameters on the end-face temperature field

4.2.1. Rotation speed. In Figure 5, with the increase of rotation speed, both the viscous shear heat and frictional heat increases, so the temperature rise of the end-face increases. At the same time, the rotation speed makes the convection heat transfer coefficient increase, which enhances the heat transfer, so the end-face tempera¬ture increases slowly.

4.2.2. Pressure difference. In Figure 6, the frictional heat in the unopened state increases with the pressure difference increases, while the convection heat transfer coefficient on the seal ring remains constant. Therefore, as the pressure difference increases, the end-face temperature increases linearly.

4.2.3. Liquid-gas ratio. In Figure 7, in the unopened state, when the liquid-gas ratio increases, the convective heat transfer effect is enhanced and the temperature drops significantly. In the open state, due to the combined effect of shear heat and convection heat transfer, the liquid-gas ratio has little effect on the temperature change.

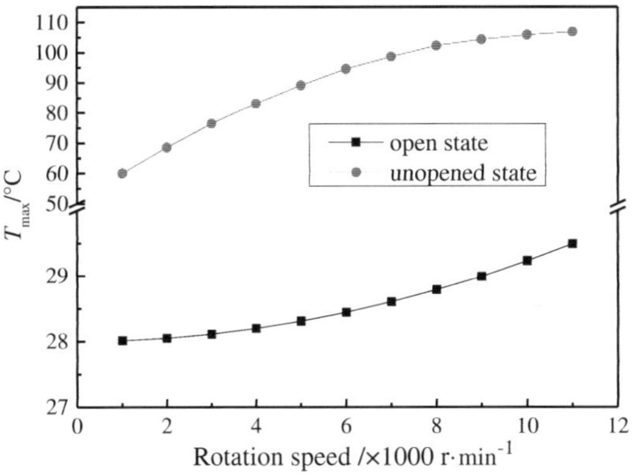

Figure 5. The effect of rotation speed on temperature.

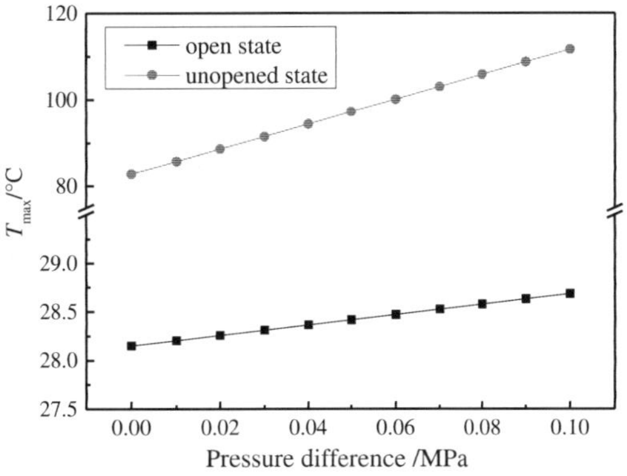

Figure 6. The effect of pressure difference on temperature.

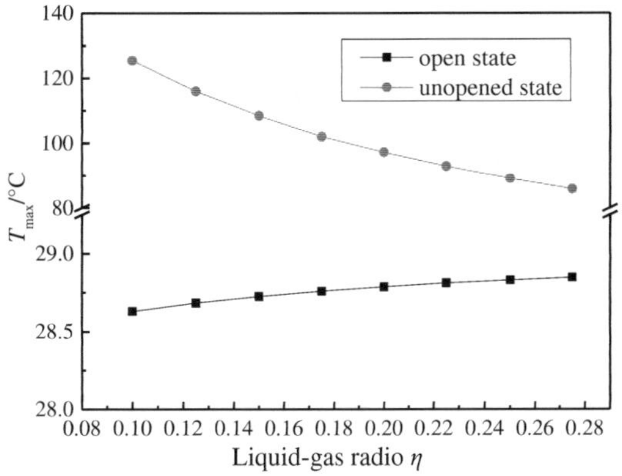

Figure 7. The effect of liquid-gas ratio on temperature.

5. Experimental verification

The variable speed, variable pressure and variable lubrication conditions tests have been carried out to verify the correctness of the numerical results.

5.1. Experimental equipment

The experiment equipment of OG-BPS is shown in Figure 8a &8b.

Figure 8a. Experiment bench.

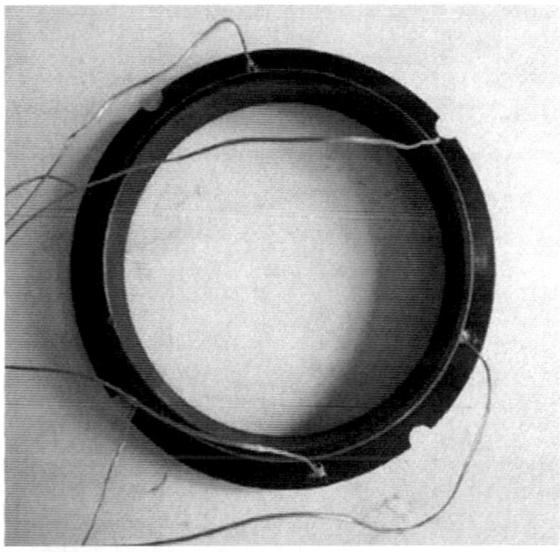

Figure 8b. The innstallation of thermocouple.

5.2. The effect of rotation speed on the end-face temperature

As shown in Figure 9, the sealing end face is opened at about 3000 rpm. The rotational speed has a greater influence on the end face temperature in the unopened state, and has less influence on the end face temperature in the open state. When the speed increases to a certain value, the end face is opened and the seal ring will have a slight temperature rise due to the dry friction before opening. The test results are basically consistent with the calculation results.

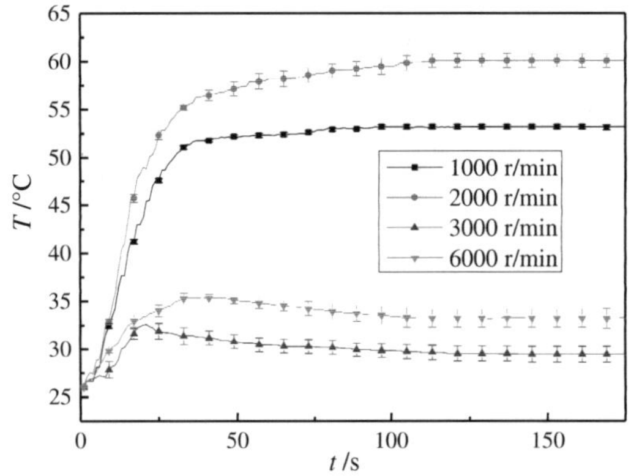

Figure 9. Temperature change curve under different rotation speed.

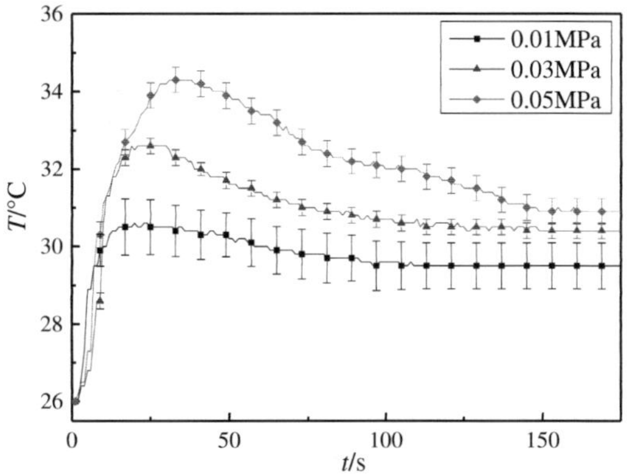

Figure 10. Temperature change curve under different pressure.

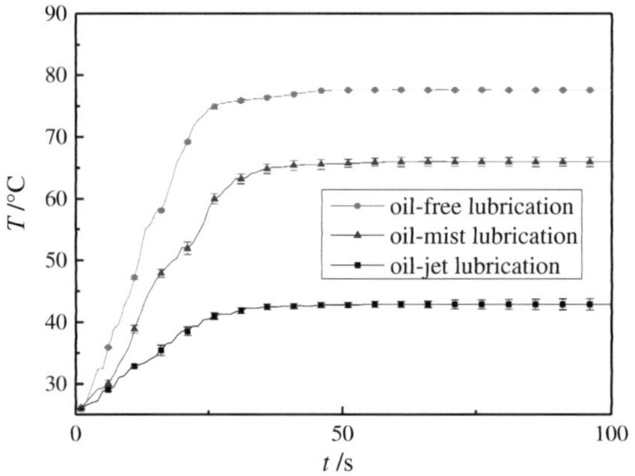

Figure 11. Temperature curve under different lubrication condition in the unopened state.

5.3. The effect of different pressure on end-face temperature

As shown in Figure 10, the differential pressure has a significant effect on the end face temperature only in the unopened state. When the seal is in the open state, the viscous shear heat of the fluid is the only source of heat, the value of which is much lower than the heat of solid friction, so the temperature rise is small. The test results are basically consistent with the numerical calculation results.

5.4. The effect of different oil-gas radio on end-face temperature

As shown in Figure 11, the test result is slightly higher than the numerical calculation result, because the seal is in the open state in the test, and the temperature rise mainly comes from the friction heat in the unopened state. When the seal is in an oil-free lubrication state, the temperature rise is the largest, followed by the oil mist lubrication, and the injection lubrication is minimal. Therefore, the lubricating oil has a great cooling effect on the seal.

6. Conclusion

The end-face temperature in the open state is much lower than that in the unopened state, so it can be used as a criterion for whether the seal end-face is open. The speed should be quickly increased/decreased in the start/stop phase, in order to avoid prolonged stay at low speed stage and prevent the seal rings thermal cracking damage. Oil-gas lubrication is conducive to the start/stop phase. Seal ring with different structural parameters has different convection heat transfer coefficients, therefore, the effect of temperature on the seal performance can be reduced by optimizing the structure of seal ring. In working condition, with the change of these three variables (rotation speed, pressure and liquid-gas ratio), the amplitude of the change is relatively small and the temperature is relatively low, OG-BPS can work stably and reliably under changing conditions. The experiment results are in agreement with the numerical results, which proves the correctness of the numerical calculation.

References

[1] Kim K T 1997 The Effect of Oil-Starvation on the Lubrication Characteristics of High-Speed Bearing: Part II-Roller Bearing *Journal of Clinical Microbiology* **34** 447-9

[2] Liu H B, Wang H Y, Lei Z, Shi Y S and Liu G P 2016 Analysis on penetration mechanism of oil jet lubrication for high speed rolling bearing *Journal of Aerospace Power*

[3] Rucker G, Weidmann W and Reisenweber K U 1983 Rotary ring seal with gas-dynamic lubrication for high-speed turbomachines. US)

[4] Xue yong L I and Fang G H 2016 Analysis on Oil-air Lubrication for Guide of Finish Rolling for High Speed Wire Rod Plant of Baotou Steel and Improvement of Oil Way *Science & Technology of Baotou Steel*

[5] Li X F, Cai J N, Zhang Q X and Li S X 2016 Analysis on Hydrodynamic Sealing Performance under Low Liquid-gas Ratio Mixed Lubrication *Lubrication Engineering* **41** 40-44

[6] Hu G Y 2012 Application Research of Seal Technologies for Aeroengine *Aeroengine* **38** 1-4

[7] Hu S T, Huang W F, Liu X F and Wang Y M 2017 Applicability Analysis of Steady-state Models for Spiral Groove Gas Face Seals *Journal of Mechanical Engineering* **53** 7-13

[8] Wang Q, Chen H L, Liu T, Liu Y H, Liu Z B and Liu D H 2012 Research on performance of upstream pumping mechanical seal with different deep spiral groove. p 2019

[9] Xu H J, Song P Y and Yu J P 2013 Analysis on the Seal Performance of Aerostatic Gas Lubricating Mechanical Seal Lubrication Engineering 41-45

[10] Zhang G, Zhao W and Tian Y X 2014 Experimental study on the water lubrication of non-contacting face seals for turbopumps *Industrial Lubrication & Tribology* **66** 314-21

[11] Ding X X, Wu H, Yan R Q, Liu Y and Yan F 2014 Analysis and calculation of thermal stressing coupled deformation of mechanical seal based on ANSYS *Journal of Lanzhou University of Technology* **40** 41-45

[12] Wang Y T 2015 The Analysis of Mechanical Seal's Leakage Rate *Hydraulics Pneumatics & Seals* **35** 40-42

[13] Zhang Z S, Yang Y M, Dai X D and Xie Y B 2014 Effects of Thermal Boundary Conditions on Thermo¬hydrodynamic Lubrication Analysis of Plain Journal Bearing *China Mechanical Engineering* **25** 1427-1432

[14] Cheng W Q 2013 Numerical Simulation of Scramjet Combustor Based on the Orthogonal Experimental Design. Nanchang Hangkong University

[15] Li N, Zhu W B, Wang H S and Wang Y F 2009 Research on Distribution Law of Steady-state Temperature for the Rotating Ring of T-shape Groove Dry Gas Seal *Chinese Hydraulics & Pneumatics* 46-49

[16] Wang Z T, Zhang Q X, Cai J N Huang B C and Li S X 2015 Analysis on Thermal Characteristic of Hot Oil Pump Mechanical Seal *Lubrication Engineering* **40** 20-25

IOP Conf. Series: Materials Science and Engineering **544** (2019) 012035 doi:10.1088/1757-899X/544/1/012035

Experimental study on hydraulic expansion of tube-fin heat exchanger

Junjie Han[1*], Lianfa Yang[2], Jinjie Huang[3], Jingyu Jiang[4] and Yulin He[5]

1,2,3,4,5School of Mechanical & Electrical Engineering, Guilin University of Electronic Technology, Guilin, Guangxi, 541004

E-mail: owlph@foxmail.com

Abstract: The tube-fin heat exchanger is the core equipment in the whole air-conditioning system, the thermal resistance of copper tube and fin contact accounts for 10%-20% of the total thermal resistance of the heat exchanger, and the thermal contact with the fin increases. Therefore, reducing the contact gap of the heat exchanger is extremely important to improve the overall heat exchange efficiency. In this paper, for the high-efficiency heat exchanger composed of seamless internal thread copper pipe, the heat exchange tube and the fin are expanded by hydraulic expansion and the liquid pressure range, reasonable pressure holding time and pull-off force are guaranteed by a large number of experiments. In this paper, the heat transfer tube and the fins are expanded by hydraulic expansion, and the relationship between the liquid pressure range, the reasonable pressure holding time and the pull-off force and the dwell time is measured through a large number of experiments. Finally, the effect of different liquid pressure expansion joints on the contact gap was further observed by a microscope.

1. Introduction

With the fierce competition between the air-conditioning industry and the multi-level development of consumer demand, people are increasingly demanding the performance of air conditioners. A large number of experimental studies have shown that the tube-fin contact thermal resistance accounts for 6%-18% of the total heat resistance of the entire heat exchanger [1]. Among them, the heat exchange tube is a key component of the heat exchanger. Besides the excellent features of the smooth copper tubes, the heat transfer effect is the smooth tube 2- 2.5 times, saving one third of copper [2]. Therefore, the seamless internal thread copper tube is widely used. Due to the structural characteristics of the seamless internal thread copper tube, the mechanical expansion tube method will cause the drawing of the inner wall thread and the defects such as work hardening, which will reduce the overall transmission. However, the non-contact uniform expansion method of hydraulic pressure to expand the heat exchanger can effectively solve the disadvantages caused by mechanical expansion. Therefore, it is extremely important to study the new expansion method to improve the development and innovation of the current heat exchanger to low power consumption and high performance.

In order to better observe the contact gap of the heat exchanger, we need to know the factors related to the contact gap, such as the magnitude of the hydraulic pressure, the specific holding time and the relationship between the pull-off force and the holding time. The contact gap between the entire heat transfer tube and the fin was observed through a microscope.

In 2003, Elsherbini et al. studied the contact of aluminum fins without flanged copper tubes, and measured the contact thermal resistance under tube fin brazing and direct expansion contact [3] Jeong is equal to the correlation between the thermal resistance and the expansion joint forming of the indirect contact between the 9.52mm and 7mm copper tubes and the aluminum fins in 2004 and 2006 respectively[4.5].In 2009, Tang Ding et al. studied the contact state of tube-fin heat exchangers, and proposed reasonable improvement of the expansion joint process by finite element method and test method [6]. However, the relationship between the contact thermal resistance and the thermal expansion of the hydraulic expansion tube is rare.

In this paper, the optimal expansion pressure range of the heat exchanger and the relationship between the dwell time and the pull-off force are obtained through a large number of experiments. Under the different hydraulic pressures obtained through experiments, the results of local contact of the heat exchanger tubes and fins were observed. A new measurement direction related to contact thermal resistance is proposed to analyze the influence of the hydraulic expansion process on the thermal resistance of the heat exchanger, which provides a new idea for the entire tube-fin heat exchanger in the case of hydraulic expansion.

2. Determination of the range of hydraulic pressure

2.1. experiment system

The whole experiment consists of three parts, the first part of the hydraulic power system, the second part of the data measurement part, and the third part is the self-designed hydraulic bulging device combination to complete the whole experiment process. the sampling rate of the data acquisition system is set to 10 Sps/s for uninterrupted sampling. Finally, different hydraulic pressure experiments were completed under the same holding time, and the high pressure liquid in the heat exchanger was released by hydraulic device to complete the expansion process of the entire heat exchanger.

2.2. Expansion pressure determination

In order to make the fins of the tube-fin heat exchanger and the heat exchange tube expand tightly, we need to know in advance the reasonable pressure of the liquid pressure expansion in the case of hydraulic expansion, using the corresponding theoretical calculation as a guide, the specific experimental phase Combine the way to determine the reasonable expansion pressure under this specification. The specific assembly and parameters of the selected hydrophilic aluminum foil fins and seamless internal thread copper tubes are shown in Figure1. The liquid belongs to a homogeneous medium. According to the calculation, the optimum expansion pressure range of the gap between the fin and the heat exchange tube and the yield of the fin just after yielding is between 8 MPa and 11 MPa.

IOP Conf. Series: Materials Science and Engineering **544** (2019) 012035 doi:10.1088/1757-899X/544/1/012035

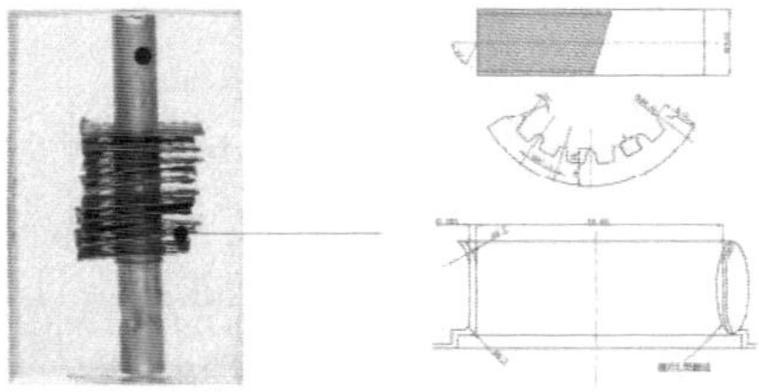

Figure 1. Schematic diagram of heat exchange tube and fin structure.

The range of the expansion pressure is tested by the previously calculated reference value under different expansion pressures, and the indentation on the surface of the heat exchange tube is used to determine whether the expansion joint phenomenon has occurred to determine whether the expansion has occurred. Finally, a liquid pressure range from 9 MPa to 11.5 MPa when the expansion joint is just expanded until the heat exchange tube is broken.

3. Determination of holding time

3.1 Expansion test and pull-off test

The experiment was arranged under the condition of 10MPa liquid pressure expansion joint, and the effective expansion time was carried out under the conditions of holding pressure for 2s, 4s, 6s, 8s, 10s, 12s and 14s respectively. The expansion of the three sets of samples was carried out under each dwell time, and the expansion of the whole experiment was completed by a hydraulic expansion joint device. The pull-off force test is carried out according to the heat exchanger's independent design of the corresponding fixture. The pull-off force test was carried out by a self-designed tube-fin fixture and a Shimadzu AX-G electronic universal testing machine to measure the pull-out force under different dwell time. After the measurement of the pull-off force reaches the maximum value, the average pull-off force value of the 0.5 mm continues to slide is the pull-out force value of the heat exchanger. The pull-off force and displacement parameters measured at different dwell times were obtained as shown in Table 1.

Table 1. Pull-down force and displacement test parameters for different dwell time.

Measurement type	Numbering	4s	6s	8s	10s	12s	14s
Pull force /N	No1	40.22	244.56	64.50	185.80	183.06	129.17
	No2	43.75	85.66	119.58	295.04	145.65	94.81
	No3	20.85	72.11	306.20	188.98	124.49	112.44
Pull force average /N		34.94	78.86	119.58	223.27	151.07	112.14
Displacement /mm	No1	0.36	14.05	3.90	15.31	7.82	5.88
	No2	4.47	8.23	6.24	31.25	6.67	3.32
	No3	0.38	5.62	10.32	3.52	4.95	9.26
Average displacement /mm		1.74	6.93	6.24	16.69	6.84	6.15

IOP Conf. Series: Materials Science and Engineering **544** (2019) 012035 doi:10.1088/1757-899X/544/1/012035

Finally, the average value was obtained to obtain the pull-out force under the corresponding dwell time. With displacement parameters. Finally, the experimental and measurement results obtained the relationship between the dwell time and the pull-off force and displacement shown in Figure 2. For the tube-fin heat exchanger, the dwell time is about 10s, which is suitable for too short. The pressing time will cause the pulling force to be insufficient. The excessive holding time will cause the heat exchanger to work less, and the holding time is suitable for 10s-12s.

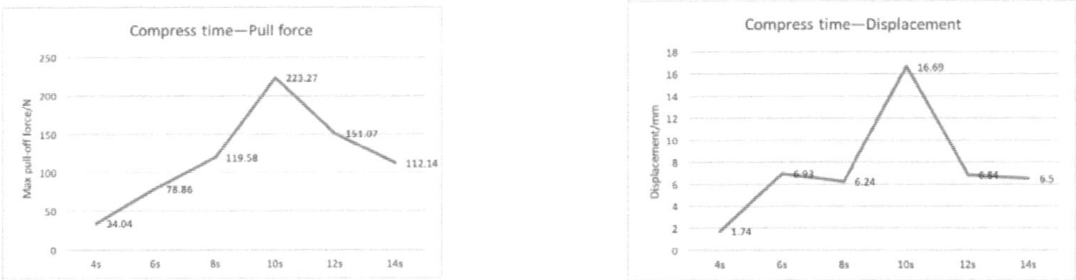

Figure 2. Relationship between packing time and pull-off force and displacement.

4. Analysis of local characteristics of hydraulic expansion process

4.1. *Sample contact observation experiment*

Through the experimental data obtained in the previous section, under the premise of the holding time of 10s, nine experiments were carried out in three groups of experiments at 9MP, 10MPa and 11MPa respectively, and the anatomy was carried out in the axial direction through the water cutting machine. Eighteen observation experiments were obtained. According to the axial direction of each heat exchange tube, it is divided into six parts, A_U\$A_D$、$B_U$\$B_D$、Cu\$C_D$, for metallographic microscope observation. The observed parts are shown in the figure 3.

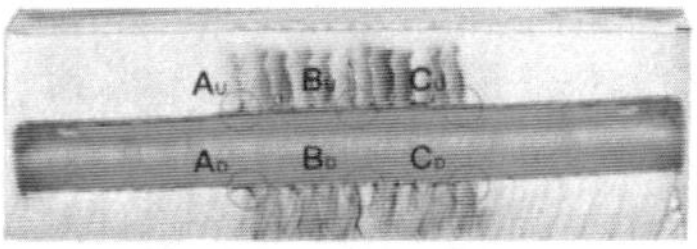

Figure 3. Experimental sample molding and size.

4.2. *Experimental research and analysis of results*

In the case, it is found that the maximum gap generated in the middle of the flat side of the fin is 0.035 mm and 0.030 mm. To this end, we made nine sets of experimental samples and cut them in half along the axis. Through actual observation, it is found that under the condition of hydraulic expansion joint, the tube-fin heat exchanger has a clear gap between the fin and the heat exchange tube under the condition of 9MPa expansion, and the heat exchange in the case of 10MPa and 11MPa. The gap of the test piece was almost completely eliminated, and a good expansion effect was obtained. According to the results obtained by the experiment, in the case of hydraulic expansion, the results of the B_U part of the 9MP, 10MPa, and 11MPa pipes were observed as shown in the figure. In the case of 9 MPa, a significant gap appeared in the expansion of the heat exchanger, and then complete expansion was achieved under other pressures.

According to these observation phenomena, the contact gap of the tube-fin heat exchanger in the hydraulic expansion joint is proposed. In the case of hydraulic expansion, the effective contact area between the flat portion of the fin and the heat exchange tube is used to express the heat transfer under the hydraulic expansion joint.

Nine sets of heat exchanger samples were subjected to water cutting and dissection under different hydraulic pressures. Eighteen groups of samples were obtained. Observation and measurement were performed by an optical electron microscope at a magnification of 50 times, and finally, statistical data as shown in the table 2 was obtained.

Table 2. Effective contact area of fin contact under different liquid pressures.

Hydraulic pressure	Numbering	Au/mm	Ad/mm	Bu/mm	Bd/mm	Cu/mm	Cd/mm
	No1	0	0	0.234	0.345	0	0
9MPa	No2	0.290	0.243	0.501	0.427	0.256	0.267
	No3	0	0	0.196	0.203	0	0
	No1	0.934	0.934	0.944	0,986	0,862	0.944
10MPa	No2	0.782	0.789	0.904	0.854	0.864	0.853
	No3	0.702	0.718	0.914	0.928	0.853	0.834
	No1	0.734	0.790	0.843	0.903	0.911	0.958
11MPa	No2	0.417	0.419	0.873	0.816	0.951	0.943
	No3	0.288	0.213	0.242	0.246	0.404	0.166

As shown in the table 2, it can be seen that there is a state of incomplete contact at both ends of the A/C of the heat exchange tube. The main reason for this phenomenon is that the entire heat exchanger has a large due to insufficient initial expansion pressure. Contact gap. In the case of 10 MPa, there is no obvious contact gap in the complete expansion of the A/B/C three parts. In the case of 11 MPa, the data in the experimental group showed large fluctuations. The possible reason was that the hydraulic pressure was unstable or the holding time was too long during the expansion, and the over-expansion occurred, resulting in plastic deformation of the fins. The decrease of contact pressure increases the corresponding contact gap.

As shown in Figure 4, the electron expansion lens is observed under the 9MPa expansion pressure. The effective expansion area of the tube-fin heat exchanger is almost 0. The intermediate ratio is 24.5%. There is a significant contact gap. The reason for this is that the insufficient expansion pressure causes the tube and the fin to not fully expand, resulting in a certain contact gap inside.

Due to the processing technology of the fin itself, we can see from the figure that there are two original angles in the flanging process of the whole fin, so we observe the fillet in the actual expansion joint contact angle during the expansion process. The range has an effect on the entire expansion effect. In the previous large number of experiments, we determined the liquid pressure value and the reasonable pressure holding. Finally, different experimental groups were made at 9 MPa, 10 MPa, and 11 MPa, respectively. The best expansion effect of the control was at 10 MPa. The proportion of the entire contact area reached 94.1%.

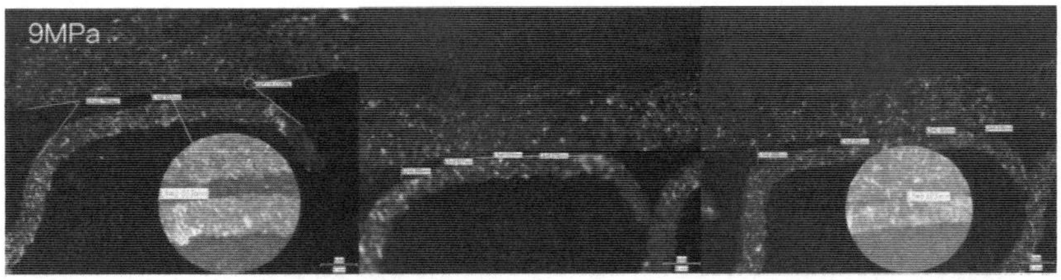

Figure 4. Contact gap of different parts under 9MPa hydraulic pressure.

5. Conclusions

The experimental research shows that the range of hydraulic expansion of the tube-fin heat exchanger is combined with theory and experiment to obtain the best expansion range in the range of 10~11MPa. The reasonable holding time is about 10s. Under this holding time, it can provide the maximum pulling force between the heat exchange tube and the fin to ensure the performance of the whole production, and at the same time ensure the effective expansion of the tube-fin heat exchanger. Forming. Through the experiment, under the condition of 10MPa hydraulic pressure, the contact gap between the tube fins is basically completely eliminated, and the substantially effective expansion joint area is above 91.4%. Since the hydraulic expansion joint can effectively solve the drawbacks caused by the mechanical expansion tube, in the cost control and the specific process, there will be a contact gap between the two ends of the heat exchange tube, how to effectively deal with the contact gap at both ends It is necessary to conduct further research in the future.

References:

[1] C.V.Madhusudana. Thermal Contact conductance[J]. Springer-Verlag, (1996).

[2] N.Nickolas, P.Moorthy,A,N.Oumer,M.Lshak.A review on improving thermal-hydraulic performance of fin-and-tube heat eachangers[J]. IOP Conference Series;Materials Science and Engineering 2017, 1757-8981.

[3] Elsherbini AI,Jacobi AM.The thermai-hydraulic impact of delta-wing vortex generators on the performance of a plain-fin-and-tube heat exchanger[J].Int JHAVAC&R 2002(8);357-370

[4] Jin Jeong, Chang Nyung Kim,Baek Youn.A study on the thermal contact conductance in fin-tube heat exchangers with 7mm

[5] Jin Jeong,Chang Nyung Kim,Baek Youn,Young Saeng Kim.A study on the correlation between the thermal contact conductance and effective factors in fin-tube hat exchangers with 9.52mm tube[J].International Journal of Heat and Fluid Flow . 2004,25;1006-1014

[6] Ding Tang,Dayong Li, A new approach in evaluation of thermal contact conductance of tube-fin heat exchanger[J].Applied thermal Engineering,2010(30);1359-4311

Acknowledgements

This work was financially supported by the National Natural Science Foundation of China (51564007), the Natural Science Foundation of Guangxi Province (2015GXNSFBA139223 and 2017GXNSFAA198133), and Guangxi Key Laboratory of Manufacturing System & Advanced Manufacturing Technology (14-045-15-005Z).

IOP Conf. Series: Materials Science and Engineering **544** (2019) 012036 doi:10.1088/1757-899X/544/1/012036

A Study on the Shape deviation rate of Hydroforming of Seamed tubes Based on Two Different Velocity

Qin Liu[1]*, Lianfa Yang[2], Jun Wei[3], Jian Guo[4] and Yulin He[5]

[1,2,3,4,5]School of Mechanical & Electrical Engineering, Guilin University of Electronic Technology, Guilin, Guangxi, 541004

E-mail: liuqin9308@163.com

Abstract: Due to the low price, the seamed tubes are used more and more widely in various industrial fields in recent years. The difference in material properties between the weld zone of the parent metal zone of the thin-walled seamed tubes, the deformation of the hydraulic bulging under the same conditions will also be different, thus resulting in an inconsistency in the overall bulging of the seamed tubes. In this paper, the hydraulic bulging experiment of the seamed tubes at low and medium velocity, and the shape deviation law of the seamed tubes at different velocity are obtained. In the first half of the bulging, the seamed tubes at the medium velocity drops faster than the low velocity shape deviation rate, and the medium velocity is closer to circle; in the second half of the bulging, the medium velocity is go up faster than the low velocity. The shape deviation rate is higher before the tubes are burst, but the medium velocity's radial average bulge height is smaller than the low velocity.

1. Introduction

Tube hydroforming (THF) technology is an advanced, net forming technology for the production of such components. It is a widely used non-traditional metal forming technology. [1] The materials used for tubes hydroforming are mainly divided into two types: seamless tubes and seamed tubes. Due to the advantages of low production cost and high production efficiency, seamed tubes have more and more applications in THF. Therefore, it is extremely important to study the influence of welds on the hydroforming process of tubes. For the seamed tubes, according to the weld line of the tube axis parallel or perpendicular, the seamed tubes is divided into a straight seamed tubes and a welded tube. The seamed tubes is greatly different in structure and preparation method from the seamless tubes. The non-uniformity of the heat transfer in the soldering process results in changes in the metallographic structure and grain size of the weld zone. which causes the mechanical properties of the material in the weld zone to be inconsistent with the parent metal zone [2], resulting in weld seams, HAZ (heat affected zone) and parent metal these three areas. The size and material properties of the three regions will be different, under the same conditions of hydraulic bulging, the deformation of the three materials will be different and will be uncoordinated, which will result in the contour of the seamed tubes to be difference with the seamless tube.

Many scholars have talked about the unevenness of tube bulging. Guannan Chu [3], it was found that the bulging state of the welded tubes was affected by the length ratio and the thickness ratio. The deformation coordination of the thick and thick welded tubes increased with the increase of the length ratio and the deformation coordination decreased with the increase of the thickness ratio. At present, the most used and most studied seamed tubes is an electric resistance welding (ERW) straight seam tubes. Some scholars assume that the axial profile of the welded tube during the bulging process is a cosine function [4], ellipse [5] and circle [6]. There are few articles on the deviation rate of radial profile in the process of seamed tubes bulging, especially the comparison of the deviation rate of radial profile and the quantitative variation of the deviation rate under different bulging rates. Zhan Mei [7] and many other scholars have proposed that it is urgent to study the non-uniform deformation behavior.

This paper studies the free bulging of high-frequency resistance welded straight seam tubes at two different forming rates, obtained the real-time bulging height of the weld zone and the parent metal zone of the welded tubes, then obtained the shape deviation rate at different bulging rates.

2. Welded tube bulging test

2.1. Test materials

The material in this experiment is electric resistance welding (ERW) SS304 stainless steel seamed tubes. The nominal diameter is 25 mm, nominal wall thickness is 0.6 mm, length is 110 mm and bulging zone length of 50 mm. The maximum diameter is measured to be 25.02 mm, the minimum diameter is 24.83 mm, the shape deviation rate λ is introduced to characterize the irregular shape of the tube:

$$\lambda = \frac{2(d_{max} - d_{min})}{d_{max} + d_{min}}$$

The initial maximum shape deviation rate of the welded tube is measured to be 0.72%.

2.2. *tube free bulging test platform*

The low-rate bulging test device consists of a hydraulic oil supply system , bulging system and 3D strain measurement analysis system , The bulging rate is 10-2s-1~1.5*10-2S-1. The medium velocity bulging experimental device add the deep drawing punch than the low velocity, so added an impact hydraulic generating system, as shown in Fig 1. The device produces a bulging rate of 1S-1 to 1.5S-1. The pressure supply system provides the hydraulic oil, which is transported to the inner part of the welded pipe through the pipe. Both ends of the welded tube are placed in the left and right locating rings, and two urethane plugs are placed inside the two ends. The tube, two locating rings, and two urethane plug are joined together by a long stud. The liquid enters the inside of the tube through the liquid filling hole at one end of the stud. With the increase of the hydraulic pressure, the urethane plug is expanded by the action of the hydraulic pressure to achieve the sealing of the tube end; since there is no other binding force at both ends and only by the frictional force, the tube can be relatively freely contracted along the axial direction, the tube gradually inflated, which is the free bulging of the tube.

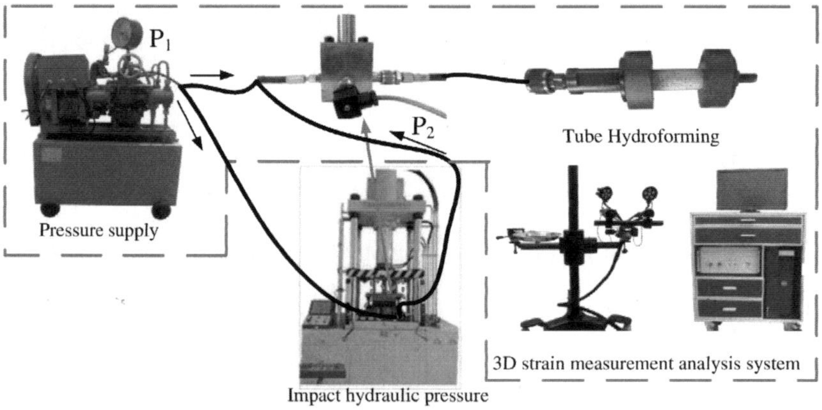

Figure 1. bulging device.

A three-dimensional strain measurement analysis system was used to collect the deformation images in the hydraulic bulging. It consists of two CCD camera, two LED lamp, a control box, a computer and a tripod. During the bulging process, the weld is directly facing the center of the camera image, and the deformation data of the material is dynamically obtained by matching the grayscale distribution information of the speckle on the image before and after the deformation.

3. Data processing of bulging experiments

First, the mesh of the seamed tubes speckle area is divided, and the grid width is set to 0.72 mm. And the 3D coordinates of each grid in each deformation time are obtained. Secondly, the center origin of the radial direction of the tube is found by fitting. The highest point of axial bulging is found according to the data of the sheet before bulging. The origin of the system is moved to the center of the radial circle of the cross section of the highest point of axial bulging. The x axis is parallel to the weld axis. Finally, using the y and z coordinates of the output, can obtain the max and min radial radius of the weld zone and the parent metal zone during the bulging process. According to the calculation formula of the shape deviation rate λ, the curve of the shape deviation rate with the radial average bulging amount during the bulging process is obtained.

4. Hydraulic bulging test results

In the experimental study, only the radial shape deviation rate at the highest point of the axial bulging shape was considered.

The initial shape deviation of the seamed tubes does not start from 0. As the bulging progresses, the shape deviation rate of the welded tube has a decreasing process, the low velocity bulging height reaches about 1.48 mm, the medium velocity bulging height reaches 1.16 mm, the shape deviation rate of the welded tube both are close to 0, indicating that the shape of the welded tube at this time is the closest to the circle. The larger the bulging rate, the closer the shape deviation rate of the welded tube is to 0.

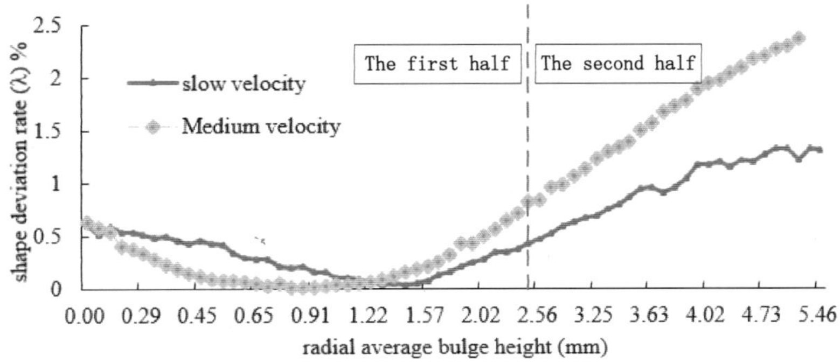

Figure 2. Shape deviation rate curve.

In the second half of the THF, the shape deviation rate of the medium velocity rises faster than the low velocity, and the shape deviation rate before the bursting is about 1% larger than the low velocity. Before bursting, the final bulging height of the medium velocity is about 0.34 mm smaller than the final bulging height of the low velocity. Because the metal deformation rate increase, the recrystallization cannot be overcome by the work hardening phenomenon, and the elongation index and the area shrinkage of the metal are decreased, the deformation resistance is increased, and the plasticity is deteriorated.

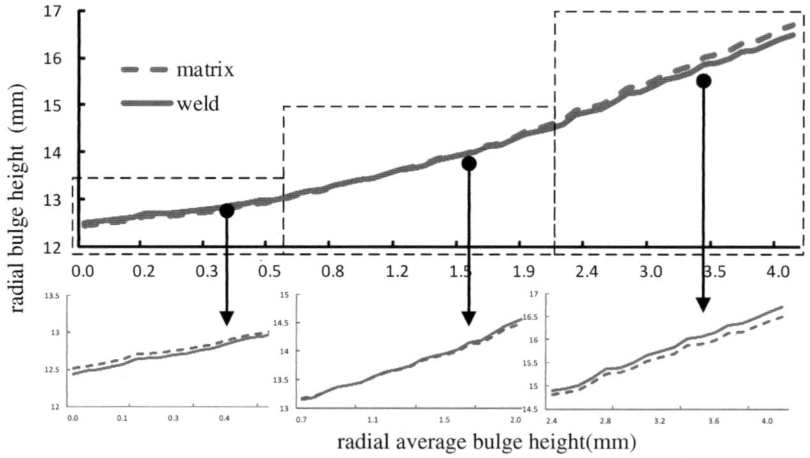

Figure 3. Shows the bulging height curve of the weld zone and the base zone at medium velocity.

Fig. 3 is the curve of the bulging height with time in the welding zone and the base zone under medium velocity. The reason why abscissa is not uniform is that the time interval of data acquisition is the same, but the bulging of welded pipe is not uniform. In the bulging front half, the cross-sectional profile of the tubular member is greater in the weld direction than in the direction of the parent metal, but in the bulging second half is opposite. The degree of work hardening in the weld zone is higher than the parent metal zone, so the deformation resistance of the weld zone is relatively large, and the amount of deformation generated in the same hydraulic environment is small. This explains and verifies the variation of the shape deviation rate in Fig. 2.

5. Conclusions

The regularity of the contour shape reflects the geometric accuracy of the bulging member and the geometric accuracy is one of the important forming quality control indicators for THF high-precision parts. In this study, the deformation data in the hydraulic test of the welded tube is obtained by DIC technology, and the deformation Radial radius of curvature is obtained. Research indicates:

1. The bulging shape of the medium and low velocity decreases in the early stage, but the shape deviation rate decreases faster at the medium velocity, and the shape deviation rate of the medium velocity increases faster than the low velocity in the late bulging.

2. The final shape deviation rate is about 1.3% at low velocity, and the final shape deviation rate is about 2.26 % at medium velocity. The bulging height of the medium velocity is smaller 0.34 mm than

that at the low velocity, indicating that the radial bulging profile of the welded tube at a lower velocity is more regular and the bulging height is higher.

3. When the radial bulging height does not exceed 2 mm, the shape deviation rate both close to the shape deviation rate of the initial welded tube. By properly controlling the bulging height, a cross-sectional shape closer to a circle can be obtained.

References

[1] Kim H S. Experimental and Numerical Analysis of Hydroformed Tubular Materials for Superconducting Radio Frequency (SRF) Cavities[J]. 2016.

[2] Panda S K, Kumar D R, Kumar H, et al. Characterization of tensile properties of tailor welded IF steel sheets and their formability in stretch forming[J]. Journal of Materials Processing Technology, 2007, 183(2–3):321-332.

[3] Chu G, Li F. Plastic deformation regularity of tailor-welded tube hydroforming[J]. Indian Journal of Engineering & Materials Sciences, 2010, 17(1):13-19.

[4] Zadpoor A A, Sinke J, Benedictus R, et al. Mechanical properties and microstructure of friction stir welded tailor-made blanks[J]. Materials Science & Engineering A, 2008, 494(1–2):281-290.

[5] Strano M, Altan T. An inverse energy approach to determine the flow stress of tubular materials for hydroforming applications[J]. Journal of Materials Processing Technology, 2004, 146(1):92-96.

[6] Hwang Y M, Wang C W. Flow stress evaluation of zinc copper and carbon steel tubes by hydraulic bulge tests considering their anisotropy[J]. Journal of Materials Processing Tech, 2009, 209(9):4423-4428.

[7] Mei Z, Guo K, Yang H. Advances and trends in plastic forming technologies for seamed tubes[J]. Chinese Journal of Aeronautics, 2016, 29(2):305-315.

Acknowledgements

This work was financially supported by the National Natural Science Foundation of China (51564007), the Natural Science Foundation of Guangxi Province (2015GXNSFBA139223 and 2017GXNSFAA198133), and Guangxi Key Laboratory of Manufacturing System & Advanced Manufacturing Technology (14-045-15-005Z).

IOP Conf. Series: Materials Science and Engineering **544** (2019) 012037 doi:10.1088/1757-899X/544/1/012037

A Design Method of Chaotic Synchronous Multi-stable Manifold

Z L Zuo[1] and X Yu[2]

1 College of Power Engineering, Naval Univ.of Engineering, Wuhan, China
2 College of Naval Architecture and Ocean Engineering, Naval Univ. of Engineering, Wuhan, China
2 Corresponding author
E-mail: zuozl_nue@163.com, yuxiang_nue@sina.com
Abstract. Line spectra in the low frequency band of the underwater radiated noise are the main factors affecting the acoustic stealth performance of submarines. The chaotification technique for line spectra reduction can realize spectra reconstruction and reduce line spectra intensity by converting line spectra in the low frequency band into chaotic spectra in the broadband frequency. Previous studies have shown that the method of the generalized synchronization of chaos can generate continuous chaos of the system under variable operating conditions, and can reduce the amplitude significantly in some cases, which provides a new idea for the theory of the chaotification technique for line spectra reduction. However, this method can only select the appropriate synchronization control strategy of chaos through "trial and error" method to achieve chaos with small amplitude. Therefore, a design method of multiple stable synchronous manifolds including stable chaotic attractors with small amplitude is proposed. This method is verified and illustrated by the coupled Duffing system, and applied to the two degree-of-freedom (DOF) nonlinear vibration isolation system(VIS).

1. Introduction

With its superior concealment and assault, the submarine has become one of the most important maritime military forces since its naissance. In physical fields such as sound, infrared, electromagnetic, and wake, only sound waves can travel long distances in seawater, which becomes the main signal for exposing the submarine's whereabouts. After World War II, through the unremitting efforts of experts and scholars in the field of vibration and noise control of submarine, spectra intensity in the high frequency band of the underwater radiated noise has been greatly reduced, and the line spectra in low-frequency bands has increasingly become a major factor affecting the safety of submarines [1].

Therefore, scholars from various countries have begun to focus on how to reduce the spectra intensity and hide the information about the submarine carried by the line spectrum. Among them, Zhu[2-7] innovatively proposed the chaotification technique for line spectra reduction, which can transform the line spectra into a broadband chaotic spectra through the nonlinear VIS in chaotic state, thereby weaken and reconstruct the spectra signal in the radiated water sound of submarine. In order to realize the continuous chaos of the VIS, a chaotification technique for line spectra reduction based on generalized synchronization of chaos is proposed by Yu [3]. Lou [8] realized the continuous chaos of the hard Duffing VIS by using the generalized synchronization of chaos based on a parameter-driven method. Sang [9] accomplished the generalized synchronization of chaos of the heterogeneous two-dimensional delay systems by using the stability theory of functional differential equations, and simplified the method of the generalized synchronization of chaos. Yang [10] actualized continuous chaos with small amplitude of 2 DOF high-static-low-dynamic-stiffness VIS by using the generalized synchronization of chaos base on a open-loop plus nonlinear closed-loop coupling method. However, at present, in order to generate continuous chaos with small amplitude by utilizing the method of the generalized synchronization of chaos, the "trial and error" is the only way to select the appropriate chaotic synchronization control strategy. Consequently, there is still no systematic method.

In this paper, a design method of multiple stable synchronous manifolds is proposed, which ensures that there are chaotic attractors with small amplitude in the phase space, thus achieving the continuous chaotic motion of the small amplitude of the VIS.

IOP Conf. Series: Materials Science and Engineering **544** (2019) 012037 doi:10.1088/1757-899X/544/1/012037

2. Study on the design method

2.1 Presentation of the method

Taking the one-way coupling system as the research object:

$$\dot{\boldsymbol{x}} = \boldsymbol{F}(\boldsymbol{x}) \tag{1a}$$

$$\dot{\boldsymbol{y}} = \boldsymbol{G}(\boldsymbol{y}) + \mathrm{k}\boldsymbol{P}(\boldsymbol{x}) \tag{1b}$$

Where $\boldsymbol{x}$ and $\boldsymbol{y}$ represent the state vector of the driving system $\boldsymbol{F}(\boldsymbol{x})$ and the response system $\boldsymbol{G}(\boldsymbol{y})$, respectively. $\boldsymbol{P}(\boldsymbol{x})$ denotes the control function and k is a scalar representing the coupling coefficient

Under the premise that the driving system $\boldsymbol{F}(\boldsymbol{x})$, the response system $\boldsymbol{G}(\boldsymbol{y})$ and the control function $\boldsymbol{P}(\boldsymbol{x})$ are known, Yu [11] obtained the functional relationship between the state vector $\boldsymbol{y}$ of the response system and state vector $\boldsymbol{x}$ of the driving system by the approximate central manifold method[12], that is, the synchronous manifold $\boldsymbol{y}$=$\boldsymbol{H}(\boldsymbol{x})$. In this paper, a reverse thinking is adopted. Under the premise that the driving system $\boldsymbol{F}(\boldsymbol{x})$ and the response system $\boldsymbol{G}(\boldsymbol{y})$ are known, and the control function $\boldsymbol{P}(\boldsymbol{x})$ is unknown, a design method of multiple stable synchronous manifold is proposed. The detailed idea is as follows: Firstly, try to construct a synchronous manifold $\boldsymbol{y}$=$\boldsymbol{H}(\boldsymbol{x})$ according to the need, and design the control function $\boldsymbol{P}(\boldsymbol{x})$ by the approximate central manifold method, and solve the simultaneous equations combined with $\boldsymbol{y}$=$\boldsymbol{H}(\boldsymbol{x})$, the driving system and the response system including the unknown $\boldsymbol{P}(\boldsymbol{x})$ to obtain the specific expression of $\boldsymbol{P}(\boldsymbol{x})$. Then, the obtained $\boldsymbol{P}(\boldsymbol{x})$ is substituted into the coupled system. According to the stability theorem of synchronous manifold proposed by Yu[13], it is judged whether $\boldsymbol{y}$=$\boldsymbol{H}(\boldsymbol{x})$ is stable in the coupled system containing the specific $\boldsymbol{P}(\boldsymbol{x})$. If $\boldsymbol{y}$=$\boldsymbol{H}(\boldsymbol{x})$ is stable, the stable synchronous manifold $\boldsymbol{y}$=$\boldsymbol{H}(\boldsymbol{x})$. is obtained. If $\boldsymbol{y}$=$\boldsymbol{H}(\boldsymbol{x})$ is unstable, then $\boldsymbol{y}$=$\boldsymbol{H}(\boldsymbol{x})$ needs to be perturbed and the above process is repeated until the resulting synchronous manifold is stable. Finally, the approximate central manifold method is used to find other synchronous manifolds of the coupled system under the control of $\boldsymbol{P}(\boldsymbol{x})$ and determine its stability.

2.2 Verification and illustration

This design method is verified and illustrated by the coupled Duffing system as below. The single DOF oscillator under harmonic excitation is selected as the driving system:

$$\begin{bmatrix} \dot{x}_1 \\ \dot{x}_2 \\ \dot{x}_3 \\ \dot{x}_4 \end{bmatrix} = \begin{bmatrix} x_2 \\ -wx_1 - vx_1^2 - x_1^3 - ux_2 + f\cos(\omega t) \\ -\omega x_4 \\ \omega x_3 \end{bmatrix} \tag{2}$$

Where the introduction of x_3 and x_4 transforms the driving system into an autonomous system, and x_3=$\cos(\omega t)$, $x_4 = \sin(\omega t)$ respectively. Set the parameters of the driving system as below: $\omega = 3.9311$, $w = 4$, $v = 3$, $u = 0.15$, $f = 9$.

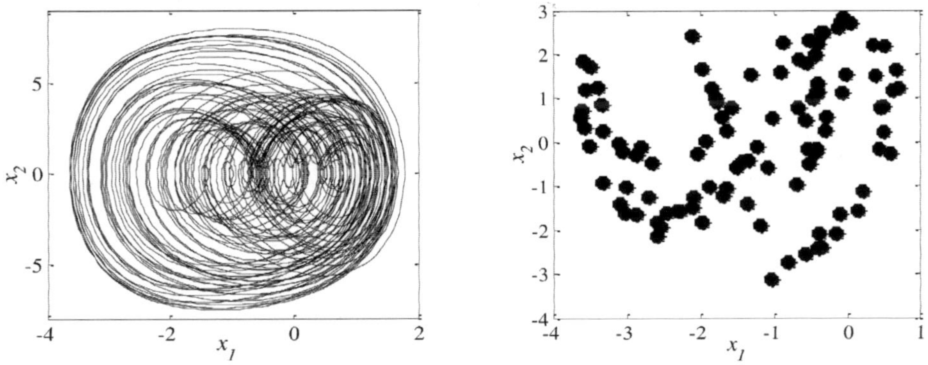

(a) (b)

Figure 1. The phase diagram(a) and the poincare map(b) of the driving system

When the initial condition is (0, 0), the phase diagram and the Poincaré map are shown in figure 1, and the power spectra are shown in figure 2. It can be seen from figure 1 and figure 2 that the movement trajectory of the driving system is entangled and spread over a certain phase space, and the power spectra exhibit the characteristics of continuity. It can be concluded that the driving system is in a chaotic state of motion under the condition of these parameters.

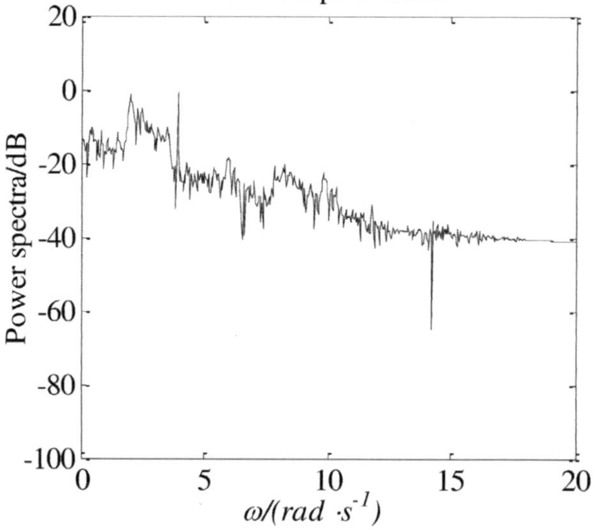

Figure 2. The power spectra of the driving system

The nonlinear Duffing VIS with control function P(x) is selected as the response system:

$$\begin{bmatrix} \dot{y}_1 \\ \dot{y}_2 \\ \dot{y}_3 \\ \dot{y}_4 \end{bmatrix} = \begin{bmatrix} y_2 \\ -\varepsilon y_2 - y_1 - 2y_1^2 - y_1^3 + d\cos(\omega t) \\ -\omega y_4 \\ \omega y_3 \end{bmatrix} + \mathrm{k}\boldsymbol{P}(\boldsymbol{x}) \tag{3}$$

The response system is under the chaotification control of the the driving system in a manner of state-driven generalized synchronization, where k=0.1, the introduction of y_3 and y_4 transforms the response system into an autonomous system, and $y_3 = x_3 = \cos(\omega t)$, $y_4 = x_4 = \sin(\omega t)$ respectively. Set the parameters of the response system as below: $\varepsilon = 0.15$, $d = 9$.

Try to construct a synchronous manifold:

$$\boldsymbol{y} = \boldsymbol{H}_1(\boldsymbol{x}) = \boldsymbol{A}_1 + \boldsymbol{B}_1\boldsymbol{x} \tag{4}$$

Where $\boldsymbol{A}_1 = [0, 0, 0, 0]^{\mathrm{T}}$,

$$\boldsymbol{B}_1 = \begin{bmatrix} 0 & 0.033 & -0.625 & -0.056 \\ -0.133 & -0.005 & 0.080 & 2.457 \\ 0 & 0 & 1 & 0 \\ 0 & 0 & 0 & 1 \end{bmatrix}$$

It is worth noting that multiple attempts are required to construct this synchronous manifold. Firstly, considering the case where the matrix $\boldsymbol{B}_1$ in the synchronous manifold retains one bit after the decimal point, the parameters are adjusted according to the obtained control function $\boldsymbol{P}(\boldsymbol{x})$. Then, the above steps are repeated by sequentially considering the case where the matrix $\boldsymbol{B}_1$ in the synchronous manifold retains two and three bit after the decimal point. The form of the resulting control function $\boldsymbol{P}(\boldsymbol{x})$should be quite simple, making it easy to implement control strategy when applied to engineering practice.

Since $\boldsymbol{y} = \boldsymbol{H}(\boldsymbol{x})$ is differentiable, then there is

$$\dot{\boldsymbol{y}} = \boldsymbol{D}_x \boldsymbol{H}(\boldsymbol{x})\dot{\boldsymbol{x}} \tag{5}$$

Substitute equation (1a) and equation (1b) into equation (5) and get

$$G(H(x)) + kP(x) - D_x H(x)F(x) = 0 \tag{6}$$

The control function $P(x)$ is designed by using the approximate central manifold method, expand $P(x)$ into a Taylor series and keep the first two terms: $P(x)=Mx+N$. Substitute $y=H_1(x)$, the equation of the driving system and the response system with the control function $P(x)$ into the equation (6), and get

$$P(x)=\begin{bmatrix} 0.01 & 0.0005 & -0.031416 & -0.000625 \\ 0.0005 & -1 & 0.007127 & -0.01938 \\ 0 & 0 & 0 & 0 \\ 0 & 0 & 0 & 0 \end{bmatrix}\begin{bmatrix} x_1 \\ x_2 \\ x_3 \\ x_4 \end{bmatrix}+\begin{bmatrix} 0 \\ 0 \\ 0 \\ 0 \end{bmatrix} \tag{7}$$

Ignore the calculation error within the allowable range and round up $P(x)$ to

$$P(x)=\begin{bmatrix} 0 & 0 & 0 & 0 \\ 0 & -1 & 0 & 0 \\ 0 & 0 & 0 & 0 \\ 0 & 0 & 0 & 0 \end{bmatrix}\begin{bmatrix} x_1 \\ x_2 \\ x_3 \\ x_4 \end{bmatrix}+\begin{bmatrix} 0 \\ 0 \\ 0 \\ 0 \end{bmatrix} \tag{8}$$

According to the stability theorem of synchronous manifold proposed by Yu [13], the following inference can be drawn: if the maximum condition Lyapunov exponent(CLE) of one-way coupled system (1) is less than 0, then the synchronous manifold is asymptotically stable. If the maximum CLE of the one-way coupled system (1) is bigger than 0, the synchronous manifold is unstable. Substitute the obtained $P(x)$ into the coupled system, and the maximum CLE of the system under the action of mapping $H_1(x)$ is -0.7449, indicating that the corresponding synchronous manifold is stable.

Another synchronous manifold of the coupled system under the control of $P(x)$ is obtained by using the approximate central manifold method:

$$y=H_2(x)=A_2+B_2x \tag{9}$$

Where $A_2 = [-1, 0, 0, 0]^T$,

$$B_2 = \begin{bmatrix} 0 & 0.025 & -0.584 & -0.035 \\ -0.100 & -0.004 & 0.088 & 2.295 \\ 0 & 0 & 1 & 0 \\ 0 & 0 & 0 & 1 \end{bmatrix}$$

In this case, the values of B_2 are exactly three decimal places. Under the action of mapping $y=H_2(x)$, the maximum CLE of the system is -0.7502. Similarly, it can be concluded that the corresponding synchronization manifold is stable.

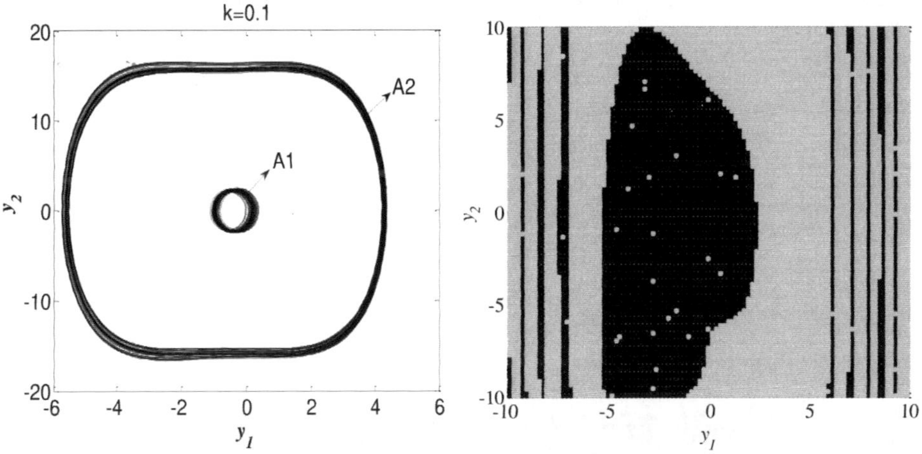

(a) (b)

Figure 3. The phase diagram(a) and the attraction basins(b) of the coexistent attractors of the coupled Duffing system when k=0.1

The validity of this design method will be proved by numerical simulation. Due to different initial conditions, the system operates in two different chaotic attractors, and the phase diagram is shown in figure 3(a).When the initial condition is (0,0,0,0), the system runs on a stable attractor A1 with smaller amplitude.When the initial condition is (0,0,4,0), the system runs on a stable attractor A2 with larger amplitude.The coexistent chaotic attractors all have their attraction basins, as shown in figure 3(b). The black region is the attraction basin of the attractor with smaller amplitude, and the gray region is the attraction basin of the attractor with larger amplitude. It can be seen that there are two stable synchronous manifolds with different amplitudes in the coupled Duffing system, which proves the validity of the proposed method.

3. Further application of the method

In order to further extend the application range of this design method, the application of the method in the 2 DOF nonlinear VIS is considered. The chaotic single DOF oscillator represented by equation(1) is still chosen as the driving system,. A 2 DOF nonlinear VIS with control function ***P***(***x***) is selected as the response system :

$$\begin{bmatrix} \dot{y}_1 \\ \dot{y}_2 \\ \dot{y}_3 \\ \dot{y}_4 \\ \dot{y}_5 \\ \dot{y}_6 \end{bmatrix} = \begin{bmatrix} y_2 \\ -\delta_1(y_2-y_4)-(y_1-y_3)+\xi_1(y_1-y_3)^2-(y_1-y_3)^3+d\cos(\omega t) \\ y_4 \\ -\mu\delta_2 y_4-\mu p_2 y_3+\mu\delta_1(y_2-y_4)+\mu(y_1-y_3)-\mu\xi_1(y_1-y_3)^2+\mu(y_1-y_3)^3 \\ -\omega y_6 \\ \omega y_5 \end{bmatrix} + \mathrm{k}\boldsymbol{P}(\boldsymbol{x}) \tag{10}$$

The response system is still under the chaotification control of the the driving system in a manner of state-driven generalized synchronization, where k=0.1, the introduction of y_5 and y_6 transforms the response system into an autonomous system, and $y_5=x_3=\cos(\omega t)$, $y_6=x_4=\sin(\omega t)$ respectively. Set the parameters of the response system as below: $\delta_1=\delta_2=0.8$, $\xi_1=2$, $d=20$, $\mu=0.2$, $p_2=2$.

Try to construct a synchronous manifold:

$$\boldsymbol{y}=\boldsymbol{H}_3(\boldsymbol{x})=\boldsymbol{A}_3+\boldsymbol{B}_3\boldsymbol{x} \tag{11}$$

where $\boldsymbol{A}_3=[0,0,0,0,0,0]^{\mathrm{T}}$,

$$\boldsymbol{B}_3=\begin{bmatrix} 0.110 & -0.050 & -1.234 & 0.399 \\ 0.200 & 0.118 & 1.118 & 4.850 \\ -0.027 & 0.054 & -0.022 & -0.186 \\ -0.218 & -0.035 & -0.239 & 0.086 \\ 0 & 0 & 1 & 0 \\ 0 & 0 & 0 & 1 \end{bmatrix}$$

The control function ***P***(***x***) is designed by using the approximate central manifold method, expand ***P***(***x***) into a power series and keep the first two terms: ***P***(***x***)=***Mx***+***N***. Here, ***P***(***x***) can be expressed as follows:

$$\boldsymbol{P}(\boldsymbol{x})=\begin{bmatrix} m_{11} & m_{12} & m_{13} & m_{14} & 0 & 0 \\ m_{21} & m_{22} & m_{23} & m_{24} & 0 & 0 \\ m_{31} & m_{32} & m_{33} & m_{34} & 0 & 0 \\ m_{41} & m_{42} & m_{43} & m_{44} & 0 & 0 \\ m_{51} & m_{52} & m_{53} & m_{54} & 0 & 0 \\ m_{61} & m_{62} & m_{63} & m_{64} & 0 & 0 \end{bmatrix} \cdot \begin{bmatrix} x_1 \\ x_2 \\ x_3 \\ x_4 \\ 0 \\ 0 \end{bmatrix} + \begin{bmatrix} n_1 \\ n_2 \\ n_3 \\ n_4 \\ n_5 \\ n_6 \end{bmatrix}$$

Substitute ***y***=***H***3(***x***), the equation of the driving system and the response system with the control function ***P***(***x***) into the equation (6) and get

$$\boldsymbol{P}(\boldsymbol{x})=\begin{bmatrix} 0 & -0.005 & 0.005089 & -0.009774 & 0 & 0 \\ -0.006 & 2.007 & 0.01435 & 0.012302 & 0 & 0 \\ 0.02 & -0.001 & -0.061846 & 0.004842 & 0 & 0 \\ 0.0004 & 2.0043 & 0.013146 & -0.003471 & 0 & 0 \\ 0 & 0 & 0 & 0 & 0 & 0 \\ 0 & 0 & 0 & 0 & 0 & 0 \end{bmatrix}\begin{bmatrix} x_1 \\ x_2 \\ x_3 \\ x_4 \\ 0 \\ 0 \end{bmatrix}+\begin{bmatrix} 0 \\ 0 \\ 0 \\ 0 \\ 0 \\ 0 \end{bmatrix} \tag{12}$$

Ignore the calculation error within the allowable range and round up $\boldsymbol{P}(\boldsymbol{x})$ to

$$\boldsymbol{P}(\boldsymbol{x})=\begin{bmatrix} 0 & 0 & 0 & 0 & 0 & 0 \\ 0 & 2 & 0 & 0 & 0 & 0 \\ 0 & 0 & 0 & 0 & 0 & 0 \\ 0 & -2 & 0 & 0 & 0 & 0 \\ 0 & 0 & 0 & 0 & 0 & 0 \\ 0 & 0 & 0 & 0 & 0 & 0 \end{bmatrix}\begin{bmatrix} x_1 \\ x_2 \\ x_3 \\ x_4 \\ 0 \\ 0 \end{bmatrix}+\begin{bmatrix} 0 \\ 0 \\ 0 \\ 0 \\ 0 \\ 0 \end{bmatrix} \tag{13}$$

Substitute the obtained $\boldsymbol{P}(\boldsymbol{x})$ into the coupled system, and the maximum CLE of the system under the action of mapping $\boldsymbol{y}=\boldsymbol{H}_3(\boldsymbol{x})$ is -0.0594, indicating that the corresponding synchronous manifold $\boldsymbol{y}=\boldsymbol{H}_3(\boldsymbol{x})$ is stable.

Another synchronous manifold of the coupled system under the control of $\boldsymbol{P}(\boldsymbol{x})$ is obtained by using the approximate central manifold method:

$$\boldsymbol{y}=\boldsymbol{H}_4(\boldsymbol{x})=\boldsymbol{A}_4+\boldsymbol{B}_4\boldsymbol{x} \tag{14}$$

where $\boldsymbol{A}_4=[1, 0, 0, 0, 0, 0]^T$,

$$\boldsymbol{B}_4=\begin{bmatrix} 0.063 & -0.034 & -1.196 & 0.316 \\ 0.137 & 0.068 & 0.934 & 4.700 \\ -0.016 & 0.051 & -0.028 & -0.168 \\ -0.206 & -0.023 & -0.199 & 0.111 \\ 0 & 0 & 1 & 0 \\ 0 & 0 & 0 & 1 \end{bmatrix}$$

In this case, the values of $\boldsymbol{B}_4$ are exactly three decimal places. Under the action of mapping $\boldsymbol{y}=\boldsymbol{H}_4(\boldsymbol{x})$, the maximum CLE of the system is -0.2033. Similarly, it can be concluded that the corresponding synchronization manifold $\boldsymbol{y}=\boldsymbol{H}_4(\boldsymbol{x})$ is stable.

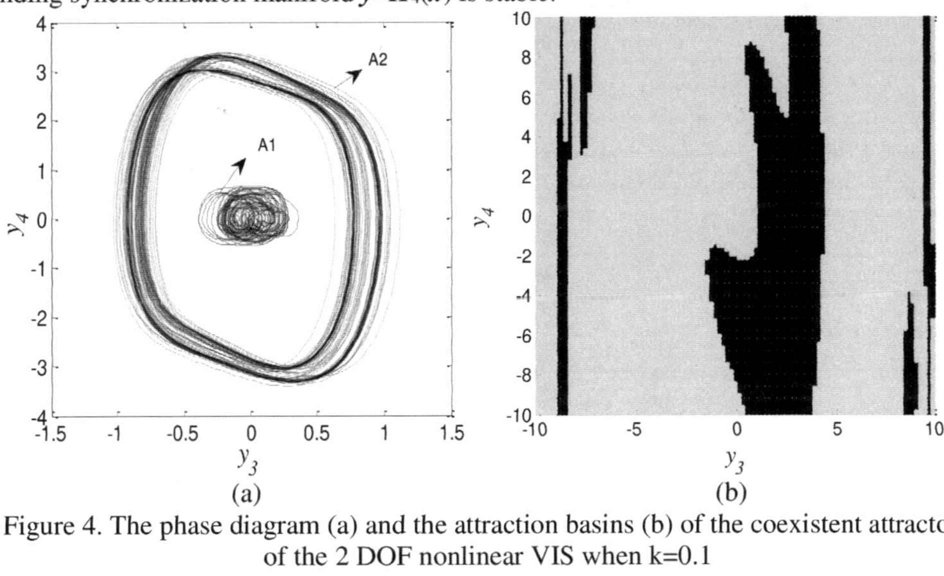

Figure 4. The phase diagram (a) and the attraction basins (b) of the coexistent attractors of the 2 DOF nonlinear VIS when k=0.1

Due to different initial conditions, the system operates in two different chaotic attractors, and the phase diagram is shown in figure 4(a).When the initial condition is (0,0,0,-5,2,0), the system runs on a stable attractor A1 with smaller amplitude.When the initial condition is (0,0,0,-5,-4,6), the system runs on a stable attractor A2 with larger amplitude.The coexistent chaotic attractors all have their attraction basins, as shown in figure 4(b). The black region is the attraction basin of the attractor A1 with smaller amplitude, and the gray region is the attraction basin of the attractor A2 with larger amplitude. It can be seen that there are two stable synchronous manifolds with different amplitudes in the 2 DOF nonlinear VIS, which proves the validity of the proposed method.

4. Conclusions

Line spectra in the low frequency band of the underwater radiated noise are the main factors affecting the acoustic stealth performance of submarines. The chaotification technique for line spectra reduction can realize spectra reconstruction and reduce line spectra intensity. The method of the generalized synchronization of chaos can generate continuous chaos of the system under variable operating conditions, and can reduce the amplitude significantly in some cases, which provides a new idea for the theory of the chaotification technique for line spectra reduction. However, at present, in order to generate continuous chaos with small amplitude by utilizing the method of the generalized synchronization of chaos, the "trial and error" method is the only way to select the appropriate synchronization control strategy of chaos. Consequently, there is still no systematic method. In this paper, therefore, a design method of multiple stable synchronous manifold is proposed to make the system have the chaotic attractor with small amplitude in phase space. This method is verified and illustrated by the coupled Duffing system, and applied to the 2 DOF nonlinear VIS.

Acknowledgment

This work is supported by the National Natural Science Foundation of China (Grant no. 51679245)

References

[1] Liu P and Liu G L 2013 *Acoust. Elec. Eng.* **112(4)** 26-27(in Chinese)
[2] Zhu S J, Jiang R J and He L 2003 *J. Nav. Uni. Eng.* **15(1)** 19-22(in Chinese)
[3] Lou J J 2006 *Nav. Uni. Eng.* PhD Thesis (in Chinese)
[4] Yu X 2008 *Nav. Uni. Eng.* PhD Thesis (in Chinese)
[5] Zhang Z H, Zhu S J and Lou J J 2011 *J. Vib. Sho.* **30(7)** 40-44(in Chinese)
[6] Zhang J, Xu D L, Li Y L and Zhou J X 2014 *J. Phys.* **63(18)** 120-30(in Chinese)
[7] Zhang Z H, Zhu S J and He Q W 2012 *J. Vib. Eng.* **25(01)** 30-37(in Chinese)
[8] Lou J J, Zhang H, Yu X and Zhu S J 2015 *J. Vib. Sho.* **34(14)** 106-09(in Chinese)
[9] Sang J Y, Wang J and Yue L J 2010 *J. Phys.* **59 (11)** 7618-22(in Chinese)
[10] Yang Q C, Chai K, Lou J J and Zhu S J 2014 *J. Vib. Eng.* **2018(04)** 620-28 (in Chinese)
[11] Yu X, Zhu S J and Liu S Y 2008 *J. Phys.* **57(05)** 2761-69(in Chinese)
[12] Brown R 1998 *Phys. Rev. Lett.* **81(22)** 4835-4838
[13] Yu X, Zhu S J and Liu S Y 2007 *J. S. V.* **306** 835-848.

IOP Conf. Series: Materials Science and Engineering **544** (2019) 012038 doi:10.1088/1757-899X/544/1/012038

Fatigue damage growth in notched laminates

A Muc[1], M Chwał[1] and P Kędziora[1]

[1] Institute of Machine Design, Cracow University of Technology, Warszawska 24, 31-155 Cracow, Poland
E-mail: olekmuc@mech.pk.edu.pl

Abstract: The issue discussed in the current paper deals with experimental and finite element analysis of fatigue problems for notched rectangular multilayered plates. The fatigue damage growth and fracture at holes in reinforced laminates loaded in tension is presented. The analysis concerns the ±45° glass/epoxy laminated plates with several types of holes such as vertical elliptical, horizontal elliptical and circular. Also a few remarks on the static analytical solutions and fatigue live predictions are presented. Here, the initial results are only demonstrated, which are fundamental for the further analysis. The experiments shown that independently on the shape of the hole the fatigue process includes three stages, i.e. the matrix cracking, the delamination between the plies and the final failure. The first stage of the failure can be described with the help of the standard finite element codes.

1. Introduction

A great deal of research has been devoted to a study of the mechanism of fatigue, and yet there is still not a complete understanding of the phenomenon and the effect of fatigue loading on the crack propagation under the complex stress of boundary and loading conditions. The behavior and fatigue degradation of notched composite laminates under static and fatigue loading exhibits many features which usually are not observed for homogeneous materials such as metals. As the most significant features of the fatigue phenomena in composite laminates, one can mention the following:

- during fatigue loading of a notched laminate, high axial shear stresses exist near the notch root; these high shear stresses will cause the material in the vicinity of the notch to be degraded much more rapidly than throughout the rest of the laminate,

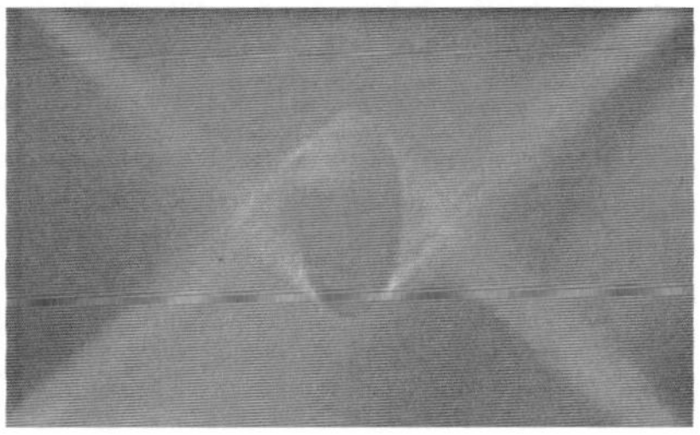

Figure 1. Fatigue failure modes in notched unidirectional glass/epoxy laminates (eight layers $\pm45^0$) - 1300000 cycles (prior to final failure) – a vertical ellipse.

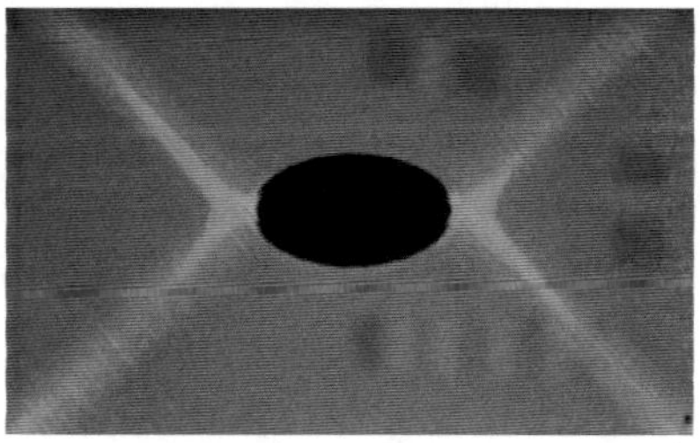

Figure 2. Fatigue failure modes in notched unidirectional glass/epoxy laminates (eight layers $\pm45^0$) – 1500000 cycles (prior to final failure) – a horizontal ellipse.

- frequently, the direction of crack growth in a fiber composite laminate is not perpendicular to the applied load, even for perpendicular slit notches (figure 1); it is worth to note that that effect is usually connected with the shape of the hole – Ref. [1] (compare figure 1 and 2),
- static failure modes may be completely different from the mode of failure observed during or after a program of cyclic loading,
- experiments demonstrate evidently the interaction of delamination and cracks (see Ref. [2] and figure 3).

Figure 3. Failure form (splitting) in notched unidirectional glass/epoxy laminates (eight layers $\pm 45^0$) – a circular hole

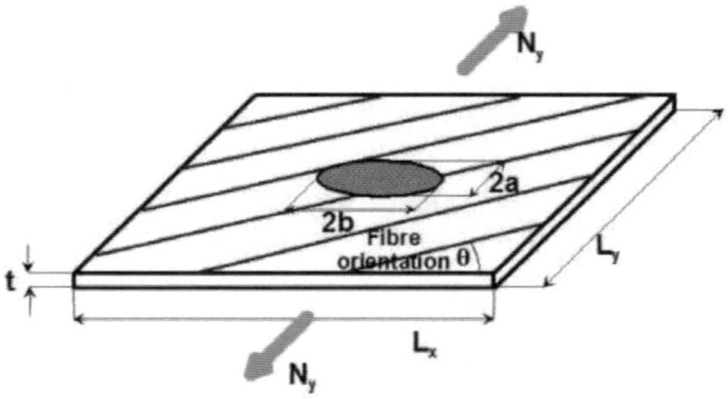

Figure 4. A multilayered laminated plate with an arbitrary shaped hole under in-plane loading.

In general, the problems considered herein deal with the rectangular multilayered plates having centrally located holes and subjected to the remote tension – figure 4. Here fatigue damage growth and fracture at holes in reinforced laminates loaded in tension have been presented. In the beginning, a few remarks on the experimental results are demonstrated. Then, the possibilities of the stress estimations around holes for static tension are briefly discussed. At the end of the paper, various models used for the fatigue life predictions are studied and they are supplemented with the finite element results.

2. A short review of the fatigue life prediction methods

2.1. Static analytical solutions

For plates having circular, elliptical or irregular shaped holes under static tension (or compression), the analytical solutions can be found with the use of the Airy function, compatibility equation and complex functions. Lekhnitskii [3] used one mapping function for isotropic plates. Lu et al. [4] introduced three different mapping functions for generally anisotropic plates. The broader discussion of those problems and solutions can be found in Muc et al. [5-8].

2.2. Fatigue life prediction

Fatigue cracks appear in components after a certain number of load cycles at nominal stress levels which are often below the tensile strength of the material. The stress-life method [9-10] was the first approach used in an attempt to understand and quantify metal fatigue. This approach requires extensive experimental work and does not take into account the actual damage mechanisms. The classical S-n correlation is given by an equation of the form:

$$\mathrm{Log}(n)=A-B\mathrm{Log}(Seq);\ Seq=S_{max}(1-R)C-D;\ R=S_{min}/S_{max} \tag{1}$$

where S_{max} is the maximum stress, S_{min} is the minimum stress and n is the life in cycles. The quantities A, B, C and D may be viewed as material properties which have to be determined through constant amplitude stress-controlled fatigue testing. It is common practice to define the S-n diagram for constant R-value. Echtermeyer [11] gave the parameters of a standard R=-1 strain-life log-log S-n curve, for mainly [0°, 90°] lay-ups of polyester and vinylester-glass laminates, and some criteria for its use. In the case of the existence of a fatigue limit, eq. (1) is sometimes formulated as:

$$S = S_{\infty} + B/nx \tag{2}$$

where S_∞ is the fatigue or endurance limit stress – see Och [12]. Sims and Brogdon [13] used a modified power law relationship to fit various datasets as follows:

$$n= [b/(S-a+cA-y)](1/x)\ ,\ A=(1-R)/(1+R) \quad (3)$$

and *a*, *b*, *x*, and *y* are fitting parameters. Caprino and Giorlea [14] formulated a two-parameter S-n formulation from a strength degradation equation:

$$US - S_{max}= aS_{max}(1-R)(nc-1) \quad (4)$$

where US means the ultimate strength. Epaarachchi and Clausen [15] have recently expanded this to a three-parameter S-n formulation, which takes into account the dependency of stress ratio influence on dominant fibre angle, and frequency effects:

$$US -S_{max}= aS_{max}(1-\psi)1.6-\psi\ |\sin\theta|[S_{max}/US]\ 0.6-\psi\ |\sin\theta|\ (\ nc-1)/fc \quad (5)$$

The parameters ψ, θ, and *f* quantify the R-value dependence, dominant fibre angle, and frequency, resp. In general, the above methods contain an assumed form of R-value dependency and should only be used with care where this dependency is not experimentally verified. In most cases, the mean S-n curve is used to describe fatigue data. However, for certification purposes it may be worthwhile to describe the S-n data regarding percentiles other than this 50% survivability line – see, e.g., Och [12]. The next group of fatigue life prediction is directly connected with the S-n method although fatigue life is estimated with the use of the fatigue failure criteria or equivalently damage parameter or parameters (in general denoted by the symbol d) [16]. The fatigue life is established with the use of the failure criteria (e.g., in the form of the Tsai-Wu criterion):

$$F_{ij}S_iS_j+F_jS_j=1,\ i,j=1,2,6 \quad (6)$$

however, the coefficients are defined with the use of the three S-n curves. The application of the above method is presented by Muc et al. [17]. In view of the above concepts dealing with the fatigue damage analysis it is worth to mention Reifsnider's approach–the remaining (residual) strength of a "critical element" governs the life of the entire structure [18]. When composing his construction from individual layers, the designer would like to have design rules that are based on strains rather than stresses. Strains are equal for all layers (in the structural reference frame of a uniformly, uni-axially loaded laminate – the notched problem). Therefore, cyclic strain-controlled tests can better characterize fatigue behavior of material than cyclic stress-controlled tests, particularly in the low cycle fatigue region and/or in notched members. The strain-life curves are often called low cycle fatigue data because much of the data are for less than 10^5 cycles. The total strain amplitude can be resolved into elastic and plastic strain components from the steady-state hysteresis loops. Both the elastic and plastic curves can be approximated as straight lines. According to this approach, the fatigue life of a member subjected to fully reversed constant amplitude strain-controlled loading is given by an equation of the form:

$$\frac{\Delta\varepsilon}{2}=\frac{\Delta\varepsilon_e}{2}+\frac{\Delta\varepsilon_p}{2}=\frac{\sigma_f'}{E}\left(2N_f\right)^b+\varepsilon_f'\left(2N_f\right)^c \quad (7)$$

where: $\Delta\varepsilon/2$ - total strain amplitude, σ_f' - fatigue strength coefficient, b - fatigue strength exponent, ε_f' - fatigue ductility coefficient, c - fatigue ductility exponent, E - modulus of elasticity, $2N_f$ - reversals to failure. The strain-life approach results in better predictions in cases where considerable plastic deformation is present in the critical regions of a component. When the deformations are predominantly in the elastic region, this approach yields the same predictions as the much simpler stress-life approach.

3. Materials and methods

The samples applied in the fatigue tension experiments were made of 8-layer of unidirectional glass/epoxy composite. The mechanical properties of the composite are given in table 1.

Table 1. Mechanical properties of unidirectional glass/epoxy composite

Material	Fiber orientation	E_1 [GPa]	E_2 [GPa]	G_{12} [GPa]	ν_{12}
Unidirectional glass/epoxy	±45°	46.4	14.9	5.2	0.27

All specimens were made in autoclave under the pressure 0.4MPa and cured at the temperature 135°C. The dimensions of the laminated plates are shown in figure 4. The total thickness of the laminate was equal to t=2.2 mm and the width and the height were equal to $L_x=L_y=250$ mm. The diameter of the circular hole was equal to 50mm, whereas the semi-axes of the ellipses were following: vertical ellipse a=17.5 mm, b=35.7 mm; horizontal ellipse a=35.7 mm, b=17.5 mm.
The experimental fatigue tests were performed on the MTS Landmark 370 servo hydraulic testing machine with a FlexTest 40 controller and the MTS 793 System Software. The experimental settings of fatigue tests are presented in table 2.

Table 2. Fatigue tests settings

	Hole		
	Circular	Elliptical horizontal	Elliptical vertical
The average force P_m [kN]	44	38	50
Amplitude (P_{max}-P_m) [kN]	4	4	4
Frequency f [Hz]	15	30	30

4. Results and discussion

The experiments have been carried out for three shapes of cut-outs: circular, horizontal elliptical and vertical elliptical – figure 4 and table 2. As it may be observed the whole process of fatigue failure is identical for all shapes of the holes (figure 5). In general, it is composed of three parts:

- the first stage I is associated with the initiation of the crack (at four points around the hole) and further (rather rapid) development of the stiffness degradation due to the matrix cracking – see figure 1 and 2;

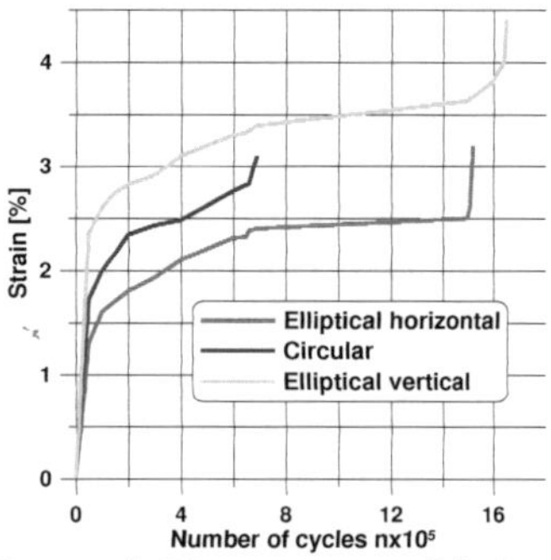

Figure 5. The process of fatigue failure – the strain corresponds to the total strain of the plate, i.e. $\varepsilon=(L_y-L_{y0})/L_{y0}$ where L_y – the actual plate length and L_{y0} – the initial plate length.

Figure 6. The final fatigue failure mode of the plate with the circular cutout.

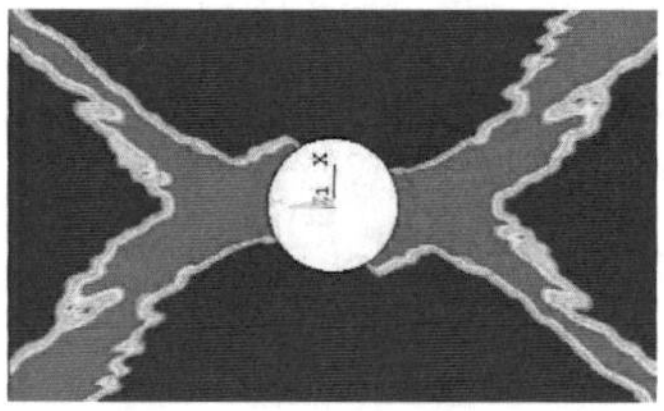

Figure 7. The propagation of fatigue failure.

- the second stage II is connected with the gradual (almost linear) increase of the strain that is caused by the delamination failure of the plies – figure 3;
- the last (third) III stage is connected with a rapid increase of strains and the final fatigue failure – see figure 6.

With the use of the FE modeling or the analytical approach, it is possible to capture the initiation of the first-ply-failure of the notched plate using, e.g., the Hashin or the Tsai-Wu criterion. In the stage I the matrix degradation can also be evaluated with the aid of the FE codes - the representative results of such an analysis are demonstrated in figure 7.
However, using the standard FE codes, it is impossible to detect the final fatigue failure of the notched laminates since the classical codes do not incorporate the evaluation and the final failure of the mixed modes that are connected with the simultaneous matrix cracking and delamination modes – see figure 5 and the discussion in Ref. [2]. Briefly speaking, the variety of fracture criteria (Paris, Wallker, Collpriest, Newmann etc. – see Ref. [18]) cannot be applied herein.

5. Summary

In the paper, the problems connected with the fatigue failure process of rectangular plates made of eight plies oriented at $\pm 45^0$, subjected to axial tension and having circular or two elliptical (vertical and horizontal) holes is discussed from both experimental and theoretical point of view. It is necessary to emphasize that the initial results are only demonstrated which will be fundamental for the further analysis. The experiments show that independently on the shape of the hole the fatigue process consists of three stages, i.e. the matrix cracking, the delamination between the plies and the final (very rapid) failure – fibres cracking and debonding. The first stage of the failure can be described with the use of the standard FE codes. The consistent characterization of the next stages of failure requires further investigations. The possible theoretical approaches and methods of the solution of the latter problem are also discussed herein.

References

[1] Durchlaub EC and Freeman RB 1974 *Design data for composite structure safe life prediction*, AFML-TR-73-225
[2] Hallett SR, Jiang WG, Khan B and Wisnom MR 2008 *Compos Sci Technol* **68** 80
[3] Lekhnitskii SG 1968 *Anisotropic plates*, New York, Gordon and Breach
[4] Lu AZ, Zhang N, Zhang XL and Li WS 2015 *Int J Geomech* **15**
[5] Muc A 2000 *Comput Struct* **76** 211
[6] Muc A and Ulatowska A 2012 *Compos Struct* **94** 1413
[7] Muc A and Romanowicz P 2017 *Compos Struct* **178** 27
[8] Muc A, Barski M, Chwal M, Romanowicz P and Stawiarski A 2018 *Compos Struct* **189** 117
[9] Bannantine JA, Corner JJ and Handrock JL 1990 *Fundamentals of metal fatigue analysis*, Prentice Hall
[10] Broek D 1989 *The practical use of fracture mechanics*, Kluwer Academic Publishers.
[11] Echtermeyer AT 1994 *Proc EWEA Conference and Exhibition*, p 391
[12] Och F 1983 *Fatigue Strength, AGARD-AG-292, Helicopter Fatigue design guide*, p 109
[13] Sims D and Brogdon V 1977 *ASTM STP* **636** 185
[14] Caprino G and Giorlea G 1999 *Compos A* **30** 299
[15] Epaarachchi JA and Clausen PD 2003 *Compos A* **34** 313
[16] Degrieck J and Paepagem V 2001 *Appl Mech Rev* **54** 279
[17] Muc A, Kedziora P and Krawiec Z 2011 *Procedia Eng* **10** 1315
[18] Reifsnider K and Stinchcomb WW 1986 *Composite Materials: Fatigue and Fracture*, ASTM STP 907, ed HT Hahn, American Society for Testing and Materials: West Conshohocken, PA. p 298

IOP Conf. Series: Materials Science and Engineering **544** (2019) 012039 doi:10.1088/1757-899X/544/1/012039

Eigenvalue problem of confined quantum dots

A Muc[1], M Chwał[1] and P Kędziora[1]

[1] Institute of Machine Design, Cracow University of Technology, Warszawska 24, 31-155 Cracow, Poland
E-mail: olekmuc@mech.pk.edu.pl

Abstract: The analysis of optoelectromechanical properties of nanostructures in band structure engineering is discussed in the paper. It is demonstrated that the design of semiconductors is based on the solution of different forms of the Schrödinger (Helmholtz) equation, dependent on the form of the Hamiltonian characterizing quantum effects. The formulation can lead to the linear or nonlinear eigenvalue problems. Then, the methods of solutions are also presented.

1. Introduction

Today the advances in chemical synthesis and manufacture techniques have led to novel nano-sized materials demonstrating unique and often unexpected properties. These new nanosystems offer the ability to control their electronic and optical behavior through the sample's size, shape and topology [1-4]. In recent decades semiconductor quantum dots (QD) have been the subject of many experimental, theoretical and technological investigations [5-9]. The design of nanoelectronic devices requires a clear understanding of the fundamental properties of nanomaterials. When nanomaterials absorb a light quantum, two charged particles are created simultaneously, an electron and a hole. Knowledge of photoinduced charge carrier dynamics in nanomaterials will help to achieve an effective functionality of prospective nanonelectronic devices. Here, the eigenproblems analysis is presented for selected QD having the conical shape embedded in a cylindrical matrix.

2. Materials and methods

2.1 Quantum dots

QDs are tiny dimensionally confined (typically semiconductor) objects where quantum effects become obvious, for example, energy spectra become discrete. Of particular interest is the class of devices that are composed of combinations of lattice mismatched materials. These material combinations, such as Si_xGe_{1-x}/Si and $In_xGa_{1-x}As/GaAs$, where x indicates the fractional content of alloying material, are selected primarily on the basis of their electronic properties and to some extent for convenience of processing [10]. Quantum effects begin to dominate as the size of semiconductor structures approaches the electron de Broglie wavelength. In low-dimensional semiconductor nanostructures (LDSN), the motion of electrons can be confined spatially, from one, two, and even three spatial directions. The operation of semiconductor quantum devices is based on the confinement of individual electrons and holes in one spatial dimension (quantum wells), two spatial dimensions (quantum wires) or three spatial dimensions (quantum dots - QDs) [2].

Yet fundamental understanding of the underlying physics responsible for carrier dynamics and the specific role that phonons play in the relaxation mechanisms in QDs is still lacking. In such nanostructures, the free carriers are confined to a small region of space by potential barriers, and if the size of this region is less than the electron wavelength, the electronic states become quantized at discrete energy levels. The ultimate limit of low dimensional structures is the quantum dot, in which

the carriers are confined in all three directions. Therefore, a quantum dot can be thought of as an artificial atom.
Usually, the QD is buried in the host matrix (figure 1) with differing elastic constants and lattice parameter. Because of the lattice mismatch, both the QD and host matrix strain and relax elastically to accommodate this mismatch and thus admit a state of stress. As is well known, the electronic structure and the consequent optoelectronic properties of QDs are severely impacted due to this lattice mismatch induced strain. The host matrix boundaries do not impact its strain state, if the distance of the QD from any free boundary is significantly larger than, e.g., the QD radius (typically > 3R).

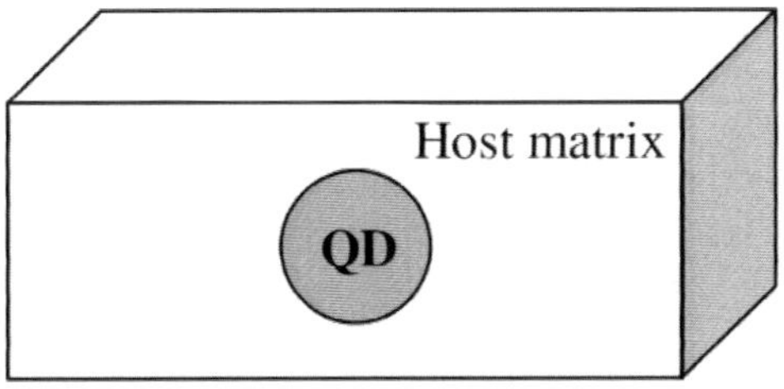

Figure 1. Spherical quantum dot embedded in host matrix

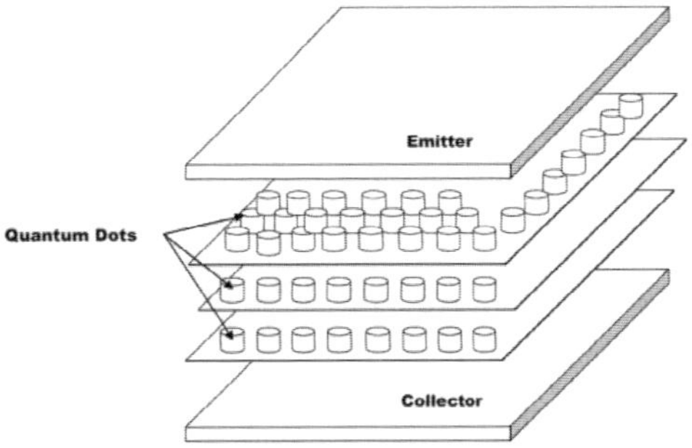

Figure 2. Schematic view of the structure (QD - InAs separated by wide gap material layers e.g. GaAs)

The typical QD infrared light system is plotted in figure 2. A lot of QDs constitute a set of separated layers.

2.2 Computational model
Many of the previous numerical modeling approaches for quantum dots used spatial discretisation methods, such as the finite element or finite difference method [5-7]. As an alternative, the boundary element method was proposed by Geldbard and Malloy [11]. Voss [12] employed the Rayleigh-Ritz method to solve the nonlinear eigenvalue problem where the eigenstates of the electron in QDs were derived with the use of the finite element method incorporated in the MATLAB package. Voon and Willatzen [13-14] found analytical solutions for paraboloidal and ellipsoidal QDs not embedded in the matrix. Various aspects of the evaluation of the eigenenergies in closed periodic systems of quantum dots were also discussed in the literature - see, e.g., [15-16].
Even in devices free of misfit dislocations, the strain induced by lattice mismatch can strongly affect electronic properties. However, this effect has not been thoroughly studied, particularly in submicron sized structures in which quantum mechanics governs the device properties and in which strains are highest and most nonuniform. In order to analyze strain effects in semiconductor quantum structures, it is necessary to adopt a model for electronic properties. Simple quantum mechanical models have long been available for describing the electronic properties of semiconductor devices based on the transport and confinement of single charge carriers. The study of quantum dots and quantum wires has renewed the interest in these models. The effects of uniform, coherent strain on electronic properties have also been well understood for many years and have been identified experimentally by Zaslavsky et al. [17]. There have been some attempts to model strain effects in quantum dots – see, e.g., [7]. Further, many quantum dot structures have a well-pronounced piezoelectric effect which does contribute to their overall properties in a non-trivial manner. These coupled electromechanical effects will become increasingly important for the current and future applications of such nanostructures [18-19]. By adopting this continuum view of confinement in semiconductor quantum devices, the spectrum of confined states available to individual electrons or holes can be characterized by the steady state Schrödinger equation, given by:

$$H_{\alpha\beta}(\vec{r})\psi_{\beta}(\vec{r})+V_{\alpha\beta}(\vec{r})\psi_{\beta}(\vec{r})=E\psi_{\alpha}(\vec{r}) \quad (1)$$

where $H_{\alpha\beta}$ is the Hamiltonian function coupling the energy of a charge carrier between energy bands α and β, ψ_β is the quantum mechanical wave function associated with energy band β, $V_{\alpha\beta}$ is an effective potential field coupling energy bands α and β, and E is the energy (the eigenvalue) of a particular quantum mechanical state.

The form of the Hamiltonian function is directly associated with the type of semiconductors, and especially with the form of the energy band. In view of that, we introduce the following division:

1. One band envelope-function formalism for electrons and holes in which the effective Hamiltonian is given by:

$$H_{\alpha\alpha} = -div\left(\frac{\hbar^2}{2m_\alpha(\lambda_E)} grad\psi_\alpha\right), \alpha = i = 1,2 \tag{2}$$

where ħ is the reduced Planck constant, and the index i corresponds to the quantum dot (i=1) and to the matrix (i=2), respectively. Such a description is valid for the direct band gap and for instance is valid for an InAs quantum dot embedded in a GaAs matrix. In the present analysis the electron effective mass m_i is assumed to be constant on the quantum dot and the matrix for every fixed energy level E and is taken as [5]:

$$\frac{1}{m_i(E)} = \frac{P_i^2}{\hbar^2}\left(\frac{2}{E+E_{g,i}-V_i} + \frac{1}{E+E_{g,i}-V_i+\delta_i}\right) \tag{3}$$

where the confinement potential V_i is piecewise constant, and P_i , $E_{g,i}$ and δ_i are the momentum matrix element, the band gap, and the spin–orbit splitting in the valence band for the quantum dot (i=1) and the matrix (i=2), respectively. The values of the above constants are given in table 1.

Table 1. The material properties of the quantum dot InAs and the matrix GaAs.

	P_i	$E_{g,i}$	V_i	δ_i
i=1 (q) InAs	0.8503	0.42	0	0.48
i=2 (m) GaAs	0.8878	1.52	0.7	0.34

2. The highest energy valence subbands. This basis consists of the two degenerate heavy hole subbands and the two degenerate light hole subbands, in reference to the relative masses of the charge carriers when treated as classical particles. The form of the Hamiltonian is given by:

$$H_{\alpha\beta} = -\frac{\hbar^2}{2m_0} L_{\alpha\beta}^{kl}(\vec{r})\nabla_{kl}^2, \alpha, \beta, k, l = 1,2 \tag{4}$$

where:

$$L_{\alpha\beta}^{kl}(\vec{r}) = \begin{bmatrix} g_1^{kl} & g_2^{kl} & g_3^{kl} \\ g_2^{kl} & g_4^{kl} & g_5^{kl} \\ g_3^{kl} & g_5^{kl} & g_6^{kl} \end{bmatrix} \tag{5}$$

This Hamiltonian can be used to model the medium in a Si_xGe_{1-x} structure g_1^{kl}, g_2^{kl}, g_3^{kl}, g_4^{kl}, g_5^{kl}, g_6^{kl} – see Singh and Muc et al. [20-21]. With the use of the relation (3) it is possible to use the identical approximations of the wavefunctions in the equations (4) and (5), since the different properties of the QD and the matrix are hidden in table 1. Using the properties of the Bessel function the numerator in the equation (3) takes the form:

$$a(\Phi,\Phi)=\sum_{i=1}^{2}\left(\frac{\hbar^2}{2m_i}D_1+V_iD_2\right) \tag{6}$$

$$D_2=\frac{r_0^2dz}{2}F(A_m)\sum_{n=0}^{\infty}\left[\frac{n\pi}{z_1-z_0}C_n\right]^2,\ D_1=-dz\sum_{n=0}^{\infty}[C_n]^2\sum_{m,k=0}^{\infty}A_mA_k\frac{\mu_m^{(l)}\mu_k^{(l)}}{r_0^2}\int_0^{r_0}drJ_{l-1}\left(\frac{\mu_k^{(l)}r}{r_0}\right)J_{l+1}\left(\frac{\mu_m^{(l)}r}{r_0}\right) \tag{7}$$

whereas the denominator:

$$b(\Phi,\Phi)=\frac{r_0^2dz}{2}F(A_m)\sum_{n=0}^{\infty}[C_n]^2,\ F(A_m)=\sum_{m=0}^{\infty}\left[A_mJ_l'\left(\mu_m^{(l)}\right)\right]^2,\ dz=\frac{z_1-z_0}{2} \tag{8}$$

The computation is carried out in the iterative way in two steps. Firstly for the assumed value of the eigenenergy E_0 the minimum of the functional $a(\Phi,\Phi)-E_0b(\Phi,\Phi)$ is searched for, with respect to the unknown coefficients A_m and C_n. Next for the calculated values of A_m and C_n a rational matrix eigenvalue problem (4) is solved to determine a new eigenvalue E_1. The procedure is repeated until the required accuracy is reached. The computations are conducted with the use of the symbolic package Mathematica.

3. Results and discussion

We have applied our method on two different problems in order to test its efficiency and effectiveness. Figure 3 demonstrates the axisymmetrical conical quantum dot embedded in the axisymmetric cylindrical matrix.

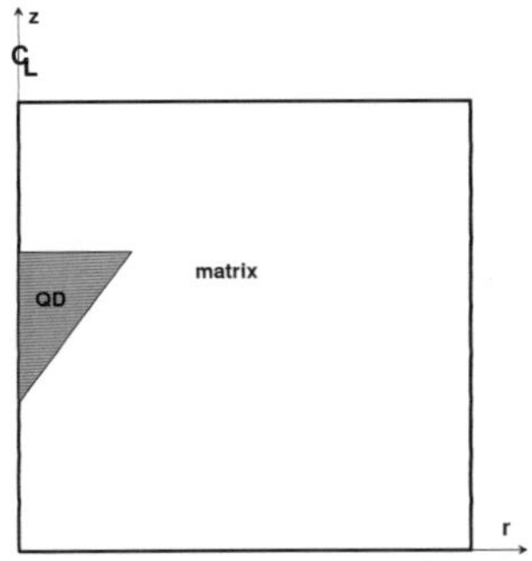

Figure 3. Axisymmetric problem - conical QD surrounded by cylindrical matrix.

Table 2. Eigenvalues for the conical quantum dot embedded in the matrix.

The electron orbital quantum number	Eigenvalue - present analysis			Eigenvalue - Voss [12]
	Fourier expansion 15 to 20 terms	Finite element 30 (linear)	Wavelet L=12	Finite element (planar) 6000
0	0.254607	0.255243	0.258931	0.254585
1	0.387332	0.386121	0.395571	0.384162
0	0.466941	0.469542	0.471121	0.467239
2	0.502889	0.503891	0.504005	0.503847
0	0.560774	0.562781	0.564451	0.561319
1	0.599138	0.604511	0.597138	0.598963
3	0.622213	0.634115	0.623356	0.617759

The radius and the height of the QD are equal to 10, whereas the radius and the height of the matrix are equal to 40 and 30, respectively. The physical properties of the materials are given in table 1. The boundary conditions are assumed to be in the Dirichlet form, i.e. the wave functions are equal to zero

on the boundaries. Table 2 shows the results of the computations. The results are presented for the first six eigenvalues and compared also with the results available in the open literature.
The computed eigenvalues shows quite good agreement with the results obtained by Voss [12]. In present the computations the maximal number of terms in the expansions of the function Z(z) is different, however, the use of the linear finite element or the wavelet approximations seems to be much more reasonable that the application of the finite element discretization with the use of the triangular elements. The approximation in the radial direction (10) was cut off on the maximal 50 terms. It is worth to note that the lowest values of the eigenvalues do not correspond to the lowest values of the electron orbital quantum number.

4. Summary

The present method shows the possibility of analytical computations of nonlinear boundary value problem characterized by the Helmholtz equations. In this way it is possible to find not only the energy spectrum and wave functions of an electron in a quantum dot but also the acoustic eigenfrequencies and eigenmodes of the pressure field inside an acoustic cavity. Eigensolutions are presented in the convenient form of the series expansions. The analysis can be conducted for the arbitrary form of the potential function, however having the axisymmetry with respect to the axis of rotations.

References

[1] Muc A, Banas A and Kedziora P 2013 *J Theor Appl Mech* **51** 387
[2] Banas A and Muc A 2013 *Key Eng Mater* **542** 1
[3] Muc A and Banas A 2014 *Proc 16th European Conference on Composite Materials*, Seville, Spain, June 22-26 , p 1217
[4] Muc A and Chwal M 2011 *Proc. IUTAM Symposium on Dynamics Modeling and Interaction Control in Virtual and Real Environments,* ed G Stepan, LL Kovacs and A Toth, Budapest, Hungary, June 7-11, 2010 (Springer) vol 30 p 239
[5] Grundmann M, Stier O and Bimberg D 1995 *Phys Rev B* **52** 119
[6] Stier O, Grundmann M and Bimberg D 1999 *Phys Rev B* **59** 5688
[7] Johnson HT, Freund LB 2001 *Int J Sol Struct* **38** 1045
[8] Lassen B, Voon LCLY, Willatzen M and Melnik R 2004 *Solid State Commun* **3-4** 141
[9] Melnik R and Mahapatra R 2007 *Comp Struct* **85** 698
[10] Muc A and Banas A 2014 *Adv Mater Res* **849** 427
[11] Gelbard F and Malloy KJ 2001 *J Comp Phys* **172** 19
[12] Voss H 2006 *Comp Phys Communications* **174** 441
[13] Voon LCLY and Willatzen M 2004 *Math Comp Simul* **65** 337
[14] Willatzen M, Voon LCLY 2005 *Comp Phys Communications* **171** 1
[15] Li Y 2005 *Math Comp Modelling* **42** 711
[16] Cattapan G, Lotti P and Pascolini A 2009 *Phys E* **41** 1187
[17] Zaslavsky A, Milkove KR, Lee YH, Ferland B and Sedgewick TO 1995 *Appl Phys Lett* **67** 3921
[18] Pan E 2002 *J Appl Phys* **91** 3785
[19] Jogai B, Albrecht JD and Pan E 2003 *J Appl Phys* **94** 3984
[20] Singh J 1993 *Physics of semiconductors and their heterostructures*, McGraw-Hill, New York
[21] Muc A and Banas A 2015 *Bulletin Polish Acad Sci Tech Sci* **63** 819

IOP Conf. Series: Materials Science and Engineering **544** (2019) 012040 doi:10.1088/1757-899X/544/1/012040

Buckling loads for sandwich structures having FRP faces and FGM core

A Muc[1], P Kędziora[1] and M Chwał[1]

[1]Institute of Machine Design, Faculty of Mechanical Engineering, Cracow University of Technology, Kraków, Poland
E-mail: olekmuc@mech.pk.edu.pl

Abstract. The aim of the present paper is to present a consistent 2D formulation of sandwiches (CS – classical sandwiches and FGMs – functionally gradient materials) that is accurate enough for a prediction of thermo-mechanical buckling load. Various kinematical hypothesis and variational formulations are studied in order to compare them with 3D finite element modelling and to estimate the effectiveness of 2D approximations. The examples deal with the compressed rectangular plates and shallow cylindrical panels. The effects of boundary conditions on both values of thermal and mechanical buckling loads as well on the variations of bifurcation loads with fibre orientations are discussed in details.

1. Introduction

Classical sandwich (CS) composites constitute a special group of structures used extensively in composite applications. They consist of thin facings made of high strength materials (e.g. FRP) and of a core made of thick and lightweight materials. The motivation for such sandwich structures elements is well-known. In the last decade a new class of sandwich structures has been developed. They are called as functionally gradient materials (FGMs) and distribute the material functions throughout the material body to achieve the maximum heat resistance and mechanical properties. Both types of structures (classical sandwiches and FGM sandwiches) are characterized in the similar manner, i.e. as structures having different properties in the thickness z direction in the form proposed by Qian et al. [1]:

$$p(z)=(p_t-p_b)V(z)+p_b \tag{1}$$

where p denotes a generic material property like modulus, p_t and p_b denote the corresponding properties of the top and bottom faces of the plate, respectively, and V(z) is a function. Here, we assume that moduli E and G, the coefficient of thermal expansion α, and thermal conductivity k vary according to Eqn (1) where the function V(z) is defined in the following way:

$$V(z)=\begin{cases}\sum_{i=1}^{N}\left[1(z-t_i)-1(z-t_{i-1})\right]\text{for CS}\\ \left(\dfrac{z}{t}+\dfrac{1}{2}\right)^n \quad z\in[-t/2,t/2] \text{ for FGM}\end{cases} \tag{2}$$

Other definitions of the function V(z) can be found in Ref [2] for the grading density of foam core for CS. The first two propositions take into account the use of relations (1) and (2 – for FGM) for degradation in a different direction than thickness. First density grading in the width direction is as follows:

$$\rho(x)=\rho_l-\frac{x(\rho_l-\rho_t)}{w} \tag{3}$$

where ρ_l and ρ_t are the mass densities of closed-cell foam of the leading edge and trailing edge, respectively; w denotes the width of the panel. Second density grading in the length direction is as follows:

$$\rho(x)=\rho_r-\frac{x(\rho_r-\rho_t)}{L} \tag{4}$$

where ρ_r and ρ_t are the mass densities of closed-cell foam of the root edge and tip edge, respectively; L denotes the length of the panel. Next definition of density diagonal grading across the xy plane (in two directions) is as follows:

$$\rho(x,y)=\frac{L-x}{L}\left(\frac{w-y}{w}\rho_u+\frac{y}{w}\rho_l\right)+\frac{x}{L}\left(\frac{w-y}{w}\rho_l+\frac{y}{w}\rho_u\right) \tag{5}$$

where ρ_u and ρ_l are the mass densities of closed-cell foam of upper and lower bounds, respectively. A model for charactering the mechanical properties of FGMs with the regular polygonal cross-section is also developed by introducing the power-law rule in Ref [3].

However, modelling of their mechanical behaviour still leads to many problems, particularly in the description of 1D or 2D structures, such as beams, plates or shells (see e.g. [4,5]). The broader discussion of those problems and solutions can be found in Muc et al. [6-10].

The aim of the present paper is to present a consistent 2D formulation of sandwiches (CS and FGMs) that is accurate enough for a prediction of thermo-mechanical buckling load. Various kinematical hypothesis and variational formulations are studied in order to compare them with 3D finite element modelling and to estimate the effectiveness of 2D approximations – see Muc et al. [6,9]. The examples deals with the compressed rectangular plates and shallow cylindrical panels. The effects of boundary conditions on both values of thermal and mechanical buckling loads as well on the variations of bifurcation loads with fibre orientations are discussed in details.

2. Formulation of the problem

In this paper we intend to propose an approach based on the symbolic computations – the use of the Mathematica package. Shear deformation effects are taken into account for different variants of laminate wise and layer wise 2D shell theories. The order of the theory is completely free but it should be fundamental kinematical relations for 3D structures are simplified to 2D equations via assumptions in the displacement field. The structure continuum vector $u_i(x,y,z)$ might be referred to the mid-surface 2D relation by using the following series expansion:

$$u_i^{(k)}\left(x^{(k)}y^{(k)}z^{(k)}\right)=U_i^{(k)}\left(x^{(k)}y^{(k)}\right)+\sum_{r=1}^{P}V_{ri}^{(k)}\left(x^{(k)}y^{(k)}\right)\Phi_r\left(z^{(k)}\right) \tag{6}$$

Using the above relation and the continuity conditions at the top and the bottom of the layers different theories may be formulate, i.e.: the Love-Kirchhoff theory, the first-order transverse shear deformation theory (FSDT) and higher-order transverse shear deformation theories (HSDT).

The formulation is based on the functional of total potential energy in the Lagrange form. An arbitrary form of 2D sandwich structure can be analysed via the required form of the Lame parameters. Then with the aid of the Rayleigh-Ritz formulation buckling load factors can be derived for an arbitrary sandwich structure. The theoretical formulation of this problem is discussed in Ref [11]. Some examples are presented in order to demonstrate the performance of such theories in the buckling analysis if sandwich plates, panels and shells.

3. Results and discussions

For compressed rectangular sandwich plates having FRP faces buckling mechanical loads are commonly dependent on fibre orientations.

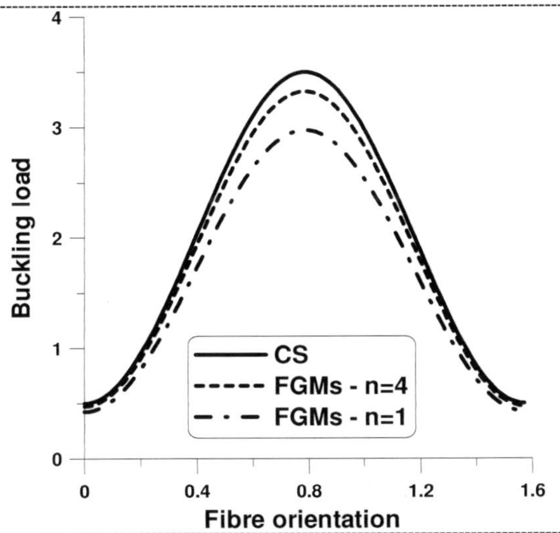

Figure 1. Buckling loads for a compressed square plate – the Love-Kirchhoff theory.

However assuming that the core mechanical properties are derived from relation (1), (2) bifurcation buckling loads are also the function of the parameter n. As it may be seen in Fig.1 the difference between buckling loads for CS and FGM sandwiches decreases as the parameter n grows.

Thermal buckling of the bifurcation type occurs in structures subjected to steadily increasing temperature field. Figure 2 is a plot of buckling loads for combined thermal and mechanical in-plane compressed loads. $N'=N/N_0$ and N_0 the buckling load value for the laminate subjected to in-plane loads (without thermal loads). As it may be seen the similar situation as previously occurs. With the increasing value of the parameter n the difference between the buckling curves for FGMs and CS disappear.

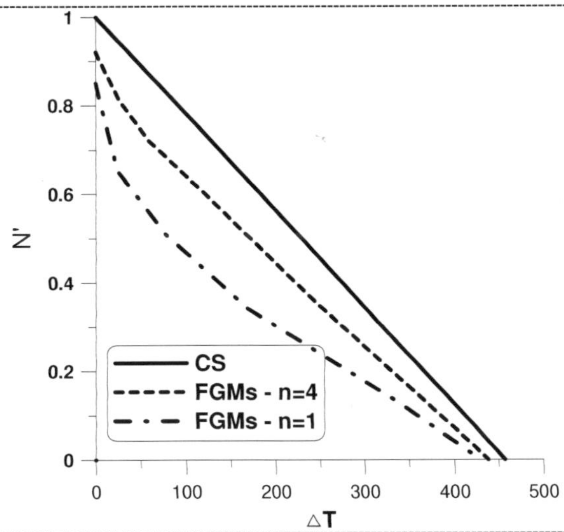

Figure 2. Combined buckling loads.

Figures 3 and 4 demonstrate the influence of the shell theory on buckling mechanical loads. As it may be observed, the growth in the accuracy in 2-D description of plates/shells results in the decrease of buckling loads. It is mainly caused by the core properties. However, as the compressibility of a normal to the plate mid-surface is taken into account (the 2-D approximation called as 11p) the location and distributions of buckling loads have been changed significantly, especially due to large deformations of the core.

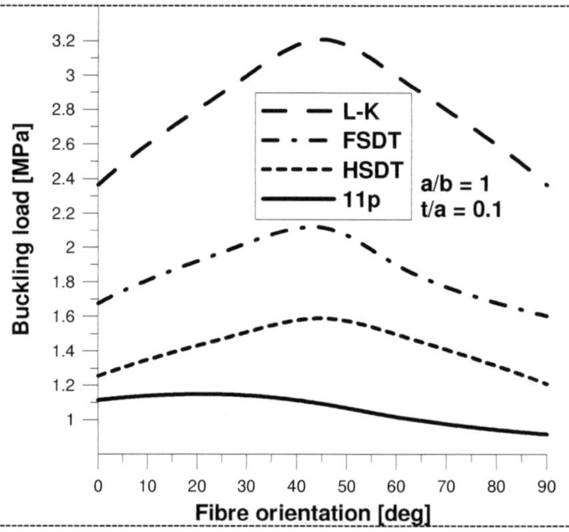

Figure 3. Comparison of buckling loads for different variants of plate theory – square plates (CS).

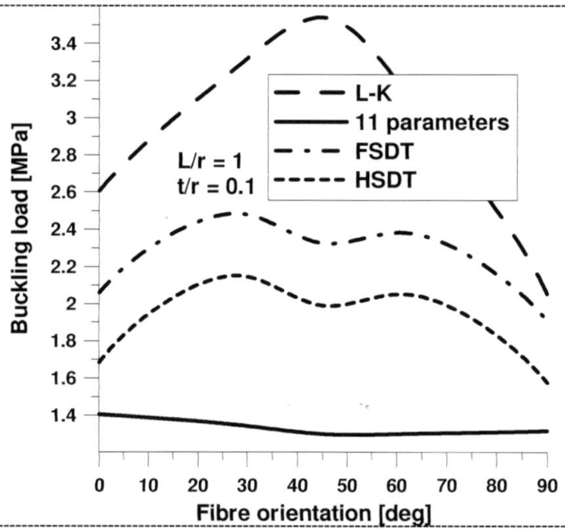

Figure 4. Comparison of buckling loads for different variants of plate theory – cylindrical shells (CS).

The type of boundary conditions affects both values of mechanical buckling loads as well as the variations of bifurcation loads with fibre orientations – see Fig.5. Similar effects have been observed to those for isotropic shells. For clamped edges buckling loads increase comparing with those obtained for simply supported plate/shell structures.

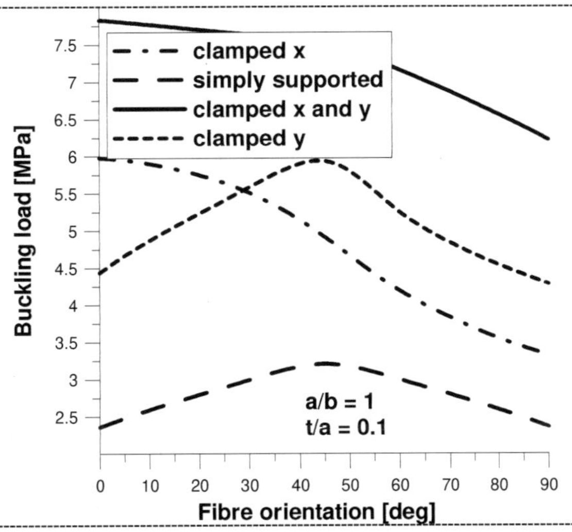

Figure 5. Variations of buckling loads for square plates having different boundary condition (CS).

4. References

[1] Qian L F, Batra R C and Chen L M 2004 *Compos: Part B* **35** 685–97
[2] Peng Jin and Xiaoping Zhong 2016 *Int J Aerospace Eng* **2016** ID 7971435
[3] Xiao-Jian Xu and Jun-Miao Meng 2018 *Compos: Part B* **145** 70–80
[4] Reddy J N and Chin C D 1998 *J. Therm Stresses* **21** 593-626
[5] Shiota I and Miyamoto Y 1997 *Functionally Graded Materials*, ed I Shiota and Y Miyamoto (Elsevier) p 137-41
[6] Muc A, Stawiarski A and Romanowicz P 2018 *Appl. Comp. Mater.* **25** 177-189
[7] Muc A Janczy S and Nogowczyk R. 2007 *16th Int.Conf. on Composite Materials* (Kyoto)
[8] Muc A and Zuchara P 1999 *Archives of Civil Engineering* **45** 289-304
[9] Muc A and Zuchara P 1999 *Comp. Struct.* **48** 145-50
[10] Muc A and Zuchara P 1998 *Mech Compos Mater* **34** 203-11
[11] Muc A 2003 *Mechanics of Fibrous Composites* (Kraków: Księgarnia Akademicka) (in Polish)

IOP Conf. Series: Materials Science and Engineering **544** (2019) 012041 doi:10.1088/1757-899X/544/1/012041

Improvement And Application Of Comprehensive Weighted Allocated Method Based On Fault Rate And Design Character

Shen Gui-xiang, Dong Lu-yao, Liu Jin-tong*, Liang Shu-bin, Ma Meng-jiao

School of Mechanical and Aerospace Engineering, Jilin University, Changchun 130000, China
E-mail: 1033922342@qq.com

Abstract. In order to solve the problem of weighting factor in the maintenance comprehensive weighted allocation model based on failure rate and design characteristics. Firstly, the fuzzy comprehensive evaluation method is introduced. Secondly, the fuzzy judgment matrix is established by using the three-scale method and the weight is obtained. Then the binary fuzzy contrast is used to establish the relationship matrix of the influencing factors and the evaluation matrix is obtained. The weighting factor is calculated by the fuzzy transformation. When the failure rates vary widely, the allocation results will weak the design characteristics by relying on the failure rate. So the weighting factor should be modified to improve the rationality of the distribution result. Finally, take a machining center as an example to verify the maintenance distribution application.
Key words: machining center ; maintenance allocation ; fuzzy comprehensive evaluation ; weighting factor

1. Introduction

Maintainability allocation refers to the design phase. At that process, the quantitative requirements of product maintenance shall be assigned to each component of the product layer by layer from the top of the system according to certain rules [1,2]. At present, the reliability level of domestic CNC machine tools has been greatly improved, but failures are inevitable. In order to improve the usability of products, maintenance research is needed. Maintenance design determines maintenance level. As an important part of design, maintenance allocation has been paid more and more attention [3]. Therefore, it is necessary to study the maintenance distribution of domestic CNC machine tools.

Commonly used maintenance allocation methods include: equal distribution method, distribution method by failure rate, comprehensive weighted distribution method by failure rate and design characteristics, distribution method by using maintenance data of similar products, guaranteed availability, and weighted distribution method by considering unit complexity [4]. The comprehensive weighted allocation method based on fault rate and design characteristics is more accurate than the allocation method based on fault rate. The application of this method can significantly improve the allocation accuracy when the failure detection, adjustable, accessible, replaceable and hazardous information of the product have been obtained. Therefore, this method is widely used in maintenance allocation.

The comprehensive weighted allocation method based on fault rate and design characteristics regards all the influencing factors are equally important. And the weighting factor coms from the statistical analysis of product structure type, which inevitably affects the accuracy of distribution. As a result, firstly, this paper introduces the theory of fuzzy comprehensive evaluation. Then the fuzzy judgment matrix is established by using the three-scale method and the weight is obtained [5,6].

Secondly, the relation matrix of influencing factors is established by applying the binary fuzzy contrast and the evaluation matrix will be obtained. Finally, the weighted factor is calculated by fuzzy transformation. In the case that the allocation results depend too much on the fault rate, the design characteristics are weakened [7]. Therefore, modifying the weighting factor to improve the accuracy of maintenance allocation.

2. Fuzzy maintainability distribution principle

The traditional comprehensive weighted distribution method based on the failure rate and design characteristics is to transform the factors that affect the design characteristics, such as fault detection, adjustable, accessibility and replaceable, into the weighted factor, and the component failure rate is used for allocation. Maintainability weighting factors are generally obtained from statistical data [4]. The weighting factor is obtained under the assumption of equal weight, which inevitably leads to error. Therefore, based on the traditional maintenance weighted distribution method based on failure rate and design characteristics, this paper introduces the fuzzy comprehensive evaluation method and proposes an improved distribution model as follows:

$$T_{CT_i} = \frac{\theta_i \cdot \sum_{i=1}^{n} \bar{\lambda}_i}{\bar{\lambda}_i \cdot \sum_{i=1}^{n} \theta_i} T_{CT} \tag{1}$$

Where: T_{CT} is the average repair time of the whole product; T_{CT_i} is the mean time to restoration (*MTTR*) of component i; $\bar{\lambda}_i$ is the average failure rate of component i; n is the number of components; θ_i is the maintainability weighting factor of component i.

2.1 Determination of maintainability weighting factor based on fuzzy evaluation

The process of establishing a maintenance weighting factor evaluation model using fuzzy comprehensive evaluation includes: using the three-scale method to establish the fuzzy judgment matrix of the influencing factors ; using the row normalization method to find the weight vector $\tilde{W}$;applying the binary fuzzy contrast to establish the relationship matrix of the influencing factors and obtaining the evaluation matrix $\tilde{R}$; getting the weight factor vector $\tilde{\theta}$ by fuzzy transformation.

$$\tilde{\theta} = \tilde{W} \circ \tilde{R} = [w_1, w_2, \cdots, w_m] \begin{bmatrix} r_{11} & r_{12} & \cdots & r_{1n} \\ r_{21} & r_{22} & \cdots & r_{2n} \\ \vdots & \vdots & \cdots & \vdots \\ r_{m1} & r_{m2} & \cdots & r_{mn} \end{bmatrix} = [\theta_1 : \theta_2 : \cdots : \theta_n] \tag{2}$$

2.1.1 Determination of the weight of the maintenance influence factor $\tilde{W}$

Due to various factors impact on the whole is more complex. A simple three-scale method is used to establish a fuzzy judgment matrix and obtain weights. The fuzzy consistency matrix obtained by the fuzzy judgment matrix satisfies the consistency condition, so no consistency test is needed [8,9].

The steps are as follows:

(1) establishing fuzzy judgment matrix by using three-scale method:

$$\mathbf{F} = \left(f_{ij}\right)_{m\times m} = \begin{cases} 0 & \text{factor } i \text{ is poorer than factor } j \\ 0.5 & \text{factor } i \text{ is the same as factor } j \\ 1 & \text{factor } i \text{ is better than facror } j \end{cases}$$

Where: m is the number of influencing factors.

(2) Transform the fuzzy judgment matrix $\mathbf{F}$ into a fuzzy consistency judgment matrix $\mathbf{Q}$

$\mathbf{Q} = \left(q_{ij}\right)_{m\times m}$, where: $q_{ij} = \dfrac{q_i - q_j}{2m} + \dfrac{1}{2}$; $q_i = \sum_{j=1}^{m} f_{ij}$.

(3) Using the row normalization method to obtain the weight vector:

$$\tilde{\mathbf{W}} = (w_1, w_2, \ldots w_m) \tag{3}$$

Where: $w_i = l_i / \sum_{i=1}^{m} l_i = \frac{2l_i}{m(m-1)}$ $i = 1, 2, \ldots, m$; $l_i = \sum_{j=1}^{m} q_{ij} - \frac{1}{2}$; $\sum_{i=1}^{m} l_i = m(m-1)/2$.

2.1.2. Determination of the evaluation matrix $\tilde{\mathbf{R}}$

Using the binary fuzzy comparison method to obtain the influencing factors B_i of maintenance distribution. The relationship matrix $\tilde{\mathbf{V}}$ is:

$$\tilde{\mathbf{V}}_i = \begin{bmatrix} \mu_{11}(B_i) & \mu_{12}(B_i) & \cdots & \mu_{1n}(B_i) \\ \mu_{21}(B_i) & \mu_{22}(B_i) & \cdots & \mu_{2n}(B_i) \\ \vdots & \vdots & \cdots & \vdots \\ \mu_{n1}(B_i) & \mu_{n2}(B_i) & \cdots & \mu_{nn}(B_i) \end{bmatrix}_{n\times n} \tag{4}$$

Where: $\mu_{jk}(B_i) = \frac{\delta_j(B_i)}{\delta_j(B_i)+\delta_k(B_i)}$, $1 \le i \le m$, $1 \le j \le n$, $1 \le k \le n$; $\delta_j(B_i)$ is the average of the expert score values of component A for the influencing factor j.

Then the evaluation matrix $\tilde{\mathbf{R}}$ of the influencing factors is:

$$\tilde{\mathbf{R}} = (r_{ij})_{m\times n} = \begin{bmatrix} r_{11} & r_{12} & \cdots & r_{1n} \\ r_{21} & r_{22} & \cdots & r_{2n} \\ \vdots & \vdots & \cdots & \vdots \\ r_{m1} & r_{m2} & \cdots & r_{mn} \end{bmatrix}_{m\times n} \tag{5}$$

Where: $r_{ij} = \frac{\sum_{k}^{n} \mu_{jk}(B_i)}{n}$.

Substituting the formulas (3) and (5) into the formula (2), the weigh factor is obtained.

2.2 Distribution result correction and verification

The status of failure rate and design characteristics are equally important when performing maintenance distribution. However, when the subsystem failure rates vary greatly (close to or greater than one order of magnitude), the distribution results will inevitably rely too much on the failure rate and weaken the design characteristics. Therefore, there is a need for maintainability redistribution of this machining center.

The modified comprehensive weighting factor is β' [10]:

$$\beta_i' = \frac{\theta_i}{\bar{\theta}} + \log_\alpha \frac{\bar{\lambda}_i}{\bar{\lambda}} \tag{6}$$

Where: $\bar{\lambda} = \frac{1}{n}\sum_{i=1}^{n} \bar{\lambda}_i$;the value of α is shown in Table 1.

Table 1. α standard table.

ξ (Ratio of maximum to minimum failure rate)	α
$1 \le \xi < 10$	0.01
$10 \le \xi < 100$	0.001
$100 \le \xi < 1000$	0.0001
$1000 \le \xi < 10000$	0.000001

Then, the restoration time that each subsystem should be assigned is:

$$T'_{CT_i} = \beta'_i T_{CT} \tag{7}$$

Bring the allocated maintenance time of each subsystem which is calculated by the above formula into the following formula:

$$T'_{CT} = \frac{\sum_{i}^{n}\left(\lambda_i \cdot T'_{CT_i}\right)}{\sum_{i=1}^{n}\lambda_i} \tag{8}$$

If T'_{CT} is less than the maintenance allocation indicator T_{CT}, the adjusted distribution is considered to accord the requirements.

3. Case Analysis

This paper takes a domestic CNC machining center as an example, considers the user demand and the actual maintenance ability, and sets the maintainability index of the whole machine as $T_{CT} = 18h$. According to the functional and structural characteristics of machining center, the machining center can be divided into the following subsystems: spindle system (A_1), cooling system (A_2), electrical system (A_3), Tool magazine system (A_4), hydraulic system (A_5), pneumatic system (A_6), feeding system (A_7), lubrication system (A8), numerical control system (A_9), auxiliary device (A_{10}), workbench (A_{11}).

According to the fault data collected and the method provided in reference [11], the average failure rate of each subsystem of the machining center is as shown in Table 2.

Table 2 Average failure rate of subsystem

Subsystem	λ_i	Subsystem	λ_i	Subsystem	λ_i
A_1	0.0009232	A_5	0.0001159	A_9	0.0006722
A_2	0.0006787	A_6	0.0001776	A_{10}	0.0004103
A_3	0.0002961	A_7	0.0006587	A_{11}	0.0002542
A_4	0.0006982	A_8	0.0004197		

In order to improve the maintenance level of the product, the subsystem with higher failure rate gives more stringent maintenance requirements, which means that lower *MTTR* is allocated to the system. On the other hand, it refers to the difficulty of product maintenance. Theoretically, the harder the product is to be repaired, higher *MTTR* should be allocated.

Combined with the maintenance characteristics of the machining center, the factors influencing maintenance allocation of the machining center are as follows: fault detection methods (B_1), replaceability (B_2), accessibility (B_3), adjustable (B_4) and perniciousness (B_5).

(1) Fault detection methods. Fault detection method in machining center can be divided into automatic, semi-automatic and manual detection. The higher *MTTR* is allocated to the subsystem which is not easy to detect the fault.

(2) Replaceability. Parts replacement in machining center can be divided into plug and pull, buckle, screw and welding. Typically, higher *MTTR* is allocated in the subsystem where the components are less likely to be replaced.

(3) Accessibility. The accessibility of machining center includes two aspects: whether the parts have cover and whether the cover can be removed quickly. In general, the higher *MTTR* is allocated in the subsystem where the cover is difficult to remove.

(4) Adjustable. Adjustment of machining center generally refers to whether adjustment is required after the replacement of new parts. Higher *MTTR* is allocated to more complex subsystems.

(5) Perniciousness. In order not to affect product quality, the more pernicious parts are, the faster the subsystem needs to be repaired after failure, so low *MTTR* is allocated to this subsystem.

Three scale method was used to compare the influence of the above five factors on the overall maintenance distribution. Then, according to the expert guidance results, the fuzzy judgment matrix of influencing factors is obtained as follows:

$$\mathbf{F}=\begin{pmatrix} 0.5 & 0.5 & 0 & 1 & 1 \\ 0.5 & 0.5 & 0 & 1 & 1 \\ 1 & 1 & 0.5 & 1 & 1 \\ 0 & 0 & 0 & 0.5 & 1 \\ 0 & 0 & 0 & 0 & 0.5 \end{pmatrix}$$

Transform the fuzzy judgment matrix into the fuzzy consistency matrix:

$$\mathbf{Q}=\begin{pmatrix} 0.5 & 0.5 & 0.35 & 0.65 & 0.75 \\ 0.5 & 0.5 & 0.35 & 0.65 & 0.75 \\ 0.65 & 0.65 & 0.5 & 0.8 & 0.9 \\ 0.35 & 0.35 & 0.2 & 0.5 & 0.6 \\ 0.25 & 0.25 & 0.1 & 0.4 & 0.5 \end{pmatrix}$$

By using formulas (3), the weight is calculated as:

$$\tilde{\mathbf{W}}=(0.225, 0.225, 0.3, 0.15, 0.1)$$

Because of the complexity of the sub-system of machining center, it is difficult to accurately evaluate the level of factors affecting maintenance. Therefore, the experts compared the importance of the different subsystems by using binary fuzzy comparison under different maintainability factors, and the score is shown in table 3.

Table 3 Score values of subsystem under each factor

Subsystem	A_1	A_2	A_3	A_4	A_5	A_6	A_7	A_8	A_9	A_{10}	A_{11}
Fault detection methods(B_1)	7.3	5.1	4.7	5.3	5.9	5.9	6.6	5.1	3.5	4.1	3.3
Replaceability(B_2)	4.3	3	4.5	3.5	3.5	3.7	5.5	2.5	2	2.5	3.3
Accessibility(B_3)	6.6	3.3	4.6	4.5	3.8	4.1	6.2	4.6	3.2	4.1	2.5
Adjustable(B_4)	3.2	1.8	1.7	3.5	1.8	2	4.5	1.8	1.1	1.2	1.6
Perniciousness(B_5)	0.8	5.3	7.4	2.5	7.6	8.9	5.8	6.9	5.7	9.2	8.3

According to formula (4), five binary fuzzy contrast relation matrices $\tilde{\mathbf{V}}$ are calculated. Due to space reasons, only $\tilde{\mathbf{V}}_1$ is listed here.

$$\tilde{\mathbf{V}}_1=\begin{bmatrix} 0.5000 & 0.5887 & 0.6083 & 0.5794 & 0.5530 & 0.5530 & 0.5252 & 0.5887 & 0.6759 & 0.6404 & 0.6887 \\ 0.4113 & 0.5000 & 0.5204 & 0.4904 & 0.4636 & 0.4636 & 0.4359 & 0.5000 & 0.5930 & 0.5543 & 0.6071 \\ 0.3917 & 0.4796 & 0.5000 & 0.4700 & 0.4434 & 0.4434 & 0.4159 & 0.4796 & 0.5732 & 0.5341 & 0.5875 \\ 0.4206 & 0.5096 & 0.5300 & 0.5000 & 0.4732 & 0.4732 & 0.4454 & 0.5096 & 0.6023 & 0.5638 & 0.6163 \\ 0.4470 & 0.5364 & 0.5566 & 0.5268 & 0.5000 & 0.5000 & 0.4720 & 0.5364 & 0.6277 & 0.5900 & 0.6413 \\ 0.4470 & 0.5364 & 0.5566 & 0.5268 & 0.5000 & 0.5000 & 0.4720 & 0.5364 & 0.6277 & 0.5900 & 0.6413 \\ 0.4748 & 0.5641 & 0.5841 & 0.5546 & 0.5280 & 0.5280 & 0.5000 & 0.5641 & 0.6535 & 0.6168 & 0.6667 \\ 0.4113 & 0.5000 & 0.5204 & 0.4904 & 0.4636 & 0.4636 & 0.4359 & 0.5000 & 0.5930 & 0.5543 & 0.6071 \\ 0.3241 & 0.4070 & 0.4268 & 0.3977 & 0.3723 & 0.3723 & 0.3465 & 0.4070 & 0.5000 & 0.4605 & 0.5147 \\ 0.3596 & 0.4457 & 0.4659 & 0.4362 & 0.4100 & 0.4100 & 0.3832 & 0.4457 & 0.5395 & 0.5000 & 0.5541 \\ 0.3113 & 0.3929 & 0.4125 & 0.3837 & 0.3587 & 0.3587 & 0.3333 & 0.3929 & 0.4853 & 0.4459 & 0.5000 \end{bmatrix}$$

According to formula (5), the evaluation matrix $\tilde{\mathbf{R}}$ of influencing factors is obtained:

$$\tilde{\mathbf{R}}=\begin{bmatrix} 0.5910 & 0.5036 & 0.4835 & 0.5131 & 0.5395 & 0.5395 & 0.5668 & 0.5036 & 0.4117 & 0.4500 & 0.3977 \\ 0.5609 & 0.4729 & 0.5718 & 0.5107 & 0.5107 & 0.5243 & 0.6192 & 0.4996 & 0.3760 & 0.4286 & 0.4962 \\ 0.6111 & 0.4430 & 0.5243 & 0.5189 & 0.4774 & 0.4960 & 0.5964 & 0.5036 & 0.4355 & 0.4960 & 0.3770 \\ 0.6109 & 0.4749 & 0.4612 & 0.6311 & 0.4749 & 0.5001 & 0.6853 & 0.4990 & 0.3602 & 0.3798 & 0.4468 \\ 0.1514 & 0.4908 & 0.5673 & 0.3286 & 0.5733 & 0.6089 & 0.5114 & 0.5513 & 0.5075 & 0.6163 & 0.5933 \end{bmatrix}$$

According to formula (2), the fuzzy evaluation result $\tilde{B}$ of maintainability weighted factor is obtained:

$$\tilde{B}=\{0.5493,0.4729,0.5206,0.5136,0.5081,0.5241,0.4387,0.5068,0.4127,0.4651,0.4406\}$$

Considering the problem that the distribution results depend too much on the fault rate and weaken the design characteristics when the failure rates vary greatly, the weighted factor is corrected. According to the value standard table of table 1 and the failure rate data of table 2, let α =0.01. By formula (6), the modified comprehensive weighting factor can be obtained as follows:

$$\beta_i'=\{0.9877,0.8976,1.1758,0.9749,1.3536,1.2938,0.8337,1.0715,0.7758,0.9908,1.0444\}$$

According to formula (7), the revised maintenance time distribution of each subsystem should be calculated as:

$$T_{CT_i}'=\{17.78, 16.16, 21.16, 17.55, 24.37, 23.29, 15.01, 19.29, 13.96, 17.83, 18.80\}$$

According to formula (8), the average repair time of the whole machine T'_{CT} is calculated. If T'_{CT} =17.40$h \leq 18h$, the adjusted distribution results are considered to meet the requirements.

Referring to the above results, higher *MTTR* should be allocated for electrical system, hydraulic system and pneumatic system, while lower *MTTR* should be allocated for numerical control system and feeding system, which is basically consistent with the actual situation. It can be seen that this method has good practicability.

4. Conclusion

(1) For the problem that weight cannot be calculated under the data of influencing factors. Fuzzy mathematics tri-scale method is introduced to establish the fuzzy judgment matrix of factors influencing maintenance allocation and to solve the weight. The method can effectively reduce the difficulty of decision-making and the subjective error of evaluation of influencing factors by different experts.

(2) Based on the comprehensive weighting method of fault rate and design characteristics, a fuzzy comprehensive evaluation method combining quantitative analysis and qualitative analysis is introduced. Fuzzy mathematics is applied to quantify the weight of each influencing factor and each subsystem, which gives full play to the role of the expert and reduces the judgment error caused by subjective judgment.

(3) Considering that while the distribution results are too dependent on the failure rate due to the failure rate varies greatly, the design characteristics will be weakened. So the weighted factor is corrected to make the maintenance allocation decision more reasonable. Providing a scientific and reasonable method for the maintenance allocation of products.

Acknowledgement

Thanks for great efforts that made by Liu Jin-tong (Corresponding author) in this article submitting process. And research in this paper was supported by Major national science and technology projects of China (Grant no.2015ZX04005005) of High grade CNC machine tools and basic manufacturing equipment, and Jilin Provincial Natural Science Foundation of China (Grant no. 20170101212JC).

Reference

[1] KANG Rui. RMS Reliability maintenance engineering foundation [M]. Beijing: National Defense Industry Press,2012.

[2] FAN Shaohua. The reliability distribution of CNC lathe based on fuzzy theory [D]. Changchun: Jilin University,2011.

[3] JI Yunfei, ZHU Huayuan, JIA Xiangjun. Research on engine maintenance optimization based on efficiency analysis [J]. China Science And Technology Information,2012(13):121+126.

[4] GJB/Z57-94 Maintainability allocation and prediction handbook[S].

[5] LIU Shihao, YE Wenhua, TANG Dunbing, et al. Fuzzy comprehensive evaluation for the performance of a numerical control machine tool based on the analytic hierarchy process[J].JOURNAL OF SHANDONG UNIVERSITY (ENGINEERING SCIENCE),2010,40(01):68-72+92.

[6] PAN Fei, PANG Hongzhao, ZHANG Xiao. Research on torpedo maintenance assignment model based on fuzzy comprehensive evaluation[J]. China Science And Technology Information, 2011(10):147-148.

[7] MAO Jiangang. Improvement of Method of Integrated Weighted Assignment by Using Failure Rate and Design Characteris [J]. Aviation Standardization And Quality,2018(01):38-41.

[8] ZHU Dexin, LIU Hengzhao. Reliability distribution of electrical spindle based on improved fuzzy analytic hierarchy process[J]. China Mechanical Engineering,2011,22(24):2923-2927.

[9] YANG Litu, CAI Hongfei, REN Jiayin, et al. Improvement of fuzzy analytic hierarchy process and its application in transformer life evaluation[J]. Journal Of Zhengzhou University (Engineering),2013,34(03):10-13.

[10] DUAN Ping, HU bin, GUO Yukun, et al. Dynamic Weighted Maintainability Allocation Method Based on Failure Rate and Design Character[J]. Fire Control & Command Contro,2014,39(12):156-159+163.

[11] MENG Shu. Research on Preventive Maintenance Strategy of CNC Lathe Based on Fault Rate Correlations[D].Jilin University,2017.

Life prediction of electric spindles with multiple degradation processes

Zhang Ying-zhi, Jiang Cheng-zi, Liu Jia-lin*, Song Min-qiao

School of Mechanical and Aerospace Engineering, Jilin University, Changchun, 130000, China
E-mail: ororange@sina.com

Abstract. In order to realize product life prediction without burst failure information, a method combining performance degradation information correlation analysis to construct a degradation process model is proposed. On the basis of obtaining the degradation information and establishing the single performance indicator degradation process model, the covariance is used to analyze the correlation of the multi-performance index degradation process, and the Copula connection function is introduced to establish the system reliability model under multi-degradation process. Achieve the reliability curve drawing and reliable life prediction. Finally, the effectiveness of the method is verified by taking the degradation information of a series of milling type electric spindles as an example.
Key words: Electric spindles; multi-degenerate process; reliability modelling; life prediction

1. Introduction

At present, it is difficult for domestically produced machine tools to form effective competitiveness. One of the reasons is that the reliability of the "heart component" electric spindle is not high [1]. By predicting the life of the electric spindle, it is possible to predict the failure time in advance, take measures to reduce the number of motor spindle shutdowns, and reduce downtime and maintenance costs during the fault repair process. Therefore, it is of great significance to carry out research on reliable life prediction of electric spindles.

At present, the life prediction methods are mainly divided into the following three categories: statistical-based methods, artificial intelligence-based methods, and model-based methods [2]. Based on the statistical principle, the life prediction needs to obtain sufficient product life information, and the reasonable statistical life model is obtained through model assumptions and parameter estimation. The model accuracy is related to the data volume. The artificial intelligence method needs to combine multi-source information fusion technology with artificial neural network, and the analysis process is relatively complicated [3]. Life prediction using models is more common in engineering practice [4].

The electric spindle is a long-life, high-reliability product, and it cannot obtain sufficient failure information in a short time. Therefore, the traditional life prediction method based on statistical principle is not applicable. The performance of the product is always degraded during use, so it can be combined with its degradation information for reliability modelling and life prediction research [5-6]. The multi-

IOP Conf. Series: Materials Science and Engineering **544** (2019) 012042 doi:10.1088/1757-899X/544/1/012042

degradation process coexists during the use of the electric spindle, and there is mutual influence between the degradation processes. Therefore, based on the multi-performance degradation process and correlation analysis of the electric spindle, this paper constructs a multi-performance degradation process model for the life prediction of the electric spindle.

2. Reliability modelling based on multiple degradation processes with mutual influence

2.1. Single degradation process modeling

The performance indicator degradation amount is calculated based on the performance measurement information. It is assumed that the degradation process obeys a common distribution function, and the Weibull distribution, the normal distribution, and the lognormal distribution are generally considered. The model parameters are estimated by the maximum likelihood estimation method. The d-test method (Kolmogorov-Smirnov test, or K-S test) is used to test the distribution model, and the error distribution is used to optimize the model distribution parameters. Then the performance index degradation data distribution function will be obtained.

2.1.1. Parameter estimation

Let $X=\{X_1 \quad X_2 \quad \cdots \quad X_n\}$ be the overall degradation value of a certain performance indicator. Set $f(x;\theta_1,\theta_2,\cdots\theta_m)$ ($\theta_1,\theta_2,\cdots\theta_m$ are parameters to be estimated with the amount of m)as the population distribution density of the random variable X. Construct a likelihood function from n observations of random variable X.

$$L(\theta_1,\theta_2,\cdots\theta_m)=\prod^{n} f(x;\theta_1,\theta_2,\cdots\theta_m) \tag{1}$$

According to the principle of maximum likelihood method, the estimated value of $\theta_1,\theta_2,\cdots\theta_m$ can be obtained by solving the likelihood equation (2).

$$L(\theta)=L\left(x;\bar{\theta}_1,\bar{\theta}_2,\cdots\bar{\theta}_m\right)=\max_{\theta\in\Theta} L\left(x;\theta_1,\theta_2,\cdots\theta_m\right) \tag{2}$$

2.1.2. Model checking

Commonly used model checking methods are the χ^2test method and the d-test method. The χ^2test method is mainly applicable to large samples. The d-test (Kolmogorov-Smirnov test, or K-S test) is finer than the χ^2test. Besides, it is applicable to small samples. Therefore, this paper chooses the d-test method for the distribution model test.

The d-test is to arrange n test data in order of small to large. According to the hypothetical distribution, calculate the corresponding $F_0(x_i)$ of each data and compared these with the empirical distribution function $F_n(x_i)$. The maximum absolute value of the difference is the observation value of the test statistic $\hat{D}_n$. It will be compared to the threshold $D_{n,\alpha}$ to determine whether to accept the null hypothesis.

$$D_n=\sup_{-\infty<x<+\infty}\left|F_n(x_i)-F_0(x_i)\right|=max\{d_i\}\leq D_{n,\alpha} \tag{3}$$

Where $F_0(x_i)$ is null hypothesis distribution; $F_n(x_i)$is empirical distribution function with sample size n.

2.1.3. Distribution model parameter optimization

The model parameters are actually the amount of change with time, further improve the accuracy of the model, and perform time series fitting on the model parameters, considering several commonly used time series fitting functions, such as linear functions, exponential functions, power functions, and logarithmic functions.

Using the sum of squared errors [7] to reflect the excellent condition of the time series function fitting model parameters, the representation is as follows

$$N = \sum_{i=1}^{n}(g_i - \hat{g}_i)^2 \tag{4}$$

Where g_i indicates the actual value, $\hat{g}_i$ indicates the fitted value

The closer the sum of squared errors is to zero, the smaller the deviation between the actual data of the statistic and the fitted curve, indicating that the degree of fit of the function is higher.

2.2. Correlation analysis of multiple degradation processes

Covariance was used to test the correlation of multiple degradation processes. The covariance formula is:

$$\mathbf{cov}(x(t), y(t)) = E\left[(x(t)-\bar{x})(y(t)-\bar{y})\right] \tag{5}$$

Where $x(t)$ and $y(t)$ are the degraded data of the two performance indicators of the product at time t, $\bar{x}$ and $\bar{y}$ are the corresponding mean.

If the covariance is equal to zero or closer to zero, it means that the degradation processes of the two performance indicators are independent of each other. Besides, there is a correlation. The covariance value reflects the close relationship of the performance degradation process.

2.3. Multi-degradation process modeling based on Copula

The single degenerate process model is used as the edge distribution. The Copula connection function is introduced to construct the joint distribution model [8]. The maximum likelihood estimation is used to calculate the model parameters. The AIC criterion is used to select the most suitable Copula connection function to complete the reliability model construction.

The more typical Copula function in the commonly used Archimedes type distribution family is selected as the connection function. Its expression is shown in Table 1.

Table 1. Copula function expression.

Copula function	Parameter range	Function expression for binary
Gumbel Copula	$\alpha \in (0,1]$	$C(F_x, F_y; \alpha) = \mathbf{exp}\left\{-\left[(-\ln F_x)^{\alpha} + (-\ln F_y)^{\alpha}\right]^{1/\alpha}\right\}$
Clayton Copula	$\theta \in (0,\infty)$	$C(F_x, F_y; \boldsymbol{\theta}) = (F_x^{-\boldsymbol{\theta}} + F_y^{-\boldsymbol{\theta}} - 1)^{-1/\boldsymbol{\theta}}$
Frank Copula	$\beta \in (-\infty,0) \cup (0, \infty)$	$C(F_x, F_y; \boldsymbol{\beta}) = -\frac{1}{\boldsymbol{\beta}} ln\left[1 + \frac{(1-e^{-\boldsymbol{\beta} F_x})(1-e^{-\boldsymbol{\beta} F_y})}{e^{-\boldsymbol{\beta}}-1}\right]$

The AIC formula is as follows. The smaller the AIC criterion value, the better the model fits.

$$\mathrm{AIC} = -2\ln L + 2k \tag{6}$$

Where L is the maximum likelihood function of the model, and k is the number of parameters of the model.

3. Examples

The electric spindle is a highly reliable and long-life product. It takes a long time to expose the motor spindle to defects or potential problems. Considering this situation, the test time is reduced by the constant stress accelerated life test when the test time is determined.

The relationship between the life t in the actual case and the corresponding life T under the acceleration test can be expressed as:

$$T = K^{-1}t \tag{7}$$

Where the acceleration factor K is taken 4.5.

According to the statistical analysis of the fault under actual use conditions, get $t = 2339.6\mathrm{h}$, the completion probability is 45.60%, and the test time for determining the accelerated life test was approximately 528h.

The eddy current sensor and the three-way acceleration sensor are used to collect the signal of the axial end runout and spindle vibration. The test data are shown in Table 2. According to the specification of the electric spindle, when the axial end runout of the electric spindle reaches 1.6 times of the initial value, or when the spindle vibration of the electric spindle reaches 2 times the initial value, it is considered that the electric spindle has failed.

Table 2. Deterioration data of motor spindle performance index.

time/h	Performance index		time/h	Performance index	
	Axial end runout (μm)	Spindle vibration ($\mu m/s$)		Axial end runout (μm)	Spindle vibration ($\mu m/s$)
0	7.1	123.2	288	8.9	138.9
24	7.8	127.3	312	8.9	140.9
48	7.8	127.6	336	8.9	139.8
72	7.8	130.5	360	9.2	141.8
96	8.1	130.9	384	9.4	143.8
120	8.2	132.2	408	9.1	146
144	8.1	130.2	432	9.6	144.2
168	8.4	132.7	456	9.5	149
192	8.2	134.1	480	9.5	151
216	8.4	134.1	504	9.6	149.3
240	8.5	136.9	528	9.9	151.3
264	8.8	135.7			

The maximum likelihood method is used to estimate the performance index degradation data, and the K-S test is performed. The results are shown in Table 3. According to the results of the K-S test, the lognormal distribution is selected as the optimal distribution of the axial end runout D_x and the spindle vibration amount D_y degradation data.

Table 3. K-S test result.

distribution	Performance index			
	Axial end runout D_x		Spindle vibration D_y	
	H	P	H	P
Normal distribution	0	0.6436	0	0.6526
Lognormal distribution	0	0.6838	0	0.7056
Weibull distribution	1	0.0148	0	0.4266

The parameters of the distribution function are optimized, and are fitted by time series, and the function is selected by the sum of squared errors. The results are shown in Table 4.

According to the fitting situation, the corresponding function with the smallest sum of squared errors is selected, and the result of parameter fitting is obtained as follows:

(1) Axial end runout: $\mu_x(t)=5.002\times10^{-4}t+2.028$ $\quad\sigma_x(t)=3.043\times10^{-4}t+1.855$

(2) Spindle vibration: $\mu_y(t)=4.833exp(7.002\times10^{-5}t)$ $\quad\sigma_y(t)=3.096exp(3.627\times10^{-5}t)$

The degradation process model of performance indicators is:

(1) Axial end runout: $F_x(t)=\Phi\left(\frac{ln\,x(t)-\mu_x(t)}{\sigma_x(t)}\right)$

(2) Spindle vibration: $F_y(t)=\Phi\left(\frac{ln\,y(t)-\mu_y(t)}{\sigma_y(t)}\right)$

The relationship between the two performance index degradation processes of the electric spindle is calculated as follows: $\mathbf{cov}(x(t),y(t))=4.346$

There is a correlation between the two degradation processes. The reliability model is established by the Copula function. List the Copula function parameter values at some time, as shown in Table 5.

Table 4 . Sum of square error.

Function	Axial end runout		Spindle vibration	
	N_μ	N_σ	N_μ	N_σ
Linear function	0.0098	0.00365	0.0074	0.000793
Exponential function	0.0100	0.00372	0.0073	0.000787
Power function	0.0479	0.01753	0.0350	0.003793
Logarithmic function	0.0517	0.01847	0.0356	0.003825

Table 5. Copula function parameter values at some times.

t	Gumbel Copula	Clayton Copula	Frank Copula
24	1.0473	0.9872	1.4253
264	0.3724	14.3267	26.4521
528	-0.5896	30.9974	57.8054

Gumbel Copula's function parameter values are not within the specified limits and are not considered. The selection of the best model among the remaining two Copula functions is done by the AIC criterion. The calculation results of the AIC are shown in Table 6.

According to the calculation results of AIC, Clayton Copula>Frank Copula can be found, therefore, the Frank Copula model is selected.

Perform a linear time series fitting on the parameters β in the Frank Copula function [9], and the parameter expression as: $\beta=0.1081t+0.0515$

Using the Frank Copula model, the reliability model of the multi-degradation process of the electric spindle is completed, and reliability model is obtained: $R(t)=-\frac{1}{\beta}ln\left[1+\frac{(1-e^{-\beta F_x})(1-e^{-\beta F_y})}{e^{-\beta}-1}\right]$

According to the reliability model of the electric spindle, the reliability curve of the electric spindle is drawn. The reliability curve of the electric spindle is shown in Figure 1.

Table 6 . Copula function AIC calculation value at a certain moment.

t	Clayton Copula	Frank Copula
24	-9.3057	-10.4209
264	-8.4781	-9.7896
528	-7.9131	-8.915

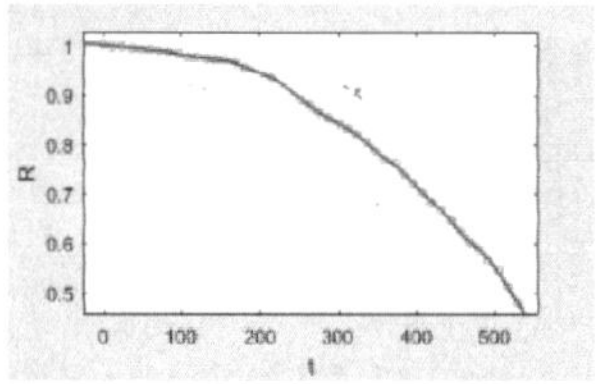

Figure1. Electric spindle multi-degenerate. process related.

When a given value R of a certain reliability of a product is given, the corresponding time t is called the reliable life t_R of the electric spindle, and this value reflects the probability that the product can operate normally at time t. The formula for the reliable life of an electric spindle can be expressed as: $R(t_R)=R$

After accelerated factor conversion, get median life: $t_{0.5}=2301.75h$, characteristic life: $t_{e^{-1}}=2617.37h$.

When $R=0.456$, $t=2407.51h$, compared the actual life time of the electric spindle under the same reliability of 2339.6h, the gap of life prediction result and the actual is 2.9%. Considering the existence of

IOP Conf. Series: Materials Science and Engineering **544** (2019) 012042 doi:10.1088/1757-899X/544/1/012042

the test error, the prediction result is acceptable, basically in line with the actual situation of the project.

The average time before failure is $MTTF = \int_0^{+\infty} R(t)dt$, the MTTF is 1885.36h, which reflects the average normal working time of the product.

4. Conclusion

(1) Assume several commonly used distribution functions, estimate the parameters by the maximum likelihood method, use the d-test to select the distribution function of the performance degradation data, further optimize the model parameters, and determine the final single degradation process model.

(2) Using the covariance to judge the relationship between the multi-performance indicators of the electric spindle, and the performance indicators are related.

(3) Based on the correlation between multi-performance indicators, the maximum likelihood method is used to estimate the parameters of Copula function, the AIC criterion is used to complete the selection of Copula function, and the reliability modeling of the electric spindle is completed to achieve a reliable life prediction.

Acknowledgement

Thanks for great efforts that made by Liu Jia-lin (Corresponding author) in this article submitting process. And research in this paper was supported by Major national science and technology projects of China (Grant no.2015ZX04005005) of High grade CNC machine tools and basic manufacturing equipment, and Jilin Provincial Natural Science Foundation of China (Grant no. 20170101212JC).

References

[1] Yang Gui-jie, Qin Dong-dong. Key technology and development trend of high speed electric spindle[J]. Servo Control, 2010(02):32+34-36.

[2] Andrew K.S. Jardine,Daming Lin,Dragan Banjevic. A review on machinery diagnostics and prognostics implementing condition-based maintenance[J]. Mechanical Systems and Signal Processing,2005,20(7).

[3] Gao Hong-li, Li Deng-wan, Xu Ming-heng. Predicting Technology of Lead Screw Life Based on Artificial Intelligence[J]. Journal of Southwest Jiaotong University, 2010,45(05):685-691.

[4] Zhou Shao-hua, Hu Chang-hua, Si Xiao-sheng et al. Product life prediction method based on nonlinear accelerated degradation model and failure rate model[J]. Chinese Journal of Electronics, 2017, 45(05): 1084-1089.

[5] Xu D, Zhao W. Reliability prediction using multivariate degradation data[C] Reliability and Maintainability Symposium, 2005. Proceedings.

[6] Zhenyu Liu; Xiaobing Ma; Jun Yang; Yu Zhao. Reliability Modeling for Systems with Multiple Degradation Processes Using Inverse Gaussian Process and Copulas [J].Mathematical Problems in Engineering,2014,Vol.

[7] Zhong Qian-hui, Zhang Zhi-hua, Liang Sheng-jie. Reliability Analysis Method Based on Multiple Degraded Data[J]. Systems Engineering Theory & Practice, 2011, 31(03): 544-551.

[8] Yang Yi-dang, Luo Yu-hua. Parameter Estimation of Copula Function[J]. Journal of Xinjiang Normal University(Natural Science Edition), 2007(02):15-18+24.

[9] Zhang Zhi-peng. Research on modeling and reliability assessment methods for system multivariate correlation degradation process [D]. University of Electronic Science and Technology of China, 2016.

IOP Conf. Series: Materials Science and Engineering **544** (2019) 012043 doi:10.1088/1757-899X/544/1/012043

Preparation of graphene coated aluminum composite powders by high-energy ball milling

Liu Heping[12], **Sun Fenger**[1], **Gao Yibo**[1], **Cheng Shaolei** [1], **Jing Xingbin**[1]

1School of Materials Science and Engineering, North University of China, Taiyuan 030051, China
2School of Materials Science and Engineering, Taiyuan University of Technology, Taiyuan 030024, China
E-Mail: peace666@126.com

Abstract: The preparation of composite powders has an important impact on the properties of the synthesized bulk materials. In this work, graphene coated aluminum composite powders were successfully prepared by high-energy ball milling. The microstructure of graphene/aluminum mixed powders was characterized by scanning electron microscope (SEM) and X-ray diffraction (XRD). The effects of the micro-morphology of the composite powders and the ratio of the ball to the particle size of the graphene-coated aluminum composite powders were analyzed in detail. The results show that graphene can be uniformly dispersed in aluminum powder without significant agglomeration in the case of ball milling parameters with a ball-to-powder ratio of 5:1. When the ratio of ball to material is 6:1, the dispersion effect was not so obvious. However, the level of agglomeration in the composite powders at ratios 7:1 and 8:1 differed respectively; some particles were deformed and the dispersion effect was reduced. The particle size of the graphene coated aluminum composite powder increased with the increase of ball to material ratio. Thermodynamic analysis of the interface reaction between graphene and aluminum composite powders shows that the maximum content of graphene microflakes on the surface of aluminum powder is 0.3wt%.

1. Introduction

Graphene is composed of one or several layers of carbon atoms, which is tightly packed into a honeycomb network. It has some unique properties, such as mechanics, heat, electricity, magnetism, optics, acoustics, corrosion resistance and so on[1,2]. This makes many scientists to carry out in-depth research on it. Up to now, graphene is considered to be the most ideal reinforcement material for metal matrix composites[3,4].

IOP Conf. Series: Materials Science and Engineering **544** (2019) 012043 doi:10.1088/1757-899X/544/1/012043

With the development of automotive and aerospace fields, higher requirements have been put forward for aluminum matrix composites, such as specific strength, specific modulus, conductivity and thermal conductivity. Traditional ceramic fibers and particle reinforced materials cannot meet the requirements of high performance materials. Now there are many reports that silicon carbide and carbon nanotubes are added to aluminum alloy as reinforcements, but the reinforcement effect cannot meet the requirements of industrial and daily use. In view of the above analysis, graphene can be used as reinforcement of aluminum matrix composites because of its various characteristics, which makes it possible to improve the properties of aluminum matrix composites[5-9]. Therefore, it has attracted wide attention from researchers.

When graphene is added to aluminum powder particles as reinforcement, the primary problem is to make graphene disperse evenly in aluminum matrix. At present, the commonly used dispersing methods are liquid and ball milling. However, ball milling has the advantages of low reaction temperature, high yield and uniform particle size distribution, which makes it play an important role in the grinding and mixing of powders. The microstructure and particle size distribution of graphene/aluminium composite powders prepared by high-energy ball milling were studied in this work, and its thermodynamic properties were analyzed.

2. Experiment

This experiment used the YXQM planetary ball mill machine for ball milling. Mixed powders containing 0.5% of graphene and 99.8% of aluminum powder were ball milled using four different ball to powder weight ratios of 5:1, 6:1, 7:1, and 8:1 respectively. Ball milling parameters were as follows: speed was 300r/min, ball milling time was 1h. In order to obtain the best ball milling effect, the proportion of stainless steel balls with a diameter of 3mm, 5mm, and 8mm was 3:5:2, respectively. The ball milling process was performed under argon atmosphere to prevent oxidation.

3. Results and Discussion

3.1 Microstructure analysis of graphene/aluminum composite powders

The SEM micrographs at different magnifications of the graphene/aluminum composite powders after milling using the different ball to powder weight ratios are shown in Figure1. As can be seen from the Figure 1, at the 5:1 ball to powder weight ratio, the vast majority of particles are approximately spherical or approximately ellipsoidal, there are a small number of particles with a large particle size distributed in them. There is no obvious agglomeration phenomenon and the dispersion is good. At a 6:1 ball to powder weight ratio, the low magnification electron micrograph is similar to the 5:1 case, but a small amount of particles are found to be deformed. Some irregularly shaped particles appear and there is a slight agglomeration and the dispersion is slightly worse. At a 7:1 ball to powder weight ratio, it can be seen that partially spherical or nearly spherical particles become irregular shapes, and some particles become rod-like particles after ball milling, and agglomeration occurs. At the 8:1 ball to powder weight ratio, most of the particles are deformed and become flat, especially in electron micrograph at high magnification. The phenomenon of agglomeration is more serious and many particles are stuck together irregularly. The dispersion is uneven and poor.

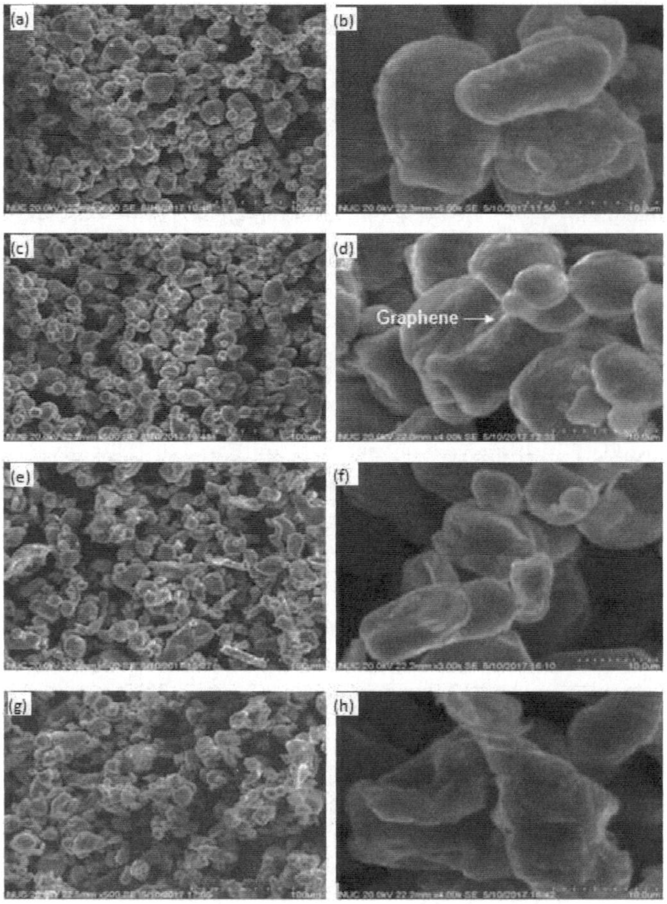

Figure 1 SEM image of graphene/aluminum composite powder with different ball material ratio after ball milling
(a) 5:1 low magnification; (b) 5:1 high magnification; (c) 6:1 low magnification; (d) 6:1 high magnification; (e) 7:1 low magnification; (f) 7:1 high magnification; (g) 8:1 low magnification; (h) 8:1 high magnification

3.2 XRD pattern data analysis of graphene/aluminum composite powder

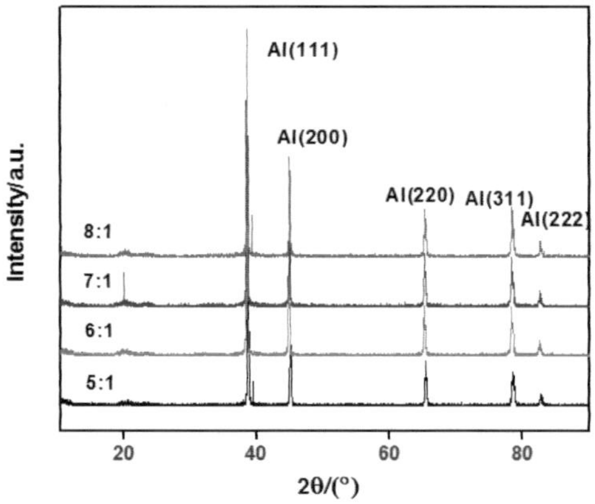

Figure 2 X-ray diffraction patterns of graphene/aluminum composite powder with different ball material ratio after ball milling

The XRD pattern of different ball to powder weight ratios is shown in Figure 2.The four types of patterns are basically similar, only the diffraction peaks of aluminum, and no characteristic peak of graphite appearing at $2\theta=26.6°$. A peak with a small peak intensity appears around 20°, indicating that there is no graphite in the composite powder, and no serious agglomeration of graphene occurs. It is shown again that graphene can be uniformly dispersed in aluminum powder by high-energy ball milling [10,11].While the above-mentioned 7:1 and 8:1 ball to powder weight ratios have agglomeration phenomenon, there is no graphite peak in the XRD pattern. It shows that the agglomeration phenomenon of graphene is not so severe; hence the signal of graphite was not detected.

3.3 Thermodynamic analysis of interfacial reaction of graphene/aluminum composite powders

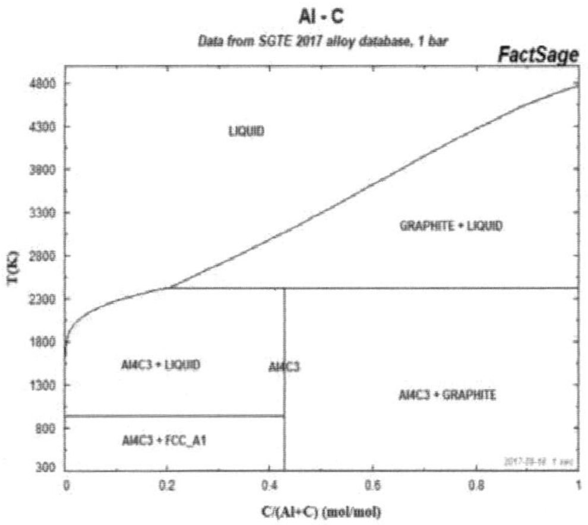

Figure 3 Al-C binary phase diagram

According to the Al-C binary phase diagram shown in Figure 3, it is possible to generate the interface product Al_4C_3, and its reaction equation is as follows:

$$4[Al]+3C(s)=Al_4C_3(s) \quad (1)$$

In the formula, Al is aluminum powder during the ball milling process, C is graphene, and Al_4C_3 is an interface product that may be generated. During the ball milling process, the grinding ball strongly impacts, grinds and agitates the aluminum powder and graphene, causing a large impact force and shearing force. The instant of collision produces a great deal of energy at the collision site, which transfers more energy to the powder, resulting in an increase in temperature at the interface between aluminum powder and graphene. Therefore, this interface reaction is a spontaneous reaction, and it can generate Al_4C_3 hard and brittle phase.

4 Conclusions

In this paper, graphene coated aluminum composite powders were successfully prepared by mixing graphene with aluminum powder by high energy ball milling. It is shown that the particle size of graphene coated aluminum composite powder increases with the increase of the ratio of ball to material. The particle size of 5:1 ball material ratio is uniform and has no obvious agglomeration, and the particle size is the smallest. Therefore, the ball ratio is the best parameter in the four ball milling parameters. The XRD analysis showed that graphite, Al_2O_3 was not found in the graphene/ aluminum composite powders, indicating that graphene/ aluminum composite powders with 7:1 and 8:1 were not particularly serious and graphite was not formed. Thermodynamic analysis shows that during high energy ball milling, the interface of graphene/aluminium composite powders will react spontaneously to form hard-brittle Al_4C_3 interface products.

Acknowledgments
This research work was financially funded by the Science and Technology Innovation Project of Shanxi Province (No.2016156), the Postdoctoral Science Foundation of China (No.2016M590214), the Key Research and Development Program of Shanxi Province(No.201803D121028), and the Natural Science Foundation of Shanxi Province, China (No. 2015011036).

References
[1] Geim A K. 2009,Graphene: status and prospects. Science, p1530-1534.
[2] Sharma A, Sagar S, Mahto R P, et al. Surface modification of Al6061 by graphene impregnation through a powder metallurgy assisted friction surfacing[J]. Surface & Coatings Technology, 2018, 337:12-23.
[3] Geim A K, Novoselov K S. 2007 The rise of graphene. Nature Materials, p183-191.
[4] Feng S, Guo Q, Li Z, et al. 2017 Strengthening and toughening mechanisms in graphene-Al nanolaminated composite micro-pillars. Acta Materialia, p98-108.
[5] Zeng X, Yu J, Fu D, et al. 2018 Wear characteristics of hybrid aluminum-matrix composites reinforced with well-dispersed reduced graphene oxide nanosheets and silicon carbide particulates. Vacuum, p364-375.
[6] Li M, Gao H, Liang J, et al. 2018 Microstructure evolution and properties of graphene nanoplatelets reinforced aluminum matrix composites. Materials Characterization, p140.
[7] Si C, Tang X, Zhang X, et al. 2017Microstructure and mechanical properties of particle reinforced metal matrix composites prepared by gas-solid two-phase atomization and deposition technology. Materials Letters, p 78-81.
[8] Wang Y, Zhou J, Cheng K, et al. 2017 Research Progress on Preparation,Microstructure and Property of Graphene Reinforced Aluminum Matrix Composite. Materials Review,p917-932.
[9] Nieto A, Bisht A, Lahiri D, et al. 2016 Graphene reinforced metal and ceramic matrix composites: a review. Metallurgical Reviews, p241-302.
[10] Liu Z Y, Xu S J, Xiao B L, et al. 2012 Effect of ball-milling time on mechanical properties of carbon nanotubes reinforced aluminum matrix composites[J]. Composites Part A,p2161-2168.
[11] Woo D J, Sneed B, Peerally F, et al. 2013 Synthesis of nanodiamond-reinforced aluminum metal composite powders and coatings using high-energy ball milling and cold spray. Carbon, p404-415.

IOP Conf. Series: Materials Science and Engineering **544** (2019) 012044 doi:10.1088/1757-899X/544/1/012044

Structural and Electrochemical Properties of Layered-Layered-Spinel Composite Cathode Materials $Li_xMn_{4/6}Ni_{1/6}Co_{1/6}O_{(1.75+0.5x)}$ ($1.1 \leq x \leq 1.3$) for Lithium-Ion Batteries

Zhu Zhenye

School of Materials Science and Engineering, Habin Institute of Technology (Shenzhen) 518055 , PR China
Corresponding author：zhuzy@hit.edu.cn

Abstract. The cathode materials $Li_xMn_{4/6}Ni_{1/6}Co_{1/6}O_{(1.75+0.5x)}$ (x=1.1, 1.2, 1.3) with different content of lithium have been synthesized by two step temperature co-precipitation in this paper. X-ray diffraction patterns confirm that there are spinel structure and layer structure in prepared composite. SEM images show that the same diameters of the assembled microspheres are at about 4 μm in $Li_xMn_{4/6}Ni_{1/6}Co_{1/6}O_{(1.75+0.5x)}$ composite. The measurement of electrochemical properties indicates that, with content of lithium varying from 1.1 to 1.3, content of spinel component declines, which induce to the increase of discharge capacity and decrease of capacity retention. $Li_{1.1}Mn_{4/6}Ni_{1/6}Co_{1/6}O_{2.30}$ composite delivers best rate capability, because there is more spinel component in $Li_{1.1}Mn_{4/6}Ni_{1/6}Co_{1/6}O_{2.30}$.

1. Introduction

The exploration and development of cathode materials in lithium-ion batteries with high capacity, good cycling stability, and low cost are of great importance, owing to the widespread applications of lithium-ion batteries in people's daily life [1-3]. However, for market requirements their available energy density is still a defect. As a result, much effort on improving energy efficiency and reducing environmental pollution should be done in the future. The commercially available $LiCoO_2$ cathode suffers from lack of further development and application because of its high cost, low capacity and toxicity [4]. Doped material $LiNi_{1/3}Co_{1/3}Mn_{1/3}O_2$ was initially synthesized in 2001 by Ohzuku and Makimura , which was considered as a promising cathode material for Li-ion batteries (LIBs) because of its milder thermal stability, lower cost, and excellent cycling performance [5-6]. Layered $LiMnO_2$ also showed structural transformation to spinel during electrochemical cycling [7,8]. A layered $LiNi_{0.5}Mn_{0.5}O_2$ material prepared by coprecipitation method showed promising capacity of around 150 mA·h/g in the voltage range of 2.5–4.3V [9]. Kang et al. [10] have reported increased discharge capacity and electrical conductivity by doping Al, Ti, and Co in $LiNi_{0.5}Mn_{0.5}O_2$. Among those materials, Co doping increased electrical conductivity most successfully.

Cathode materials with different stoichiometry containing lithium manganese oxide (Li_2MnO_3) and lithium metal oxide [$LiMeO_2$ (Me = Co, Ni, Mn, etc.)] have been studied in the recent years and are considered as potential candidates to replace the commercial lithium cobalt (III) oxide ($LiCoO_2$). These mainly include $Li_{1.2}Ni_{0.2}Mn_{0.6}O_2$ [$0.5Li_2MnO_3 \cdot 0.5LiNi_{0.5}Mn_{0.5}O_2$] and $Li_{1.2}Ni_{0.13}Co_{0.13}Mn_{0.54}O_2$ [$0.5Li_2MnO_3 \cdot 0.5LiNi_{0.33}Mn_{0.33}Co_{0.33}O_2$] [11,12]. The use of the cobalt reduces the electrode polarization. Li_2MnO_3 is known to be electrochemically inactive because of the presence of Mn in its +4 oxidation state. However, when charged above 4.5 V, this material can apparently become electrochemically active due to the extraction of Li and O, which in turn leads to the formation of themanganese dioxide (MnO_2) host structure in the compound, which can then reversibly intercalate lithium ions [13-15]. In recent years, the lithium-ion layered solid solutions in a chemical formula of $yLi_2MnO_3{\cdot}(1-y)LiMO_2$ (0<y<1; M=Ni, Co, Mn, Fe, Cr, $Ni_{1/2}Mn_{1/2}$,$Ni_{1/3}Co_{1/3}Mn_{1/3}$, ect.) have attracted more and more attention from specialists and have been recognized as one of the most promising candidates for the next generation of lithium-ion batteries [16-19].

2. Experimental

2.1 Synthesis: two-step co-precipitation method

Layer-layer-spinel structure $Li_{1.2}Mn_{4/6}Ni_{1/6}Co_{1/6}O_{2.35-x}F_x$ (x=0, 0.05, 0.10, 0.15) cathode material was synthesized by two-step co-precipitation method. Figure 1 and Figure 2 show the preparation procedure of the precursor $[Mn_{4/6}Ni_{1/6}Co_{1/6}]CO_3$ and the cathode material $Li_xMn_{4/6}Ni_{1/6}Co_{1/6}O_{(1.75+0.5x)}$(x=1.1, 1.2, 1.3).

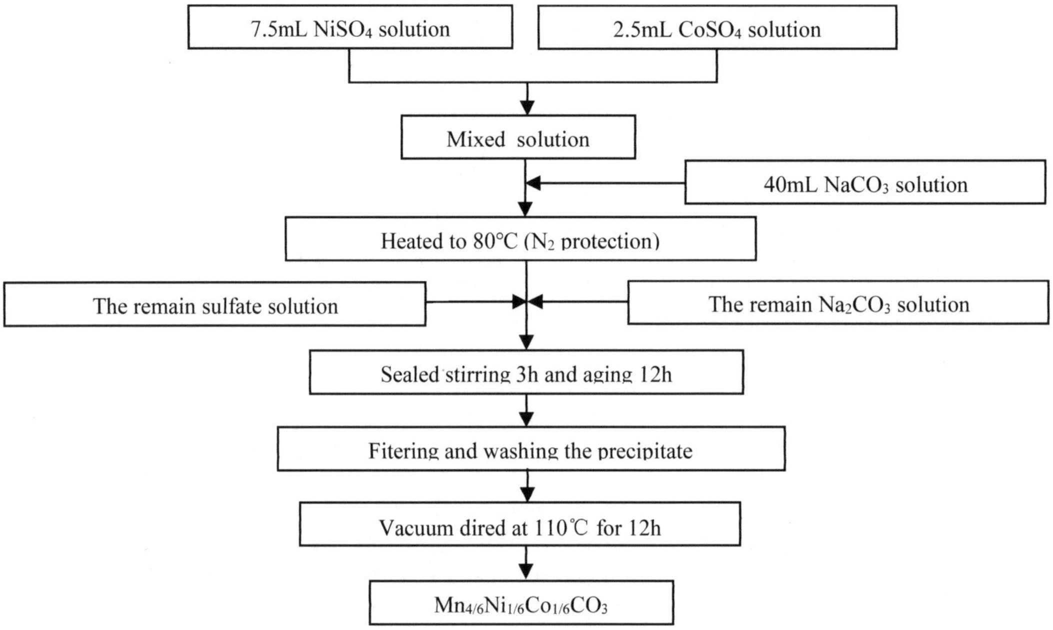

Figure 1. Schematic diagram of the synthesis of precursor $[Mn_{4/6}Ni_{1/6}Co_{1/6}]CO_3$

Firstly, the aqueous solution was prepared by dissolving the salts of $NiSO_4{\cdot}7H_2O$ (Shanghai Aladdin, 99.9%) , $CoSO_4{\cdot}7H_2O$ (Shanghai Aladdin,99.999%) and $MnSO_4{\cdot}H_2O$ (Shanghai Aladdin, 99%) in distilled water with a concentration of $1molL^{-1}$, and dissolving the $NaCO_3$ in distilled water with a concentration of $0.3mlL^{-1}$. Secondly, $NiSO_4{\cdot}7H_2O$ and $CoSO_4{\cdot}7H_2O$ with a mount of 7.5ml an 2.5ml were pumped into the reaction tank together and 40mL $NaCO_3$ aqueous solution was added dropwise to reaction tank, and the mixed solution was stirred and maintained at 40°C. Let the reaction temperature raise to 80°C after completion of the dropwise-addition, and then the mixed solution (2.5mL $NiSO_4{\cdot}7H_2O$, 7.5mL $CoSO_4{\cdot}7H_2O$ and 40mL $MnSO_4{\cdot}H_2O$) were added dropwise into the continuous stirred (900r/min) solution in the reaction tank, while the required $NaCO_3$ was also dropped into the reactor drop by drop. After the completion of the precipitation reaction, the mixed solution was stirred for 3h at the constant temperature and then aged over night. the precursor was washed for several times with the distilled water and dried in vacuum at 110°C for 12h. Then, 5%

excess $LiCO_3$ were mixed thoroughly with the obtained $[Mn_{4/6}Ni_{1/6}Co_{1/6}]CO_3$, and first fired at 500°C for 5h and then 850°C for 15h in air with a heating rate of 5°C/min.

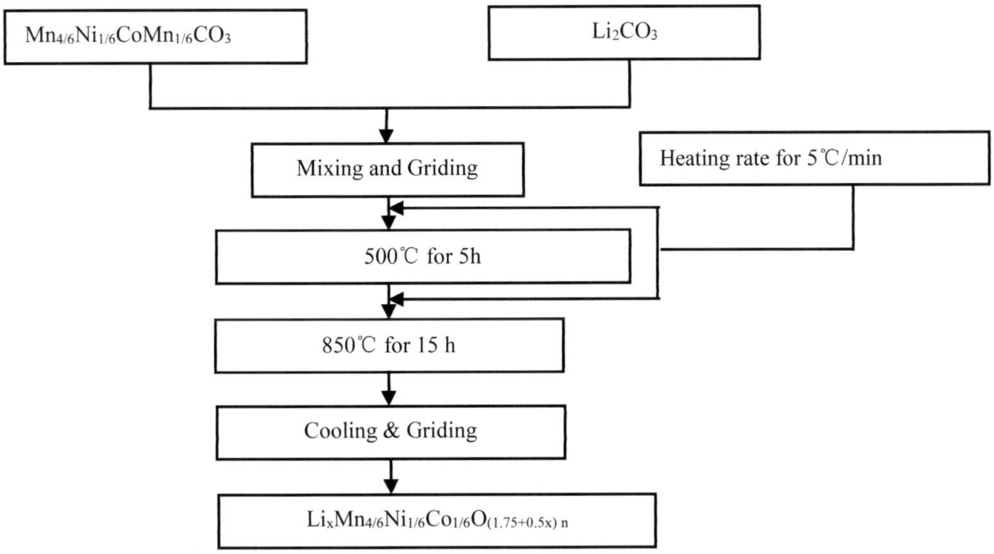

Figure 2. Schematic diagram of the synthesis of cathode material $Li_xMn_{4/6}Ni_{1/6}Co_{1/6}O_{(1.75+0.5x)}$.

2.2 Structure and morphology

A powder X-ray diffractometer (XRD, Rigaku RINT2000 with Cu-Kα radiation) in a range of 2θ=10°-80° with a scan rate of $6°min^{-1}$ was employed to analyze the structure and phase composition of synthesized samples. The scanning electron microscopy (SEM, Hitachi S4700) images were taken to get the morphology and size of the samples.

2.3 Electrochemical measurements

The electrochemical properties of the synthesized samples were examined using a coin cells (CR2032) . The coin type half-cell contains cathode and Li metal anode, which were separated by a porous polypropylene film (Celgard 2400) and electrolyte consisting 1M LiPF6 (EC:DMC=1: 1 in volume). The positive electrode was prepared by spreading mixture of prepared powders (80wt.%), carbon conducting additive (Super-P Li 10wt.%) and polyvinylidene flouride (PVDF 10wt.%) dissolved in N-methyl-2-pyrrolidone (NMP) onto a smooth aluminum foil. The positive electrode was allowed to dry at 75°C for 3h in the air and then at 120°C for 10h in a vacuum oven.The resulting electrode film was then punched into circular discs. The charge-discharge tests were set in the voltage range of 2.0-4.8V ($1C=250mAhg^{-1}$).The electrochemical measurements were carried out by using the Solartron Electrochemical workstation.

3. Results and discussion

3.1 Structure of $[Mn_{4/6}Ni_{1/6}Co_{1/6}]CO_3$ precursor

Figure 3 shows the XRD diffraction peak of the precursor, and there is no extra hetero peak, which is consistent with characteristic diffraction peaks of standard card (PDF#44-1472). Figure 3 shows that our cathode materials synthesized precursor $[Mn_{4/6}Ni_{1/6}Co_{1/6}]CO_3$ has MnCO3 type phase, which is trigonal calcite crystal phase.

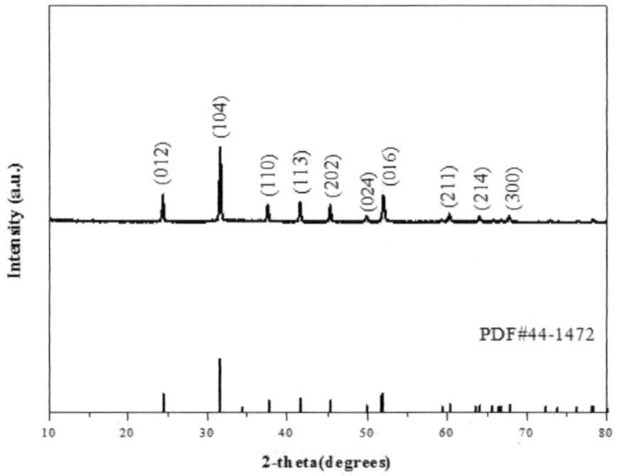

Figure 3. XRD patterns of $Mn_{4/6}Ni_{1/6}CO_3$ precursor materials.

3.2 Structure and morphology of $Li_xMn_{4/6}Ni_{1/6}Co_{1/6}O_{(1.75+0.5x)}$(x=1.1, 1.2, 1.3)composites

XRD patterns of synthesized $Li_xMn_{4/6}Ni_{1/6}Co_{1/6}O_{(1.75+0.5x)}$(x=1.1, 1.2, 1.3) composites are presented in Figure 4. In Figure 4, the diffraction peaks are very clear and sharp, and synthesized materials have relatively good crystallinity and relatively complete crystal form. Compared with the standard cards, the diffraction peaks are consistent with the standard peak $LiNi_{0.5}Mn_{1.5}O_4$ cubic spinel materials (PDF # No.80-2162) with Fd-3m space group, which indicate that the synthesized materials contain spinel material $LiNi_{0.5}Mn_{1.5}O_4$. In addition, several diffraction peaks are inconsistent with standard cubic spinel peak of $LiNi_{0.5}Mn_{1.5}O_4$ materials (PDF # No.80-2162) , and other phase exist in synthesized materials. The diffraction peaks at 20°~25° and 63°~67° refer to Li_2MnO_3 phase (PDF # No.98-016-6861) and layered structure $LiMn_{1/3}Ni_{1/3}Co_{1/3}O_2$ phase, respectively.

Figure 5 shows SEM images of synthesized $Li_xMn_{4/6}Ni_{1/6}Co_{1/6}O_{(1.75+0.5x)}$(x=1.1, 1.2, 1.3) composites. The diameters of the assembled microspheres of $Li_xMn_{4/6}Ni_{1/6}Co_{1/6}O_{(1.75+0.5x)}$ are almost the same at about 4 μm, and the average particle size of primary particles is 200 nm. Therefore, the content of lithium in the composites does not affect the morphology.

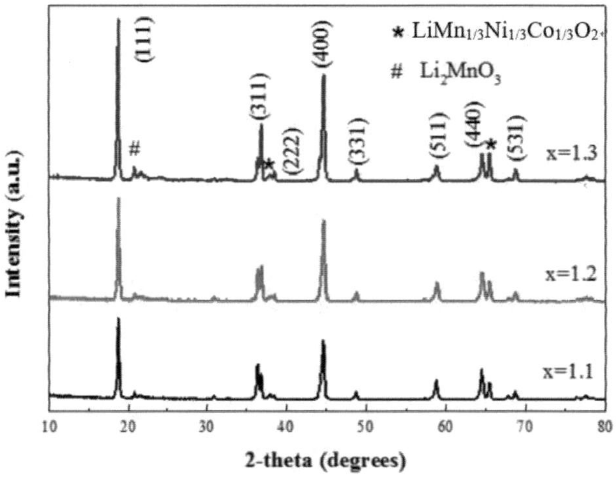

Figure 4. XRD patterns of $Li_xMn_{4/6}Ni_{1/6}Co_{1/6}O_{(1.75+0.5x)}$ (x=1.1, 1.2, 1.3) materials.

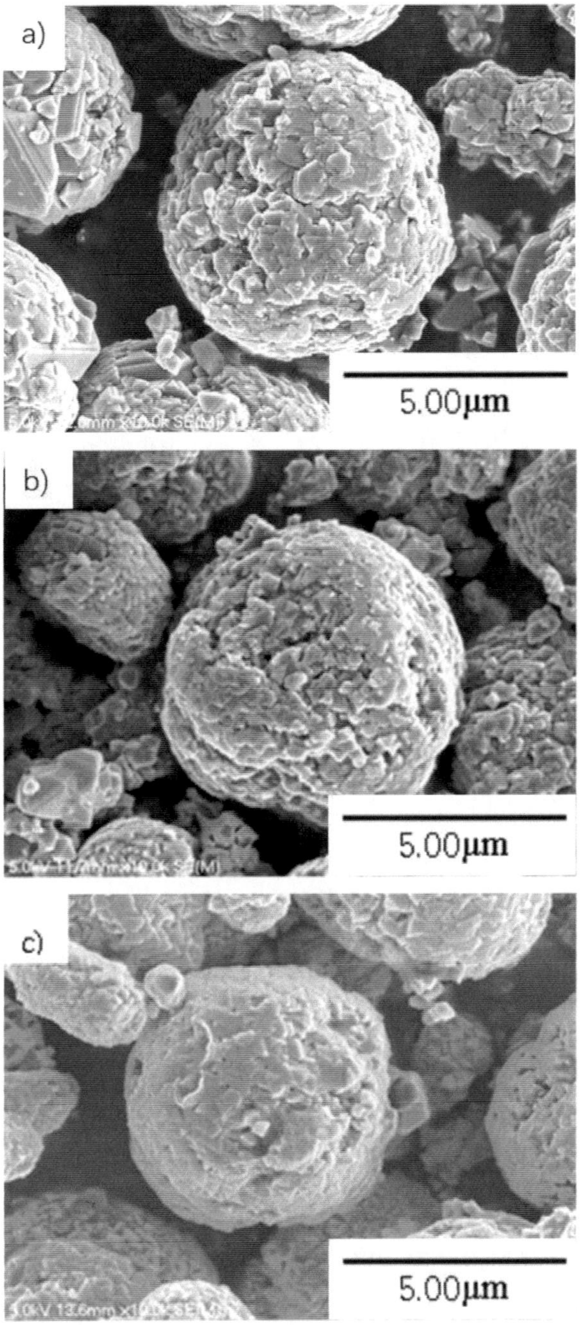

Figure 5. SEM images of $Li_xMn_{4/6}Ni_{1/6}Co_{1/6}O_{(1.75+0.5x)}$ (x=1.1, 1.2, 1.3)
a)x=1.1, b)x=1.2, c)x=1.3.

3.3 Electrochemical performance

Figure 6 indicates that the discharge capacity of synthesized $Li_xMn_{4/6}Ni_{1/6}Co_{1/6}O_{(1.75+0.5x)}$(x=1.1, 1.2, 1.3) composites for 50 cycles at the current density of 0.5C (125mAhg^{-1}) in a voltage range of 2.0-4.8V. As shown in Figure 6 a), with the content of lithium varying from 1.1 to1.3, discharge capacity gradually increases, Figure 6 b) shows the initial discharge capacity and cycle capacity retention rate of 50 times in $Li_xMn_{4/6}Ni_{1/6}Co_{1/6}O_{(1.75+0.5x)}$(x=1.1, 1.2, 1.3) composites. In Figure 6 b), the initial discharge capacity rises and the capacity retention decreases as the lithium content is increased. Spinel structures in the $Li_xMn_{4/6}Ni_{1/6}Co_{1/6}O_{(1.75+0.5x)}$ composite become less and less as the content of lithium is creased. $Li_{1.3}Mn_{4/6}Ni_{1/6}Co_{1/6}O_{2.4}$ deliver largest initial discharge capacity and smallest capacity retention. Since the spinel structure is stable and shows small discharge capacity, reduction of spinel structure will certainly affect the capacity retention rate of composite.

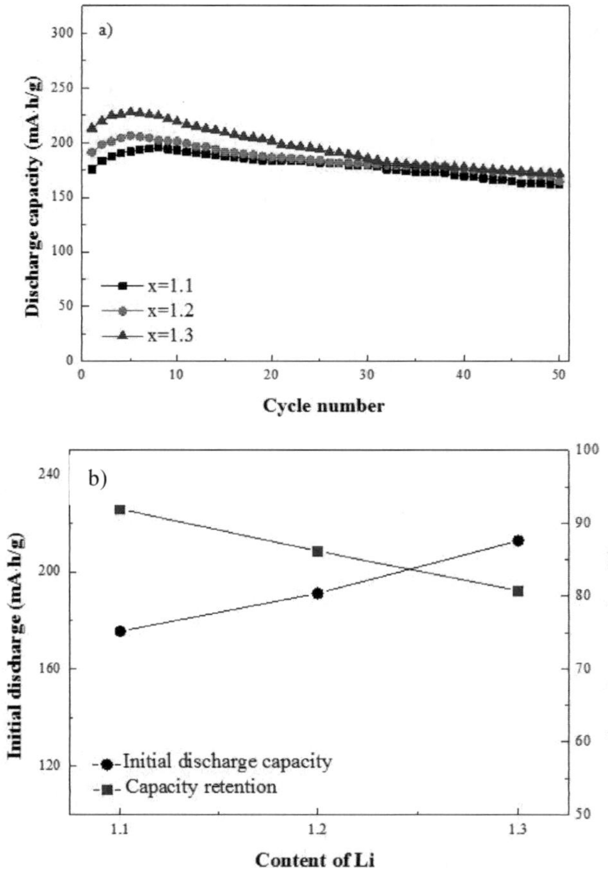

Figure 6. a) Cycle performance of $Li_xMn_{4/6}Ni_{1/6}Co_{1/6}O_{(1.75+0.5x)}$ (x=1.1, 1.2, 1.3) at a discharge rate of 0.5C (1C=250mAhg-1), and b) the corresponding initial discharge capacity and capacity retention values.

Figure 7 shows the rate capability of the $Li_xMn_{4/6}Ni_{1/6}Co_{1/6}O_{(1.75+0.5x)}$ (x=1.1, 1.2, 1.3) between 2.0V and 4.8V at charge-discharge rates from 0.1C to 5C (1C=250mA/g). In Figure 7, With the applied current density increasing, all discharge capacity of all composites decline gradually, to some degree, , which can be ascribed to the increasing polarization of the electrodes at high current densities. All discharge capacity of $Li_xMn_{4/6}Ni_{1/6}Co_{1/6}O_{(1.75+0.5x)}$ (x=1.1, 1.2, 1.3) at different current density are listed in Table 1. It is indicated that initial discharge capacity of $Li_{1.3}Mn_{4/6}Ni_{1/6}Co_{1/6}O_2$.40 material is (259.6mA·h/g) was significantly better than that of the other two materials (241mA·h/g and 242mA·h/g). However, with charge-discharge current density increasing from 0.1C to 5C, the discharge capacity of $Li_{1.3}Mn_{4/6}Ni_{1/6}Co_{1/6}O_{2.40}$ composite decrease greatly, achieving 53.4mA·h/g at 5C. When the charge-discharge current density is 5C, $Li_{1.1}Mn_{4/6}Ni_{1/6}Co_{1/6}O_{2.30}$ has a maximum discharge capacity (78.6 mA·h/g). $Li_{1.1}Mn_{4/6}Ni_{1/6}Co_{1/6}O_{2.30}$ composite shows the best rate performance, because $Li_{1.1}Mn_{4/6}Ni_{1/6}Co_{1/6}O_{2.30}$ composite contains more spinel component, which is beneficial to structural stability during charge-discharge process.

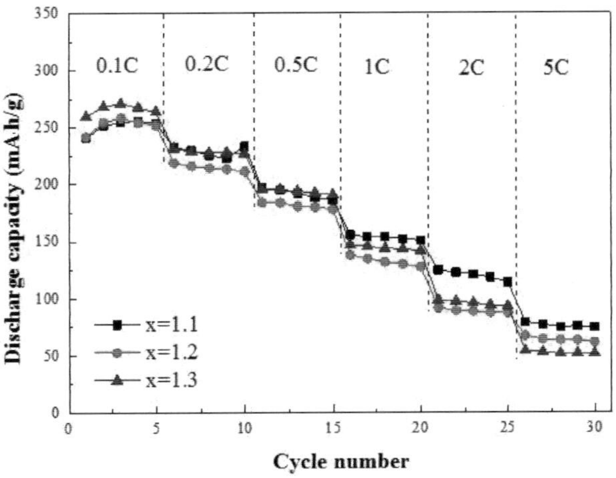

Figure 7. Rate capacity of $Li_xMn_{4/6}Ni_{1/6}Co_{1/6}O_{(1.75+0.5x)}$ (x=1.1, 1.2, 1.3) in the voltage range of 2.0-4.8V.

Table 1. First discharge specific capacity at different rates

$LixMn_{4/6}Ni_{1/6}Co_{1/6}O_{(1.75+0.5x)}$	First discharge specific capacity at different rates (mA·h/g)					
	0.1C	0.2C	0.5C	1C	2C	5C
$Li_{1.1}Mn_{4/6}Ni_{1/6}Co_{1/6}O_{2.30}$	241	232.2	196.7	155.7	124.6	78.6
$Li_{1.2}Mn_{4/6}Ni_{1/6}Co_{1/6}O_{2.35}$	242	218.8	174.1	137.9	91.1	66.8
$Li_{1.3}Mn_{4/6}Ni_{1/6}Co_{1/6}O_{2.40}$	259.6	231.3	195.2	146.8	98	53.4

4. Conclusion

The Layered-Layered-Spinel cathode materials $Li_xMn_{4/6}Ni_{1/6}Co_{1/6}O_{(1.75+0.5x)}$ (x=1.1, 1.2, 1.3) with different lithium content have been synthesized by two-step co-precipitation method in this paper. Spinel phase and layer phase coexist in the $Li_xMn_{4/6}Ni_{1/6}Co_{1/6}O_{(1.75+0.5x)}$ (x=1.1, 1.2, 1.3) composites. With the content of lithium increasing from 1.1 to 1.3,the initial discharge capacity of composite gradually increases and the retention of capacity declines, due to decrease of the spinel component. $Li_{1.1}Mn_{4/6}Ni_{1/6}Co_{1/6}O_{2.30}$ composite deliver best rate capability, because there is more spinel component in $Li_{1.1}Mn_{4/6}Ni_{1/6}Co_{1/6}O_{2.30}$.

Acknowledgements

This work is supported by Shenzhen Basic Research Free Exploration Project (Nos.JCYJ20 180306171650007).

References

[1] Tarascon J M and Armand M B 2001 Issues and challenges facing rechargeable lithium batteries *Nature* **414(6861)** pp 359-367.

[2] Whittingham M S 2004 Lithium batteries and cathode materials *Chemical Reviews* **35(50)** pp 4271-4302.
[3] Arumugam D and Kalaignan G P 2010 Synthesis and electrochemical characterizations of nano-La_2O_3-coated nanostructure $LiMn_2O_4$ cathode materials for rechargeable lithium batteries *Materials Research Bulletin* **45(12)** pp 1825-1831.
[4] Goodenough J B and Kim Y 2010 Challenges for rechargeable Li batteries *Chemistry of Materials* **22(3)** pp 587-603.
[5] Yabuuchi N and Ohzuku T 2003 Novel lithium insertion material of $LiCo_{1/3}Ni_{1/3}Mn_{1/3}O_2$ for advanced lithium-ion batteries *Journal of Power Sources* **119(6)** pp 171-174.
[6] Lee M H, Kang Y, Myung S T and Sun Y K 2005 Synthetic optimization of $Li[Ni_{1/3}Co_{1/3}Mn_{1/3}]O_2$ via co-precipitation *Electrochimica Acta* **50(4)** pp 939-948.
[7] Ammundsen B and Paulsen J 2001 Novel lithium-ion cathode materials based on layered manganese oxides *Advanced Materials* **13(12)** pp 943-956.
[8] Soo P P, Huang B, Jang Y I, et al 1999 Rubbery Block Copolymer Electrolytes for Solid-State Rechargeable Lithium Batteries *Journal of the Electrochemical Society* **146(1)** pp 32-37.
[9] Yabuuchi N and Ohzuku T 2003 Novel lithium insertion material of $LiCo_{1/3}Ni_{1/3}Mn_{1/3}O_2$, for advanced lithium-ion batteries *Journal of Power Sources* **119(6)** pp 171-174.
[10] S H Kang, J Kim, M E Stoll, D Abraham, Y K Sun, K Amine 2002 Layered $Li(Ni_{0.5-x}Mn_{0.5-x}M_{2x'})O_2$ (M'=Co, Al, Ti; x=0, 0.25) cathode materials for Li-ion rechargeable batteries *Power Sources* **112** pp 41-48.
[11] Liu J, Wang Q, Reeja Jayan B and Manthiram A 2010 Carbon-coated high capacity layered Li[Li0.54/0.13/0.132] cathodes *Electrochemistry Communications* **12(6)** pp 750-753.
[12] Kang S, H Kempgens, Greenbaum Kropf and A J 2007 Interpreting the structural and electrochemical complexity of $0.5Li_2MnO_3.0.5LiMO_2$ electrodes for lithium batteries (M = $Mn_{0.5-x}Ni_{0.5-x}Co_{2x}$, $0 \leqslant x \leqslant 0.5$). J.*mater.chem*, 17, pp 2060-2077.
[13] Z Lu and J R Dahn 2002 Understanding the anomalous capacity of Li/Li $[Ni_xLi_{(1/3-2x/3)}Mn_{(2/3-x/3)}]O_2$ cells using in situ x-ray diffraction and electrochemical studies *Electrochem. Soc* **149** pp A815-A822.
[14] Chenhao Zhao, Xinxin Wang, Xinru Liu, et al 2014 Mn-Ni Content-Dependent Structures and Electrochemical Behaviors of Serial $Li_{1.2}Ni_{0.13+x}Co_{0.13}Mn_{0.54-x}O_2$ as Lithium-Ion Battery Cathodes *American Chemical Society* **6** pp 2386-2392.
[15] S H Park, S H Kang, C S Johnson, K Amine 2009 *J. Mater, Chem.* **19** pp 4510-4516
[16] Kitaura H, Hayashi A, Tadanaga K, Tatsumisago M 2010 Electrochemical performance of all-solid-state lithium secondary batteries with Li–Ni–Co–Mn oxide positive electrode *Electrochim Acta* **55** pp 8821-8828.
[17] Kim J M and Chung H T 2004 Effect of lithium boron oxide glass coating on the electrochemical performance of $LiNi_{1/3}Co_{1/3}Mn_{1/3}O_2$ *Electrochim Acta* **49** pp 937-944.
[18] Y J Hong, S H Choi, C M Sim, J K Lee and Y C Kang 2012 Effect of boric acid on the properties of $Li_2MnO_3 \cdot LiNi_{0.5}Mn_{0.5}O_2$ composite cathode powders prepared by large-scale spray pyrolysis with droplet classifier *MRS Bull* **47** pp 4359-4364.
[19] C H Zhao, W P Kang, Q B Xue and Q Shen 2012 Polymerization pyrolysis-assisted nanofabrication of solid solution $Li_{1.2}Ni_{0.13}Co_{0.13}Mn_{0.54}O_2$ for lithium-ion battery cathodes *Nanopart.Res.* **14(1240)** pp1-9.

IOP Conf. Series: Materials Science and Engineering **544** (2019) 012045 doi:10.1088/1757-899X/544/1/012045

Impact of the pressure between the casting and water-cooled mode on the interfacial heat transfer coefficient under LPDC

Y D Zeng ,Q H Yao,L T He,J Zhang

College of Materials Science and Engineering, Inner Mongolia University of Technology, Huhehaote,010051, China

E-mail: zyidan@sina.com

Abstract. Low pressure die casting（LDPC） is the main method for aluminum alloy castings. In order to obtain no porous castings, a top-down progressive solidification is required. Owing to the complicated casting shape and the uneven wall thickness, it is difficult to achieve the above requirement, which limits the application of the method. The water-cooled mode, with controllable cooling capacity, can solve this problem, and great importance has been attached to the heat exchange between water-cooled mode and castings. In the present paper, the inverse solution method is adopted to study the influence of the casting mold/casting interface pressure and circulating water flow rate on the interface heat transfer coefficient（IHTC）. The results show that the increasing interface pressure and the circulating water flow rate can lead to the highest IHTC of 781.6~1002.4 W/(m^2·K) and 1002.4~1657.6 W/(m^2·K) respectively. The calculation result of IHTC provides a basis for the design of low pressure casting with water-cooled mode.

1. Introduction

Low pressure die casting is an ultra-precision forming process with a metal utilization rate of 90 to 98%. The castings produced through this technical process are characterized by no porous structures and high mechanical properties. Therefore, nearly 50% of the castings are produced by adopting this technique, and low pressure die casting has become the main method for aluminum alloy castings.

In the exertion of pressure, low pressure die casting aims to press the melt at the bottom of the mold into the model cavity from the stalk. The casting solidifies from the top to the gate, and the melt in the stalk provides the feeding for the casting. Therefore, in the low pressure die casting, the top-down progressive solidification must be ensured in order to obtain a non-porous structure of the casting.

However, in general, the shape of the casting is complicated, and the wall thickness is not uniform, and it is difficult to ensure the up-bottom solidification sequence. For this reason, forced cooling is required to improve the heat exchange condition between the partial mold and the casting to ensure the casting is sequentially solidified. Water-cooled mode with controllable heat transfer strength are valued by casting workers at home and abroad, and heat exchange between water-cooled mode and castings becomes the key to low pressure die casting process design.

Foreign researches on water-cooled mode are relatively conducted early, and it is possible to produce complex parts such as large gear box [1]. Domestic researches have been carried out in recent years, but there is still a certain gap in the types of castings compared with foreign countries. An important reason lies in the lack of researches on heat exchange between water-cooled mode and castings, and basically designs on empirical formula. Thus, in order to improve the quality of the casting, it is necessary to study the interfacial heat transfer of the casting/water-cooled mode.

The heat transfer of the casting/water-cooled mold is usually described by IHTC. For low pressure die casting, the casting/mold pressure and the mold cooling capacity are the main factors which affect IHTC. The former is related to the pressure of the low pressure casting(hereinafter referred to as "pressure"), and the latter is related to the cooling water flow rate(hereinafter referred to as "water flow rate"). Therefore, the study of the heat exchange of castings/molds is attributed to the study of the influence of pressure and water flow rate on IHTC.

The difficulties in the study of IHTC are manifested in the following aspects. Due to the influence of many factors, IHTC has no accurate theoretical solution, and it is difficult to measure IHTC directly under complicated interference of various factors.

The combination of the measured temperature field of the casting and the inverse calculation is an effective way to obtain IHTC. Extensive investigation has been made in terms of calculation methods and models [2,3], as well as the solution of casting / mold IHTC from perspectives of casting materials and casting temperature [4], initial mold temperature and material [5,6], geometry of casting[7], the mold surface coating material and thickness[8], the mold surface roughness and the air gap [9,10], etc.

In this paper, IHTC of casting/water-cooled die is taken as the core case, along with the method of inverse solution. Firstly, a new experimental model is proposed to accurately control the interface pressure of casting/mold and water flow rate, and to measure the cooling curve of the casting feature points. Then, the inverse calculation method is used to calculate IHTC, and the influence of pressure and water flow rate on IHTC is studied. The results of the study will provide a more accurate basis for the design of low pressure casting process.

2. Experiment

2.1. Experimental model

The experimental model is a combined mold (Fig. 1). The inner cavity is a rectangular paralleled pipe, and the upper and lower bottoms and three walls of the mold are made of heat insulating material. The remaining is a metal water-cooled wall. The water-cooled wall has a water channel inside, which can be cooled by circulating water. The water-cooled wall can move relatively to the other five walls. The push rod on the water-cooled wall is equipped with a pressure sensor to accurately grasp the contact state of the water-cooled wall and the casting and the interfacial pressure. Changing the water flow rate controls the cooling capacity of the water-cooled wall.

In the experiment, after the molten metal is poured into the mold, the water-cooled wall is kept closely with the casting. When the solidification thickness reaches the preset value, the water-cooled wall quickly moves to the left side to reach a preset interface pressure.

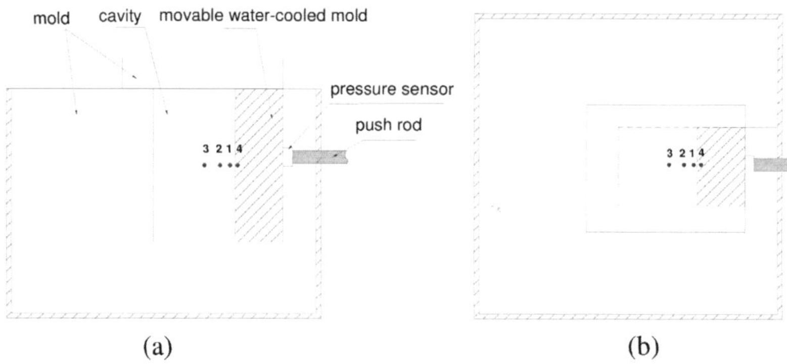

Figure 1. Experimental apparatus: (a) Front sectional view. (b) Horizontal sectional view.

2.2. Experimental Protocol

In order to study the heat exchange of the castings under low-pressure / water-cooled mode, the water flow rate and pressure are selected as the controllable factors in the experiment. In actual production, the pressure of low pressure die casting is applied according to the filling and pressurizing sections. According to the actual production, the pressure is set to 0, 70, 200, 300kg; the water flow rate is set to 23.4, 38.4, 67.2kg/ Min.

2.3. Temperature Measurement

The thermocouple temperature measurement point distribution is shown in Figure 1. The 1, 2, and 3 points are 5mm, 10mm, and 40mm away from the casting/water-cooled interface. The temperature of the water-cooled wall is measured at 4 points and placed in the water-cooled wall, 1 mm from the surface. K-type exposed thermocouples with a diameter of 0.1 mm is used to measure the temperature. The data is collected by the DAQ-USB-2401 module at a sampling frequency of 100 Hz. The temperature change at a temperature measurement point during the experiment is recorded.

2.4. Experimental Materials

Abc The experimental casting material is A356 aluminum alloy, and the pouring temperature is controlled at about 730℃, and the top gating is adopted in this process. PROCAST software is used to calculate the thermal properties of castings and mode materials.

2.5. Inverse Calculation of IHTC

The nonlinear inverse algorithm proposed by Beck et al. [11]was used to compute the IHTC. A random initial value h was set under the assumption that IHTC would remain constant within a certain time interval Δ t. Casting and mold temperature fields were computed with this initial value. Squared differences between the simulated temperature field and the measured temperature field (point 1~3) were calculated, and h was modified repeatedly to minimize the square difference, and the final h value obtained was then considered IHTC.

3. Experimental results and analysis

3.1. Curves of measured temperature

Fig. 2 shows the temperature curve of the measuring point 1~3 when the water flow rate is 23.4 kg/s and the pressure is 0, 70, 200, 300 kg. When the pressure is 0kg (Fig. 2a), the temperature drops slowly at the measuring points of 1~3. At 400s, the temperature at 1-3 points is 564.5, 568.1, 577.4℃.

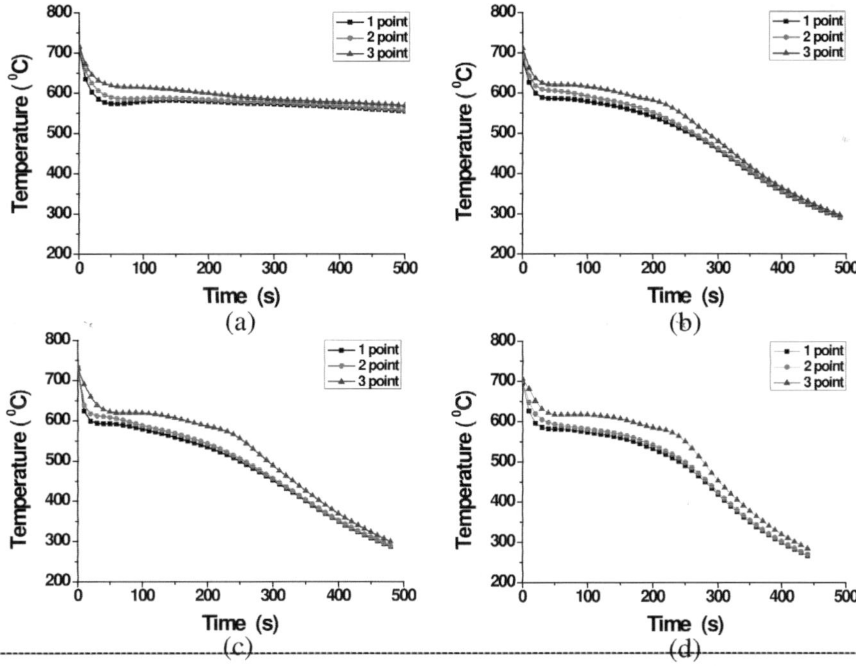

Figure 2. Measured temperature at different pressures

(a)pressure 0kg (b) pressure 70kg (c) pressure 200kg (d) pressure 300kg

The pressure is 70kg (Fig. 2b), and the temperature drop of the measuring point 1~3 is accelerated. At 400s, the temperature of 1~3 points is 354.2, 356.5, 365.1℃. Based on the comparison of two temperatures, it can be found that the pressure change will lead to a change in the cooling rate in the casting, which in turn affects the solidification process and the final microstructure. Comparing Fig. 2b, c, and d, indication can be made that as the pressure increases, the temperature drop rate increases, as is seen from 1 to 3 points; At 400 s, the temperatures are 354.2, 356.5, 365.1℃, 349.0, 352.1, 368.7℃ and 330.8, 336.0, 348.4℃, respectively.

Fig.3 is the temperature curve of 1~3 measuring points when the pressure is 300kg and the water flow rate is 23.4, 38.4, 67.2kg/min respectively. The water flow rate increases, and the temperature drop of 1~3 is accelerated. At 400s, the temperatures are 330.8, 336.0, 348.4℃, 324.7, 330.3, 347.1℃, and 323.8, 332.3, 338.7℃, respectively. Therefore, the water flow rate is increased, and the cooling capacity of the mold is improved. Under the high mold pressure, the cooling effect of the mold on the casting is enhanced.

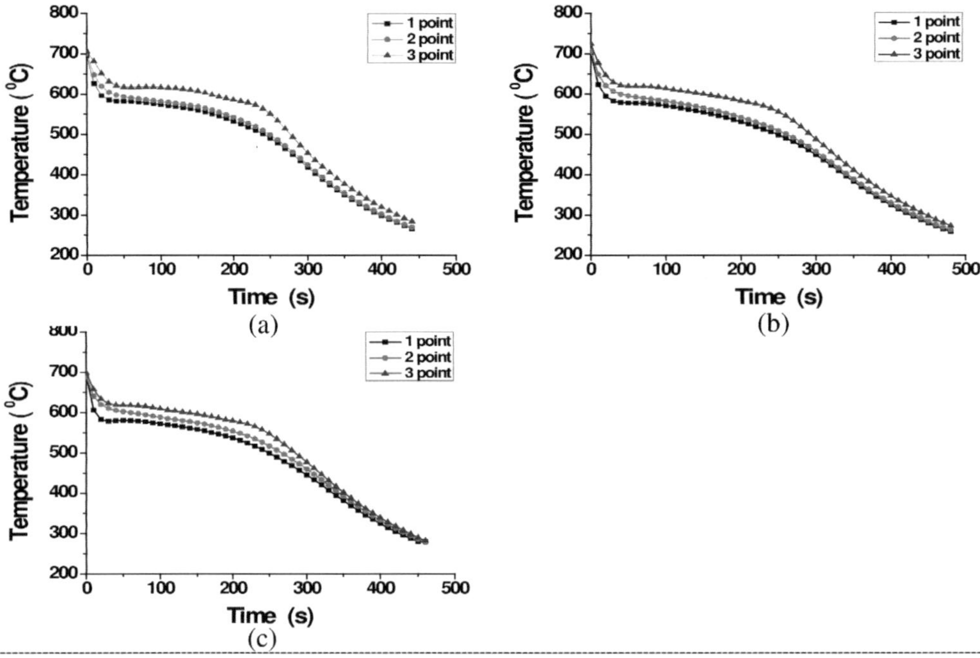

Figure 3. Measured temperature of different water flow rates

(a) flow rate 23.4 kg/min (b) flow rate 38.4kg/min (c) flow rate 67.2kg/min

3.2. *Solution of the IHTC*

It can be seen from the temperature measurement results that the pressure and water flow rate affect the cooling of the casting, and IHTC is used to quantify the interface heat transfer.

Fig.4 shows the calculation results of IHTC when the water flow rate is 23.4 kg/min and the pressure is 0, 70, 200, and 300 kg, respectively. When the pressure is 0kg (Fig. 4a), IHTC reaches the highest value and then decreases rapidly after the pouring. IHTC decreases slowly after 150s, and then remains basically unchanged. When the solidification time is 400s, IHTC drops to 291.8 W/ (m^2·K). IHTC curve of the pressure applied after pouring is shown in Fig. 4b~d, which shows a rebounding of IHTC, and then decreases slowly over time. Through the three IHTC, it is obvious that the stronger the pressure, the greater IHTC recovery. Under the experimental conditions, the pressure increased from 70 to 300 kg at a flow rate of 23.4 kg/min, and IHTC increased from 781.6 to 1002.4 W/(m^2·K). In addition, IHTC of 400s is greater than that of no pressure applied. We can easily conclude that IHTC can more accurately reflect the influence of pressure on heat transfer than the temperature change of the casting.

IOP Conf. Series: Materials Science and Engineering **544** (2019) 012045 doi:10.1088/1757-899X/544/1/012045

Fig. 5 shows the calculation results of IHTC when the pressure is 300 kg and the water flow rate is 23.4, 38.4, and 67.2 kg/min, respectively. As the water flow rate increases, IHTC increases, which is 1002.4, 1335.9 and 1657.6 W/(m^2·K), respectively, and IHTC of 400s also increases.

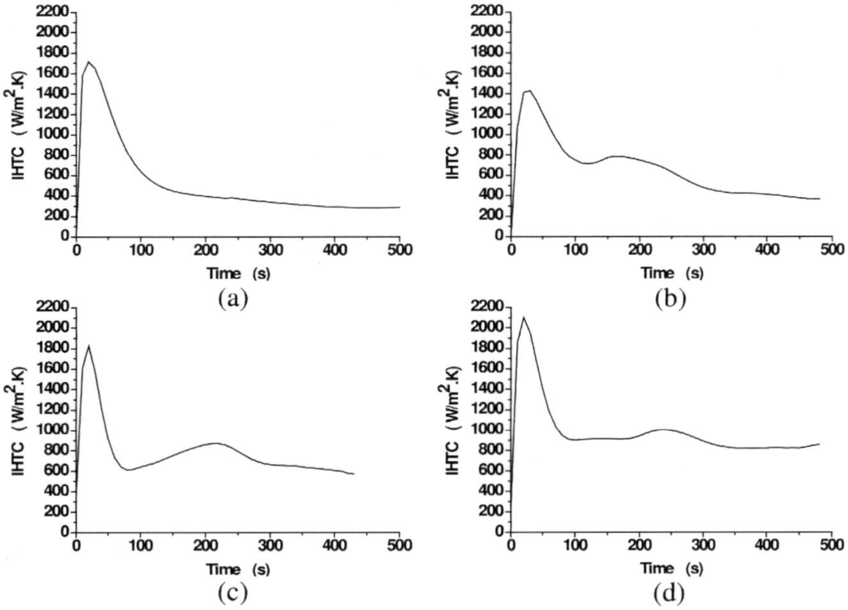

Figure 4. IHTC at different pressures

(a)pressure 0kg (b) pressure 70kg (c) pressure 200kg (d) pressure 300kg

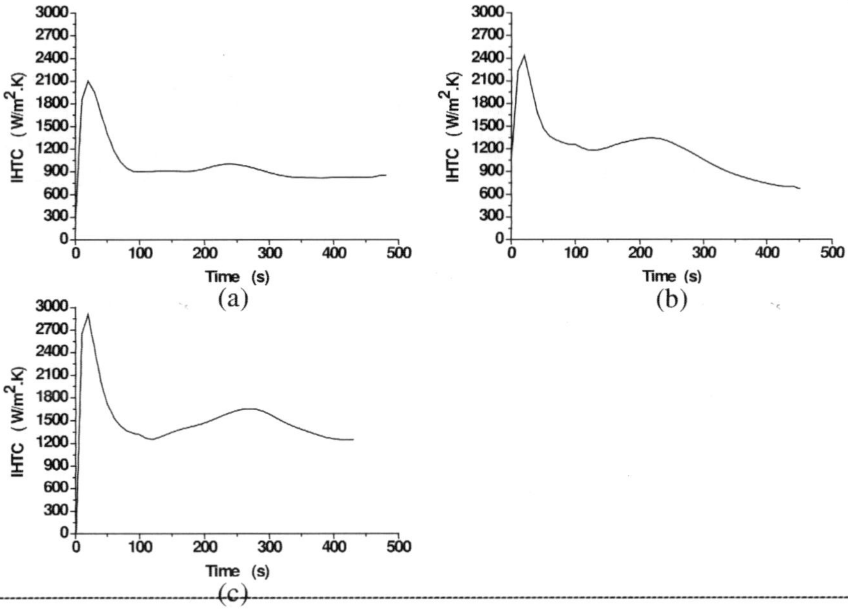

Figure 5. IHTC of different water flow rates

(a) flow rate 23.4 kg/min (b) flow rate 38.4kg/min (c) flow rate 67.2kg/min

Fig. 6 presents the relationship between the highest and lowest IHTC and pressure combined with water flow rate in the above experiment. The portion between lines 1 and 2 of Fig. 6a is the change range of IHTC after the water flow rate changes. The portion between lines 1 and 2 of Fig. 6b indicates the change range of IHTC after pressure changes. This range of values can provide reference for low pressure casting mold design.

In summary, the pressure and cooling water flow rate will cause the change of heat exchange rate of the casting/mode interface. In the low pressure die casting process, the joint regulation of these two parameters can effectively control IHTC and further control the partial cooling rate and the solidification sequence of the casting to ensure that the castings are progressively solidified from the top to the gate, and finally a high-quality die-casting part with complicated shape is achieved.

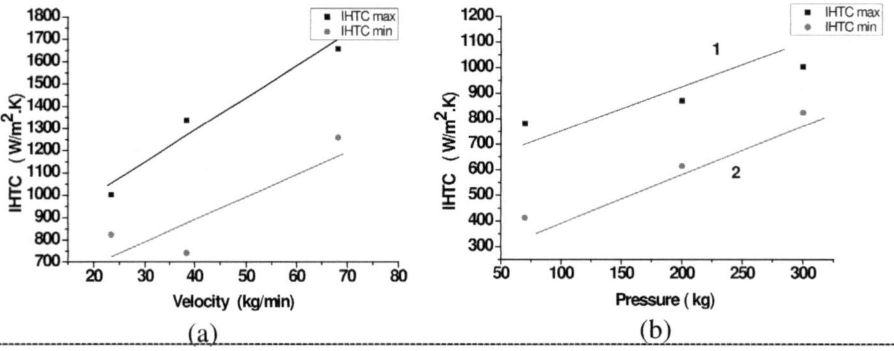

Figure 6. Variation range of IHTC corresponding to different pressures and water flow rates (a) water flow rate 23.4~67.2kg/min (b) pressure 70~300kg

4. Conclusions

(1) IHTC increases with the interface pressure. In this test, the highest IHTC increased from 781.6 to 1002.4 W/(m^2·K). When the solidification time is 400s, IHTC is still 41.3% higher than that without pressure.

(2) The water flow rate has a great influence on IHTC. When the water flow rate increases, IHTC grows. The water flow velocity increases from 23.4 kg/min to 67.2 kg/min, and IHTC rises from 1002.4 to 1657.6 W. /(m^2·K).The pressure enhances the cooling capacity of circulating water cooling mold.

(3) The experimental model which can accurately control the casting/mode pressure and circulating water flow rate is designed to study IHTC of low pressure die casting. The calculation result of IHTC provides the basis for the design of low pressure die casting water-cooled mode.

References

[1] Pereborshchikov S I and Khabarov A N 1997 Russian Castings Technology **8-9** 44-46
[2] Blackwell B and Beck J V 2010 *International Journal of Heat and Mass Transfer* **53** 753-759.
[3] Duda P 2015 *International Journal of Heat and Mass Transfer* **84** 1101-1112.
[4] Ferreira I L, Spinelli J E and Nestler B, et al 2008 *Materials Chemistry and Physics* **111** 444-454.
[5] Prabhu K N, Kumar S T and Venkataraman N 2002 *Journal of Materials Engineering and Performance* **11(3)** 265-273
[6] Guo Z P and Xiong S M 2007 *Acta Metallurgica Sinica* **43(11)** 1155-1160.
[7] Kumar T S P and Kamath H C 2004 *Metallurgical and Materials Transactions* **35(B)** 575-585
[8] Hamasaiid A,Dargusch M S and Davidson C J, et al, 2007 *Metallurgical and Materials Transactions* **38(A)** 1303-1316.
[9] Coates B and Argyropolulos S A 2007 *Metallurgical and Materials Transactions* **38(B)** 243-255.
[10] Zeng Y D,Yao Q H and Wang F 2016 *Materials Science Forum* **893** 174-180
[11] Beck J V and Blackwell B F 1985 *Inverse heat conduction Ill-posed problems*, Wiley, New York.

IOP Conf. Series: Materials Science and Engineering **544** (2019) 012046 doi:10.1088/1757-899X/544/1/012046

Vibration characteristic analysis and control of hospital ship

Yanhe Shan[1], Xuhong Miao[1,2,*], Chengshun Li[3] and Zhigang Liu[3]

[1]College of Shipbuilding Engineering, Harbin Engineering University, Harbin, 150001, PR China
[2]Naval Research Academy, Beijing, 100161, PR China
[3]Tianjin Xingang Shipyard Heavy Industry, Tianjin, China
E-mail: shanyanhe@hrbeu.edu.cn

Abstract. Aiming at the problem of high requirement of hospital ship for hull vibration, this paper takes a hospital ship as an example to analyse its vibration characteristic. The three-dimensional finite element model of hospital ship is established. Take the pulsating force of propeller as the exciting force, the vibration response of the whole ship is calculated by acoustic-structure coupling method. Refer to related standard of ISO6954-2000(E), the numerical result shows that the vibration of some region of stern is over standard. An optimization scheme is proposed for the area whose vibration over standard, the result suggests that the whole ship vibration meets the standard after optimization design. This research provides a reference for design and construction of hospital ship.

1. Introduction

In recent years, as ships become larger and faster, the vibration of ships has attracted more and more attention. For the problem of large ship vibration, Yanhui Mao et al.[1] taking non-self-propelled cutter dredger as research object, the difference of ship vibration characteristics under different working conditions is discussed; the influence of different calculation methods of attached water quality on the inherent characteristics of ship total vibration is analyzed. Yu Song et al.[2] in view of the strong vibration of superstructure of a dump sand carrier during its voyage, the response values of forced vibration of ship hull under different working conditions are obtained by frequency response analysis method, and the improvement method of strengthening structure is put forward. Li Yu et al. [3] have carried out research on the vibration performance of a large multi-purpose ship, The calculation results show that the ship has good vibration performance and meets the vibration standard requirements.

In previous studies of ship vibration characteristics, Most of them focus on the vibration of conventional ships, but few on the vibration of hospital ships. Aiming at the problem of vibration characteristics of hospital ship, this paper takes a hospital ship as the research object. Take the fluctuating pressure of propeller as the exciting force, the vibration response of the whole ship is calculated by acoustic-structure coupling method. Referring to ISO6954-2000 (E) [4] related standards, the optimum design scheme is put forward for the vibration over standard area, which has reference significance for the design and construction of hospital ships.

IOP Conf. Series: Materials Science and Engineering **544** (2019) 012046 doi:10.1088/1757-899X/544/1/012046

2. Dynamics Equation of Acoustic- structure Coupling

The vibration system of hull structure is simplified as a multi-degree-of-freedom vibration system, and the differential equation of forced vibration with damping is given as:

$$M\ddot{X} + C\dot{X} + KX = F \tag{1}$$

Where M is the mass matrix; C is damping matrix; K is stiffness matrix; F is exciting load; X is displacement response.

Assuming that the fluid is ideal acoustic medium, the acoustic wave equation can be expressed as:

$$\nabla^2 p = \frac{1}{c^2}\frac{\partial^2 p}{\partial t^2} \tag{2}$$

Where p is instantaneous acoustic pressure, c is sound velocity.

For the acoustic-structure coupling problem, the finite element method is used to calculate the structural vibration and sound field in the same coupling environment. The dynamic equation of the acoustic-structure coupling is given as:

$$\begin{bmatrix} K_s + j\omega C_s - \omega^2 M & \bar{Q} \\ \rho_a \omega^2 \bar{Q}^T & K_a + j\omega C_a - \omega^2 M_a \end{bmatrix} \begin{bmatrix} u \\ p \end{bmatrix} = \begin{bmatrix} F_s \\ F_a \end{bmatrix} \tag{3}$$

Where $\bar{Q}$ is acoustic-structure coupling matrix; M_s and M_a are the mass matrix of structural and acoustic medium; C_s and C_a are the damping matrix of structural and acoustic medium; K_s and K_a the stiffness matrix of structural and acoustic medium; u is displacement vectors of structural nodes; p is sound pressure vector; F_s is structural load vector; F_a is sound field excitation load vector.

3. Numerical simulation

3.1. Numerical simulation model

3.1.1. Basic parameters. The light ship weight of the hospital ship about 14,000 tons. The vessel is propelled by four main engines and two propellers. The maximum power of main engine is 3000Kw. Propeller is a 4-bladed propeller, its rated speed is 160rpm.

3.1.2. Finite element model. According to the drawings of hospital ships, establishing a finite element model of the vessel. Simulation of hull plate structure by shell element, using beam elements to simulate stiffeners such as longitudinal frames, beams, and pillars. There are more than 40 kinds of plate thickness and section sizes in the model, for ensure that the structural parameters of the model are the same as those of the design. In order to simulate the effect of entrained water on ship structure vibration, the acoustic-structure coupling finite element method is used in this paper, that is to establish a semi-infinite flow field coupled with the hull. The radius of the flow field is two times breadth, the outer surface of the flow field is defined as a non-reflective boundary condition. The schematic diagram of the acoustic-structure coupling model is shown in Figure 1.

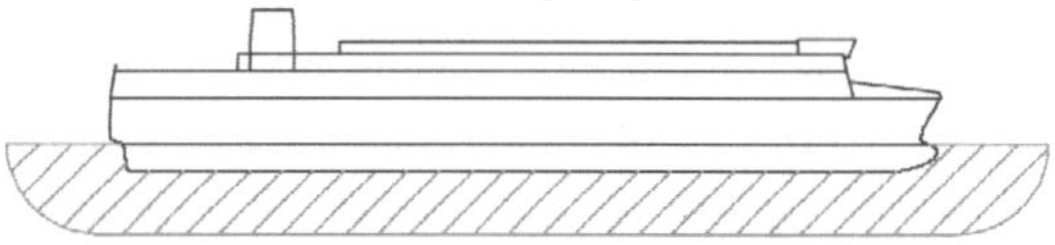

Figure 1. The schematic diagram of the acoustic- structure coupling model.

3.1.3. Exciting force. This paper studies the forced vibration characteristics of hospital ship under the propulsive force of propeller pulsating pressure, due to propeller is one of the main exciting sources of hull vibration and noise. The pulsating pressure of propeller is mainly driven by blade frequency. The formula for calculating blade frequency is given as:

$$f = \frac{n}{60} \times Z \tag{4}$$

Where f is exciting frequency; n is shaft speed; Z is number of blades.

Propeller pulsation pressure is generally estimated using approximate method. The maximum pulsating pressure P_{max} of the propeller when there is no air bubble can be given as [5]:

$$P_{max} = 4.77 K_{P_0} \frac{SHP}{nD^3} \quad (5)$$

Where SHP is power of propeller shaft; n is shaft speed; D is diameter of a propeller; K_{P_0} is dimensionless factor.

3.2. Vibration analysis and control of hospital ship

3.2.1. Vibration standard. This paper refers to ISO69540-2000(E) standard. It proposed a proposal for the suitability of different areas of the ship for the vibration speed as shown in Table 1.

Table 1. Suitability of different areas of the ship for the vibration speed (mms^{-1}).

	Areas classification		
	Passenger cabins	Crew accommodation areas	Working areas
Values above which adverse comments are probable	4	6	8
Values below which adverse comments are not probable	2	3	4

3.2.2. Forced vibration response of hospital ship. Based on acoustic-structure coupling model of hospital ship, taking pulsating pressure of propeller as exciting force, calculated the vibration response of the whole ship. The calculated frequency band is 1Hz~80Hz. Figure 2 shows the distribution of vibration velocity response of typical decks at the propeller blade frequency. It can be seen that the vibration response of the stern of hospital ship structure is larger than anywhere else, and the vibration response from stern to bow shows a decreasing trend.

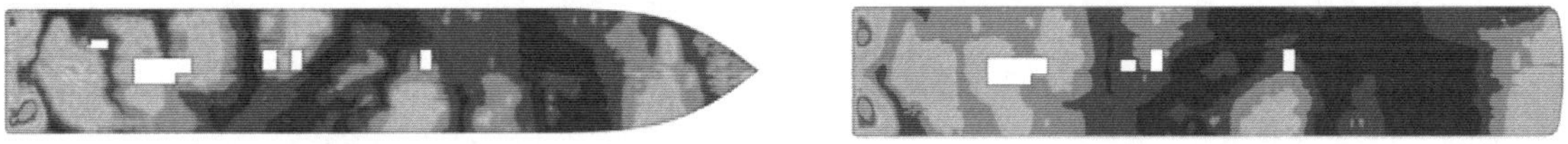

(a) (b)

Figure 2. The distribution of vibration velocity response of typical decks (a) deck 3, (b) deck 5.

According to ISO6954-2000 (E) standard, the overall frequency-weighted r.m.s. velocity of local area are calculated and compared with the standard. Limited to space requirements, only partial results are given, as shown in Table 2. The vibration response curves of deck in typical cabin is shown in Figure 3. The results show that the vibration response of most areas of the hospital ship meets the standard, some areas have potential over-standard risk, vibration response in some areas is over standard, the over-standard area are mainly located in the stern.

Table 2. Evaluation of local vibration response.

Deck	Number of cabin	Name of cabin	vibration velocity（mm/s）	standard（mm/s）	Evaluation
3	314	Operating room	2.2	4	Acceptable
3	319	General office	4.5	4	Unacceptable
4	447	Treatment room	3.9	4	Potential dangerous area
5	503	Standard ward	2.3	4	Acceptable
5	508	Standard ward	2.5	4	Acceptable

5	552	Standard ward	3.9	4	Potential dangerous area
6	601	Waiter room	8.1	4	Unacceptable
6	602	Waiter room	6.2	4	Unacceptable
6	651	Doctor room	1.8	4	Acceptable
8	849	Wheelhouse	2.4	4	Acceptable

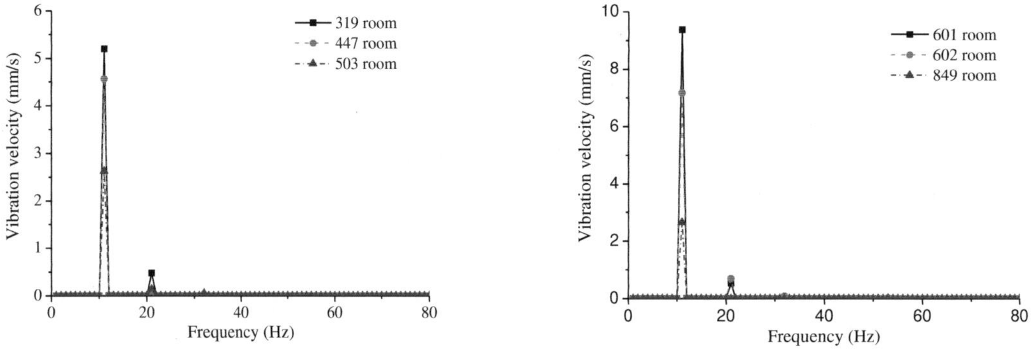

(a) (b)

Figure 3. Vibration response curves of deck in typical cabin.

3.2.3. Vibration control of hospital ship. For the problem that the propeller causes vibration over standard, main control measure is to attach high damping material to the bottom of the ship above and around the propeller. This paper aims at the over-standard areas and potential dangerous area, the proposed optimization measure is to lay constrained damper plates in the bottom area from 0# to 21#. The laying thickness is 12mm, in the simulation calculation, the loss factor is 0.05 when the restrained damper plate is laid. The schematic diagram of the constrained damper plate is shown in Figure 4.

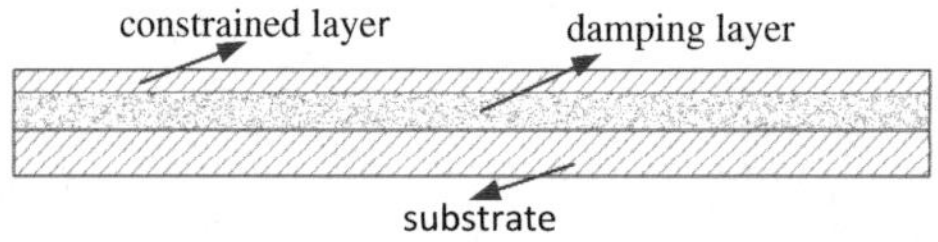

Figure 4. The schematic diagram of the constrained damper plate.

After laying damping material, Limited to space requirements, only partial results are given, as shown in Table 3. The results show that the optimized vibration response of the whole ship meet the standard. Optimized vibration response curves of deck in typical cabin is shown in Figure 5.

Table 3. Evaluation of local vibration response.

Deck	Number of cabin	Name of cabin	vibration velocity（mm/s）	standard（mm/s）	Evaluation
3	314	Operating room	1.8	4	Acceptable
3	319	General office	0.4	4	Acceptable
4	447	Treatment room	3.3	4	Acceptable
5	503	Standard ward	0.9	4	Acceptable
5	508	Standard ward	0.5	4	Acceptable
5	552	Standard ward	3.2	4	Acceptable
6	601	Waiter room	1.4	4	Acceptable
6	602	Waiter room	1.7	4	Acceptable
6	651	Doctor room	1.0	4	Acceptable
8	849	Wheelhouse	2.0	4	Acceptable

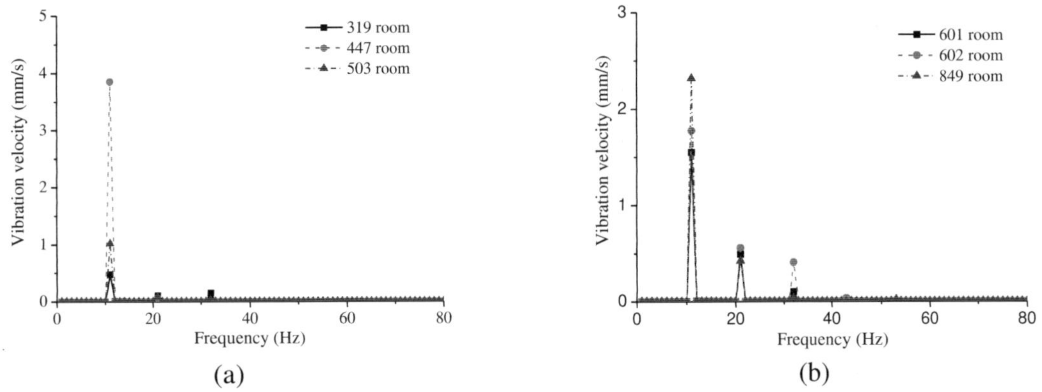

Figure 5. Optimized vibration response curves of deck in typical cabin.

4. Conclusion

This paper take the fluctuating pressure of propeller as exciting force, the vibration response of the hospital ship is calculated by acoustic-structure coupling method. Referring to ISO6954-2000 (E) related standards, the optimum design scheme is put forward for the vibration over standard areas. Through the above research, we can draw the following main conclusions :

(1) Under the action of propeller fluctuating pressure, structural vibration of hospital ship mainly concentrates on the stern part. The vibration response shows a downward trend from stern to bow.

(2) Referring to ISO6954-2000 (E) standards, under the action of propeller fluctuating pressure, the vibration response of most areas of the hospital ship meets the standard, only some areas is over standard. These over-standard areas are mainly located in the stern.

(3) Aiming at the over-standard areas and potential dangerous area, the proposed optimization measure is to lay constrained damper plates in the bottom area from 0# to 21#. The laying thickness is 12mm. The optimized vibration response of the whole ship meet the standard.

Reference

[1] Yanhui Mao, Lijuan Xia 2018 Vibration characteristic analysis of 5 000 m3 /h non self-propelled cutter suction dredger *Jiangsu University of Science and Technology(Natural Science Edition)* **38** 465-71.

[2] Yu Song, Hong Fan, Yuxiao Zou, Xucong Ding 2018 Vibration analysis and control of self-discharging sand carrier *Ship & Ocean Engineering* **47** 60-3+7.

[3] Li Yu, Lu Zhao, Liangsheng Xiong, Hongrui Li 2017 Research on vibration and response of large multi-purpose vessel *Mine Warfare & Ship Self-Defence* **25** 75-80.

[4] Jiameng Wu, Lijuan Xia, Xianding Jin, Shaoyan Chi 2012 Comparison between new and old versions of vibration standards ISO 6954 *Journal of Vibration and Shock* **31** 177-82.

[5] Xiongliang Yao 2004 *Hull Vibration* Harbin Engineering University Press.

Acknowledgments

This study was funded by Major innovation projects Of High Technology Ship Funds of Ministry of Industry and Information of P. R. China, Ph.D. Student Research and Innovation Fund of the Fundamental Research Funds for the Central Universities (HEUGIP201801), National key Research and Development program (2016YFC0303406).

IOP Conf. Series: Materials Science and Engineering **544** (2019) 012047 doi:10.1088/1757-899X/544/1/012047

The Application of Transfer Matrix Method in Acoustic Performance Analysis of Composite Materials and Acoustic Protection Design of Ship

Ruidong Huo[1*], Lin Chen[2], Jiangtao Liu[1] and Dewei Peng[1]

[1]College of Shipbuilding Engineering, Harbin Engineering University, Harbin, 150001, PR China
[2]Marine Design and Research Institute of China, Shanghai, 200011, PR China

E-mail: huoruidong@hrbeu.edu.cn

Abstract. This research introduced the transfer matrix method to calculate the acoustic performance of composite materials, which solving the problem of traditional methods existing in calculating acoustic performance of complex composite structure. Based on the impedance tube method and the reverberation chamber method, the experiment of measuring sound absorption and insulation parameters of typical composite materials was conducted. And then, the validity of theoretical calculation of the transfer matrix method was verified. In addition, under the application of the statistical energy method, the model simulation was carried out by taking a ship as an example, and the transfer matrix method in the noise protection design of the ship was discussed.

1. Introduction

Compared with traditional metal materials, composite materials have some advantages in acoustic performance, damping performance and designability. They have been so widely used in noise control of aerospace, underwater ships, automobile and other industries that it has great significance to study the acoustic properties of composite materials.

At present, there are quite a lot of theoretical researches and applications on the acoustic performance calculation of single-layer and double-layer composite materials, but the theoretical calculation of acoustic performance of multi-layer complex composite structures as a hot issue that has not been solved well. The acoustic performance of the composite material is affected by the important parameter of porosity, thickness, Young's modulus and the lamination sequence of the composite structure. Due to the above characteristics, the traditional methods of composite performance mainly include the equivalent circuit method[1], the finite element method[2], etc., which are difficult to deal with the acoustic performance of complex structures.

The transfer matrix method has a superiority in the calculation of chain structure[3], so it has been introduced into the calculation of acoustic performance of composite materials. The impedance performance and the reverberation chamber method are used to measure the acoustic performance and verify the effectiveness of the transfer matrix method in dealing with such problems. In addition, a statistical energy model simulation is carried out by taking a ship as an example. The application of this method in noise prediction of ship is discussed, which provides the support for the application of noise protection design of ship.

2. Analysis of the acoustic performance of composite based on the transfer matrix method

The principle of acoustic performance analysis of composite materials based on transfer matrix method is as follows. It is assumed that the incident wave is incident at the incident angle θ_0, and the theoretical model of the composite material is shown in figure 1.

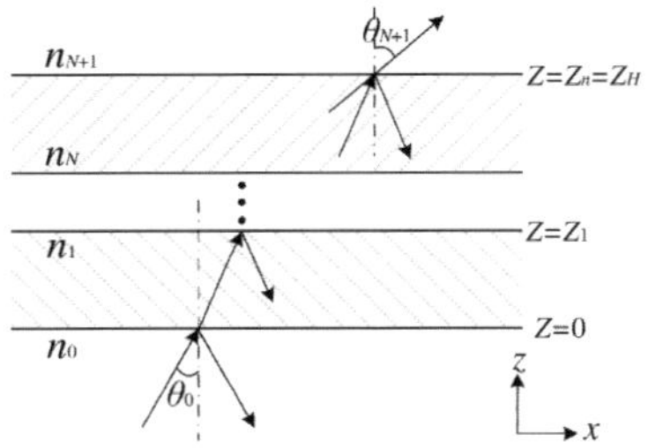

Figure 1. Multilayer composite material schematic

It is also assumed that the composite interface is a plane parallel to each other, the interface is marked as $z=z_n$, and the outermost layer of the structure is a semi-infinite uniform medium. A plane wave(frequency=f) is incident at an incident angle θ_0, defining in the x-o-z coordinate system. Meanwhile, the sound pressure in each layer of the composite material is p_n, and the sound velocity of normal direction is v_{nz}, wherein the subscript n represents the number of composite material layer. The sound pressure p_n can be expressed as:

$$p_n = a_n e^{ik_n x\sin\theta_n + ik_n z\cos\theta_n} + b_n e^{ik_n x\sin\theta_n - ik_n z\cos\theta_n} \tag{1}$$

The limited condition is that the sound pressure and the sound velocity of normal direction are continuous at the interface:

$$p_n = p_{n+1}, v_{nz} = v_{(n+1)z}(z = z_n) \tag{2}$$

The transfer matrix is a relationship matrix between p_n and v_{nz} in two adjacent interfaces, and in the subsequent derivation process, the factor $e^{ik_n x\sin\theta_n}$ is omitted for the operation of the x coordinate is not involved, then the following matrix is obtained:

$$q_n = \begin{pmatrix} p_n \\ v_{nz} \end{pmatrix} = \begin{pmatrix} a_n e^{ik_n z\cos\theta_n} + b_n e^{-ik_n z\cos\theta_n} \\ \dfrac{1}{z_n}(a_n e^{ik_n z\cos\theta_n} - b_n e^{-ik_n z\cos\theta_n}) \end{pmatrix}, z_n = \frac{d_n c_n}{\cos\theta_n} = \frac{E_n}{\cos\theta_n} \tag{3}$$

Where Z_n is the acoustic impedance of normal direction, d_n is the density of the n-th layer composite and E_n is the elastic modulus of the n-th layer composite. The matrixes of the upper and lower surfaces of the n-th composite material are respectively denoted as q_n^{up} and q_n^{down} , and the equation 4 is applied to the interface of the incident layer (z=0) and the interface of the transmission layer (z=H):

$$q_0^{up} = \begin{pmatrix} 1+R_p \\ \frac{1}{z_0}(1+R_p) \end{pmatrix}, q_{(N+1)}^{down} = \begin{pmatrix} T_p \\ \frac{T_p}{z_{N+1}} \end{pmatrix}, \text{and } q_{(N+1)}^{down} = T \cdot q_0^{up} \tag{4}$$

Where T is the transfer matrix between q_n^{up} and q_n^{down}, which is a second-order square matrix. It is proved below that T is the transfer matrix between the incident surface and the exit surface of the composite. The sound pressure reflection coefficient can be obtained as follow:

$$R_p = \frac{(T_{11} + \frac{1}{z_0}T_{12}) - z_{N+1}(T_{21} + \frac{1}{z_0}T_{22})}{z_{N+1}[(T_{21} - \frac{1}{z_0}T_{22}) - \frac{1}{z_{N+1}}(T_{11} - \frac{1}{z_0}T_{12})]} \tag{5}$$

Then the acoustic energy reflection coefficient R_e and sound absorption coefficient α can be obtained as follow:

$$R_e = \left|R_p\right|^2 \cdot \frac{z_0}{z_{N+1}}, \alpha = 1 - R_e \cdot R_e^* \tag{6}$$

According to the equation 3, the matrix of the n-th layer can be expressed as:

$$q_n^{down} = \begin{pmatrix} e^{ik_n z\cos\theta_n} & e^{-ik_n z\cos\theta_n} \\ \frac{e^{ik_n z\cos\theta_n}}{z_n} & \frac{-e^{-ik_n z\cos\theta_n}}{z_n} \end{pmatrix}_{z=z_{n-1}} \begin{pmatrix} a_n \\ b_n \end{pmatrix}, \quad q_n^{up} = \begin{pmatrix} e^{ik_n z\cos\theta_n} & e^{-ik_n z\cos\theta_n} \\ \frac{e^{ik_n z\cos\theta_n}}{z_n} & \frac{-e^{-ik_n z\cos\theta_n}}{z_n} \end{pmatrix}_{z=z_n} \begin{pmatrix} a_n \\ b_n \end{pmatrix} \tag{7}$$

According to boundary conditions of the composite sound absorbing system, the matrixes q_n above the n-th layer material and under the (n+1)-th layer material has the following relationship: $q_n{}^{up}=q_{n+1}{}^{down}=E_n$, it can be obtained: $E_n=t_n \cdot E_{n+1}$. T_n is the transfer matrix of the n-th layer material, and the specific form is:

$$t_n = \begin{pmatrix} \cos O_n & iz_n \sin O_n \\ \frac{i}{z_n}\sin O_n & \cos O_n \end{pmatrix} \tag{8}$$

Where $O_n=k_n h_n \cos\theta_n$. From the equation 4 to equation 8, the relationship between the total transfer matrix T and the transfer matrix t_n of each layer is obtained:

$$T = t_n \cdot t_{n-1} \cdot \ldots \cdot t_2 \cdot t_1 \tag{9}$$

3. Comparative analysis of test results and theoretical calculations

In this section, we apply the impedance tube method and the reverberation chamber method to respectively test the sound absorption coefficient and sound insulation of composite structures.

3.1. The principle of the impedance tube method and the reverberation chamber method

3.1.1. The impedance tube method for measuring sound absorption coefficient. The basic principle of the impedance tube method for measuring the sound absorption coefficient is that the normal-incident reflection factor r can be determined by the transfer function H_{12} measured by the two microphones in front of the sample. The schematic of the measurement principle is shown in the figure 2.

IOP Conf. Series: Materials Science and Engineering **544** (2019) 012047 doi:10.1088/1757-899X/544/1/012047

The sound pressure at the position of the two microphones is expressed as:

$$P(x_1) = P_i \cdot e^{jkx_1} + P_r \cdot e^{-jkx_1}, P(x_2) = P_i \cdot e^{jkx_2} + P_r \cdot e^{-jkx_2} \tag{10}$$

Where p_i and p_r are the amplitudes of the incident sound pressure and the reflected sound pressure, k is the complex wave number. The transfer function H_{12} of the total sound field is:

$$H_{12} = \frac{p(x_2)}{p(x_1)} = \frac{e^{jkx_2} + r \cdot e^{-jkx_2}}{e^{jkx_1} + r \cdot e^{-jkx_1}} \tag{11}$$

This gives the normal-incident reflection factor r and the sound absorption coefficient α:

$$r \cdot e^{j\psi} = \frac{H_{12} \cdot e^{j\theta} \cdot e^{jkx_2} - e^{jkx_1}}{e^{-jkx_1} - H_{12} \cdot e^{j\theta} \cdot e^{jkx_1}}, \text{ and } \alpha = 1 - |r|^2 \tag{12}$$

It can be seen that the sound absorption coefficient can be obtained from the inter-microphone transfer function H_{12} and the mounting position parameters x_1, x_2.

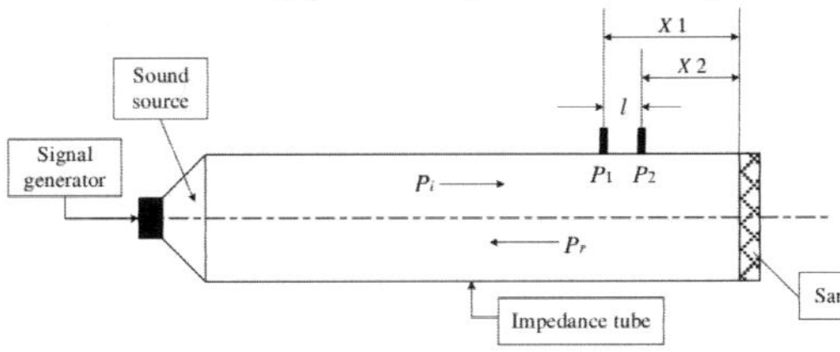

Figure 2. Schematic diagram of the impedance tube method

3.1.2. The reverberation chamber method for measuring sound insulation. The sound insulation measurement of large sample is carried out in two connected chambers. The schematic diagram is shown in figure 3. The sound pressure level measured by the reverberation chamber is SPL1 while the sound pressure level measured by the receiving chamber is SPL2, In addition, it is necessary to correct the sound absorption effect of the receiving room. Therefore, the measurement formula of the sound insulation amount is:

$$SPL = SPL_1 - SPL_2 + 10\lg(\frac{1}{4} + \frac{A}{\alpha S}) \tag{13}$$

Where A is the surface area of the sample, and S is the total area of the receiving chamber.

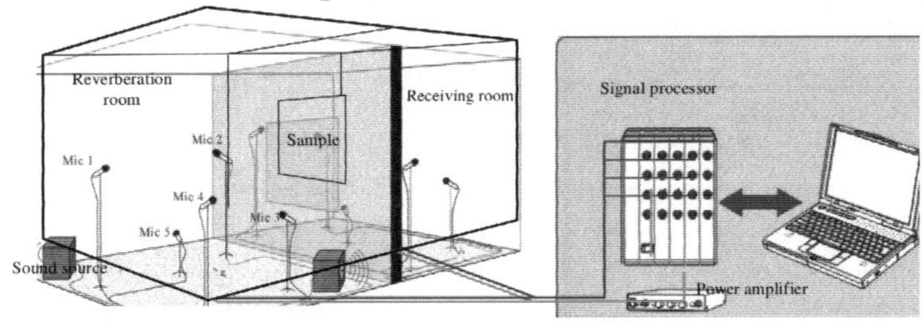

Figure 3. The sound insulation measured by the reverberation chamber method

The layout of sample is shown in figure 4.

Figure 4. The layout of the specimen

3.2. The result

Figure 5-figure 6 shows the comparison between measured values and the theoretical value calculated by the transfer matrix method. The error of the sound absorption coefficient is within 0.1, but it gets

IOP Conf. Series: Materials Science and Engineering **544** (2019) 012047 doi:10.1088/1757-899X/544/1/012047

bigger and bigger with the increase of the frequency, as seen in figure 5. In the other hand, it can be seen from the figure 6 that the error of the calculated value of the sound insulation is within 5dB, which meets the requirements of engineering application. The theoretical value has less error in the middle frequency band, and the error increases with the increase of frequency. Altogether, the effectiveness of the composite matrix based on the transfer matrix method is verified.

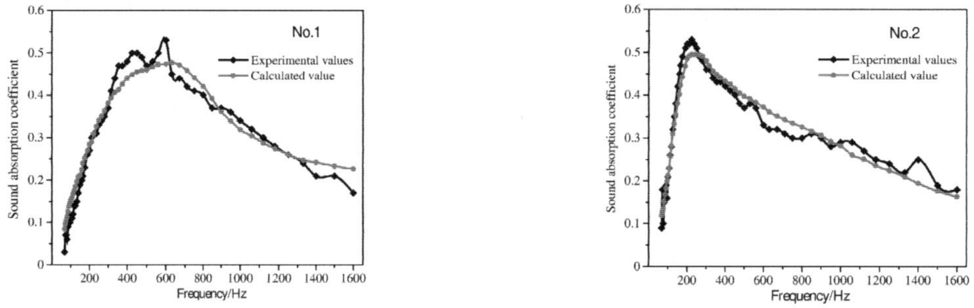

Figure 5. Comparison of experimental and theoretical values of Sound absorption coefficient

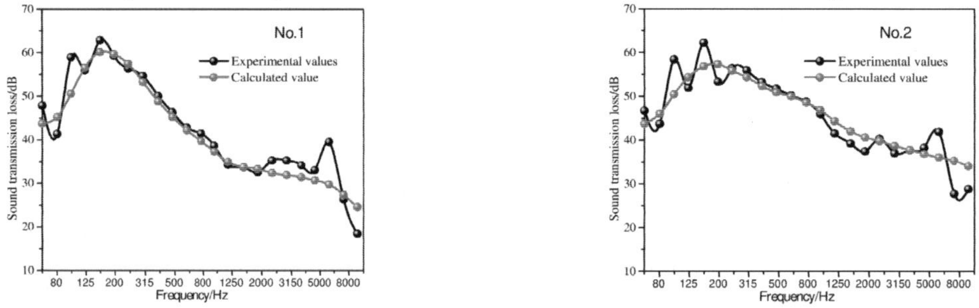

Figure 6. Comparison of experimental and theoretical values of Sound transmission

4. Application of transfer matrix method in noise protection design of ship

This section is based on the statistical energy method to simulate the noise protection design of a ship as an example. It is used to investigate the application of the calculation of acoustic parameters by transfer matrix method in the acoustic protection design of ships. The geometric model of ship is introduced into VAOne and the statistical energy model is established. Then, the acoustic loads are applied in the ship's statistical energy model. Figure 7 shows the layout of the load application position in 1800mm deck while figure 8 shows the nephogram of cabin noise.

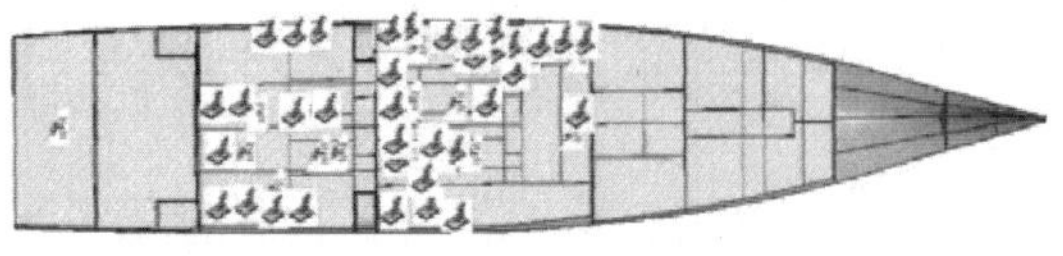

Figure 7. The load layout diagram

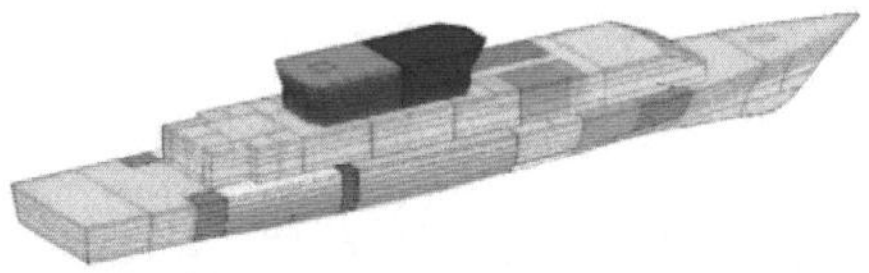

Figure 8. The nephogram of cabin noise

Take the running of all equipment as the calculation conditions, and the conference room in the living compartment is selected as the assessment cabin. There are two types of the outfitting materials including sound insulation materials and sound absorption materials. The former laid in the assessment cabin, and the later laid in the cabin near the noise source. Through the transfer matrix method, the acoustic parameters of two materials can be calculated. Then the results are input into the statistical energy model, and then the noise values of the assessment cabin, whether laying the sound insulation composite material or not, are calculated and compared. Finally, the effect of cabin noise reduction by laying composite material can be obtained and shown in figure 9(a). It can be seen that the noise reduction of the cabin is more obvious, and the total noise reduction reaches 11.2dB.

The noise values of the cabin near the noise source and the assessment cabin after laying the sound absorbing composite material are shown in figure 9(b) and figure 9(c). The noise reduction effect of the evaluation cabin is obvious by using the sound absorbing composite materials and the noise reduction amount is 16.1dB, as seen in figure 9(b). But the noise reduction on the cabin near the sound source is only 3.1dB, as seen in figure 9(c), which indicates that the sound absorption composite materials has a good effect on blocking noise transmission, but the noise caused by the noise source inside the cabin cannot be effectively treated.

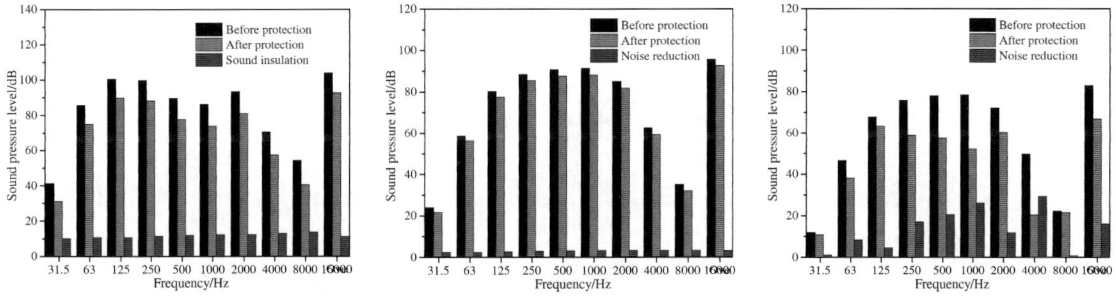

(a) (b) (c)

Figure 9. The noise reduction effect of outfitting materials
(a) sound insulation; the evaluation cabin, (b) sound absorbing composite materials; the cabin near the sound source, (c) sound absorbing composite materials; the evaluation cabin.

5. Conclusion

In this paper, the transfer matrix method is used to theoretically deduct the acoustic performance of composite materials. The excellent accuracy and reliability of this method are confirmed by comparing the caculated results with experimental values. Using the transfer matrix method to calculate the acoustic performance of composite materials has the characteristics of high calculation efficiency, great reliability and wide applicability.Then the model simulation was carried out by taking a ship as an example, and the transfer matrix method in the noise protection design of the ship was discussed.

References

[1] MA D Y 1997 *General theory and design of microperforated-panel absorbers J. Acta Acustica*, **22** pp 385-393.
[2] BarbieriR 2006 Finite element acoustic simulation based shape optimization of a muffler J. Applied Acoustics, 67 pp 346-357
[3] ZHANG J G 2007 Interval Analysis for Natural Frequencies of an Uncertain Structure Based on Transfer Matrix Method J . Journal of vibration and shock, 26 pp 100-103.

Acknowledgments

This study was funded by Major innovation projects Of High Technology Ship Funds of Ministry of Industry and Information of P. R. China, Ph.D. Student Research and Innovation Fund of the Fundamental Research Funds for the Central Universities (HEUGIP201801), National key Research and Development program (2016YFC0303406), High Technology Ship Funds of Ministry of Industry and Information of P.R. China, Assembly Advanced Research Fund Of China (6140210020105), China Postdoctoral Science Foundation (2014M552661), National Natural Science Foundation of China (51209052).

IOP Conf. Series: Materials Science and Engineering **544** (2019) 012049 doi:10.1088/1757-899X/544/1/012049

Research on Measurement of Loss Factor based on Transient State Decay Method

Dewei Peng[1], Na Wang[2*], Guonan Li[3] and Ruidong Huo[1]

[1]College of Shipbuilding Engineering, Harbin Engineering University, Harbin, 150001, PR China
[2]Yantai CIMC Raffles Ocean Engineering Co., Ltd., Yantai, 264001, PR China
[3]China Ship Research and Development Academy, Beijing, 100192, PR China
E-mail: pengdewei@hrbeu.edu.cn

Abstract. As an important parameter in application of the Statistical Energy Analysis (SEA), loss factor can measure damping characteristics of a system and determine the ability of vibration energy dissipation. First, the experimental models of steel plate structures are established. And then, the loss factor test is carried out based on transient state decay method. Finally, the variation characteristics of loss factor with medium, thickness and size of structures are explored. The results show that: (1) The loss factor of steel plates has the characteristics that larger at low frequency and small at high frequency, it decreases gradually with the increase of frequency. (2) For the same structure, the loss factor tested in water is greater than the loss factor tested in air. (3) The loss factor increases with the raise of thickness of plates when its cross-sectional dimension is constant.

1. Introduction

The vibration and noise of structures are directly related to the stability and stealth in the work, vibration will lead to fatigue problems of equipment, and noise will have a greater impact on the working environment. It is very important to establish high-precision model for structure design and noise prediction, and accurate damping loss factor is one of the most important factors. Therefore, obtaining the loss factor accurately of structures and master its law is essential to predict the vibration and noise[1-5].

Domestic and foreign scholars generally believe that the main component of noise is the structural acoustic radiation generated by the vibration of equipment, and propose to use the finite element method and the statistical energy method to predict the low-frequency mechanical noise and medium-high frequency mechanical noise. When the statistical energy analysis method is used for noise prediction and evaluation, the key is to obtain the fundamental parameters of structural system, such as modal density, coupling loss factor and internal loss factor[6]. The common and reliable way to determine structural loss factor is experimental measurement, including steady-state energy flow method, transient state decay method, and so on. The input power needs to calculate accurately by using steady-state energy flow method, this is undoubtedly inconvenient to measure the internal loss factor, and the error is also large. The transient state decay method is suitable for estimating average internal loss factor of the structure and acoustic volume, it doesn't need to measure input power, the statistical error of experimental results is relatively small[7-8].

In this paper, the loss factor test is carried out based on transient state decay method, the law of loss factor changing with frequency is analysed. and the effects of changing the test environment and steel structure parameters on the loss factor are discussed. The purpose of this study is to provide parameter input basis for prediction and evaluation of structural noise, and provide preliminary accumulation for loss factor database of steel plate structure.

2. Test principle

Assuming that the acceleration response signal of the structure under excitation is a real function $a(t)$, the corresponding Hilbert transform is as follows[9]:

$$\hat{a}(t) = H[a(t)] = a(t) * \frac{1}{\pi t} = -\frac{1}{\pi}\int_{-\infty}^{+\infty} \frac{a(\tau)}{t-\tau} d\tau \tag{1}$$

the inverse transformation is:

$$a(t) = \hat{a}(t) * \frac{1}{\pi t} = -\frac{1}{\pi}\int_{-\infty}^{+\infty} \frac{\hat{a}(t)}{t-\tau} d\tau \tag{2}$$

Where * represents convolution.

Then the parsing signal of the real function is:

$$u(t) = a(t) + j\hat{a}(t) \tag{3}$$

Where the plural expression of $u(t)$ is:

$$u(t) = A(t)e^{j\theta t} \tag{4}$$

Where $A(t) = \sqrt{a(t)^2 + \hat{a}(t)^2}$ is the envelope of the response signal $a(t)$, and the instantaneous phase is $\theta(t) = \arctan \hat{a}(t) / a(t)$. So, the real function can be re-represented as $a(t) = A(t)\cos\theta(t)$. The logarithm of the envelope of the response signal is used to obtain the attenuation curve, the absolute value of the slope is the structural damping η, the damping ratio $\zeta = \eta/\omega$ is dividing η by frequency, the damping ratio divided by 2 is the loss factor of structures.

3. Test scheme

3.1. Test model

The test models include three different sizes of steel plates structure, they were tested in air and water respectively. The plates are fixed by elastic rope lifting. The size of the test models is shown in Figure 1.

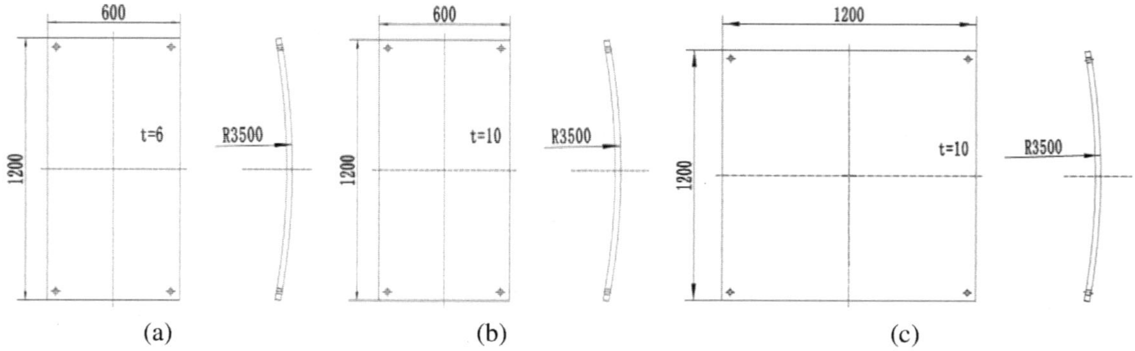

Figure 1. Schematic diagrams of the test models(a) 1# plate, (b) 2# plate, (c) 3# plate

In the process of testing, the hammer is used to knock the structural parts in different positions, vibration acceleration of plates is collected by using acceleration sensors that placed on the surface of the test model, the acceleration response signal is recorded from knocking plates until the response to minimal. The schematic diagram and data acquisition flow is shown in Figure 2.

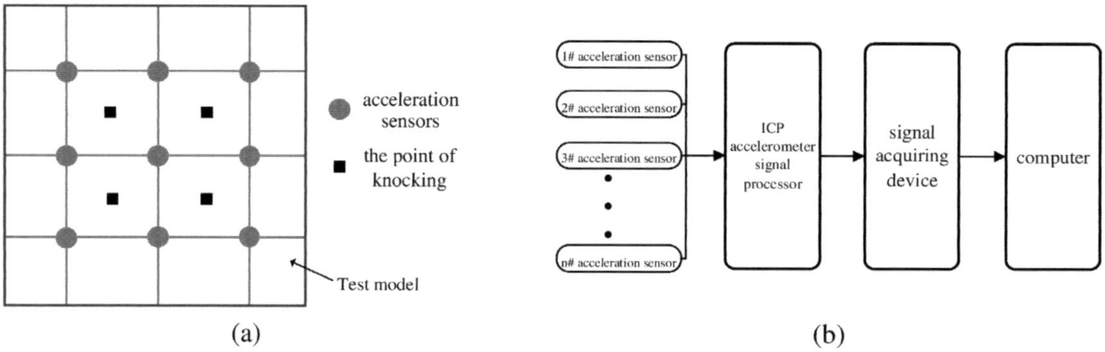

Figure 2. The schematic diagrams of test mode and data acquisition(a) Schematic diagram of measuring model acceleration, (b) Flow chart of collecting test data

3.2. Data Processing and Results

After processing and transforming the collected data, the response signal is obtained. The acceleration curves measured by a certain channel of 1# plate is shown in Figure 3. The attenuation curve and envelope slope of the response signal at 1/3 octave of a 1# plate at 160 Hz are shown in Figure 4.

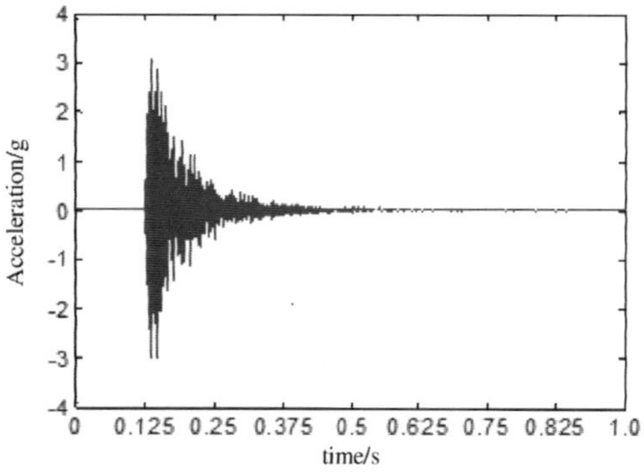

Figure 3. the curve of Acceleration

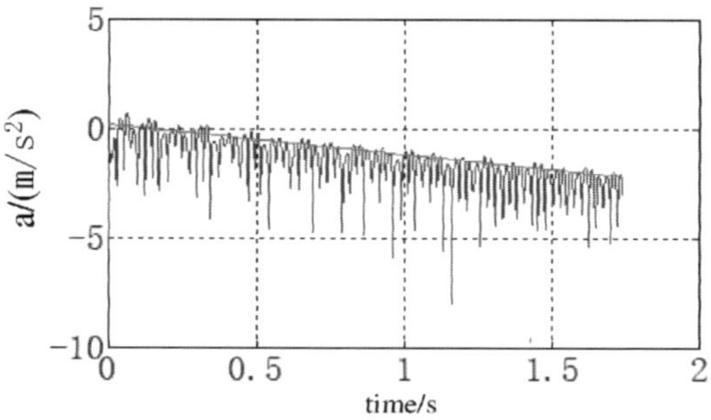

Figure 4. The diagram of response signal attenuation curve and slope

3.2.1. The effect of test environment on loss factor. To explore the impact of the environment in which the structure is on the loss factor, according to the test results, the comparison curves of 1/3 octave loss factor of each plate under different test environments is shown in Figure 5.

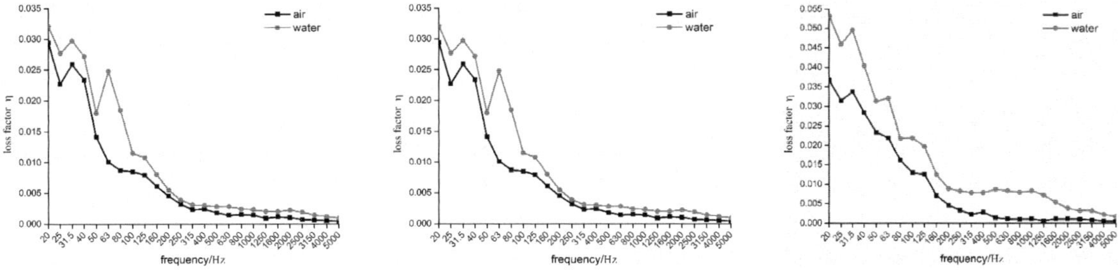

(a) (b) (c)

Figure 5. The contrast curves of 1/3 octave loss factor of test models (a) 1# plate, (b) 2# plate, (c) 3# plate

It can be seen from Figure 6 that the loss factor of three different structural size steel plates is on the order of 10^{-3}~10^{-2}, which is characterized by larger at low frequency and small at high frequency. In the 20~125Hz frequency band, except for the fluctuation of individual frequency points, the loss factor decreases with the increase of frequency. In the 125~5000 Hz band, the fluctuations tend to be flat and gradually decrease. For the same steel plate structure, the loss factor measured in water is greater than the loss factor measured in air in the full frequency band.

3.2.2. The effect of structural parameters on loss factor. To explore the influence of structural parameters on loss factor, the comparison curves of steel loss factors for different structural parameters in the same experimental environment is shown in Figure 6.

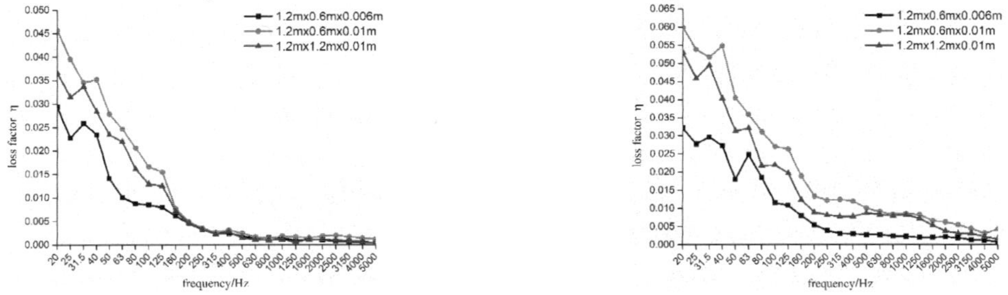

(a) (b)

Figure 6. The contrast curves of 1/3 octave loss factor of test models in different testing environment (a) In air, (b) In water

It can be seen from figure 6, that in the same test environment, the loss factor of steel plate with the same cross-section size increases significantly in the whole frequency band with the increase of thickness, especially in water. For the steel plate with the same thickness, the loss factor decreases slightly with the increase of cross-section area, but it is larger than the steel plate with smaller thickness in the whole frequency band.

4. Conclusion

The loss factor test is carried out based on transient state decay method. Finally, the variation characteristics of loss factor with medium, thickness and size of structures are explored. Through the research of this paper, the following main conclusions can be obtained:

(1) The loss factor of three different structural size plates is on the order of 10^{-3}~10^{-2}, which is characterized by larger at low frequency and small at high frequency. In the 20~125Hz frequency band, except for the fluctuation of individual frequency points, the loss factor decreases with the increase of frequency. In the 125~5000 Hz band, the fluctuations tend to be flat and gradually decrease.

(2) For the same plate structure, the loss factor measured in water is greater than the loss factor measured in air in the full frequency band.

(3) In the same test environment, the loss factor of plates with the same cross-section size increases significantly in the whole frequency band with the increase of thickness, especially in water. For the plates with the same thickness, the loss factor decreases slightly with the increase of cross-section area, but it is larger than the steel plate with smaller thickness in the whole frequency band.

References

[1] J M Cuschieri, J C Sun 1994 Determination of dissipation and coupling loss factors using energy ratios *J. Journal of Sound and Vibration* **170** pp181～91.

[2] Weng C J, Wu W G 2000 Vibration and noise control in high-speed ship *J. Journal of Wuhan University of Technology* **1** pp 16-9.

[3] Sun X R, Zhu X 2005 Survey and tendency of study on the underwater noise of ship structure *J. Journal of Vibration and Shock* **24** pp 106–13.

[4] F J Fahy, H M Ruivo 1997 Determination of statistical energy analysis loss factors means of an input power modulation technique *J. Journal of Sound and Vibration* **203** pp 763～79.

[5] Sheng M P, Wang M Q, Sun J C 2001 On fairly accurately determining the loss factor of two-mode system in steady vibration with the attenuation method *J. Journal of Northwestern Polytechnical University* **19** pp 130～5

[6] Cheng G L, Zhu S J, Wu X J 2004 A summary of statistical energy analysis method and its loss factor' s determination *J. Ship Engineering* **26** pp 10–5.

[7] Lyon R H, Maidanik G 1962 Power flow between linearly coupled oscillators *J. Journal of the Acoustical Society of America* **34** pp 623–39.

[8] Gu J T, Sheng M P 2015 Estimation of steady loss factor with decay rate method *J. Acta Physica Sinica* **64** pp 339–46.

[9] Ning F H, Zhang J 2002 Hilbert transform in loss factor measurement *J. Journal of Shandong Institute of Technology* **16** pp 17–20.

Acknowledgments

This study was funded by Major innovation projects Of High Technology Ship Funds of Ministry of Industry and Information of P. R. China, Ph.D. Student Research and Innovation Fund of the Fundamental Research Funds for the Central Universities (HEUGIP201801), National key Research and Development program (2016YFC0303406), High Technology Ship Funds of Ministry of Industry and Information of P.R. China, Assembly Advanced Research Fund Of China (6140210020105), Fundamental Research Funds for the Central University (HEUCFD1515, HEUCFM170113), China Postdoctoral Science Foundation (2014M552661), National Natural Science Foundation of China (51209052).

IOP Conf. Series: Materials Science and Engineering **544** (2019) 012048 doi:10.1088/1757-899X/544/1/012048

Study on Vibration and Acoustic Radiation Characteristics of Composite Material Superstructures of Underwater Vehicles

Cong Gao[1] , Xichun Huang [2*],Guobing Huang[2] and Huiling Dai[1]

[1]College of Shipbuilding Engineering, Harbin Engineering University, Harbin, 150001, PR China
[2]China Ship Development and Design Center, Wuhan, 430064, China
E-mail: conggao@hrbeu.edu.cn

Abstract. Composite materials have been widely used in underwater ships for advantages of high specific strength, excellent acoustic performance and well designability. As an important indicator to measure combat effectiveness and vitality of underwater ships, structural radiation noise is the main target of acoustic detection weapons and the main factor that interferes with sonars. Underwater vibration and acoustic radiation characteristics of superstructure using composite sandwich panels directly affects the acoustic performance of underwater vehicles and the comfort and safety of personnel, therefore, research on vibration and acoustic radiation characteristics of underwater vehicle superstructure with composite material has important theoretical value and engineering prospect.

1. Introduction

Superstructure structure is an important part of a underwater vehicle. When composite material is applied to superstructure of the underwater vehicle, vibration and acoustic radiation performance of the structure are changed relative to traditional steel material. At present, composite material superstructure use sandwich panel structures, which are mainly composed of two thin layers of fiber skin and a relatively low density core material in middle. Composite sandwich panels that have been widely used have simple structures and mature preparation processe [1]. At the same time, their mechanical properties are relatively simple, various dynamic and static mechanical properties have been extensively studied, such as extrusion shear, bending deformation and fatigue damage[2]. In addition, influences of factors such as thickness of the skin layer, layup order and core material on the mechanical properties of sandwich structure are extensive[3]. Therefore, this paper takes a composite material superstructure of underwater vehicle as a research object, and carries out the analysis of underwater vibration and acoustic characteristics of the structure, aiming at providing reference for low noise design of underwater vehicle structures.

2. Theoretical basis

In this paper, a composite material superstructure is taken as a research object which size is 1.5m×2.0m, and thickness of the composite sandwich panel is 50mm. Numerical calculation model of composite superstructure is showed in Figure 1. The ratio of length of short side to the thickness of the panel is greater than 10 so the first constraint condition of the first-order shear deformation theory is satisfied; and because surface layer of the composite sandwich panel is made of FRP material, and core layer is made of foam material, elastic modulus of the two materials along thickness direction is not more than two orders of magnitude, satisfying the second constraint of the first-order shear deformation theory. Therefore, the composite material superstructure can be calculated by the first-order shear deformation theory. Compared with the boundary element method, calculation scale of acoustic-solid coupling finite element analysis method is relatively slow with the increase of the number of meshes, and its prediction accuracy is relatively high. Therefore, this paper uses the acoustic-solid coupling finite element analysis method to calculate and analyze underwater vibration and acoustic radiation characteristics of the composite superstructure.

Figure 1. Numerical calculation model of composite superstructure

3. Research of Vibration characteristics

3.1 Calculation function

The mean square vibration velocity is defined as:

$$v^2 = \frac{1}{S}\iint_s |V|^2 ds \tag{1}$$

According to the definition of vibration speed level or vibration acceleration level, vibration speed level and the vibration acceleration level can be expressed as follows:

$$L_a = 20\times\log(a/a_0),\ L_v = 10\times\log(v^2/v_0^2) \tag{2}$$

Where the acceleration reference value $a_0=1\times10^{-6}$ m / s ^{2}, and the speed reference value $v_0= 1\times10^{-9}$ m / s;

Vibration acceleration level or mean square vibration speed level of each frequency point is solved by energy superposition, and the total level of vibration acceleration or mean square vibration speed can be obtained:

$$L_T = 10\log(\sum_{i=1}^{n} 10^{(L_i/10)}) \tag{3}$$

3.2 Analysis of vibration characteristics in air

Under action of vertical unit force at center of the superstructure, vibration acceleration level at the loading position in air and vibration acceleration level of shell surface of the superstructure are shown in figure 2 and figure 3.

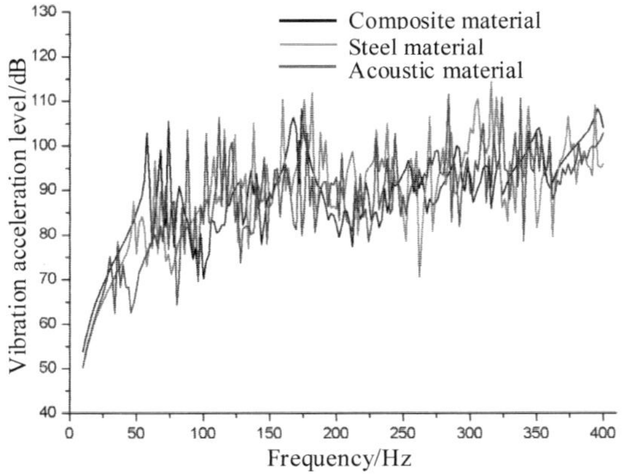

Figure 2. Comparison curve of Vibration acceleration leve in air

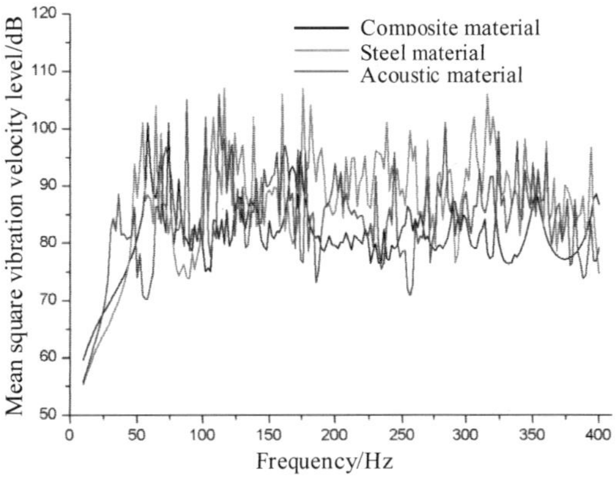

Figure 3. Comparison curve mean square vibration speed level in air

Vibration acceleration of loading position of the superstructure of three materials is basically consistent with frequency. In 10Hz-400Hz frequency band, vibration acceleration response peak of the composite superstructure is significantly less than the other two materials. When the superstructure structure adopts different materials, influence of mean square vibration velocity on surface of the structure is greater. When the superstructure uses composite reinforced sandwich panels, mean square vibration velocity in 10Hz-100Hz frequency band is not much different from the other two material parameters, and after 100Hz, the vibration response is significantly lower than the other two materials.

3.3 Analysis of underwater vibration characteristics

Considering coupling between flow field and superstructure of composite material, vertical unit force is also applied at the center of the superstructure shell. Vibration acceleration at the loading position of the shell surface are shown in the figure 4.

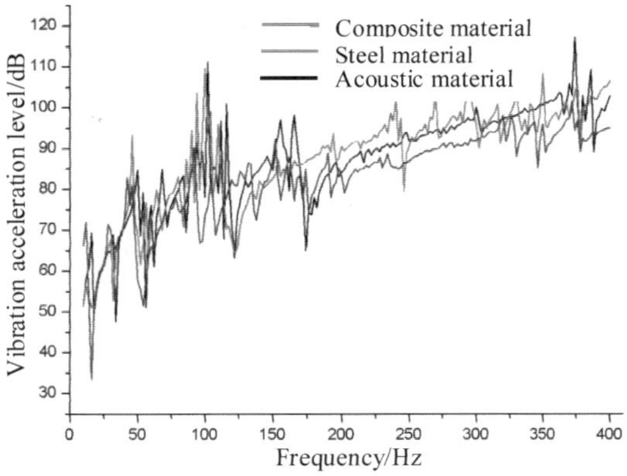

Figure 4. Comparison curve of vibration acceleration level under water

Vibration acceleration at the loading position of the superstructure of three materials is basically consistent with frequency. Vibration acceleration response of the composite superstructure is significantly less than that of the other two materials, especially after 200 Hz. Acoustic material and the superstructure of steel material have a large peak of vibration response and a concentrated distribution in frequency band of 80 Hz-120 Hz. When the superstructure uses a composite reinforced sandwich structure, response of vibration acceleration level is significantly smaller than the other two materials in 150Hz-400Hz frequency band.

4. Study on characteristics of acoustic radiation

4.1 Calculation function

Underwater radiated acoustic power and acoustic source level of the superstructure structure can reflect its acoustic characteristics as a noise source, and the two do not differ with change of measurement position. Sound intensity refers to work done by the sound wave per unit area in direction of advancement adjacent to the coal mass, and radiated acoustic power refers to average sound energy radiated outward by area S perpendicular to direction of sound propagation per unit time. Therefore, the radiated acoustic power can be expressed as:

$$W = I * S = \frac{S}{T}\int_0^T \mathrm{Re(p)\,Re(v)\,dt} \tag{4}$$

The definition of acoustic power level is as follows:

$$L_W = 10 \times \log\left(\mathrm{w} / \mathrm{w}_0\right) \tag{5}$$

Among them, acoustic power reference value $w_0 = 1 \times 10^{-12}$Pa.

4.2 Analysis of underwater acoustic radiation characteristics

Under action of vertical unit force at center of the shell plate, underwater acoustic radiation of the superstructure structure changes with change of material parameters. Far-field calculation results are transformed, and radiation of the superstructure of three materials can be obtained. Comparisons of acoustic radiated power level of the three materials are shown in figure 5.

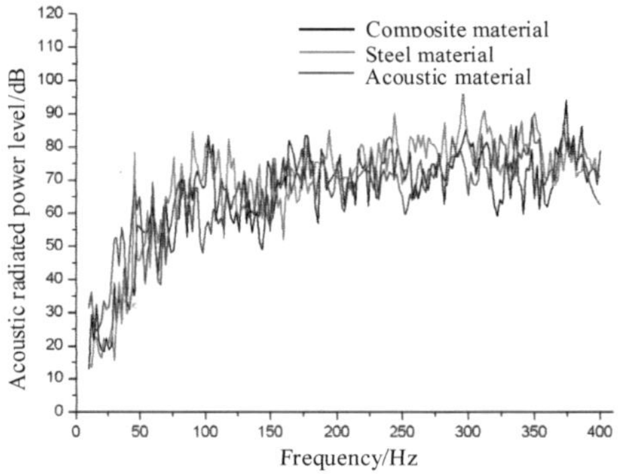

Figure 5. Comparison curve of acoustic radiated power level

The change of material parameters has a great influence on underwater acoustic radiation of the superstructure structure. Radiation acoustic power of the superstructure of the three material parameters is basically consistent with frequency. In 10Hz-100Hz frequency band, acoustic material superstructure has a large acoustic radiation power level. When the composite building shell structure adopts composite materials, acoustic radiation power level is obviously smaller than the other two material parameters, and acoustic radiation power level of the superstructure is the largest when using steel material.

5. Conclusion

Underwater vibration and acoustic characteristics of underwater vehicle superstructure with composite material greatly improved related to traditional material.

Reference

[1] Inayathullah J, Nagarajan V A, Kumar S H 2016 Development of hybrid composite radar wave absorbing structure for stealth applications *J. Bulletin of Materials Science*, **39** pp 279-84.

[2] Lei H, Yao K, Wen W 2016 Experimental and numerical investigation on the crushing behavior of sandwich composite under edgewise compression loading *J. Composites Part B: Engineering*, **94** pp 34-44.

[3] Ahmed A, Bingjie Z, Ikbal M H 2015 Experimental Study on the Effects of Stacking Sequence on Low Velocity Impact and Quasi-Static Response of Foam Sandwich Composite Structures *J. Advances in Structural Engineering*, **18** pp 1789-806.

Acknowledgments

This study was funded by Major innovation projects of High Technology Ship Funds of Ministry of Industry and Information of P. R. China, Ph.D. Student Research and Innovation Fund of the Fundamental Research Funds for the Central Universities (HEUGIP201801), National key Research and Development program (2016YFC0303406), High Technology Ship Funds of Ministry of Industry and Information of P.R. China, Assembly Advanced Research Fund Of China (6140210020105), Fundamental Research Funds for the Central University (HEUCFD1515, HEUCFM170113), China Postdoctoral Science Foundation (2014M552661), National Natural Science Foundation of China (51209052).

Numerical Study of Hydrodynamic Noise of a Cone-column Combined Structure

Chen Yu[1], Zhihua Shen[2*] Guobing Huang[2] and Jiangtao Liu[1]

[1]College of Shipbuilding Engineering, Harbin Engineering University, Harbin, 150001, PR China
[2]China Ship Development and Design Center, Wuhan, 430064, China
E-mail: huoruidong@hrbeu.edu.cn

Abstract. Hydrodynamic noise is one of the main sound source of underwater vehicle. Flow noise and flow-excited noise are important part of hydrodynamic noise. In this paper, large eddy simulation is applied to simulate the flow field around a cone-column combined structure to obtain dipole sound source and turbulence fluctuating pressure. The surface pressure distribution is validated by experiment measured data. Structure dynamic response can be obtained by solving fluid-structure coupling equation. The flow noise and flow-excited noise of the structure are calculated according to acoustic analogy theory. From the numerical result, it can be seen the spectra characteristic of flow-excited noise is different from flow noise. The flow-excited noise obviously increases when system resonance occurs.

1. Introduction

Underwater vehicles are distinguished from other crafts, surrounded by viscous water. Irregular motion on the surface of the underwater vehicle will generate noise, often called hydrodynamic noise. Researchers usually divide hydrodynamic noise into three categories, including flow noise, flow-excited noise and cavitation noise. Both of them have become a research hotspot.

In order to simplify the study of hydrodynamic noise, researchers often regard underwater vehicle as a rigid body. Based on the acoustic analogy theory proposed by Lighthill, it is possible to predict the flow noise generated by the dipole sound source on the surface of the underwater vehicle. However, the actual situation is that the underwater vehicle will have a fluid structure interaction (FSI) with the water flow, and the structural surface will generate dynamic response and radiated noise under the action of turbulence flow. This will cause the predicted noise smaller than the real value. Currently, related research has been conducted on hydrodynamic noise with consideration of FSI using the boundary element method (BEM). Yao and Zhang[1] predicted the flow noise and flow-excited noise of fully appended submarine by using CFD and BEM methods. Wei and Wang[2] used the CFD to obtain the hydrodynamic load during the rotation of the propeller. Based on this, the BEM method was used to calculate the radiated noise.

In this paper, the underwater vehicle is simplified into a cone-column combined structure. The flow field around the cone-column combined structure is simulated by applying the large eddy simulation (LES). Finally we use BEM to calculate the flow noise and flow-excited noise generated by the dipole sound source and turbulent pulsating pressure. A series of hydrodynamic noise experiments were conducted to verify the effectiveness of the above calculation methods.

2. Mathematical formulations

2.1. Large eddy simulation

LES is used to simulate the turbulence field around the cone-column combined structure .Based on incompressible Navier-Stokes Equation and spatial filtering function, we can divide the turbulence flow into large-scale and small-scale two parts. The processed N-S Equation and filter function can be written as:

$$G(x,x',\overline{\Delta})=\begin{cases}\dfrac{1}{\Delta x_1\Delta x_2\Delta x_3}, & |x_i'-x_i|\le\dfrac{\Delta x_i}{2}\quad i=1,2,3\\ 0, & |x_i'-x_i|>\dfrac{\Delta x_i}{2}\quad i=1,2,3\end{cases}\tag{1}$$

$$\frac{\partial\overline{u}_i}{\partial t}+\frac{\partial}{\partial x_j}\left(\overline{u}_i\overline{u}_j\right)=-\frac{1}{\rho}\frac{\partial\overline{p}}{\partial x_i}+\nu\frac{\partial^2\overline{u}_i}{\partial x_j^{\,2}}-\frac{\partial}{\partial x_j}\tau_{ij}^S\tag{2}$$

In Eq.1 $G(x,x',\overline{\Delta})$ is a boxy filter function, Δx_i is the grid scale. We apply a filtering operation to N-S Equation ,then we can obtain Eq.2. In Eq.2 $\overline{u_i}$ means the filtered average velocity components, $\tau_{ij}^S=\overline{u}_i\overline{u}_j-\overline{u_iu_j}$ is so-called sub-grid scale stress (SGS).

2.2. Acoustic analogy theory and FSI

Ffowcs-Williams and Hawkings[3] extended Lighthill's acoustic analogy theory and raised the famous FW-H Equation which can be written as follows:

$$\frac{1}{c^2}\frac{\partial^2 p}{\partial t^2}-\nabla^2 p'=\frac{\partial^2}{\partial x_i\partial x_j}\left[T_{ij}H(f)\right]-\frac{\partial}{\partial x_i}\left[P_{ij}n_j\delta(f)\right]+\frac{\partial}{\partial t}\left[\rho_0u_n\delta(f)\right]\tag{3}$$

Where p' represents far field sound pressure, $\delta(f)$ and $H(f)$ are Dirac delta function and Heaviside function. In Eq.3, its right source term contains three components, and both of them represent different physical meanings. The first component contains Lighthill stress tensor, representing the quadrupole source's contribution to flow noise; the second component refers to wall pressure fluctuation, representing the contribution of dipole source to flow noise; the third component is related to the velocity of structure surface, it can be regarded as monopole source. Due to the low velocity of flow, quadrupole source and monopole source are negligible. We can just take dipole source into consideration to calculate the flow noise.

As for flow-excited noise, we should regard wall pressure fluctuation as external load to calculate system dynamic response. The governing equation can be written as:

$$\begin{bmatrix}[M_s] & [0]\\ [M^{fs}] & [M_f]\end{bmatrix}\begin{Bmatrix}\{\ddot{U}\}\\ \{\ddot{P}\}\end{Bmatrix}\ [illegible]\ \begin{bmatrix}\cdots & K^{fs}\\ \cdots & K_f\end{bmatrix}\begin{Bmatrix}\{U\}\\ \{P\}\end{Bmatrix}=\begin{Bmatrix}[F_s]\\ [0]\end{Bmatrix}\tag{4}$$

In Eq.4 $\left[M^{fs}\right]=\rho[R]^T$, $\left[K^{fs}\right]=-[R]$, $[F_s]$ means external load, $[R]^T=\int_s [N']^T[N]^T\{n\}d(s)$, $\{N\}$ represents pressure shape function and $\{N'\}$ is the structure displacement shape function. By solving Eq.4, we can obtain the displacement and sound pressure of structure surface. Then the sound pressure of the whole acoustic field can be obtained by solving the following equation according to the normal speed and sound pressure of structure surface.

$$p(x)=\int_s\left[p(y)\frac{\partial G(x,y)}{\partial n_y}-\frac{\partial p(x,y)}{\partial n_y}G(x,y)\right]dS(y) \quad (5)$$

3. Numerical simulation and experiment

3.1. Flow field simulation

The cone-column combined structure was chosen for the study of flow noise and flow excited noise. The overall length is 0.95m, the diameter of the structure is 0.1m, and the thickness of the structure is 6mm. Both ends of the structure are two hemispheres. The velocity inlet boundary is one structure length away from the cone-column combined structure and the inflow velocity is 2.46m/s. The pressure outlet boundary is two times structure length from the model. The Reynolds number of the structure model is 1.25E6. The surface of the model is set as no- slip boundary condition.

In order to decrease the total number of the grid, the grid in the computational flow field near the structure surface is quite fine, whereas the grid in the non-interesting region is sparse.

The computation domain is discretized by applying multi-block strategy and the number of hexahedral grids of the computational domain is nearly 1.8 million. In order to ensure the non-dimensional spacing wall y+ on the hull surface is within 10, we set the height of first layer grid as 5‰ of the overall length. Due to the sampling frequency was up to 12.5kHz, the time-step was set as 4E-5 and the total calculation time was 10s. The structure design and details of the mesh used in this paper are separately shown in Figure 1 and Figure 2.

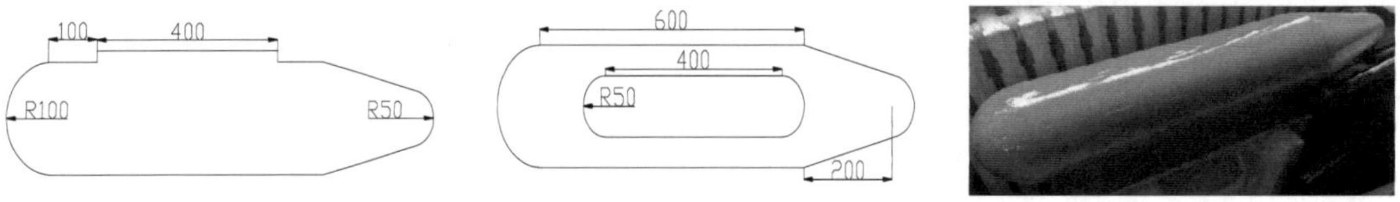

(a) (b) (c)

Figure 1. Designed structure (a) Front view, (b) Bottom view and (c) Model photograph.

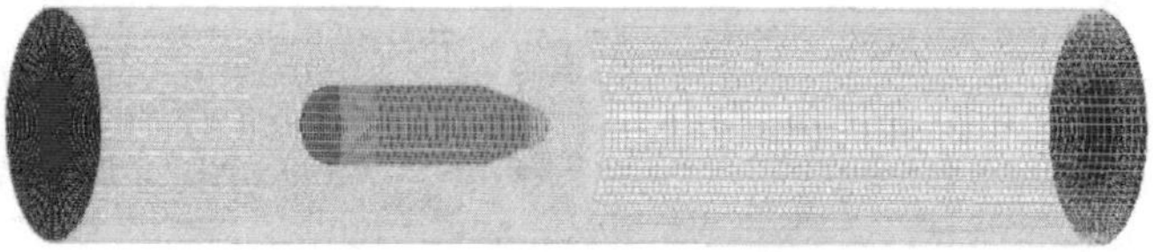

Figure 2. Computational domain.

For excluding the influence of grid scale, grid independent verification should be implemented. Four sets of grids with different mesh scale are simulated to check the accuracy of simulation. Along longitudinal direction, 11pressure sensor are set to obtain the surface pressure distribution. Due to the space limitation, we just choose point 5 and 8 to compare the simulation data with measured data . The arrangement of measuring point and calculated result are shown in Figure 3 and Figure 4.

According to the simulation result, it is clear that when the grid cell is set as 2mm, we can get minimal error between simulation and experimental results. So the following fluid field simulation will use the same mesh resolution.

Finally we get the fluid field information of inflow velocity equals 2.46m/s by using above mesh resolution and calculation method.

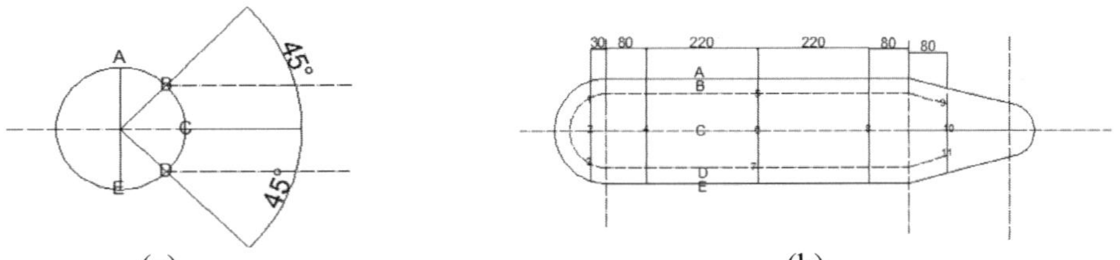

(a) (b)

Figure 3.Arrangement of pressure sensor (a) Front view, (b) Side view.

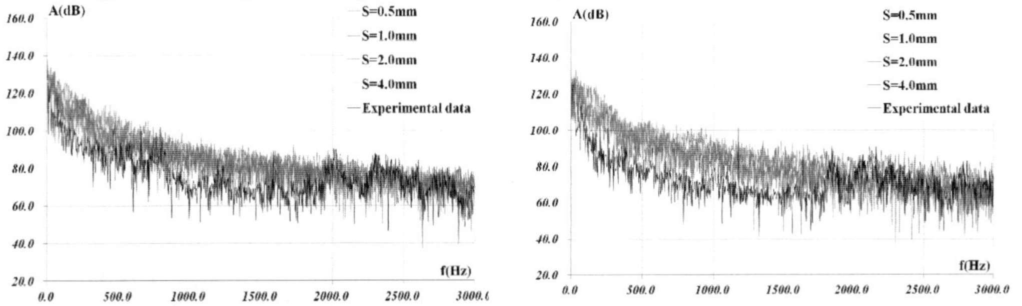

(a) (b)

Figure 4. Grid independent verification (a) Point 5, (b) Point 8.

3.2. Flow noise and flow-excited noise simulation

The flow noise and flow-excited noise of cone-column combined structure are calculated by BEM. The distributions of the overall sound pressure level (OASPL) of flow noise and flow-excited noise along the longitudinal direction are presented in Figure 5. According to the Figure 5, the OASPL of flow-excited noise is smaller than that of flow noise in the near field. The flow-excited noise decreases by 43.8% while flow noise decreases by 28.6% after spreading 8.5m. It can be seen that flow-excited noise decreases rapidly in the near-field which need us to pay attention to. In the far-field, the difference between the OASPL of flow noise and flow-excited noise is nearly 30dB.

The sound directivity diagrams in axial direction at the center of the structure are shown in Figure 6. From the comparison result, it is clear that the sound directivity patterns of flow noise and flow-excited noise are different from each other. The sound directivity of flow noise is almost the same at every direction of the acoustic field and its OASPL is nearly 74.5dB. But the OASPL of flow-excited noise in height direction is larger than that in lateral direction, its maximum OASPL appears in the directions of 90 and 270 degrees.

The spectrum of flow-excited noise received at the location of hydrophones is compared with that of the flow noise in Figure 7. It's clear that flow noise doesn't have obvious peak value while flow-excited noise has its maximum scale SPL at about 400Hz. Through modal analysis in Figure 8, we know that the first natural frequency of the structure is 390Hz and it is close to the peak frequency of spectrum curve of flow-excited noise.

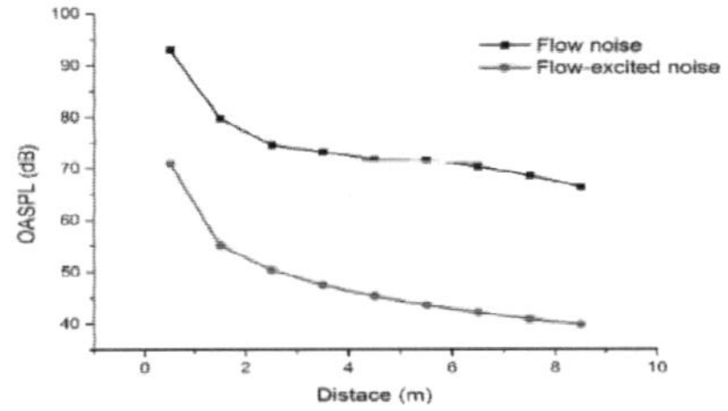

Figure 5. Distributions of the OASPL of flow-excited noise and flow noise along the longitudinal direction of the cone-column combined structure.

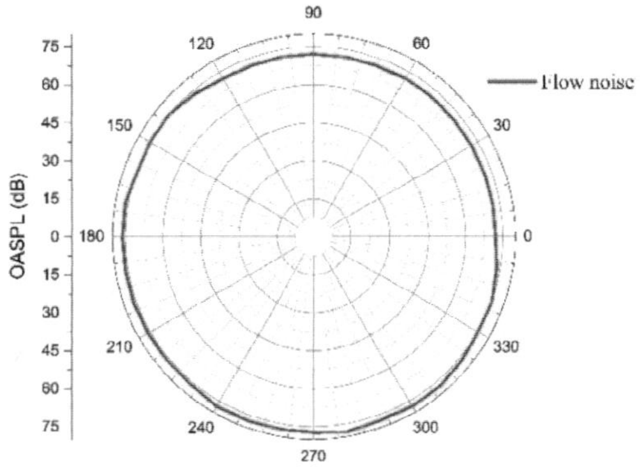

(a)

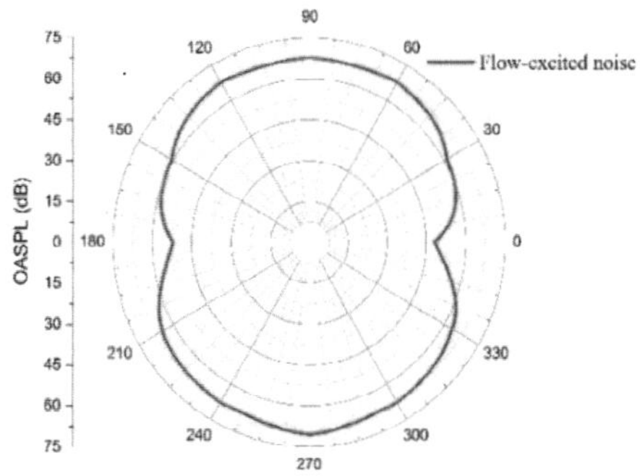

(b)

Figure 6. Sound directivity in axial direction (a) Flow noise, (b) Flow-excited noise.

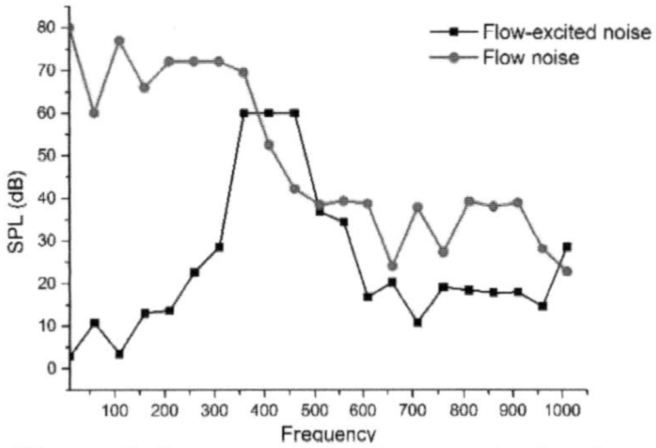

Figure 7. Spectra of the flow-excited noise and flow noise at the same location.

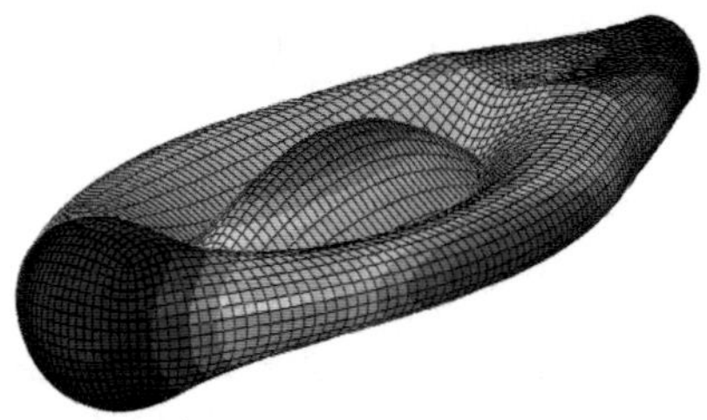

Figure 8. The first natural frequency of the structure.

4. Conclusion

In this paper, LES is applied to simulate the flow field around the cone-column combined structure and BEM is applied to calculate the flow noise and flow-excited noise. The flow-excited noise is smaller than flow noise along the longitudinal direction and rapidly decreases in the near-field. The sound directivity of flow noise is different from that of the flow-excited noise. The flow-excited noise has its own peak frequency which is similar to the structure natural frequency at 390Hz.

Acknowledgments

This study was funded by Major innovation projects Of High Technology Ship Funds of Ministry of Industry and Information of P. R. China and High Technology Ship Funds of Ministry of Industry and Information of P.R. China.

Reference

[1] Yao H, Zhang H, Liu H, and Jiang W 2017 Numerical study of flow-excited noise of a submarine with full appendages considering fluid structure interaction using the boundary element method Engineering Analysis with Boundary Elements 77 pp 1-9

[2] Wei YS and Wang YS 2012 Structure noise prediction of submarine by propeller excitation *Appl Mech Mater* **105** pp 43-6

[3] Fwowcs Williams J E and Hawkings D L 1969 Sound generation by turbulence and surface in arbitrary motion *Philosophical Transaction of the Royal Society of London Series A* **264** pp 321-342

Study on experiment and finite element modelling of small punch test

Xianlong Zhang[1], Zhaonan Ding[1], Yuguang Chen[1], Liqing Zhang[1] and Chonghong Zhang[1]

1 Institute of Modern Physics, Chinese Academy of Sciences. Lanzhou, China.
E-mail: zhang.xl@impcas.ac.cn

Abstract. The small scale specimen techniques have been a fast-growing research field for the past three decades. Small punch test is the most commonly used small specimen test technology and numerous mechanical properties are obtained from the test. Due to the highly complex stress state of the sample during the experimental process, many empirical formulas are used to obtain mechanical properties of the specimen. The curve obtained by finite element modelling of the small punch test is quite different from that obtained from the experiment, this paper present a systematic study on the problem. A revised model is proposed to solve the issue and we conclude that the rod must be modelled in the finite element simulation and the deformation of the rod is the causes of the misaligned.

1. Introduction

Small-scale specimen techniques attract considerable attention nowadays. Over the past three decades, numerous techniques for non-standard small samples have been established to characterize mechanical and physical properties. Small-scale specimen techniques are firstly used in the nuclear industry due to the limitation of irradiation space in the reactor and the dose to testers for post-irradiation testing[1, 2]. Small punch test (SPT) is the most wildly used to extract the fracture properties among all the small-scale test techniques, since SPT only requires a thin slice ($\varnothing 8 * 0.5$mm or even $\varnothing 3 * 0.3$ mm) of material to assess materials properties of an in-service component.

Researchers proposed empirical relation between tensile yield stress and the inflection point of the SPT load-deflection curve as well as between the maximum load of the SPT curve and the ultimate tensile stress based on their own experimental results. SPT has also been employed to non-nuclear industries such as pressure vessel and fossil fuel energy production plants[3]. SPT has been extended to evaluate mechanical properties including ductility, fracture toughness, creep behaviour and ductile to brittle transition temperature. Four different stage were identified in a typical SPT load-deflection curve, the four consecutive regime are: (1) elastic bending of the specimen, (2) plastic bending of the disk followed by membrane stretching (3) and, finally, (4) local crack formation. However, it is important to mention that the boundaries between different stages are difficult to identify.

Assessing SPT apparatus parameters on the load-deflection curve is necessary for understanding this technique, Lucas et al[4] systematically investigated the effect of specimen thickness, clamp apparatus size and materials through experiment. Moreno investigate the influence of different displacement measurement method on the experimental results [5]. Campitelli[6, 7] investigated the effect of yield stress, friction coefficient and compliance of the load system by finite element method (FEM) combing with experiment. Due to the higher complexity of the stress state of SPT, the mechanical properties can not be obtained directly from the load-deflection curve of the test and FEM was used to simulate the experiment in the present work.

However, a significant issue is often ignored, this is the load-deflection curve obtained by the FEM simulation disagree with the experimental curve. There are still no systematic studies on this issue and only few authors corrected the FEM simulation curve by adding the compliance of the system. In the present work, a through study is performed with the aim of clarifying this issue. It includes revision of traditional FEM modelling of the rigid loading system of SPT, the loading system compliance and discussion on the permanent plastic deformation of the loading system.

2. Experiment details

2.1. Material employed

The material employed in this study is low carbon content austenitic stainless steel 316L, commercial steel provided by Goodfellow Company, its nominal composition is listed in Table 1.

Table 1. Composition of austenitic SS316L. (wt%)

Cr	Ni	Mo	Mn	Si	S	C	Fe
17	12	2.5	2	1	0.03	0.03	65.44

2.2. Uniaxial tensile test

Uniaxial tensile test is performed at room temperature. Sheet specimen with a width of 12.5mm, length of parallel section of 65mm, thickness of 0.5mm, and gauge length of 150mm is used for uniaxial tensile test, presented in figure 1. The electro-mechanical testing machine is controlled at a constant crosshead velocity to make sure the strain rate of 0.015 s^{-1}, which is nearly the strain rate of the first stage of the small punch test. Note that the strain rate varies of different stage of the SPT experiment. The elongation of the interest section is measured by clip gauge attached to the sample. The nominal and the true stress-strain curve are presented in figure 2.

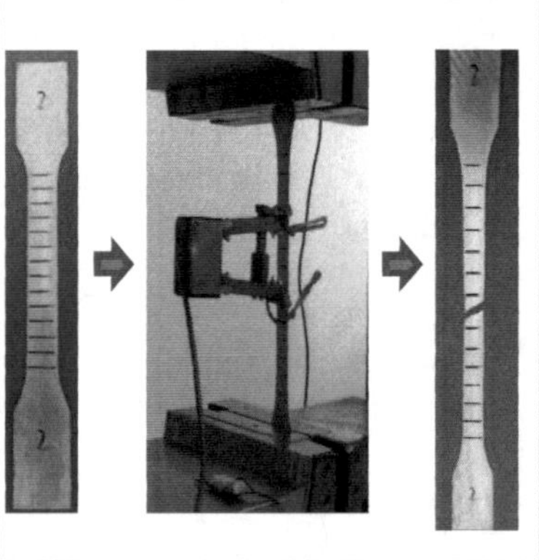

Figure 1. Experimental set-up of the uniaxial tensile test.

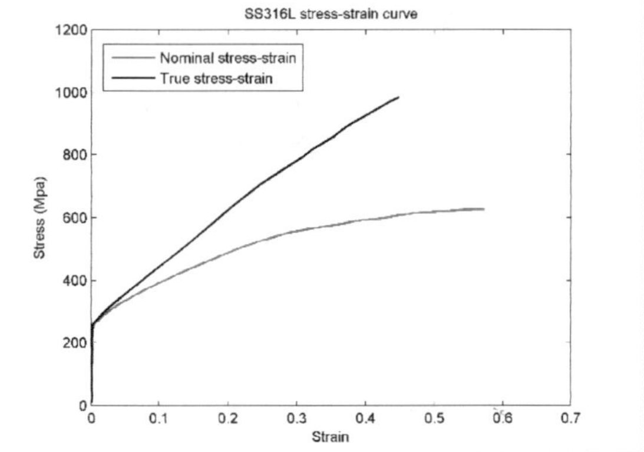

Figure 2. The experimental nominal and true stress-strain curve of the uniaxial tensile test.

2.3. Small punch test

The small punch test is performed with an MTS-E43 electro servo testing machine. The rod and the ball is driven by the testing machine and deforming the clamped disk sample, while the load and the deflection of the rod are recorded. The schematic of the SPT device is shown in figure 3, the punch is a 0.5mm radius ball, the upper die hole diameter is 1.05mm, the lower die hole diameter d is 1.5mm and the specimen is a TEM sample (3mm diameter disc and its thickness B is varying from 0.1 to 0.3 mm). The rod is made of low carbon steel with Elastic modules of 210GPa and the punch ball is a ceramic ball with Elastic modules of 300GPa. The top of the rod is flat and have a surface contact with the ball during the test. A constant punch velocity of 0.2mm/min is used to perform the test. In order to remove the effect of the roughness on the experiment, the surfaces of the sample are polished.

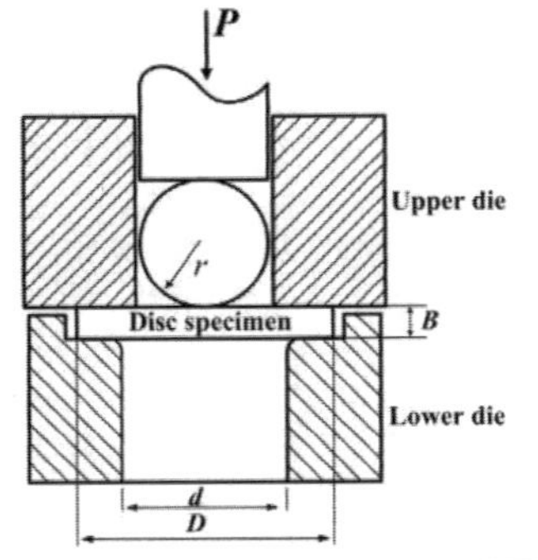

Figure 3. Schematic of small punch test

3. Finite element modelling of SPT

3.1. Traditional FEM model

The finite element model of small punch test is developed using ABAQUS 6.14-2 standard. The ball punch, lower die and upper die are modelled as rigid bodies shown in figure 4. The sample disc is modelled with CAX4R element, and the von Mises yield criterion is adopted. A friction coefficient of 0.2 is employed between the upper (lower) die and disc sample, while the no friction is assumed between the puncher and the sample. The input material properties from the uniaxial tensile test are used, Young's modulus E is 205GPa and the Poisson's ratio ν is 0.33. The plastic strain and the corresponding flow stress is calculated by the equation 1.

$$\varepsilon_{pl} = \varepsilon - \sigma/E \tag{1}$$

Where ε_{pl} is true plastic strain, ε is the true strain, and σ is the true stress. A series of ε_{pl} and σ pairs are input of the ABAQUS plastic parameters. The experimental load-deflection curves and FEM modelling result with and without loading system compliance corrected are shown in figure 5.

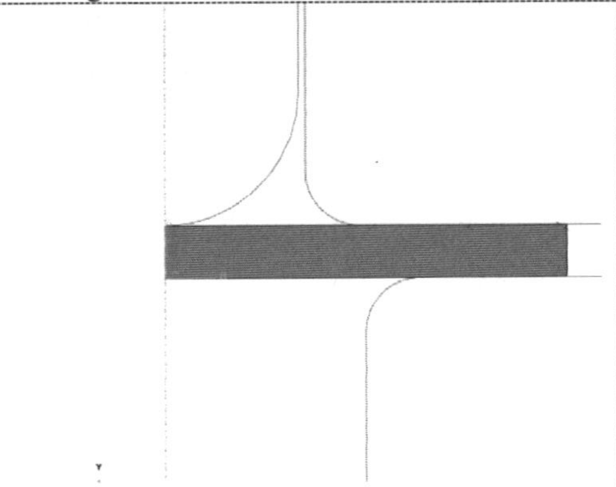

Figure 4. Traditional FEM set-up of the SPT

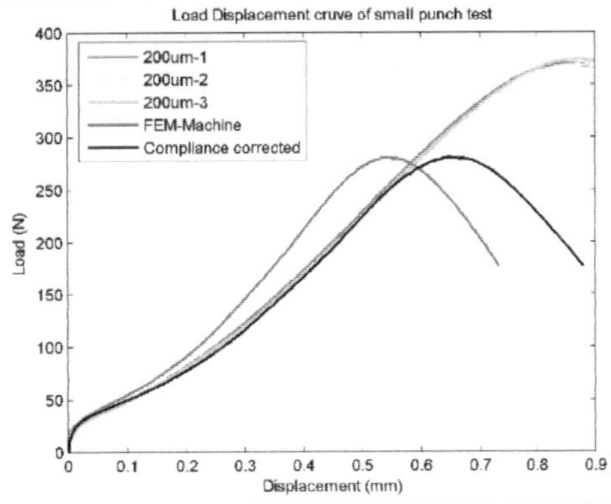

Figure 5. Experimental and FEM SPT load-deflection curves.

Three experimental results show good repeatability. The misalignment of the experimental and FEM results become larger as the displacement (Load) increase, the FEM (red line) load-deflection curve is much steeper than the experiment curves indicate that some system deformations are not included in the finite element calculation. The U_{max} for the experimental curves is higher than that of the FEM results, but we only focus on the horizontal misaligned. Compitelli[5] corrected the FEM simulation curve by adding the elastic displacement of the system through equation 2.

$$d_{sp} = d_{measure} - P/K \tag{2}$$

Where d_{sp} is the target displacement, the $d_{measure}$ is the displacement measured by the extensometer, P is the load and the K is the compliance of the loading system. They got good alignment with this method shown in black curve in figure 5, but K differs in different experiment. Note that no constitute relation extrapolation is performed, the maximum height of the experimental and FEM results differs in maximum height and this phenomenon is beyond the scope of this paper.

3.2. Revised FEM model-1

We revised the FEM modelling by adding sphere puncher into SPT set-up shown in figure 6, the ceramic ball is modelled as elastic ball with Young's modulus E=300GPa and Poisson's ratio $v = 0.27$, the plastic properties of the ball is $\varepsilon_{pl} = 0.2, \sigma = 800$ and $\varepsilon_{pl} = 0.4, \sigma = 1200$. The results of the revised FEM model-1 and the experiment load-deflection curve is shown in figure 7.

Figure 6. Revised FEM model-1 set-up of the SPT.

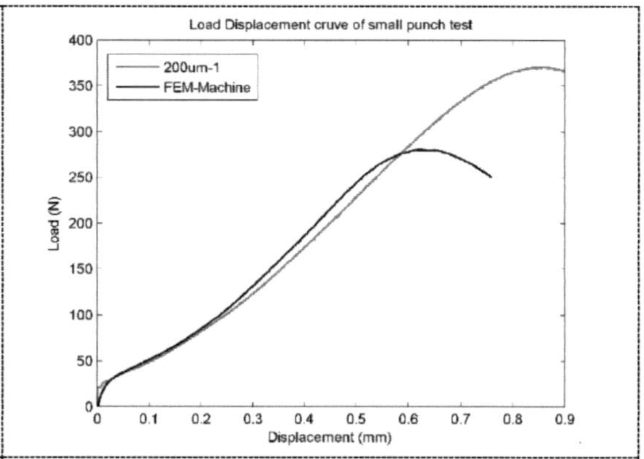

Figure 7. The experimental and revised FEM model-1 SPT load-deflection curves.

The revised model-1 still differs from the experimental result and is not much different from the original model, since the ceramic ball is harder than the material and there is hardly any deformation during the test.

3.3. Revised FEM model-2

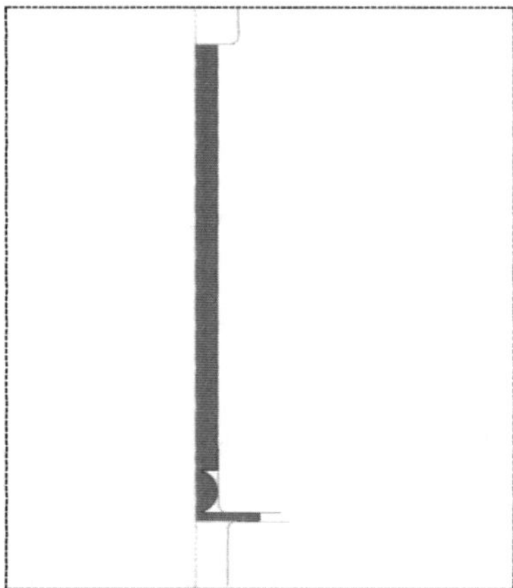

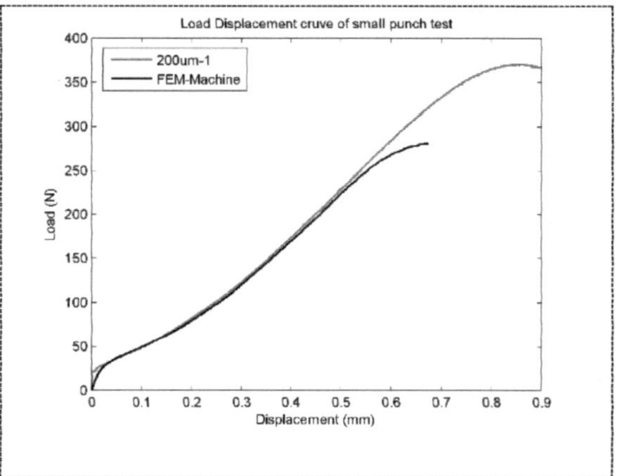

Figure 8. Revised FEM model-2 set-up of the SPT.

Figure 9. The experimental and revised FEM model-1 SPT load-deflection curves..

We revised the FEM model-1 by adding steel rod into SPT set-up shown in figure 8, the ceramic ball is modelled as revised FEM model-1 and the rod is 1mm in diameter and 35mm in height, modelled as Q345 low carbon steel with Young's modulus E=206GPa, Poisson's ratio $\nu = 0.3$ and the yield stress $\sigma_y = 345$MPa.

The results of the revised FEM model-2 and the experiment load-deflection curve is shown in figure 9. The revised model-2 alignment well with the experimental result except the misalignment with the height of the curve, this phenomenon might be solved by extrapolation of the constitutive relation of the testing material, will be discussed elsewhere. The ceramic ball puncher do not deform during the test and the loading system error is mainly from the elastic and plastic deformation of the rod. It is important to notice that the elastic compression of the rod is much smaller than the maximum loads difference of the FEM model and the experimental results. The difference come from the plastic deformation of the rod shown in figure 10, and the issue is observed in our experiment.

In order to avoid the plastic deformation we recommend the use the rod after plastic deformation instead of rounded rod surface, since the rounded rod made of steel might be worn after several experiment and the ceramic ball has good friction and wear property.

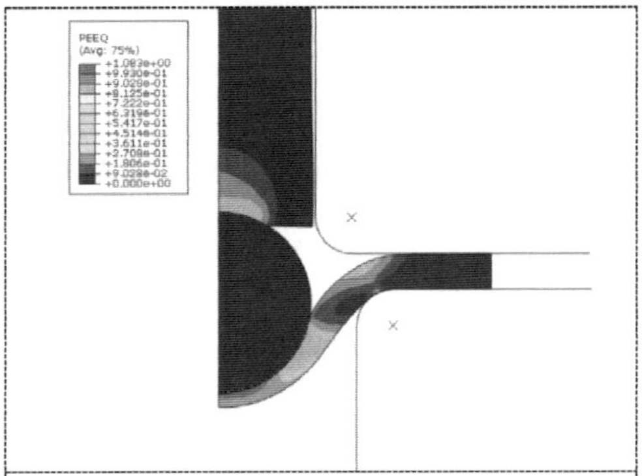

Figure 10. Equivalent plastic strain of the revised FEM model-2.

4. Conclusions

The rod plastic deformation during the small punch test can not be ignored, where the compliance of the loading system is relatively small compared to the plastic deformation of the rod. The rod and the sphere ceramic ball must be modelled of our loading system.

5. References

1.Mao, X. and H. Takahashi. Journal of Nuclear Materials, 1987. **150**(1): p. 42-52.
2.Murty, K.L. and I. Charit. Journal of Nuclear Materials, 2008. **383**(1–2): p. 189-195.
3.Ha, J.S. and E. Fleury. International Journal of Pressure Vessels & Piping, 1998. **75**(9): p. 707-713.
4.Lucas, G.E.,1983. **117**(8): p. 327-339.
5.Moreno, M.F., Materials & Design, 2018. **157**: p. 512-522.
6.Campitelli, E.N., et al., Materials Science & Engineering A, 2005. s **400–401**(1): p. 386-392.
7.Campitelli, E.N., et al., Journal of Nuclear Materials, 2004. **335**(3): p. 366-378.

Acknowledgments
This work is support by The National Key Basic Research Program (2017YFB0702202) and National Natural Science Foundation of China (U1532262).

IOP Conf. Series: Materials Science and Engineering **544** (2019) 012052 doi:10.1088/1757-899X/544/1/012052

Static Characteristics Analysis and Research on the Energy-Storage Flywheel Rotor Based on Finite Element

Xiu-hua ZHANG [1,2,*], Kang-jie SONG [2], Jiang PU [1], Ya-feng SU [2] and Zhi-guo FENG[2]

[1]School of Mechatronics Engineering of GuiZhou MinZU University, Guiyang 550025, China
[2]School of Mechanical Engineering of Guizhou University, Guiyang 550025 , China
E-mail: 756845681@qq.com

Abstract: The energy-storage flywheel system has been becoming a kind of energy storage device with broad application prospect. It has the advantages of high energy storage density, high power, high efficiency, short charging time , long life, etc. In order to prevent energy storage flywheel system components deformation or failure under high speed, the static characteristic analyses on the flywheel rotor were done respectively under two types of the permanent magnet don't act on the rotor internal wall and the permanent magnet acts on the rotor internal wall, the deformation, equivalent stress, tangential stress, radial stress of the rotor were researched. The work can provides certain reference for the structure design , optimization design of the flywheel system.

1 Introduction

The energy-storage flywheel system is a device to realize energy conversion and storage by physical methods[1]. The kinetic energy of the flywheel is increaseed and stored when the flywheel speed is increaseed; the kinetic energy of the flywheel is decreaseed and released when the flywheel speed is decreaseed, so the flywheel can be used to reduce the speed fluctuation of the mechanical operation process. The mechanical vibration of flywheel will be intensified when acted by the unbalanced forces, external dynamic load etc. This makes some parts easy to wear, fatigue and shortened life spans, even will cause the excessive deformation of flywheel components, and then whole system cleft, causing serious accidents[2-6]. Therefore, it is necessary to analyze deeply the energy-storage flywheel system , studying its stress and deformation under working conditions, so as to take protective measures.

In this paper, the static characteristics of the rotor of energy-storage flywheel system being designed were analyzed by finite element method, and the distribution of equivalent stress,

IOP Conf. Series: Materials Science and Engineering **544** (2019) 012052 doi:10.1088/1757-899X/544/1/012052

circumferential stress and radial stress of the flywheel rotor were studied, which can provides certain reference for the structure design , optimization design of the flywheel system.

2 Finite element model of outer rotor of flywheel

When stress analysis is carried out on the flywheel rotor model, the axial symmetry of the structure is taken into account and a quarter of the circumference is taken for analysis. The finite element model is shown in figure 1, and the main properties of the outer rotor material [7] are shown in table 1.

Table1. the Main Properties of the Outer Rotor Material

Material	Elasticity Modulus	Density	Poisson's ratio	yield strength
30CrMnSi	E=200GPa	7750kg/m^3	0.3	885MPa

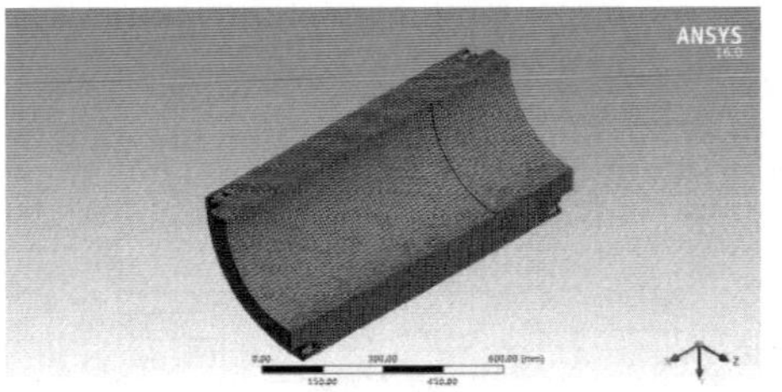

Figure 1. A quarter of flywheel rotor model

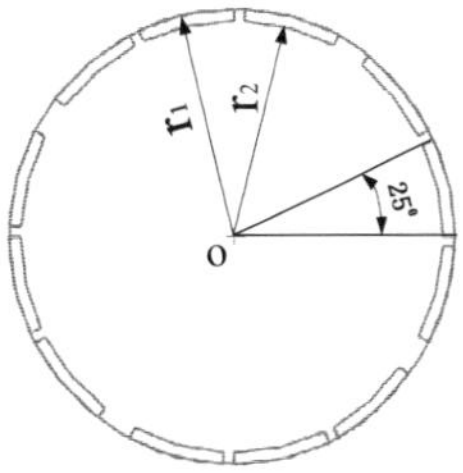

Figure 2. Assembly diagram of rotor and permanent magnet

3 Calculation and Addition of Load

(1)Centrifugal load

The outer rotor is a rotating body, which bears centrifugal load when it rotates, and the rotating speed is set at 6000RPM for stress analysis

(2) Inner wall pressure of outer rotor

The permanent magnets (each piece length L1 = 80 mm) are installed on the outer rotor inner wall (length in axial L = 400 mm), shown in figure 2, the permanent magnet outer raduis r1 is 266 mm, inner raduis r2 is 214 mm, every piece of permanent magnet is 25 ° circumference of Angle.The centrifugal force generated by the permanent magnet acts on the inner wall of the outer rotor with a form of external load.

The working area of each permanent magnet is:

$$S = 2\pi \times \mathrm{r}_1 \times L \times \frac{25}{360} \approx 7.6\times10^{-3}\mathrm{m}^2 \tag{1}$$

The volume of each permanent magnet is:

$$V = \pi(r_1^2 - r_2^2)\times L_1 \times \frac{25}{360} \approx 9.2\times10^{-5} m^3 \tag{2}$$

The mass of each permanent magnet is(Permanent magnet density is 7800 kg/m^3):

$$\mathrm{M} = \rho v \approx 0.72kg \tag{3}$$

When the rotating speed is 6000RPM, the centrifugal force generated by each permanent magnet is $F = M\omega^2\overline{R}$, in which:

$$\omega = \frac{2 \times \pi \times 6000}{60} = 628 rad/s \tag{4}$$

$$\overline{R} = \frac{r_1 + r_2}{2} = 220mm \tag{5}$$

Then the centrifugal force generated by each permanent magnet is: $F = 6.2 \times 10^4\ N$

And then the pressure acting on the inner wall of the outer rotor is:

$$P = F / S \approx 8.2MPa \tag{6}$$

4 Static simulation analysis results

4.1 the permanent magnet don't act on the rotor internal wall

The boundary conditions is shown in figure 3. The fixed constraints on both ends are added, and the rotation speed are at 6000rpm.

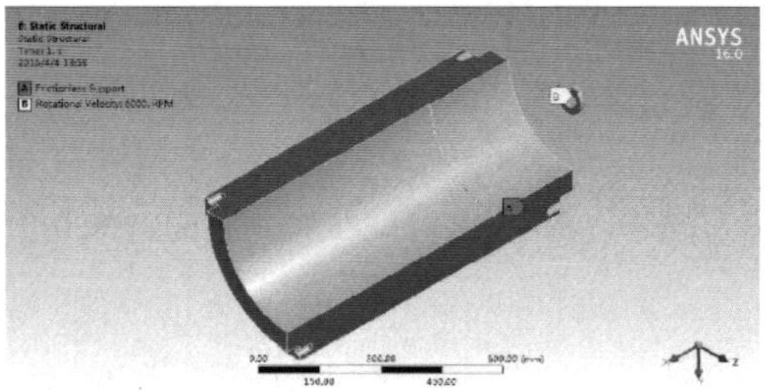

Figure 3. Boundary conditions

The static analysis results are shown in figure 4:

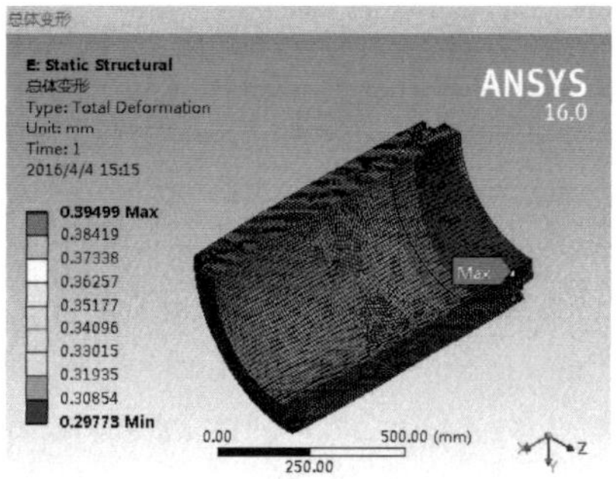

(a) Overall deformation of rotor

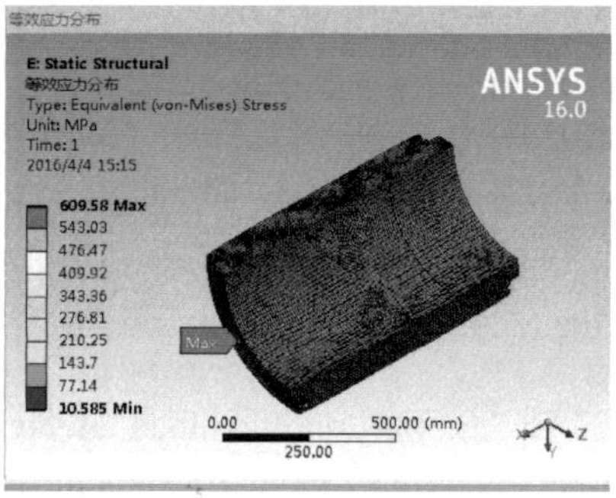

(b) Equivalent stress distribution of rotor

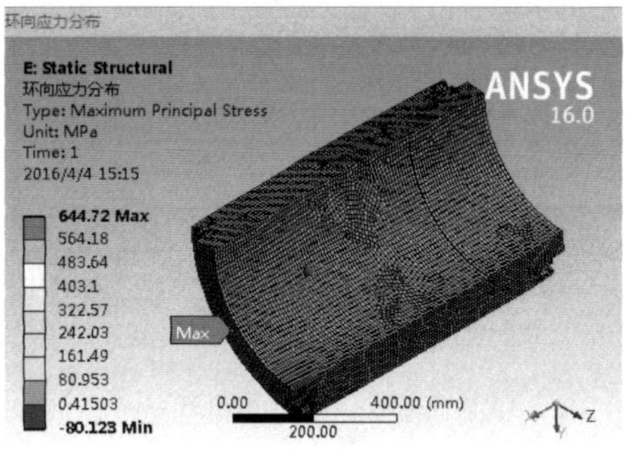

(c) Circular stress distribution of rotor

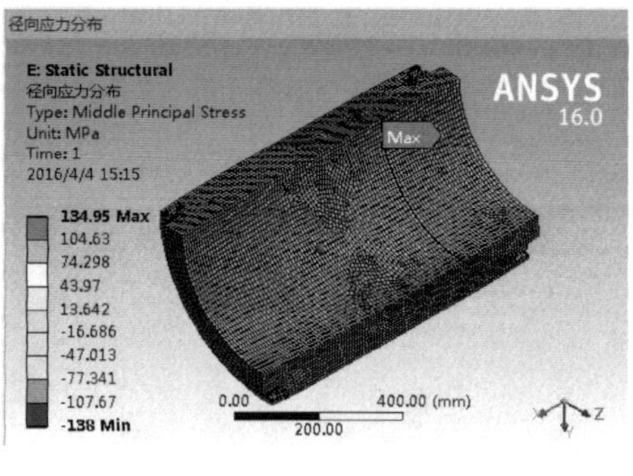

(d) Radial stress distribution of rotor

Figure 4. The static analysis results in which permanent magnets do not act on the inner wall

It can be seen from figure 4 that the maximum deformation of the rotor is about 0.395mm at the upper end.The equivalent stress is distributed in 10.58~609.58 Mpa, and the annular stress is distributed in -80.12~644.72 Mpa. The distribution of the two is not much different, here is the stress concentration. Radial stress is distributed in -138~134.95MPa, part is compressive stress, part is tensile stress.

4.2 the permanent magnet act on the rotor internal wall

The boundary conditions is shown in figure 5. In other words, 8Mpa pressure was added to the inner wall of the rotor on the basis of figure 3.

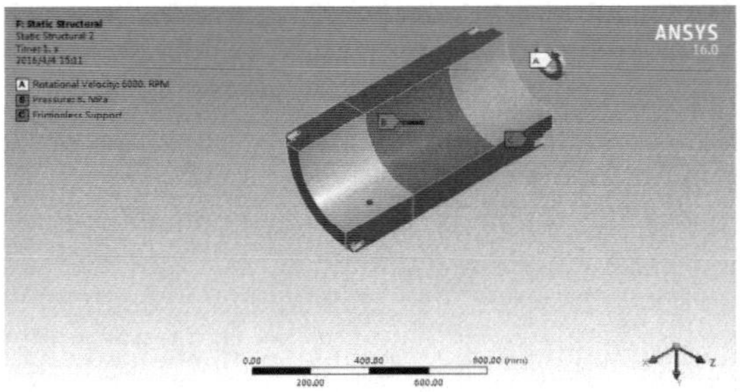

Figure 5. The internal pressure produced by the permanent magnet

The static analysis results are shown in figure 6:

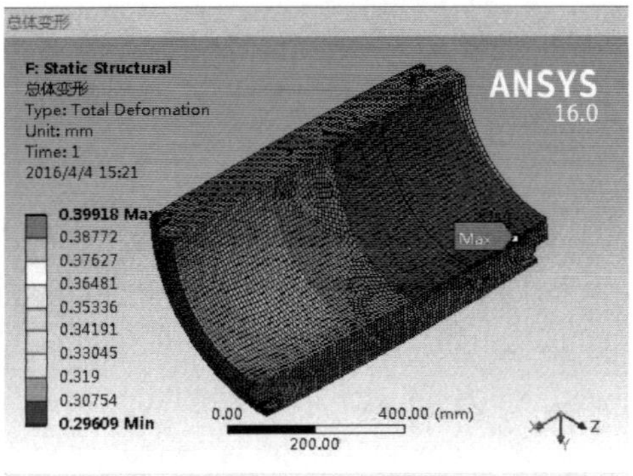

（a）Overall deformation of rotor

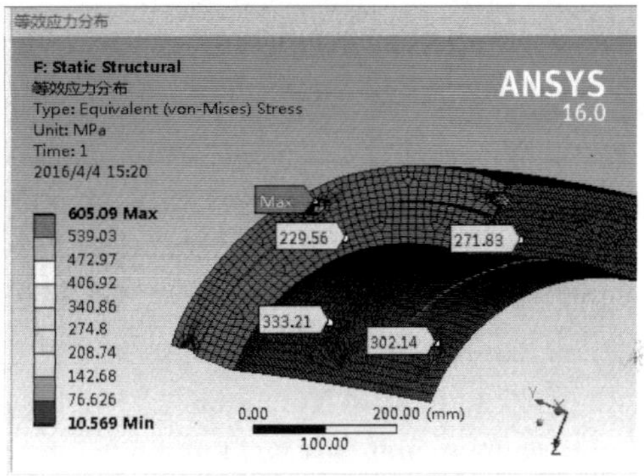

（b）Equivalent stress distribution of rotor

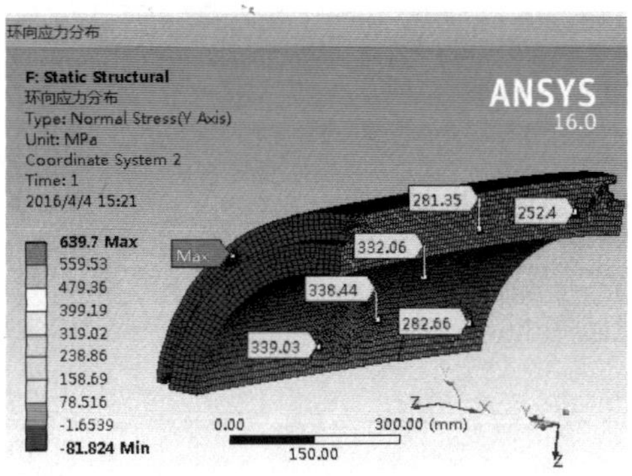

（c）Circular stress distribution of rotor

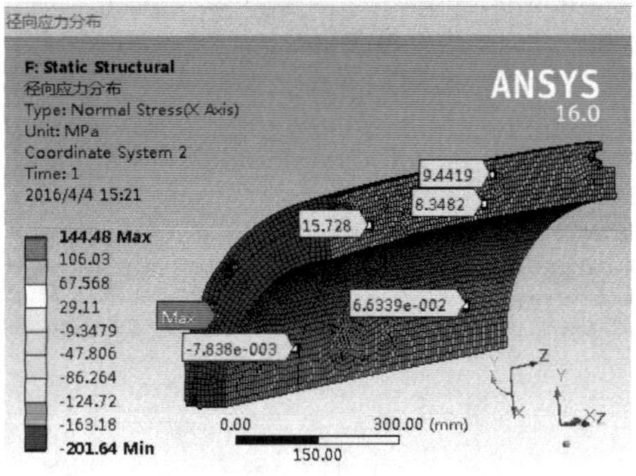

（d）Radial stress distribution of rotor

Figure 6. The static analysis results in which permanent magnets act on the inner wall

It can be seen from figure 6 that the maximum deformation of the rotor is about 0.399mm at the upper end.The equivalent stress is distributed in 10.569~605.09 Mpa, and the annular stress is distributed in -81.82~639.7 Mpa. The distribution of the two is not much different, here is the stress concentration. Radial stress is distributed in -201.64~144.48MPa, part is compressive stress, part is tensile stress.

5 conclusion

When the flywheel is rotating at high speed, the centrifugal stress acted on the flywheel cannot be neglected.In this paper, finite element stress analysis of flywheel rotor at 6000rpm is performed.By comparison, the maximum displacement deformation is concentrated at the upper end of the flywheel, and the maximum deformation is 0.399mm.The maximum equivalent stress is 609.58 Mpa, the maximum annular stress is 644.72 Mpa and the maximum radial stress is 144.48 Mpa.When design the flywheel rotor, the strength of the stress concentration of the rotor should be considered and proper rotor material should be selected.The research results have certain guiding significance for structure design , optimization design of the flywheel system.

Reference

[1] TANG Chang-liang, DAI Jian, WANG Jian, LI Yi-liang. Dynamic analysis and experimental study of shaft system of 20kW/1kWh flywheel energy storage system [J]. Vibration and impact,2013,32(01):38-42.

[2] Li X, Anvari B, Palazzolo A, et al. A Utility Scale Flywheel Energy Storage System with a Shaft-less, Hub-less, High Strength Steel Rotor[J]. IEEE Transactions on Industrial Electronics, 2017, PP(99):1-1.

[3] S. M M G, Faraji F, Majazi A, et al. A comprehensive review of Flywheel Energy Storage System technology[J]. Renewable & Sustainable Energy Reviews, 2017, 67:477-490.

[4] Gayathri N S, Senroy N, Kar I N. Smoothing of wind power using flywheel energy storage system[J]. Iet Renewable Power Generation, 2017, 11(3):289-298.

[5] ZHANG Xin-bin, JIANG Wei, LI Hong-liang, SUN Zhong-xin, RUAN Wen-zhong. Key technologies and research status of flywheel energy storage system [J]. Energy storage science and technology,2015,4(01):55-60.

[6]REN Yong-qiang, LI Xin-heng. Optimization design and finite element analysis of flywheel system on test bench [J]. Combined machine tools and automatic processing technology,2013(08):47-50.

[7] Chen Qiang, Yuan Xi-bing. Rotor statics analysis of flywheel energy storage system [J]. Manufacturing automation,2017,39(12):58-61.

Acknowledgments

1. Guizhou Science and Technology Support Program([2016]2002);
2. Guizhou University introduced talent project [(2015)-51].

IOP Conf. Series: Materials Science and Engineering **544** (2019) 012053 doi:10.1088/1757-899X/544/1/012053

Physical properties of Ti_3SiC_2/Zr heterojunction under hydrostatic pressure

Y Li[1], C Mao[2], R Tu [1], L Xue[2], H Li [1], J Wang[2], Z Chen[2], H Ni[2], Y Hu[1, 2, *]

1 Division of nuclear material and fuel, State Power Investment Corporation Research Institute, Beijing, China
2 Hubei University of Science and Technology, Xianning, China
*E-mail: hchyh2001@tom.com

Abstract. The geometric structure, elastic constants and electronic properties of Ti_3SiC_2/Zr heterojunctions under high hydrostatic pressure (0-10GPa) were investigated by first-principles calculations. It's found that the increase of hydrostatic pressure loaded on the Ti_3SiC_2/Zr heterojunctions induces the monotonically decrease of their lattice parameters. Under hydrostatic pressure below 10GPa, the Ti_3SiC_2/Zr heterojunction is stable and its elastic properties change linearly with pressure. Ti_3SiC_2/Zr heterojunction under hydrostatic pressure is found to be ductile material, and its anisotropy is weak. Finally, Ti_3SiC_2/Zr heterojunction under hydrostatic pressure retains metallic properties, and hydrostatic pressure slightly changes the band structure of Ti_3SiC_2/Zr heterojunction.

1. Introduction

Recently, MAX phase compounds have attracted much attention, due to their unique properties: high-temperature stability, good thermal and electrical conductivity, high thermal shock resistance, strong damage tolerance and particular microscale ductility[1,2]. For n=2, Ti_3SiC_2 is identified as a typical MAX material, which is widely used in high temperature and chemical anticorrosive conditions, such as in the fields of nuclear power and aviation [3-5]. In the past, theoretical studies on Ti_3SiC_2 have been carried out, calculating the electronic structure, elastic and optical properties etc. Owing to their outstanding properties, such as low neutron absorption cross section etc, zirconium-based alloys are the best water-cooled reactor nuclear materials [6,7]. In recent years, as good radiation and corrosion resistant material, Ti_3AC_2 (A=Si, Al) are usually the best choice of coating material for zirconium-based alloy to prevent hydrogen embrittlement [8-10]. Recently, we designed and studied the physical properties of Ti_3SiC_2/Zr heterojunctions, which focuses on the elastic and thermal properties of Zr metal and Ti_3SiC_2. It's reported that the mechanical and thermal properties of Zr metal and Ti_3SiC_2 matches well [11].

As is well known that in nuclear condition, materials always encounter high pressure, i.e. 10 GPa in pressurized water reactor. However, the effect of high pressure on the geometric and elastic constants of Ti_3SiC_2/Zr heterojunction has not been investigated in detail. In this article, we have investigated the geometirc structure, electronic and elastic propertics of Ti_3SiC_2/Zr heterojunction by ab initio calculations using Generalized Gradient Approximation (GGA) method, and studied the effect of pressure on the geometric, electronic and elastic properties. Our present work indicates that Ti_3SiC_2/Zr heterojunction under 10GPa has good stability and suitable elastic properties, which means that Ti_3SiC_2 is good coating material of Zr metal under high hydrostatic pressure.

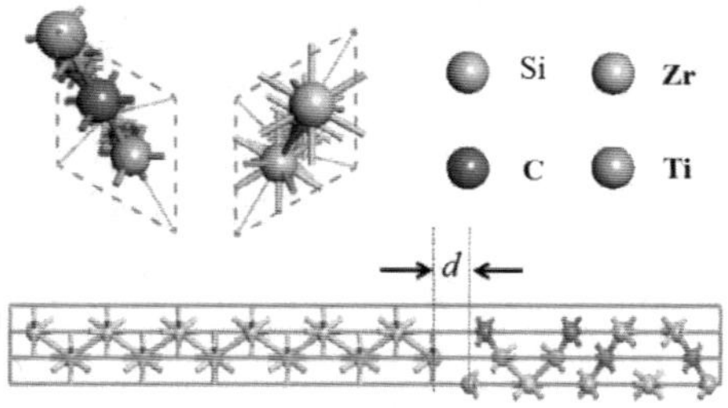

Figure 1. Perspective spatial model of Ti_3SiC_2/Zr heterojunction.

2. Calculation details

First-principles calculations were completed by the Materials Studio software[12]. The ultrasoft pseudopotential was adopted to simulate the interaction between ions and electrons. The GGA method [13] is used to deal with the exchange correlation between atoms. Ti_3SiC_2 supercell and Zr metal supercell are assembled in the (001) direction, and the mismatch of lattice parameters a and b in Ti_3SiC_2/Zr are both below 4.8%, as is shown in figure1. We fully optimized the Ti_3SiC_2/Zr heterojunction structures, under which the external pressure changes from 0 to 10 GPa. In this work, the energy deviation is less than 5.0×10^{-6}eV/atom, and convergence is assumed when the forces on atom is less than 0.01eV/Å. The plane-wave cutoff energy of 400 eV is employed. And the self-consistent field (SCF) tolerance is set to be 5.0×10^{-7}eV/atom. The calculations are carried out using a (10,10,2) Monkhorst-Pack grid[14].

3. Results and discussion

3.1Geometric structure

Ti_3SiC_2 and Zr metal both have hexagonal structure, whose space group are P63/mmc. However, after fully relaxation, the symmetry of Ti_3SiC_2/Zr heterojunction supercells become lower, and the symmetry belongs to P-3M1. The lattice constants of optimized Ti_3SiC_2/Zr heterojunction under different pressure are listed in table 1. It's found from table 1 that the structural parameters and total energy of Ti_3SiC_2/Zr heterojunction all change monotonically with pressure. With the increase of pressure from 0 to 10 GPa, the lattice parameters (a, c, d and V) all decrease linearly with pressure. In figure 2, we show the pressure dependence of a/a_0 and c/c_0 as well as V/V_0 (where a_0,c_0 and V_0 are the lattice parameters under zero pressure). We notice in figure 2 that the changes of the a/a_0 and c/c_0 as well as the V/V_0 are approximately linear when the pressure rises from 0 to 10 GPa. These changes are in consistent with each other. It can be seen from table 1 that the total energy of the Ti_3SiC_2/Zr heterojunction increases with pressure, while the volume decreases with pressure. Moreover, the bonds in the Ti_3SiC_2/Zr heterojunction become closer, which indicating that the interactions between the atoms become stronger. Under higher and higher pressure, the contractions of inter-atomic distances results in bigger and bigger change of electronic structure of the Ti_3SiC_2/Zr material.

Table 1. Lattice parameters and total energy of system under pressure.

P (GPa)	a (Å)	c (Å)	d (Å)	V (Å^3)	E (eV)
0.0	3.13	51.93	2.67	440.58	-27439.4
2.5	3.11	51.42	2.61	430.70	-27432.9
5.0	3.09	51.19	2.59	423.27	-27425.9

7.5	3.07	50.89	2.55	415.36	-27419.4
10.0	3.05	50.60	2.53	407.63	-27413.0

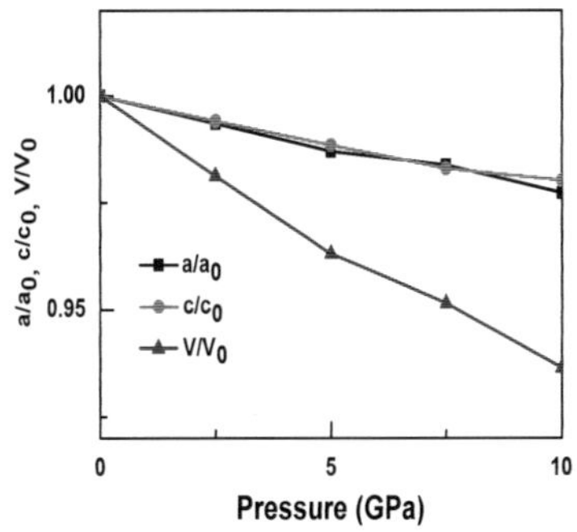

Figure 2. The V/V_0, a/a_0, c/c_0 for Ti_3SiC_2/Zr heterojunction under different pressure.

3.2 Elastic constants

Elastic properties reflect material's ability to deform and recover under different forces. Under certain force range, objects all have elasticity. Ti_3SiC_2/Zr and Zr metal have six elastic constants which is C_{ij} $(i,j=1,2,3,4,5,6)$. The elastic constants of Ti_3SiC_2/Zr are calculated in detail, which are shown in table 2. The elastic constants of Ti_3SiC_2/Zr in table 2 are taken into the inequality in reference [11]. It is found that the elastic constants of Ti_3SiC_2/Zr all satisfy the inequality when the external pressure is changed from 0 to 10GPa, which means the Ti_3SiC_2/Zr heterojunction has good stability under the given pressure. In addition, the ratios of elastic parameters of Ti_3SiC_2/Zr are C_{11}/C_{33}, C_{12}/C_{13} and C_{44}/C_{66}. All ratios deviate from 1, which means that Ti_3SiC_2/Zr heterojunctions under different pressure belong not to pseudocube crystal.

Table 2. Constants C_{ij}, B (GPa), G (GPa) and ν of material under pressure p (GPa).

p	C_{11}	C_{12}	C_{13}	C_{33}	C_{44}	C_{66}	B	G	G/B	ν
0.0	212.4	60.36	78.90	192.82	44.19	77.06	117.13	59.83	0.51	0.281
2.5	225.26	73.79	83.49	217.54	52.36	76.29	127.45	68.09	0.53	0.273
5.0	232.22	74.33	95.22	219.54	42.73	80.08	134.68	53.64	0.39	0.324
7.5	247.10	77.43	102.79	235.36	46.57	86.13	143.82	62.18	0.43	0.311
10.0	257.47	84.31	111.00	247.84	59.59	86.39	152.43	70.44	0.46	0.300

In figure 3, the variations of elastic coefficients of Ti_3SiC_2/Zr heterojunction with pressure are plotted respectively. It's found that most elastic coefficients except for C_{44} increase monotonically with pressure from 0 to 10 GPa.

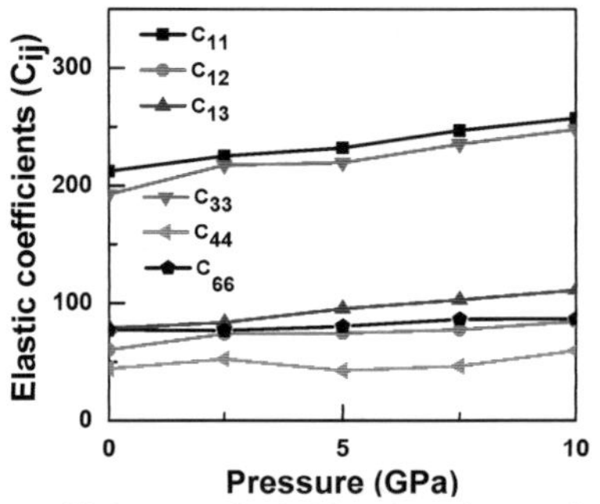

Figure 3. The elastic coefficients of Ti_3SiC_2/Zr heterojunction under pressure.

According to reference [11], we get the expressions of B_V and G_V, which are shown below:

$$B_V = (2C_{11} + C_{33} + 2C_{12} + 4C_{13})/9$$
$$G_V = (2C_{11} + C_{33} - C_{12} - 2C_{13} + 6C_{44} + 3C_{66})/15 \quad (1)$$

Further using the Reuss approximation, we get the expressions of B_R and G_R [17], which are as follows:

$$B_R = [(C_{11} + C_{12})C_{33} - 2C_{13}^2]/(C_{11} + C_{12} + 2C_{33} - 4C_{13})$$
$$G_R = 15/(8s_{11} + 4s_{33} - 4s_{12} - 8s_{13} + 6s_{44} + 3s_{66}) \quad (2)$$

Further, we get the expressions of B and G, which are given in below:

$$B = (B_V + B_R)/2\text{,} \quad G = (G_V + G_R)/2 \quad (3)$$

The body modulus and shear modulus calculated according to relations (1-3) are shown in table 2. From the data in table 2, the G/B values for the Ti_3SiC_2/Zr heterojunction under 0, 2.5, 5.0, 7.5 and 10.0GPa are 0.51, 0.53, 0.39, 0.43 and 0.46 respectively. Thus, the Ti_3SiC_2/Zr heterojunction materials under pressure (0-10GPa) are ductile materials, for all the $G/B < 0.57$. Based on the above results, ν , A, and Y can also be calculated, their expressions are as follows:

$$\nu = (B - 2/3G)/(2B + 2/3G)\text{,} \quad A = 2C_{44}/(C_{11} - C_{12})\text{,} \quad Y = 9GB/(G + 3B) \quad (5)$$

Our calculation results of ν , A, and Y according to equations (5) are listed in table 3. The value of ν for Ti_3SiC_2/Zr under pressure are close to 0.33, which are the normal value for a metallic material, which means that the bonding between the atoms in the the Ti_3SiC_2/Zr heterojunctions are dominated by ionic bonds. The anisotropic factors for Ti_3SiC_2/Zr decrease with pressure, and they are all less than 1, therefore, it can be concluded that the anisotropy of all these materials are weak. The anisotropic factors for Ti_3SiC_2 and Zr are 1.36 and 1.06, respectively, which are larger than the value of ν for Ti_3SiC_2/Zr. Therefore, it's found that after assembling of Ti_3SiC_2 and Zr into Ti_3SiC_2/Zr heterojunction, the elastic anisotropy of materials become weaker than those of pristine Ti_3SiC_2 and Zr metal. Young's modulus for materials usually indicate their deformations under different stress. In table 3, it is shown that the value of Y for Ti_3SiC_2/Zr changes non-linearly with pressure.

Table 3. Parameters ν, A, and Y (GPa) of material under various pressure (GPa).

Pressure	ν	A	Y
0.0	0.28	0.58	153.37
2.5	0.27	0.54	173.39
5.0	0.32	0.45	142.05
7.5	0.31	0.37	163.05
10.0	0.30	0.18	183.11

3.3 Electronic properties

In this section, we mainly calculated the band structures of Ti_3SiC_2/Zr heterojunctions under hydrostatic pressure. The band structures of Ti_3SiC_2/Zr heterojunctions under three different pressure (0, 5 and 10GPa respectively) are shown in figure 4. It can be seen from figure 4 that the Ti_3SiC_2/Zr under different pressure all exhibit conduction bands crossing the Fermi level, which indicates that all these systems have metallic properties. Obviously, the high hydrostatic pressure doesn't change the metallic properties of Ti_3SiC_2/Zr heterojunctions, but changes their band structure (As shown in figure 4) [11].

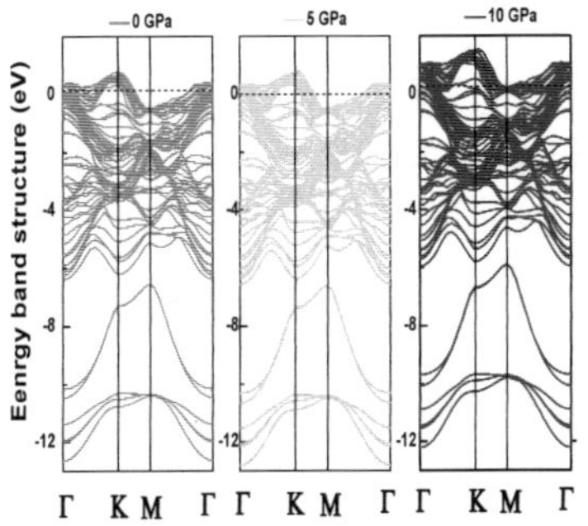

Figure 4. Band structures of Ti_3SiC_2/Zr heterojunction under different pressure.

The total density of state (TDOS) and partial density of state (PDOS) of Ti_3SiC_2/Zr heterojunction under different hydrostatic pressure were calculated by GGA-PBE functions. For brevity, taking the case of p=5 GPa as an example, the total density of state (TDOS) and partial density of state (PDOS) of Ti_3SiC_2/Zr heterojunction under hydrostatic pressure of 5GPa are shown in figure 5. From the figure 5, we can see that N (EF) values are all greater than zero, which means that the Ti_3SiC_2/Zr heterojunction under hydrostatic pressure are metallic. The peak near Fermi level in figure 5 is mainly attributable to the hybridization of Ti-3d and Zr-4d. The hybridization between atoms indicate that the bonds in Ti_3SiC_2/Zr heterojunction is partly covalent. These results are in agreement with our previous report [11].

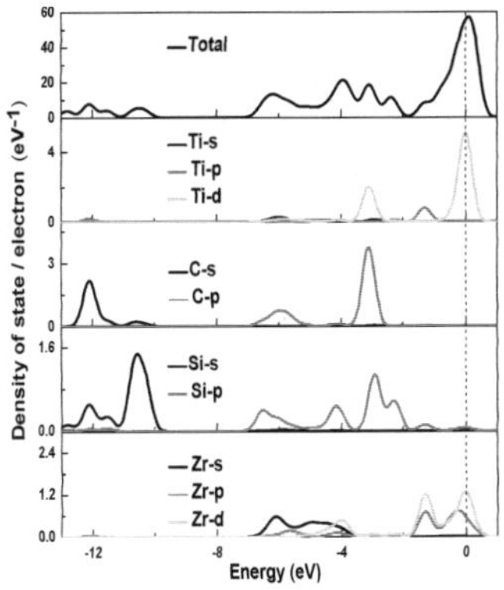

Figure 5. The TDOS and PDOS of Ti_3SiC_2/Zr heterojunction under p=5GPa.

4. **Conclusion**

In this paper, a few physical properties of Ti_3SiC_2/Zr under different hydrostatic pressure were analyzed by the first-principles method. First of all, the geometric structures of Ti_3SiC_2/Zr heterojunctions under static pressure were optimized, and it's found that the increase of hydrostatic pressure caused the monotonically decreases of their lattice parameters. And the total energy of the Ti_3SiC_2/Zr heterojunction increase with hydrostatic pressure. Secondly, when the external hydrostatic increase from 0 to 10GPa, the Ti_3SiC_2/Zr heterojunction is stable and its most elastic coefficients change linearly with pressure. Moreover, it's found that Ti_3SiC_2/Zr heterojunction materials under hydrostatic pressure are ductile material, and its anisotropy is weak. Finally, the energy band structure and DOS of Ti_3SiC_2/Zr heterojunction under different hydrostatic pressure were investigated. Our results show that hydrostatic pressure doesn't change their metallic properties, but slightly changes their band structure.

Acknowledgments
This project has received the funding from the National Science and Technology Major Project of China (2015ZX06004001-002), the Natural Science Foundation of Hubei Province of China (2017CFB526), the Nurturing Scientific Research Projects of Hubei University of Science and Technology (2016-18X048), and the teaching research project of Hubei university of science and technology (2016-XA-004; 2017-ZYB-00).

References

[1] Keast V, Harris S, Smith D, 2009 Prediction of the stability of the $M_{n+1}AX_n$phases from first principles *Phys. Rev. B* **80** 214113
[2] Jia L, Wang Y, Ou X, Shi L, Ding W 2012 Decohesion of Ti_3SiC_2induced by He impurities *Materials Letters.* **83** 21
[3] Togo A, Chaput L, Tanaka I, Hug G 2010 First-principles phonon calculations of thermal expansion in Ti_3SiC_2, Ti_3AlC_2,and Ti_3GeC_2 *Phys. Rev. B* **81** 174301
[4] Bai Y, He X, Sun Y, Zhu C, Li M, Shi L 2010 High Sensitivity Viologen for a Facile and Versatile Sensor of Base and Solvent Polarity in Solution and Solid State in Air Atmosphere *Solid State Sci.* **12** 1220
[5] Xiao J, Wang C, Yang T, Kong S, Xue J 2013 Theoretical investigation on helium incorporation in Ti_3AlC_2 *Nucl. Instrum. Meth. B* **304** 27
[6] Kutty T, Ravi K, Ganguly C 1999 Studies on hot hardness of Zr and its alloys for nuclear reactors *J. Nucl. Mater.* **265** 91
[7] Cox B J 2005 How the crystallography and nanoscale chemistry of the metal/oxide interface develops during the aqueous oxidation of zirconium cladding alloys *Nucl. Mater.* **336** 331
[8] Allen C 1996 New instrumentation in Argonne`s HVEM-Tamdem Facility: Expanded capability for in situ ion beam studies *Materials Research Society* **396** 641
[9] Riley D, Kisi E 2007 Low-temperature sintering of (K,Na)NbO_3 -based lead-free piezoceramics with addition of LiF *J. Am. Ceram. Soc.* **90** 2231
[10] Ou X, Wang Y, Shi L, Ding W, Wang M, Zhu Y 2011 Effect of hydrogen-doping on bonding properties of Ti_3SiC_2 *Physica. B* **406** 4460
[11] Li Y, Mao C, Hu Y et al. 2018 First-principles study on the physical properties of Ti_3AC_2/ Zr (A = Si, Al) van der Waals heterojunction *J. Phys. Chem. Solid.* **121** 298
[12] Ceperley D, Alder B 1980 First-principles investigation of the structural, mechanical and electronic properties of the NbO-structured 3d and 5d transition metal nitrides *J. Phys. Rev. Lett.* **45** 566
[13] Perdew J, Burke K, Erenzerhof M 1996 Generalized Gradient Approximation Made Simple *Phys. Rev. Lett.* **77** 3865
[14] Finkel P, Barsoum M W, Elraghy T 2000 Low temperature dependencies of the elastic properties of Ti_4AlN_3, $Ti_3Al_{1.1}C_{1.8}$,and Ti_3SiC_2 *J. Appl. Phys.* **87** 1701

IOP Conf. Series: Materials Science and Engineering **544** (2019) 012054 doi:10.1088/1757-899X/544/1/012054

Transformer Selection Calculation for the design of Flyback Switching Power Supply

Hao Ni[1], Aixia Wu[2*], Xin Zhang[3], Ruolin Ruan[4], Yonghong Hu[5] and Wei Zhang[6]

1 Associated Professor, Hubei University of Science and Technology, Xianning, PRC
2* Lecturer, Hubei University of Science and Technology, Xianning, PRC
3 General Manager, Hubei Tongda Power Supply Company Limited, Xianning, PRC
4 Professor, Hubei University of Science and Technology, Xianning, PRC
5 Professor, Hubei University of Science and Technology, Xianning, PRC
6 Senior Engineer, Hubei University of Science and Technology, Xianning, PRC

E-mail: 278928643@qq.com

Abstract. Pulse-width modulating (PWM) is one of the most popular methods to design the switching power supply. The framework to design a novel flyback switching power supply with a PWM integrated circuit is described in this paper. In order to meet the design demand, the significant step of transformer selection calculation is discussed in details. A_P method is used in the transformer calculation. Then some other parameters of the transformer, such as the transformer ratio, Primary current of transformer, the primary inductance of transformer, the transformer turn and air gap and the wire diameter of the winding, are computed step by step. With the selected transformer, the proposed switching power supply has simpler circuit and higher accuracy. Different input voltage ranges and output voltages required by different electric appliances can be well satisfied.

1. Introduction

In the design progress of switching power supply, efficiency is one of the most important indicators. So far the efficiency of domestic switching power supply is mostly about 70%. In order to respond to the call of energy saving and emission reduction in China, the design concept of future switching power supply should be focused on reducing the loss and improving efficiency while designing. The loss of switching power supply is composed of input rectifier loss, switching loss, loss of buffer circuit, conduction loss, loss of transformer and inductance, etc. Besides the loss of input rectifier, other losses can all be reduced. These years the relatively new technology is to adopt zero voltage/zero current switching synchronous rectifier to reduce the switching loss and grid drive loss of synchronous rectifier[1-2]. In the future, period-hopping control could be used to reduce light load and standby loss. There are many ways to improve efficiency, such as passive lossless snubber circuit, synchronous rectifier, low power control chip and so on. Flyback topology is proved to be an effective solution for making switching power supply due to its low cost and high efficiency [3]. For example, the AC-DC adapters and chargers of laptops are realized by pulse-width modulating (PWM) power conversion. Therefore, in order to improve the efficiency and reduce the loss, a novel flyback switching power supply for the electronic devices with low power is proposed in this paper.

The rest of this paper is organized as follows. First, the working principles will be proposed in Section 2. Second, the design indicators and the system framework will be described in Section 3. In this section, the hardware framework and the function of the modules will be proposed. Furthermore, the calculation to select transformers will be introduced in Section 4, as well as some other parameters of the transformer. Then the conclusion will be stated in the last Section.

2. Working principles

To adapt to different output power, switching power supply has a variety of topologies: Boost, Buck, Flyback, forward, half bridge, full bridge and so on[4-5]. In this design, the major indices are as follows: the output voltage and current, power and switching frequency are 12V, 3.4A, 40W and 65 KHz respectively. Hence the flyback topology is chosen in this paper to meet the specification of simple circuit and power level. The so-called flyback repeasents that the primary side polarity of transformer is opposite to the secondary side polarity. When the switch is on, the energy is stored in the transformer, and when the switch is off, the output rectifier diode is turned on, providing energy to the output capacitor and load. At the same time, flyback switching power supply can output multiple sets of voltages which can be isolated from each other. As a good choice for the switching power supply less than 100W, they provide power to the chip and energy to the output simultaneously. If the upper end of the primary side of the transformer is positive, the upper end of the secondary side is negative [6]. Flyback converter is widely used because of its high efficiency, simple circuit and multichannel output. Unfortunately, the ripple voltage is relatively high. We can use larger output filter capacitor and inductance to solve this problem. But this will make the power supply larger.

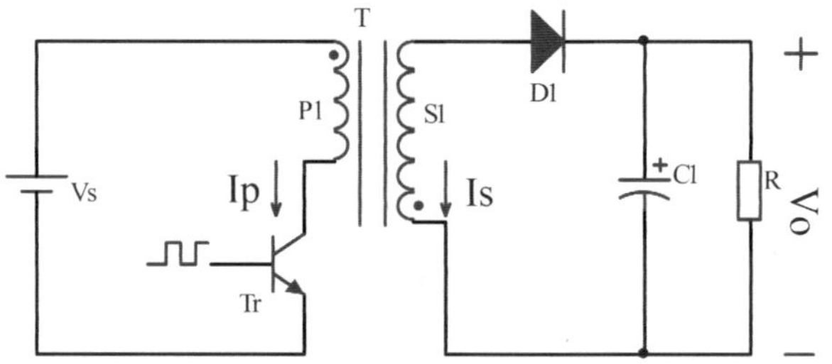

Figure 1. Flyback converting circuit

Recently, developers have found that small LC noise filters is effective. Flyback converter has two working mode. One is complete energy conversion. All the energy stored in the storage cycle of the transformer is transmitted in the flyback cycle. The other is incomplete energy conversion. Some of the energy stored in the transformer during the energy storage cycle is kept in the flyback cycle until the next energy storage cycle. While using complete energy conversion mode in PWM switching converter, the width of the starting pulse of the control circuit can be reduced. But there are some problems such as waveform distortion and modulation difficulties [7-8]. Only the incomplete energy conversion method is discussed here.

The flyback converting circuit is described in Figure.1. When the switch transistor Tr is off, the energy accumulated on the primary side of the transformer is transferred to the secondary side. At this time, the lower end of the secondary winding is negative while the upper end is positive. The diode D1 is positive conduction. The conduction voltage is filtered by the capacitor C1 to supply power to the load R. When the stored energy on the primary side of the transformer is released to a certain extent, the power supply voltage charges the collector of the transistor through the primary winding of the transformer. The primary winding begins to store energy. When V_s rises to a certain extent, the transistor is off and a new round of discharge begins. During the charging cycle, the output voltage of the converter is $Vo = \frac{N_2}{N_1} * V_S * D$.

3. System design framework

In order to make the power supply work stably in practical applications, it is often necessary to set high voltage protection, conduction interference suppression, green energy saving and other functions.

So switching power supply should have some important indicators. For example, the Input voltage range should be 100-240Vac. The input voltage frequency range should be 47-63Hz. The input and output have good There should be good isolation effect between the input and the output. Output ripple should be limited below 120mV. Other functions such as over-voltage, over-current, over-temperature protection should be taken into consideration.To achieve the above indicators, the design is mainly composed of the following circuits shown in the block diagram of Figure 2:

- *Input Part*: including EMI filter module and rectifier module
- *Power Conversion Part*: including drive module, PWM module, high-frequency transformer module, output filter module and some protection circles
- *Feedback Part*: including voltage sampling module and photoelectric coupling module

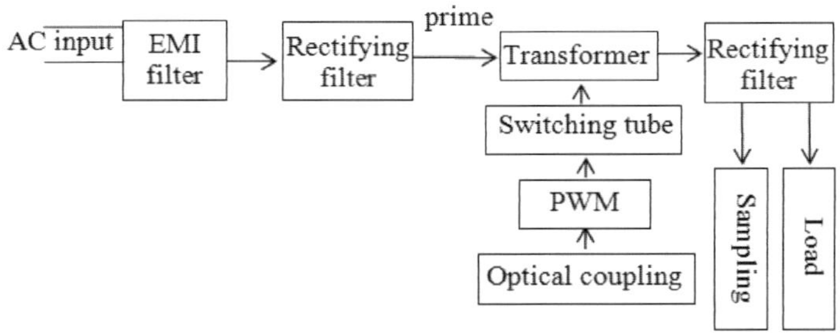

Figure 2. The System framework of flyback switching power supply

4. Caculations to select transformer

4.1. Selection of transformer core

Generally, there are two methods for transformer selection. One is to calculate the required A_P of the core in the system according to the input and output power. Then the appropriate core can be selected after looking up in the core specification table with the solved A_P .The other is to find the serial number of the core in the specification table according to the values of the geometric parameter for design. The former is called A_P method while the latter is called K_G method [9]. In this paper, A_P method is used in the transformer calculation.

4.1.1. Total apparent power. Because the output power is 40.8W, supposed the efficiency $\eta = 0.9$, the input power is $P_{in} = \frac{P_{out}}{\eta} = 45W$. Then the apparent power is $P_T = P_{out} + P_{in} = 85.8W$.

4.1.2. A_P value. Supposed K_o =0.4, K_f =4.0 (square wave), B_w =0.2T, f_s =65KHz. According to some materials, when the temperature is 25℃, K_j =323, x=-0.14, then

$$A_P = \left(\frac{P_T \times 10^4}{K_o K_f f_s B_w K_j}\right)^{\frac{1}{1+x}} = \left(\frac{85.8 \times 10^4}{0.4 \times 4 \times 65 \times 10^3 \times 0.2 \times 323}\right)^{\frac{1}{1-0.14}} = 0.09(cm^4) \quad (1)$$

In order to leave enough allowance and make the diameter of the wire coarser, we choose PQ3220 core. The A_p and A_e value are $1.37\,cm^4$ and $170\,mm^2$ respectively [9].

4.2. Calculation of transformer parameters

The input voltage of this design is 90-264 Vac. Some other parameters are as follows: switching frequency f_s =65 KHz, pulse maximum duty cycle D_m =0.45. The power conversion efficiency is set to n = 0.8. There are two output groups. One is the main output 12 V, 3.4 A. The other is the auxiliary output, providing power supply for the power chip. The voltage set to 20V while the chip's normal current is less than 2 mA. Therefore the power of the auxiliary winding is negligible.

4.2.1. Transformer ratio. When the low frequency ripple is 27V, the minimum input voltage $V_{in(\min)} = 90*\sqrt{2} - 27 = 100V_{dc}$. With the voltage-second balance $V_{in(min)}*D_m = (V_{out}+V_f)*(1-D_m)*N$, we can get the transformer ratio $N = \frac{V_{in(min)}*D_m}{(V_{out}+V_f)*(1-D_m)} = 6.294$. Finally, we set the integer N=7.

4.2.2. Primary current of transformer. When the output power of the power supply is P_{out} =40.8W, the efficiency η =0.8, the input power is $P_{in} = \frac{P_{out}}{\eta} = 51W$. The average input current is $I_{avg} = \frac{P_{in}}{V_{in(min)}*D_m} = \frac{51}{100*0.45} = 1.13A$. Supposed $I_{P_2} = k*I_{P_1}$, k=0.4, then $I_{P_1} = \frac{2P_{in}}{(1+k)*V_{in(min)}*D_m} = 1.619A$, $I_{P_2} = 0.648A$ and $\Delta I = I_{P_1} - I_{P_2} = 0.971A$.

4.2.3. Primary inductance of transformer. From $L_P*\Delta I = V*\Delta t$, we can get the primary inductance of the transformer $L_P = \frac{V*\Delta t}{\Delta I} = \frac{100*0.45}{0.971*65*10^3} = 713uH$. Then $I_{P_2} = 0.648A$ and $\Delta I = I_{P_1} - I_{P_2} = 0.971A$.

4.2.4. Transformer turn and air gap. Due to $N_P = \frac{L_P*\Delta I}{A_e*B_w}$, we can get the transformer turn $N_P = \frac{713*0.971}{170*0.2} = 20.36$. We set N_P =28. Then N_s =4 according to $N = \frac{N_P}{N_s} = 7$. Because the thyristor can only be opened by 3V, we can get N_{s2} =1 Then the auxiliary winding is $N_{aux} = \frac{N_s*(Vcc+V_f)}{V_o+V_f} = 6.462$. Finally, we set the integer N_{aux} =7. According to $L_P = u_o*u_r*A_e*N_P^2/l_g$, we can get the length of the air gap $l_g = \frac{u_o*u_r*A_e*N_P^2}{L_P} = \frac{4*3.14*10^{-7}*1*170*28^2}{713} = 0.235mm$.

4.2.5. The wire diameter of the winding. Because the wire diameter of the winding depends on the effective value of current, the effective value of the primary current is $I_{prms} = \sqrt{\frac{Dm}{3}(I_P1^2 + I_P2^2 + I_P1*I_P2)} = 0.783A$. Similarly, the effective current of the secondary winding is $I_{srms} = \sqrt{\frac{1-D_m}{3}((\frac{I_P1}{N})^2 + (\frac{I_P2}{N})^2 + (\frac{I_P1}{N})*(\frac{I_P2}{N}))} = 0.124A$. We set the current density $I_j = 6A/mm^2$. Then the primary wire diameter is $d_P = \sqrt{\frac{4*I_{prms}}{\pi*I_j}} = 0.408mm$, while the secondary one is $d_s = \sqrt{\frac{4*I_{srms}}{\pi*I_j}} = 0.162mm$. Therefore, we should set the wire diameters of the primary and secondary winding as $d_P = 0.4$mm and $d_s = 0.17$mm, respectively. For power supply winding, it is enough to set the wire diameter as 0.12mm because the current is very small. In order to improve the efficiency and reduce the leak inductance, we use sandwich winding method to wind the transformer. The primary winding comes first. The secondary one comes next. Then it is the auxiliary winding. The primary one comes last.

5. Conclusion

Flyback topology is adopted to design a switching power supply in this paper. In order to realize the given design indicators, the framework is finished with a PWM integrated circuit as the core. It is the key procedure to select the transformer while designing the power supply. Therefore the selection method of the transformer, which is called A_P method, is discussed in detail. Then we decide to choose PQ3220 core according to the solved total apparent power and A_P value. After the calculation of some other significant parameters, we get the details of the transformer as follows: the transformer ratio of 7, the primary current of 1.619A and 0.971A, the primary inductance of 713 μH, the air gap of 0.235 mm etc. These calculations and parameter settings can guarantee the design demand and the efficiency of the flyback switching power supply.

Referneces

[1] Redondo, L. M., and J. F. Silva. "Flyback Versus Forward Switching Power Supply Topologies For Unipolar Pulsed-Power Applications." IEEE Transactions on Plasma Science 37.1(2008):171-178.

[2] Nambiar, Vishnu P., A. Yahya, and T. R. Selvaduray. "SPICE modelling of a valley switching flyback power supply controller for improved efficiency in low cost devices." IEEE International Conference on Circuits and Systems IEEE, 2013:10-14.

[3] Yazdani, Mohammad Rouhollah, H. Farzanehfard, and J. Faiz. "Conducted EMI modeling and reduction in a flyback switched mode power supply." Power Electronics, Drive Systems and Technologies Conference IEEE, 2011:620-624.

[4] Padiyar, U. Suprabha, and V. Kamath. "Design and implementation of a universal input flyback converter." International Conference on Electrical, Electronics, and Optimization Techniques IEEE, 2016:3428-3433.

[5] Perrin, Remi, et al. "High-Temperature GaN Active-Clamp Flyback Converter With Resonant Operation Mode." IEEE Journal of Emerging & Selected Topics in Power Electronics 4.3(2017):1077-1085.

[6] Kaliyaperumal, Deepa, H. M. Saju, and M. V. Kumar. "High Efficiency Single Output ZVS–ZCS Voltage Doubled Flyback Converter." Journal of the Institution of Engineers 97.2(2016):139-146.

[7] Hapidin, Dian Ahmad, et al. "Design and Development of a Series-configuration Mazzilli Zero Voltage Switching Flyback Converter as a High-voltage Power Supply for Needleless Electrospinning." Procedia Engineering 170(2017):509-515.

[8] Hu, Liangdeng, et al. "Research on Driver for High Voltage Input Double-Transistor Flyback Auxiliary Power Supply." Transactions of China Electrotechnical Society 30.4(2015):161-170.

[9] Padiyar, U. Suprabha, and V. Kamath. "Design and implementation of a universal input flyback converter." International Conference on Electrical, Electronics, and Optimization Techniques IEEE, 2016:3428-3433.

Acknowledgments

This work was supported in part by the Nurturing Scientific Research Projects of Hubei University of Science and Technology(2016-18X048), the Team Plans Program of the Outstanding Young Science and Technology Innovation of Colleges and Universities in Hubei Province(T201513) ,the Educational Commission of Hubei Province of China (D20172803), the teaching research project of Hubei university of science and technology (2016-XA-004, 2017-ZYB-00) and the Teaching Research Project of Hubei University of Science and Technology(2017-XB043).

Study on Basic Properties of Inorganic Isolation Partitioned Polystyrene Insulation Board

Jusong Zhang[1] Ruge Chen[1] and Xin Wang[2]

1.College of materials science and engineering,Shenyang jianzhu University,Shenyang Liaoning 110168;
2 .Liaoning super insulation technology Co. Ltd,liaoning dandong 118000
E-mail:739007960@qq.com

Abstract. In order to ensure the safety of the inorganic isolation partitioned polystyrene insulation board in applying, this paper studied its heat retention,flame retardancy, water absorption, fatigue performance, and microscopic analysis.The test results showed that the apparent density of the board is about 120kg/m^3 would meet the demands of great flame retardancy and good thermal insulation; the volume water absorption rate is curve growing in the early stage, it drops slightly and keeping stable after reach the maximum, the maximum is less than 8%; after 1500 times 12.1KN/ ㎡ impact on the board, the internal inorganic material is destroyed basically and remain unchanged in the substrate level; x-ray diffraction analysis result showed that there were $Mg(OH)_2$ and hydrated magnesium sulfate in the solidified inorganic material .

Key words. Inorganic isolation partitioned polystyrene insulation board; Heat retention; Flame retardancy;Water absorption; Fatigue

1. Introduction

Inorganic isolation partitioned polystyrene insulation board(shorter form TEPS) is a composite insulating board, which is made up of mixed inorganic cementitious materials slurry and the base plate through impregnation and vacuum adsorption processes.It has good thermal insulation, chemical corrosion resistance and excellent mechanical strength.TEPS not only retains the good physical properties of the traditional benzol board, but also significantly improves the flame retardancy of the benzol board.

In order to improve the flame retardancy of EPS, Ruijuan Liu and Zhensen Liu[1] made a flame-retarded, economical and light composite polystyrene insulation board by painting the surface of mature benzol board and steaming curing slurry (mainly consisting of fast-hardening cement and modified resin).Hongyi Jiang and Zhonghai Yin[2] compacted silicate fibers, expanded perlite and polystyrene particles to prepare inorganic rubber powder polystyrene particles thermal insulation materials. This material has low self-weight, low thermal conductivity and recycled waste fly ash and polystyrene board. Miao Zhang[3] used EPS as substrate, EG as flame retardant and PF as coating agent to prepare a new thermosetting insulation board, which improved the compressive strength and flame retardancy of the insulation board. However, previous studies focused on the flame retardant performance of thermal insulation panels, and less on the performances of modified materials in application.

TEPS is a new type of modified flame retardant thermal insulation material. Firstly there is a contradiction between flame retardant and thermal insulation. Secondly, water absorption is one of the important properties to measure the durability of thermal insulation materials that has not been clearly studied at present. Thirdly, the modified insulation board is brittle, and its ability to withstand negative wind pressure has not been studied. Therefore, this paper studied the thermal insulation, flame retardancy, homogeneity and water absorption of TEPS, also studied the fatigue resistance of its mechanical properties by simulating the effect of wind pressure. The related problems are explored by micro-means.

2 Raw materials and measures

2.1 Raw materials

The base plate, EPS, TEPS made in Liaoning super insulation technology Co. Ltd; sizes are 1000mm × 600mm × 100mm. The segregated material is magnesium oxysulfide-based inorganic materials. The main components of the plastering mortar for exterior wall insulation materials are cement, fine sand, crack resistant fibers and polymer additives.

2.2 Measures

2.2.1 Thermal Conductivity and Combustion Calorific Value Test

According to GB/T 10295-2008 and GB/T 14402-2007, TEPS with apparent densities of 80 kg/m^3, 100 kg/m^3, 120 kg/m^3, 140 kg/m^3 and 160 kg/m^3 were selected for thermal conductivity test and combustion calorific value test.

2.2.2 Mechanical Water Absorption Test

According to GB/T 8810-2005, four TEPS specimens were immersed below 5 mm of water level. Begin to measure the weight of specimens every hour, after 5 hours weigh the specimens every 2 hours, and then weigh the specimens every 3 hours⋯⋯With the prolongation of soaking time, the weighing interval time is adjusted accordingly until the weight does not change. Responding the water absorption character of TEPS by testing volume water absorption of difference time point.

2.2.3 Fatigue test design

Wind loading is the main external force acting on insulation board. By consulting relevant data[4], the maximum negative wind pressure in China is 12.1kN/㎡.Wind force is 121N equals wind pressure multiplied by specimen area. Using 121N force to impact the specimen 0 times, 500 times, 1000 times and 1500 times respectively.Then do pull-out test to measure the tensile strength.

2.2.4 X-ray diffraction analysis

In order to explore the solidified phase of inorganic materials,grinding inorganic materials and testing in D8 X-ray diffractometer to determine the phase composition of inorganic materials.

3 Test results and analysis

3.1 Thermal conductivity and combustion calorific value

The thermal conductivity and combustion calorific value corresponding to different apparent densities of TEPS obtained by the above test method are shown in Fig.1.

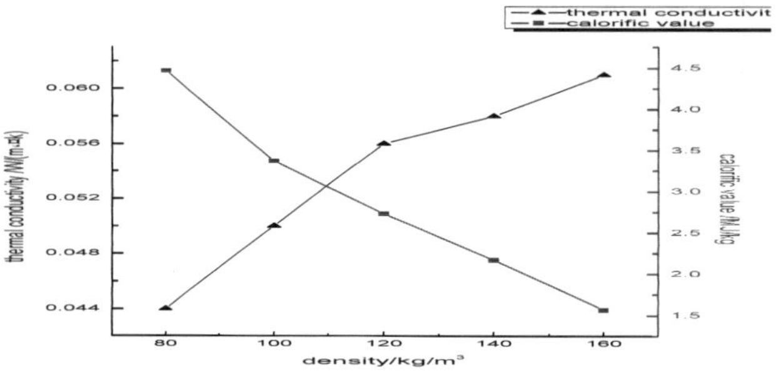

Fig.1 The thermal conductivity and calorific value corresponding of different density TEPS

As can be seen from Fig. 1, with the increase of apparent density, the thermal conductivity of TEPS increases and the combustion calorific value decreases.This is because the higher the apparent density of TEPS is, the more inorganic materials are added, which results in the increase of thermal conductivity and the decrease of calorific value.When the apparent density of TEPS is 120 kg/m3, its thermal conductivity is 0.56W/(m.k) and its calorific value is 2.6MJ/kg.To meet the engineering requirements for thermal insulation and flame retardancy. Therefore, it is suggested that TEPS with density of 120 kg/m3 be used in actual construction.

3.2Water absorbency

The volumetric water absorption of TEPS at different time points measured by the above test method is shown in Fig. 2.

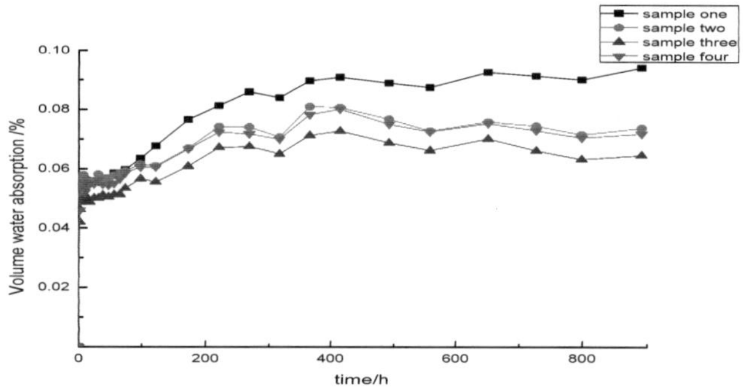

Fig. 2 The curve of volume water absorption with time

As can be seen from Figure 2, the volumetric water absorption increased dramatically in the first 1 h.Volume water absorption rate of 1-400 h increases first by curve and then by wave to maximum value. The maximum volume water absorption rate of insulation board at 400 h, then wave reduced to steady, add up to 900 h.

The specimen dried in the first hour absorbed water rapidly and abundantly. 1-400h specimen still absorbs water mainly but speed slows down.Water immersion the benzene particle gaps that without inorganic filling dense. At the same time, MgO hydration into $Mg(OH)_2$, $MgSO_4$ sopping up become hydrated magnesium sulfate may increase the quality of specimens. To ensure the accuracy of test data, When weighing the quality of specimens would wipe the surface moisture of specimens with wet towels according to the specifications. This process may remove $Mg(OH)_2$ that generated and wrapped on the surface. Additionally, partially soluble crystals may dissolve with the prolongation of immersion time. These lead to the fluctuation of sample quality after immersion.

The maximum volume water absorption rate of individual TEPS samples is about 8% at 400 h. After test verification and according to relevant standards, the board is suitable for external insulation materials.

3.3 Fatigue property

Obtaining the tensile strength of TEPS, benzol board, substrates immersed in water and placed naturally after different impact times by the above test method, the test results are shown in Fig. 3.

As can be seen from the TEPS natural polyline in Figure 3, the initial strength of the plate is fully specified in the requirement that the tensile strength perpendicular to the plate surface should not be less than 0.1MPa. The tensile strength decreases continuously with the increase of impact times.After one thousand and five hundred times shock, the strength reaches 0.07 MPa. From soaking TEPS line diagram, it can be seen that immersion has an obvious effect on the strength of the board, after one thousand times of impact strength, the immersed board has basically reached 0.07 MPa.

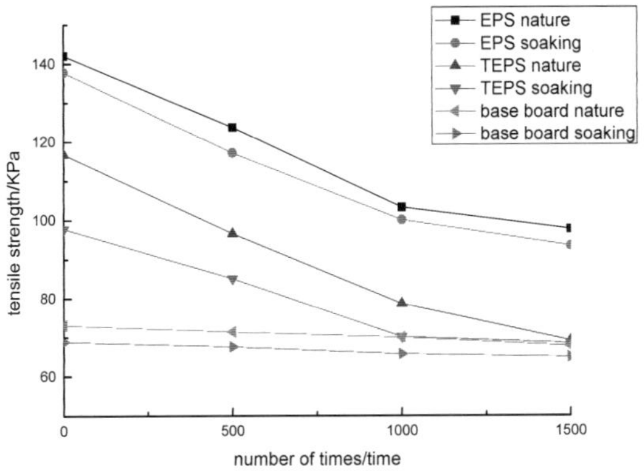

Fig.3 The tensile strength

The overall tensile strength of TEPS is between the substrate and EPS. Naturally TEPS and EPS polylines are almost parallel, intersect with the base plate broken line finally. The tensile strength of immersed TEPS decreased to the level of substrate ahead of time. This phenomenon shows that the inorganic materials inside the TEPS plate are destroyed continuously during the impact process, inorganic materials destroyed basically by one thousand and five hundred times shock, and immersion accelerated fatigue of TEPS. From the strength broken line of the substrate, it can be seen that the impact of the substrate is not significant. Therefore, it can be inferred that the tensile strength of TEPS can be maintained at 0.07 MPa whether undergoes how many times shock. If the construction operation is standardized, the board is safe in use,there will be no internal fracture of insulation board.

3.4 X-ray diffraction analysis

The XRD diffraction pattern is shown in Fig. 4.

As shown in Figure 4, the main components of inorganic materials are MgO,$Mg(OH)_2$, SiO_2, $(Mg,Fe)CO_3$, hydrated magnesium sulfate, basic magnesium sulfate and so on. It is proved that $Mg(OH)_2$, hydrated magnesium sulfate, basic magnesium sulfate and other substances are formed after the reaction of inorganic materials.

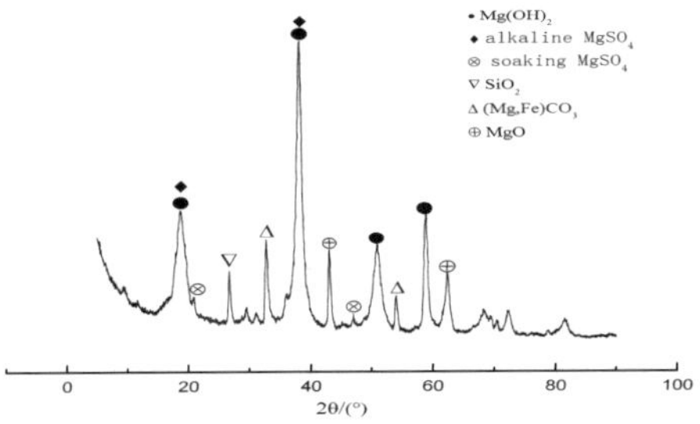

Fig. 4 X-ray diffraction analysis

4 Conclusion

(1)When the apparent density of TEPS is about 120 kg/m^3, it can meet the requirements of class A flame retardant and good heat preservation performance at the same time.

(2)The volume water absorption of TEPS increased by a curve in the early stage, then the wave decreased and remained stable after the wave reached its maximum value. The maximum volumetric water absorption does not exceed 8%, it is suitable for use in cold areas.

(3)The inorganic materials in TEPS were basically destroyed after 1500 shocks of 12.1kN/strength, keeping around 0.07 MPa. It is safe in application.

(4)The reaction of inorganic material components produces $Mg(OH)^2$, hydrated magnesium sulfate, basic magnesium sulfate, which may cause fluctuation of water absorption quality.

reference documentation

[1] Liu Ruijuan and Liu Zhensen 2000 *Development of Composite Polystyrene Thermal Insulation Board* (Concrete and Cement Products) pp 45-46

[2] Jiang Hongyi and Yin Zhonghai 2002 *Research on Inorganic Rubber Powder Polystyrene Particle Thermal Insulation Material* (Journal of Wuhan University of Technology) pp 1-3

[3] Zhang Miao 2017 *Preparation and properties of halogen-free inorganic coated styrene sheets* (North University of China)

[4] Beijing Zhenli Energy Saving and Environmental Protection Technology Co. 2011 *Ltd Theory and Application of External Thermal Insulation Technology* [M]. (Beijing: China Building Industry Press) pp 149-155

[5] Zhu Yanfang 2002 *Durability Test of External Thermal Insulation System* (New Building Materials) pp 30-31

[6] Li Bo and Shao Lingling 2008 *XRD Identification of Alumina and Aluminium Hydroxide* (Inorganic Salt Industry) pp 54-57

Fund projects:National Natural Science Foundation Project (50274025);Science and Technology Tackling Project of Liaoning Education Department (11211004)

Author brief introduction:Zhang jusong (1962-) Male Professor Doctor's degree Mainly engaged in new building materials research

IOP Conf. Series: Materials Science and Engineering **544** (2019) 012056 doi:10.1088/1757-899X/544/1/012056

Effect of Electrospun PVDF-Fibers Orientation for Vibration Sensing

Federico Fabriani, Krishna Chytanya Chinnam, Arnaldo Casalotti and Giulia Lanzara [1]

1 Engineering Department, University of Rome, Roma TRE, Rome, 00146, ITALY
E-mail: giulia.lanzara@uniroma3.it

Abstract. Advanced Structural Health Monitoring (SHM) systems require the use of high performing, ultra-thin and light weight flexible sensing devices to reduce impact in a hosting material or structure. In this paper piezoelectric polyvinylidene difluoride (PVDF) electrospun fibers were investigated to realize advanced vibration sensors. The microscale structure of the fabricated sensors is of a porous nature and consists of an aggregation of microfibers that are either randomly distributed or oriented. Being piezoelectric these fibers are capable to detect mechanical stress or strain over time in terms of a voltage signal. To investigate the reliability of piezoelectric fibers as vibration sensors, tests of a dynamic nature were conducted. It is found that the developed sensors are capable to detect a cantilever vibration over time with great accuracy and that the microfibers orientation is an important factor to maximize the sensor performance.

1. Introduction

Structural health monitoring (SHM) systems are very promising to improve reliability and integrity of structures during their life-time. These systems typically consist of sensors that are surface mounted or embedded into a structure. Sensors signal is gathered by a conditioning system and analysed by ad-hoc designed algorithms that are capable to reinterpret the acquired signal as a structural health status. Various physical changes, such as displacement, velocity, acceleration, tension, deformation, force, can be used for this scope. Piezoelectric-based devices are among the most widely used for SHM applications. These devices are made of piezoelectric materials that are characterized by an electromechanical response: they produce a voltage when they are strained or, vice versa, they strain when a voltage is applied. In other words, piezoelectric-based devices can work as both sensing and actuating elements within an SHM system. The working principle is that sensors and actuators are mounted on a structure in a specific and predefined array. Actuators send a diagnostic signal (strain waves) into the hosting structure and this signal, that is influenced by the structural integrity, is then detected by the surrounding sensors. The problem is that the most common commercially available piezoelectric sensors are relatively large in size, require time-consuming assembling procedures and can only be applied in a small number on large structural surfaces (low sensors density). Device size is critical not only in terms of an overall weight increase that can impact advanced light weight structures, but also because it can compromise structural integrity when sensors/actuators are conceived as embedded devices. In addition, piezoelectric ceramics are rigid in their nature and this limits their application to rigid structures that are affected by ultra-low strain levels.
The above limitations could be overcome by reducing the sensors/actuators size, weight and rigidity. To this end, the excellent traits of polyvinylidene difluoride (PVDF), low production cost, high

[1] Address all correspondence to this author.

IOP Conf. Series: Materials Science and Engineering **544** (2019) 012056 doi:10.1088/1757-899X/544/1/012056

chemical resistance and excellent piezo and pyroelectric properties, makes it very attractive for fabricating lighter and flexible film devices for SHM applications [1]. In the past, PVDF has been widely used for pressure [2], strain [3] and vibration sensing [1,4] as well as to implement SHM systems for composite structures [5].
PVDF has four crystalline phases, one of which, the β phase, is characterized by the highest electro-active properties. In nature this material is present in the *non-polar* phase and to attain its piezoelectric properties it must undergo into a *polarization process*, which involves subjecting the material to an elongation while applying a high voltage. This process leads to an orientation of the internal material dipoles. However, despite the excellent properties of this material, its usage for SHM systems carries most of the above-mentioned limitations except for the acquired device flexibility and partial weight reduction. To make PVDF a major candidate for SHM embedded systems, it is mandatory to drastically reduce the material size and density. Both of which have in fact the potential to further reduce the overall device weight.
In this paper we propose to change the PVDF film microstructure taking advantage of an electrospinning process. Electrospinning is a technique by which a polymer, in solution form, can be spun into smaller-diameter fibers (micro-nanometres) due to a very high electric field. One of the appreciable features of electrospinning is the simplicity of the process apparatus consisting of a syringe pump, syringe, high-voltage source and a collector. The polymeric solution contained in a syringe fitted to the syringe pump passes through the needle with a regulated and a steady flow. A very high electric field is applied between the needle and the collector. As the voltage increases, the meniscus of the solution at the needle tip deforms, creating a Taylor cone. When the electric field reaches a critical value, the repulsive forces overcome the forces of surface tension and from the Taylor cone a jet of polymer emanates and dries up during flight. The formed fibers get deposited on a collector as aggregated fiber [6]. The use of this technique to electrospin a PVDF solution [7,8,9] has the advantage that the mechanical elongation and the strong electrical field applied during electrospinning, tend to align the PVDF dipoles producing a direct *in-situ* polarization of the material [10]. Therefore, electrospinning transforms the non-polar alpha phase of PVDF into the electro-active polar beta phase with an ease.

2. Materials and methods

The PVDF solution was created by mixing together PVDF powder, acetone and N, N-dimethylacetamide (DMAc), produced by the Sigma-Aldrich company. The powder was dissolved in a mixture of acetone/DMAc. A syringe, that was filled with a polymer solution, was pressurized with a syringe pump to set the solution flow rate. A high voltage (12 KV) between the needle tip and a metal plate that collects the nanofibers (collector), was applied with a power supply. The needle to collector distance was of 20 cm. The electrospinnig setup is shown in figure 1. Changes to the collector configuration were done to produce different types of fibers. In one case, fibers with a random orientation were obtained using a rectangular aluminium foil as a collector (Fig 1a). In the second case, aligned fibers were obtained using two co-planar aluminium electrodes, placed at a specific distance, as collectors (Fig 1b).

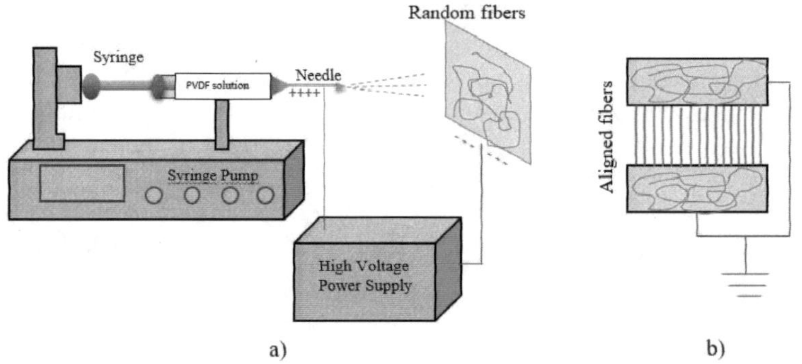

Figure 1: 1a) Set-up to electrospin random PVDF fibers; 1b) Collector set-up to obtain aligned PVDF fibers

The sensing device was fabricated by sandwiching the electrospinned fibers between two copper electrodes (Figure 2) onto which electrical wires were soldered for electrical measurements.

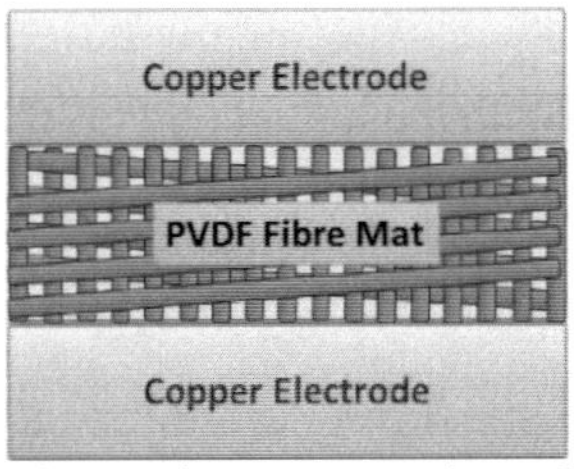

Figure 2: Schematic cross section of the sensor.

3. Results and Discussion

Figure 3 shows an optical microscope image of the resulting PVDF fibers. The fibers showed an average diameter of 3μm. As expected, the fibers that were electrospinned using a single collector, were characterized by a random distribution while the fibers that were electrospinned using the co-planar collector configuration, showed a preferential direction of alignment (oriented fibers). The microscopic characterization highlights that with this fabrication method the resulting PVDF film shows a porous microstructure and an overall film thickness of 200 μm.
The functional capability of the microfibers and their piezoelectric response were studied by testing the devices under dynamic loading conditions.

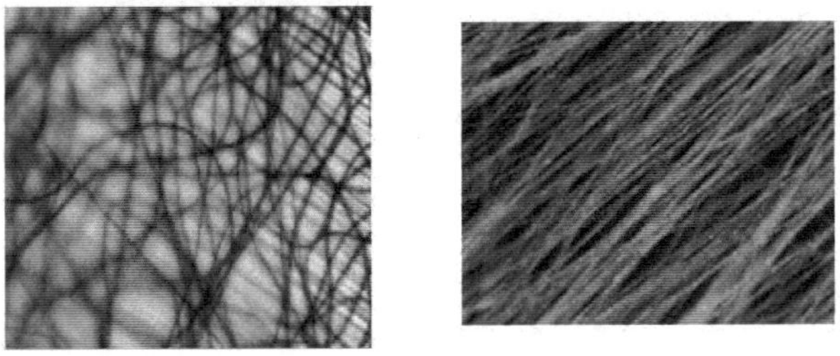

Figure 3: Microscope images of the PVDF porous films formed by random (top image) or aligned (bottom image) fibers.

The fabricated sensors were bonded to a fiber-glass composite cantilever using an epoxy resin (epoxy resin L20). Three sensors of each type (random and aligned) were tested.
The functional capability of the sensors was tested with the setup schematically represented in Figure 4. The cantilever was impulsively stressed by imposing a constant vertical displacement at the free end. The cantilever was then let free to oscillate over time while the electrical output of the sensor signal was recorded with a Digital Storage Oscilloscope (2556, BK Precision). The From the graphs it sensor response was compared with that acquired by a Polytec Laser Vibrometer that was focused at the free end of the cantilever throughout the test duration.

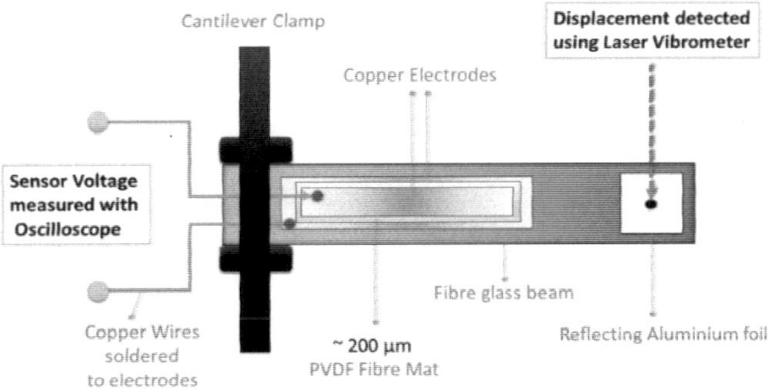

Figure 4: Schematic representation of the test setup.

The sensor time history and its comparison with the laser output is shown in Figure 5 and Figure 6 for the random and oriented fibers, respectively. These data clearly highlight the reliability of the sensor. The signals of the two detection systems are found to be in perfect phase over time. The frequency of the first vibration mode of the cantilever was calculated by means of the Fourier Transforms of the recorded signals. Table 1 shows the frequencies that were calculated from the sensor output voltage and from the laser output.

Table 1: Dynamical test results

	#	f_{laser} [Hz]	f_{sensor} [Hz]	Δf [%]	V_{max} [mV]
Random Fibers	1	90.89	90.24	-0.714	113.128
	2	93.09	93.99	0.967	84.421
	3	95.87	95.04	-0.866	120.142
Aligned Fibers	4	102.97	104.17	1.161	104.275
	5	84.21	84.86	0.761	138.991
	6	94.26	93.99	0.285	109.000

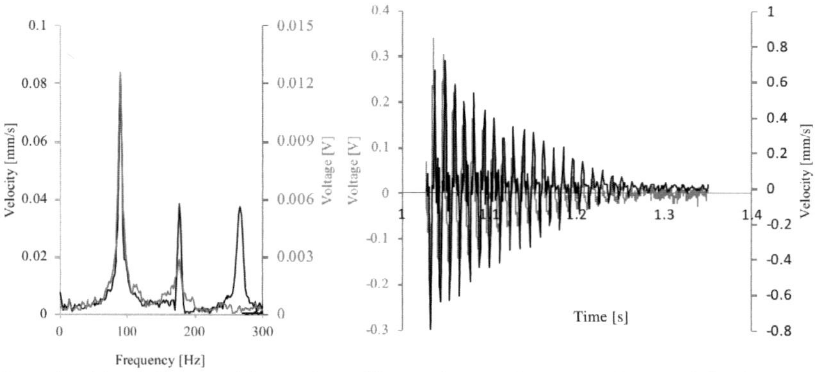

Figure 5: top) Fourier Transform with random fibers; bottom) comparison of the laser response and of the electrospun sensor.

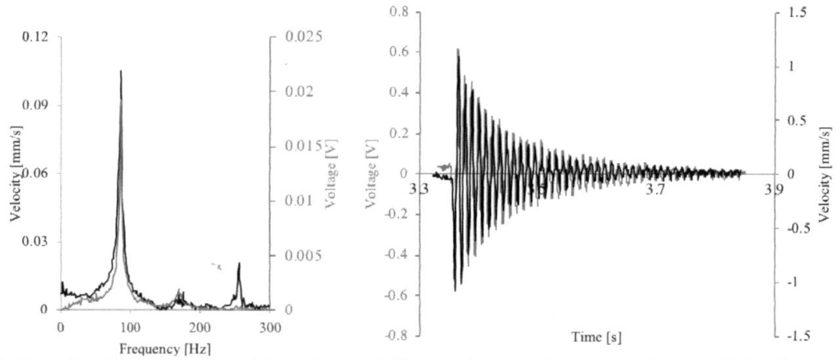

Figure 6: top) Fourier Transform with oriented fibers; bottom) comparison of the laser response and of the electrospun sensor.

As seen in all the tests carried out in this study, the various sensors have a frequency of the first mode of vibration very close to that of the cantilever beam, this indicates their correct operation. Additionally, it is observed that the sensor that has aligned fibers shows a better response compared to that made with random fibers. This can be observed from the time history response as well as the Fourier transform whose peak amplitude is closer to that derived from the laser signal. The percentage frequency variation of the detected signal with that of the laser also highlights the superiority of the sensor with aligned fibers. The latter fibers are in fact characterized by an average 0.7% variation against the average 0.85% of the randomly distributed fibers.

This study shows for the first time that electrospun piezoelectric fibers can be used to monitor dynamic mechanical deformation with a high accuracy. In particular the frequency detected by the PVDF sensor was found to be identical to the frequency that was measured with a sophisticated laser system designed for this purpose. The frequency variation of the proposed sensor was calculated to be less than 1% than that of the laser. In conclusion the developed sensors represent a promising design solution for advanced structural health monitoring systems.

4. Conclusion

This study shows for the first time that electrospun piezoelectric fibers can be used to monitor dynamic mechanical deformation with a high accuracy. In particular the frequency detected by the PVDF sensor was found to be identical to the frequency that was measured with a sophisticated laser system designed for this purpose. The frequency variation of the proposed sensor was calculated to be less than 1% than that of the laser. In conclusion the developed sensors represent a promising design solution for advanced structural health monitoring systems.

ACKNOWLEDGMENTS

The research leading to these results was supported by the European Research Council under the European Union's Seventh Framework Program (FP/2007-2013)/ERC Grant Agreement n. 308261.

REFERENCES

[1] F. Fabriani, K. C. Chytanya, G. Iucci and G. Lanzara: "Characterization of Electrospun PVDF Fibres for Sensing and Actuation", doi:10.3390/ICEM18-05404, 28 June 2018;

[2] D. Mandal, S. Yoon and K. J. Kim: "origin of Piezoelectricity in an Electrospun Poly (vinylidene fluoride-trifluoroethylene) Nanofiber Web-based nanogenerator and Nano-Pressure sensor", Macromolecular Journals, DOI: 10.1002/marc201100040, 2011, 32, 831-837.

[3] J. Sirohi and I. Chopra: "Fundamental Understanding of Piezoelectric Strain Sensor", DOI 10.1106/8BFB-GC8P-XQ47-YCQ0, April 2000;

[4] T. Lei, L. Xu, Z. Zhan, J. Du, Y. Jiang, G. Zheng, L. Wang and D. Sun, "Direct Fabrication of Polymer Nanofiber Membrane for Piezoelectric Vibration Sensor", 2011 IEEE;

[5] H. P. Konka, M.A. Wahab, K. Lian:" Piezoelectric fiber composite transducers for health monitoring in composite structures", Sensor and Actuators A: Physical, 2013;

[6] F. Mokhatari, M. Latifi and M. Shamshirsaz: "Electrospinning/electrospray of polyvinylidene fluoride (PVDF): piezoelectric nanofibers", DOI: 10.1080/00405000.2015.1083300, 18 September 2017;

[7] G. Luo, K.S. Teh, Y. Liu, X. Zang, Z. Wen & L. Lin (2015): "Direct.write, self-aligned electrospinning on paper for controllable fabrication on three-dimensional structures", Department of Mechanical Engineering, University of California, Berkeley, California;

[8] J. Zheng, A. He, J. Li and C.C. Han: "Polymorphism Control of Poly (vinylidene fluoride) through Electrospinning", DOI: 10.1002/marc.200700544, 2007;

[9] Z.H. Liu, C. T. Pan, L.W. Lin, J.C. Huang and Z.Y. Ou: "Direct-write PVDF nonwoven fiber fabric energy harvesters via the hollow cylindrical near-field electrospinning process": DOI:10.1088/0964-1726/23/2/025003, 13 December 2013;

[10] C. Chang. Y. K. Fuh and L. Lin:" A direct-write piezoelectric PVDF nanogenerator", 15th International Conference on Solid-State Sensors, Actuators and Microsystems, Denver, June 2009.

IOP Conf. Series: Materials Science and Engineering **544** (2019) 012057 doi:10.1088/1757-899X/544/1/012057

Multifunctional Glass-Fiber Composite with Self-Monitoring Capability

Federico Fabriani, Roberto Previti and Giulia Lanzara [1]

1 Engineering Department, University of Rome, Roma TRE, Rome, 00146, ITALY
E-mail: giulia.lanzara@uniroma3.it

Abstract. Composites that are capable to provide multiple functionalities represent a fundamental step toward the realization of light weight, high performing and autonomous materials. In this paper a piezoelectric polyvinylidene difluoride (PVDF) electrospun fabric is investigated as sensing layer for making a self-monitoring composite material. The micro-scale structure of the fabric consists of an aggregation of micro-fibers that are randomly distributed forming a porous film that allows a full absorption of the epoxy matrix as required for multi-layered composites. The application of a force on the composite induces a charge separation within the atomic structure of the embedded PVDF fabric generating an electrostatic output voltage. These unique features allowed to continuously monitor vibrations under dynamic conditions.

1. Introduction

Piezoelectrics are commonly used in smart structural systems as both sensors and actuators [1, 2], and they can be classified into ceramic or polymeric materials.
When this material is used as a sensor the direct piezoelectric effect is exploited: the application of an external force produces a charge separation within the atomic structure of the material generating an electric output voltage. Vice-versa, a piezoelectric can be actuated by applying a high electric field which induces a deformation in the material (reverse piezoelectric effect).
Lead Zirconate Titanates (PZT) is the most commonly used piezoceramic while Polyvinylidene Fluoride (PVDF) is the first polymer whose electromechanical properties were discovered. PZT devices exhibit the typical properties of ceramics like: high elastic modulus, low tensile strength and brittleness. The latter property implies that these ceramics undergo a relatively minor permanent deformation prior to final failure which limits their application to rigid structures that are affected by ultra-low strain levels.
On the other hand, PVDF films are lighter and have a more flexible structure, as they have a Young's modulus lower than PZT, which allows their usage to a wider strain range. Therefore, they are much more suitable for sensing applications in which non-invasive and conformable sensing elements are necessary. In the past, PVDF has been widely used for pressure, strain [2,4] and vibration sensing [1,2,5] as well as to implement Structural Health Monitoring systems for composite structures [6].
PVDF films are not piezoelectric in their nature but require an ad-hoc polarization process (poling) to exploit their piezoelectric properties. This step is complex and time consuming. Recently an electrospinning technique has been used to produce polarized fabrics made of piezoelectric fibers [8,9,10,11]. The beauty of this approach is that it allows to polarize the material in situ, directly during

[1] Address all correspondence to this author.

the manufacturing process, by exploiting simultaneously a high electric field and mechanical stretching [12].

In this paper a self-monitoring composite material is proposed. This is achieved by placing an electrospun PVDF fibers layer between several glass fibers layers, all embedded in an epoxy resin matrix.

The electromechanical behaviour of the composite material is studied in dynamic regime by applying constant frequency cycles. The goal of this work is to verify the capability of the embedded sensing layer to detect the vibrations of the composite structure.

2. Materials and methods

A PVDF powder, with a molecular weight of 534 g / mol, acetone and N, N-dimethylacetamide (DMAc) are produced by the Sigma-Aldrich company and are used without further purification. The powder (wt % 17) is dissolved in a mixture of acetone/DMAc (80:20). Figure 1 shows the electrospinning setup that is used to deposit the fibers. The polymeric solution is put into a syringe and, thanks to a syringe pump, the electrospun amount is selected. A high voltage (12 KV) between the needle tip and a metal plate that collects the nanofibers (collector), is applied with a power supply. The needle to collector distance was of 15 cm.

Fibers with a random orientation are obtained using a rectangular aluminium foil as a collector (Fig 2).

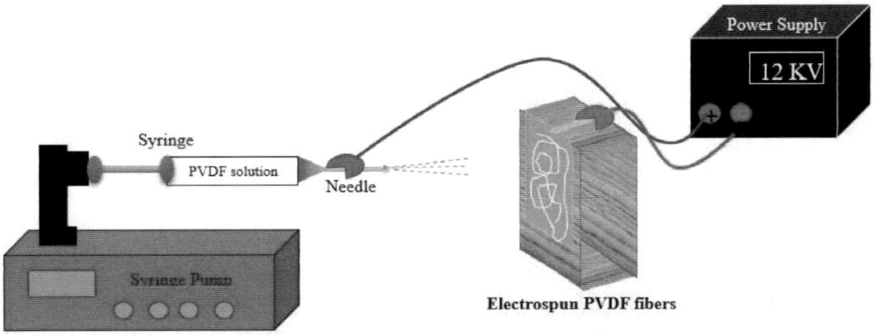

Figure 1: Electrospinning setup.

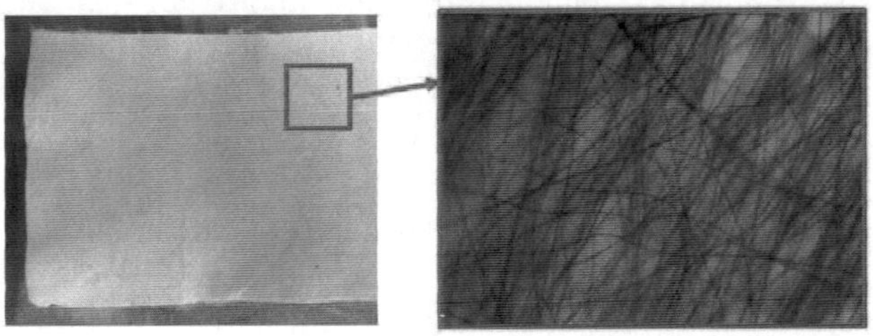

Figure 2: left) fibers mat; right) microscope image of the PVDF fibers.

This porous fibers mat, that works as a sensing layer in the composite, is embedded within a glass-fiber layup during the manufacturing process.

In particular, a composite beam made of four glass-fiber layers and a PVDF fibers layer, is obtained by using the wet lay-up technique. The sensing layer is placed between the first and the second glass-fiber layer to maximize its strain level (Fig. 3).

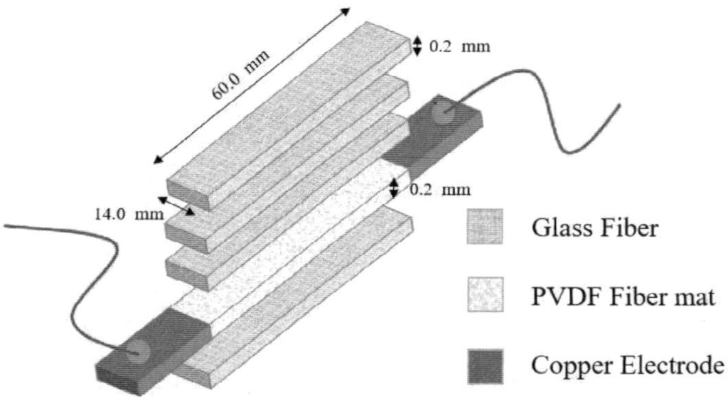

Figure 3: Self-monitoring composite material configuration.

Samples with a variable PVDF layer thickness are made to investigate the effect of this parameter on the composite functionality. In one case study the thickness of the sensing layer equals that of the glass fibers layers (about 200 µm per layer). This is done to have a thickness uniformity in the structure. Then a thickness reduction of the sensing layer is implemented through the integration of 150, 50 and 25 µm thick fabrics within the composite layup. Electromechanical tests are then performed under dynamic conditions by using the Dynamic Mechanical Analyzers (DMA q800, TA Instruments). The sample is placed on a three-point bending clamp and a frequency sweep, three cycles at 100 Hz, is applied (Fig 4a). The aim of this test is to investigate the capability of the sensing layer to detect the frequency imposed by the instrument.

3. Results and Discussion

Mechanical oscillations are imposed by the DMA clamp to the composite and the embedded sensing layer consequently produces a charge separation within the atomic structure of the material, generating an output voltage. The electric signal, generated by the embedded fibers, is acquired with a Digital Storage Oscilloscope (2556, BK Precision). Fourier Transform of the recorded electrical signal is used to calculate the frequency of the composite beam.

This test is carried out for all the samples characterized by a variable thickness of the functional layer. The results show that in all the tested samples the sensing layer is able to perfectly detect the frequency applied by the DMA. An example is reported in figure 4b. This data clearly highlights the functional reliability of the proposed composite material.

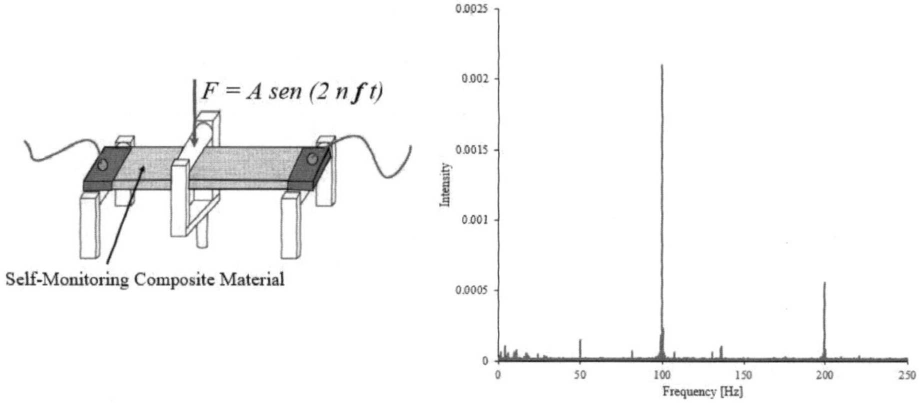

Figure 4: a) Schematic test setup; b) Fourier Transform of the self-monitoring composite material signal

IOP Conf. Series: Materials Science and Engineering **544** (2019) 012057 doi:10.1088/1757-899X/544/1/012057

In Figure 5 the acquired voltage variation over time is reported for each sample type. It is observed that, the voltage recorded by the oscilloscope, calculated as the peak-to-peak distance of the signal (V_{pp}), increases with increasing thickness, reaching a maximum value of 2.6 mV for the thicker sample.
The graphs also show that the electrical signal remains constant throughout the cycle at a frequency of 100 Hz and returns to the initial configuration as soon as the applied stress is removed.

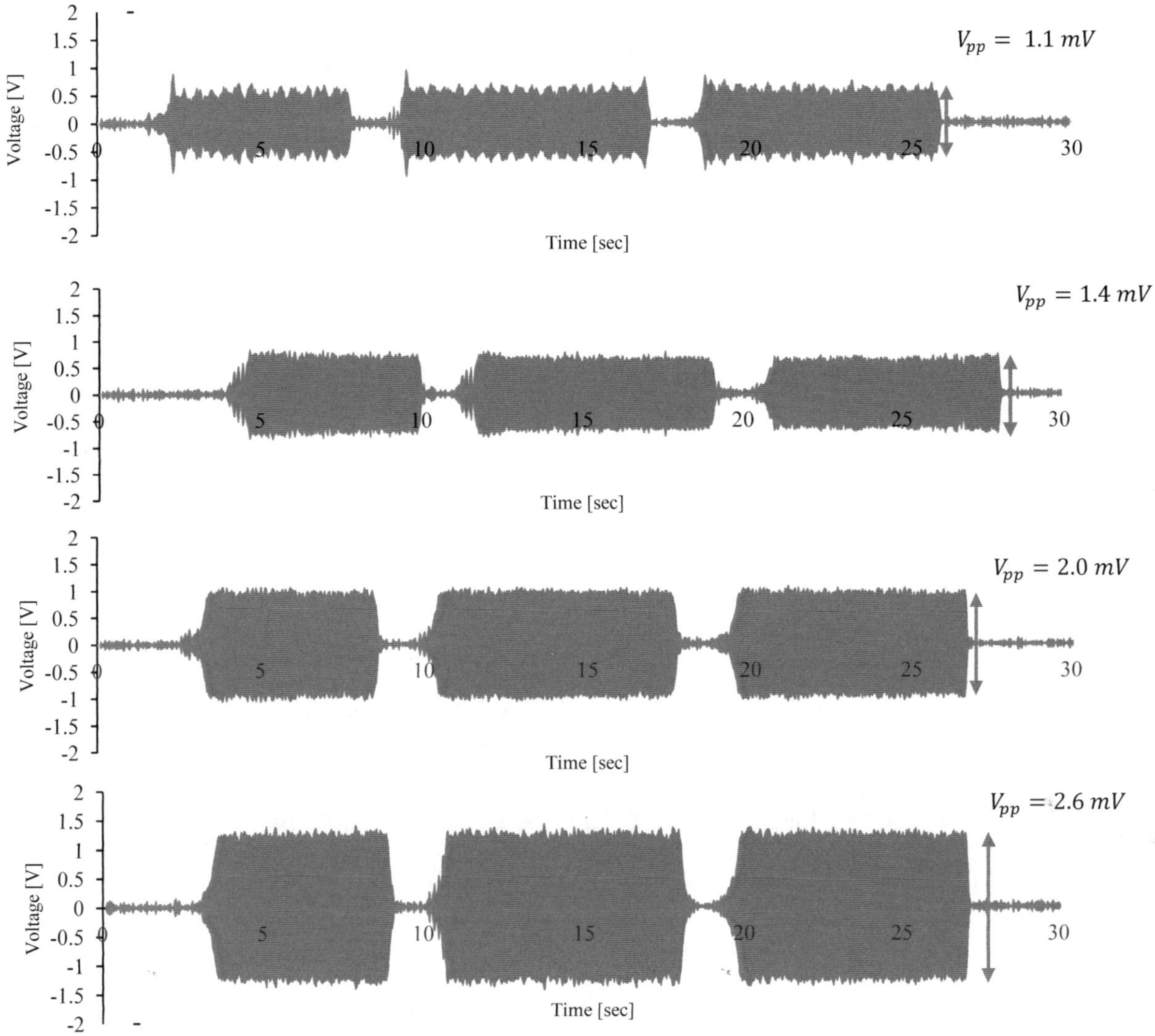

Figure 5: Electrical signal of the composite material at different thicknesses (25, 50, 150 and 200 μm respectively).

4. Conclusion

This study shows how electrospun piezoelectric fibers can be used as sensing layer in a composite material to monitor dynamic mechanical deformation with a high accuracy. The frequency detected by the PVDF fibers mat is found to be very close to the frequency applied by the Dynamic Mechanical Analyzers (DMA). Furthermore, this sensing layer is easily embeddable, does not affect the properties of the structure, such as stiffness and mass, and allows monitoring large areas.

In conclusion this Multifunctional Glass-Fiber Composite represents a promising design solution for advanced materials design, able to monitor continuously stress, strain or vibration in dynamic regimes, to avoid a sudden failure.

ACKNOWLEDGMENTS

The research leading to these results was supported by the European Research Council under the European Union's Seventh Framework Program (FP/2007-2013)/ERC Grant Agreement n. 308261.

REFERENCES

[1] F. Fabriani, K. C. Chytanya, G. Iucci and G. Lanzara: "Characterization of Electrospun PVDF Fibres for Sensing and Actuation", doi:10.3390/ICEM18-05404, 28 June 2018;

[2] K. C. Chytanya, F. Fabriani, G. Iucci and G. Lanzara: "Electrspun PVDF Miniaturized Muscles for Bio Inspirend Morphing Materials", doi:10.1115/SMASIS2018-8054;

[3] A. Shirinov and W. Schomburg: "Pressure sensor from a PVDF film", Sensor and Actuators A: Physical, vol 142, 10 March 2008;

[4] J. Sirohi and I. Chopra: "Fundamental Understanding of Piezoelectric Strain Sensor", DOI 10.1106/8BFB-GC8P-XQ47-YCQ0, April 2000;

[5] T. Lei, L. Xu, Z. Zhan, J. Du, Y. Jiang, G. Zheng, L. Wang and D. Sun, "Direct Fabrication of Polymer Nanofiber Membrane for Piezoelectric Vibration Sensor", 2011 IEEE;

[6] H. P. Konka, M.A. Wahab, K. Lian:" Piezoelectric fiber composite transducers for health monitoring in composite structures", Sensor and Actuators A: Physical, 2013;

[7] F. Mokhatari, M. Latifi and M. Shamshirsaz: "Electrospinning/electrospray of polyvinylidene fluoride (PVDF): piezoelectric nanofibers", DOI: 10.1080/00405000.2015.1083300, 18 September 2017;

[8] G. Luo, K.S. Teh, Y. Liu, X. Zang, Z. Wen & L. Lin (2015): "Direct.write, self-aligned electrospinning on paper for controllable fabrication on three-dimensional structures", Department of Mechanical Engineering, University of California, Berkeley, California;

[9] S.W. Choi, J.R. Kim, Y.R. Ahn, S.M. Jo & E.J. Cairns; "Characterization of electrospun PVDF fiber-based polymer electrolytes": Materials Science and Engineering Program, University of Texas at Austin, Austin, Texas;

[10] J. Zheng, A. He, J. Li and C.C. Han: "Polymorphism Control of Poly (vinylidene fluoride) through Electrospinning", DOI: 10.1002/marc.200700544, 2007;

[11] Z.H. Liu, C. T. Pan, L.W. Lin, J.C. Huang and Z.Y. Ou: "Direct-write PVDF nonwoven fiber fabric energy harvesters via the hollow cylindrical near-field electrospinning process": DOI:10.1088/0964-1726/23/2/025003, 13 December 2013;

[12] C. Chang. Y. K. Fuh and L. Lin:" A direct-write piezoelectric PVDF nanogenerator", 15th International Conference on Solid-State Sensors, Actuators and Microsystems, Denver, June 2009.

IOP Conf. Series: Materials Science and Engineering **544** (2019) 012058 doi:10.1088/1757-899X/544/1/012058

Study of Mechanical Property and Cellular structure Based on the Controllable Crosslinking Polyethylene Foaming Materials

Bihua Xia[1,2*], Wenqiang Xu[1*], Ke Wang[1], Qigu Huang[2], Wenbin Liang[1], Xiaojie Sun[1]

1 National Institute of Clean and Low-carbon Energy Technology, Advanced Materials Center, Beijing 102211;
2 Beijing University of Chemical and Technology, Materials Science and Engineering College, Beijing 100029)
E-mail: xiabihua@nicenergy.com; xuwenqiang@nicenergy.com

Abstract. In this paper, the controllable crosslinking polyethylene foaming materials were established though adding foaming agent additives zinc stearate firstly. Then, the density, expansion ratio, cellular structure, gel content and compression strength of crosslinking polyethylene foaming materials were studied by changing zinc stearate content. Furtherly, those properties were researched after added nucleating agent. Experimental results showed that zinc stearate could facilitate foaming reaction and broad foaming window. Density, compression strength and gel content was decreased with zinc stearate content increased, and the expansion ratio increased with zinc stearate content increased. In addition, the compression strength was improved and cellular structure became uniform after adding nucleating agent. Generally, this material not only has the controllable property, but also has the higher mechanical property and more uniform structure.

1. Introduction

Because of the cellular structure and lightweight advantages, polyethylene foaming materials are widely employed in different and diverse applications, in fields like thermal insulation, cushioning or packaging, polyethylene foaming materials are currently necessary materials[1-4]. When used for these purposes, mechanical properties and cellular structure are usually the most important properties which need considered, where the plastic matrix and gas diffusion from the cells are the fundamental determinant for those properties[5]. While, gas diffusion path could variate with polyethylene modify, for instance, by introducing crosslinks between the polymer chains, a network is formed which inhibit the gas release freely. In addition, network structure also improve heat and chemical resistance, stress cracking and shrinking[6].

Presently, there are several methods for providing network formation in polyethylene foaming material via a crosslinking reaction: with peroxides, by use of radiation and by the use of silanes[7-9]. Because the convenient and cheap of chemical initiator, chemical crosslinking was the more widely using method than other crosslinking method[10]. Masayuki et al. used peroxide crosslinking liner low density polyethylene as addition to reduce heterogeneous deformation during foaming process, and finally a foam material with a uniform cell size distribution was obtained[11]. The investigation demonstrated that chemical crosslinking could improve polyethylene mechanical property and reduce cell size.

Due to add chemical crosslinking agent, the obtaining crosslinking polyethylene foaming material have better mechanical property than polyethylene foaming material, but its expansion ratio was decreased, which is due to the enhancement of the elastic modulus[12]. In addition, the foaming window was fixed and narrowed in this kind of chemical crosslinking polyethylene foaming system (usually, crosslinking reaction occurred in the front and foaming reaction occurred in the after)[12]. A broad foaming window facilitates foam processing and it is a useful way to evaluate the foamability of polymers, while, a narrow foaming window tends to make higher demands on the foaming system, particularly on the temperature control system[13].

In order to improve the expansion ratio, keep good mechanical property and broad foaming window, a controllable crosslinking polyethylene foaming materials were designed in this paper, in which, azodicarbonamide (AC) as the foaming agent, 2,5-dimethyl-2,5-double (tert-butyl peroxide) hexane (DBPH) as the crosslinking agent and zinc stearate (ZnSr) as the foaming agent addition that could facilitate foaming agent decomposition. Rubber Process Analyzer (RPA) was used for the in situ measurement of variation of viscosity and pressure after different amount of ZnSr added. Density and expansion ratio were also tested and analyzed. SEM was used to characterize the cellular structures. Compression strength as the typically mechanical property was also analyzed, especially in comparing with polyethylene foaming materials.

2 Experimental

2.1 Materials

Linear Low Density Polyethylene (LLDPE) DMDA7042 was purchased from Shenhua Baotou Co., Ltd., with melt index 2.6 g/10min and density 0.918 g/cm^3. 2,5-dimethyl-2,5-double (tert-butyl peroxide) hexane (DBPH) from Arkema Co., Ltd was used as the crosslinking agent, it was purchased from Aladdin Chemistry Co., Ltd. Azodicarbonamide (AC) and zinc stearate (ZnSr) were employed as the chemical foaming agent and foaming agent additive, respectively, they were purchased by J&K Chemical. Talcum powder was used as the uncleating agent, which was purchased from Sinopharm Chemical Reagent Beijing Co., Ltd.

2.2 Samples preparation

HAAKE mixture: in this experiment, there were four groups of samples, they were all used 7042 as the matrix resin. The first group of samples' mass ration of 7042, DBPH and AC were maintained a consist value, which is 100:1:3, and the amount of ZnSr were 0, 0.1, 0.2, 0.4, 0.8 and 1.6, respectively. The second group of samples' mass ration of 7042 and DBPH were maintained a certain value, which is 100:1, and the amount of ZnSr were 0, 0.1, 0.2, 0.4, 0.8 and 1.6, respectively. The third group of samples' mass ration of 7042 and AC were maintained a fixed value, which is 100:3, and the amount of ZnSr were 0, 0.1, 0.2, 0.4, 0.8 and 1.6, respectively. And the four group of samples' mass ration of 7042, DBPH, AC and talcum powder were maintained a fixed value, which is 100:1:3:1, and the amount of ZnSr were 0, 0.1, 0.2, 0.4, 0.8 and 1.6, respectively. Then mixed polyethylene resin and those chemical agents, finally, put the obtained mixture into twin screw extruder, the temperature was set at 130℃ and rotation speed was 110r/min. In order to distinguish these haake mixture clearly, their sample code was listed in table1.

Table 1. Sample code list

Serial number	Mixture formula	Sample code
1	7042:DBPH:AC:ZnSr=100:1:3:0	0 ZnSr
2	7042:DBPH:AC:ZnSr=100:1:3:0.1	0.1 ZnSr
3	7042:DBPH:AC:ZnSr=100:1:3:0.2	0.2 ZnSr
4	7042:DBPH:AC:ZnSr=100:1:3:0.4	0.4 ZnSr
5	7042:DBPH:AC:ZnSr=100:1:3:0.8	0.8 ZnSr

6	7042:DBPH:AC:ZnSr=100:1:3:1.6	1.6 ZnSr
7	7042:DBPH:ZnSr=100:1:0	0 ZnSr (DBPH)
8	7042:DBPH:ZnSr=100:1:0.1	0.1 ZnSr (DBPH)
9	7042:DBPH:ZnSr=100:1:0.2	0.2 ZnSr (DBPH)
10	7042:DBPH:ZnSr=100:1:0.4	0.4 ZnSr (DBPH)
11	7042:DBPH:ZnSr=100:1:0.8	0.8 ZnSr (DBPH)
12	7042:DBPH:ZnSr=100:1:1.6	1.6 ZnSr (DBPH)
13	7042:AC:ZnSr=100:1:0	0 ZnSr (AC)
14	7042:AC:ZnSr=100:1:0.1	0.1 ZnSr (AC)
15	7042:AC:ZnSr=100:1:0.2	0.2 ZnSr (AC)
16	7042:AC:ZnSr=100:1:0.4	0.4 ZnSr (AC)
17	7042:AC:ZnSr=100:1:0.8	0.8 ZnSr (AC)
18	7042:AC:ZnSr=100:1:1.6	1.6 ZnSr (AC)
19	7042:DBPH:AC:talcum:ZnSr=100:1:3:1:0	0 ZnSr (tal)
20	7042:DBPH:AC:talcum:ZnSr=100:1:3:1:0.1	0.1 ZnSr (tal)
21	7042:DBPH:AC:talcum:ZnSr=100:1:3:1:0.2	0.2 ZnSr (tal)
22	7042:DBPH:AC:talcum:ZnSr=100:1:3:1:0.4	0.4 ZnSr (tal)
23	7042:DBPH:AC:talcum:ZnSr=100:1:3:1:0.8	0.8 ZnSr (tal)
24	7042:DBPH:AC:talcum:ZnSr=100:1:3:1:1.6	1.6 ZnSr (tal)

Sheet preparation: Set the press machine temperature at 130℃, take out 40g haake mixture and put in the mold with size of 16cm × 14cm × 4cm, then compression-molded under vacuumed condition for 5min at a pressure of 5MPa.

Oven foaming: The pressed sheets were put into an oven and heated for 12min at 240℃ for foaming and crosslinking. Then the samples were taken out and cooled at room temperature for 24h.

2.3 Characterization

Rubber Process Analyzer (RPA) test: A Rubber Process Analyzer 2000p (Alpha Technologies) was used to measure the crosslinking and foaming behavior of polyethylene mixtures.[9,10] The viscosity obtained from RPA instrument for cross-linkable polyethylene compounds is an index of melt strength for the melting resin. In this study, the viscosity is also used for simplicity to represent the degree of crosslinking of the crosslinked articles. The test process is as follows: take 2g polyethylene mixture samples, then put the sample into the test wheel, test conditions were maintained at an angle of 0.5° and an oscillation frequency of 100 CPM (1.667 Hz), 230□ last for 10min.

Expansion Ratio Test: The densities of foamed samples, ρ, were determined by deducing the Archimedes law involving weighing polymer foam in water with a sinker using an electronic analytical balance (HANG-PING FA2104), calculated using equation 1.

$$\rho = \frac{a}{a+b-c} \times \rho_w \tag{1}$$

where a, b, and c are the weights of the sample in air without sinker, the totally immersed sinker, and the specimen immersed in water with sinker, respectively, and ρ_w is the density of water. The volume expansion ratio of the crosslinking polyethylene foaming material, *Rv*, was the ratio of the bulk density of initial mixture density (ρ_0) to that of the crosslinking polyethylene foaming material (ρ), calculated as follows:

$$R_v = \frac{\rho_0}{\rho} \tag{2}$$

IOP Conf. Series: Materials Science and Engineering **544** (2019) 012058 doi:10.1088/1757-899X/544/1/012058

Compression strength test: The compression behavior of the cross-linking polyethylene foaming materials were also analyzed using the universal mechanical testing machine (5965, INSTRON Company, American). The compressive tests at room temperature were performed in air. For this purpose, samples with 50cm × 50cm foursquare size were employed. Stress-strain curves were obtained at a rate of 50 mm/min and the samples were compressed to 50% of its original height for according to the Standard ASTM D695-02.

Scanning Electron Microscope (SEM) Test: Those foam samples were immersed in liquid nitrogen for 10min, and then fractured; the surface was vacuum coated with gold examined using a NOVA Nano SEM450 (oxford instruments). The size and size distribution of the cells were measured using Nano-measure software.

Gel content Test: The gel content was determined by using xylene as the solvent according to ASTM-D2765, oil bath temperature was 170°C. A polyethylene sample of 0.400 ± 0.015 grams by weight was cut into small pieces and placed in a 100-mesh stainless steel wire cloth. The sample was dropped into xylene, boiled and reflux condensation for 8 hours. After the extraction cycle, the sample was dried in a vacuum oven at 160°C for 2 hours. The gel content (%) was calculated as follows:

$$X_{gel} = (m_3 - m_1)/(m_2 - m_1) \times 100\% \quad (3)$$

where m_1 is the weight of stainless steel wire cloth, m_2 is the weight of initial sample and stainless steel wire cloth, m_3 is the total weight of sample and stainless steel wire cloth after extraction.

3 Results and Discussion

3.1 Effect of ZnSr on pressure

In this paper, ZnSr was employed as the foaming agent addition to facilitate the AC decomposition, and there have been some reports about the ZnSr as the catalyst in AC decomposition[14]. In this experiment, Rubber Process Analyzer (RPA) 2000p was used to character the foaming reaction process of AC and ZnSr mixture which inner polyethylene matrix. Fig.1 was the viscosity curve of 0ZnSr and pressure curves of 0ZnSr, 0.1ZnSr, 0.8ZnSr and 1.6ZnSr.

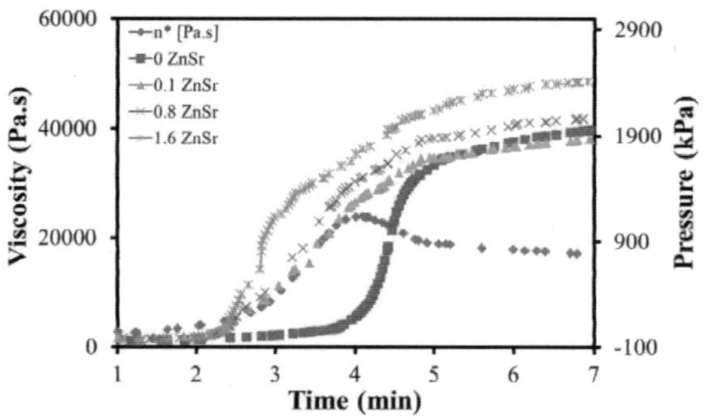

Fig.1 viscosity curve of 0ZnSr and pressure curves of 0ZnSr, 0.1ZnSr, 0.2ZnSr, 0.4ZnSr, 0.8ZnSr and 1.6ZnSr

As shown in Fig.1, the red and the blue curves were the pressure and viscosity changing curves of 0ZnSr, which corresponding to the foaming reaction and crosslinking reaction, respectively. Obviously, when there was no ZnSr in polyethylene mixture, the crosslinking reaction occurred before foaming reaction, and crosslinking reaction began at 3min, and foaming reaction began at 4.3min, the foaming reaction time lasted for 2.7min. After added 0.1% ZnSr, the foaming reaction started forward to 3min, which started at the same time with crosslinking reaction, and the foaming reaction lasted for 4min. When ZnSr content added to 0.8%, the foaming reaction started at 2.8min, the foaming reaction lasted for 4.2min. Increased ZnSr amount to 1.6% sequentially, the foaming reaction started at 2.5min, which even started earlier than crosslinking reaction, and the foaming reaction lasted for 4.5min. This phenomenon indicated that the more the ZnSr content, the earlier the foaming reaction occurs. And it also means that the presence of ZnSr broadened the foaming window of polyethylene mixture. In addition, there was another phenomenon, that was the final pressure value raised with the increased of ZnSr content, the main reason for this phenomenon was that the foaming reaction was less constrained by the crosslinking reaction when ZnSr content increased and resulted in foaming reaction started earlier than crosslinking reaction[15,16].

3.2 Effect of ZnSr on viscosity and gel content

According to those above experimental results, we can know that ZnSr could affect the pressure of crosslinking polyethylene foaming material obviously. However, viscosity is another important characteristic of crosslinking polyethylene foaming material. In this experiment, RPA was used to test the viscosity variation with different ZnSr amount, and the test results were showed in Fig.2a.

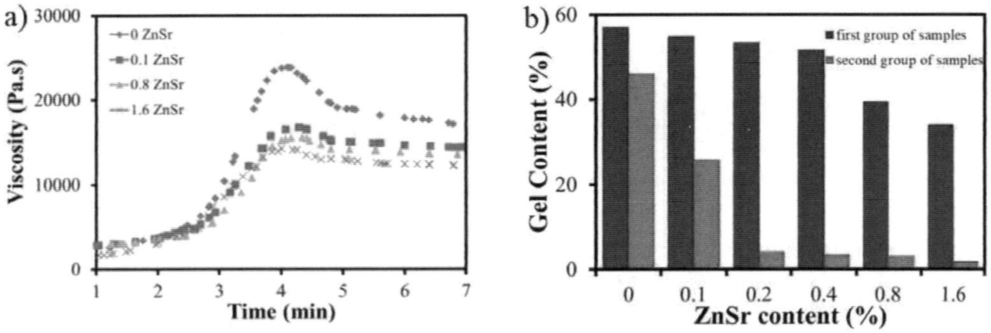

Fig.2 (a)Viscosity variation curves with different ZnSr amount; (b) Gel content of first group of samples and second group of samples with different ZnSr amount

As shown in Fig.2a, when there was no ZnSr in polyethylene mixture, the maximum viscosity was about 25000Pa.s; after adding 0.1% ZnSr, the maximum viscosity decreased to 18000Pa.s; when increased the ZnSr amount to 1.6%, the maximum viscosity decreased to 16000Pa.s continually. These experimental results demonstrated that ZnSr could reduce the crosslinking degree of polyethylene.

In order to verify the effect of ZnSr on crosslinking reaction of polyethylene, gel content was also tested. The experimental results were shown in Fig.2b.The gel content of first group of samples (0 ZnSr, 0.1 ZnSr, 0.2 ZnSr, 0.4 ZnSr, 0.8 ZnSr and 1.6 ZnSr) were shown in Fig.2b. When there was no ZnSr in polyethylene mixture, the gel content was 45.99%; after 0.1% ZnSr was added, the gel content was dropped to 25.81% obviously; when ZnSr amount was increased to 0.2%, the gel content was dropped to 4.15% sharply; when ZnSr amount increased to 0.4%, the gel content was dropped to 3.48%; when ZnSr amount increased to 0.8%, the gel content was dropped to 3.13 % sequentially; and when ZnSr amount increased to 1.6%, the gel content dropped to only 1.83%. With the ZnSr content increased, the gel content was decreased gradually. This experimental result was consisted with the Fig.2 viscosity variation.

In order to avoid the influence of foaming agent AC, the gel content of second group of samples (0 ZnSr (DBPH), 0.1 ZnSr (DBPH), 0.2 ZnSr (DBPH), 0.4 ZnSr (DBPH), 0.8 ZnSr (DBPH) and 1.6 ZnSr (DBPH)) were also tested, which without AC in polyethylene mixture. Test results showed that when there was no ZnSr, the gel content was high up to 57%; after added 0.1% ZnSr, the gel content decreased to 54.83% slightly; when ZnSr content increased to 0.8%, the gel content decreased to 39.42% obviously; when ZnSr amount increased to 1.6%, the gel content decreased to only 33.97%. This decreasing trend of gel content variation with increasing ZnSr content was similar with the first group of samples, and it demonstrated that ZnSr inhibited the crosslinking reaction[17]. The main reason for this phenomenon was that ZnSr captured the decomposition peroxide of DBPH and inhibited the process of crosslinking reaction[18]. In addition, there was another phenomenon that was the gel contents of second group of samples were higher than the first group of samples. This difference demonstrated that foaming agent AC also could inhibit the crosslinking reaction[19,20].

3.3 Effect of ZnSr on density and expansion ratio

ZnSr not only affect the pressure and viscosity of crosslinking polyethylene foaming materials, but also affect the density and expansion ratio. In this experiment, the density and expansion ratio were tested, and the test results were listed in Fig.3.

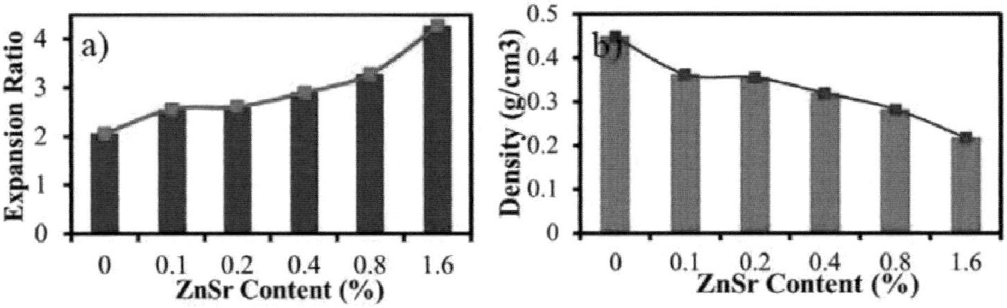

Fig.3 (a) Expansion ratio and (b) density variation with ZnSr content

As shown in Fig.3a, expansion ratios of first group of samples (0ZnSr, 0.1ZnSr, 0.2ZnSr, 0.4ZnSr, 0.8ZnSr and 1.6ZnSr) presented increasing trend with the increased the ZnSr content. When there was no ZnSr, the expansion ratio was only 2.06, after added 0.1% ZnSr, the expansion ratio raised to 2.56; it even raise to 4.26 when the content of ZnSr was up to 1.6%. Oppositely, the density appeared the decreasing trend. When there was no ZnSr, the density was $0.45g/cm^3$, after added 0.1% ZnSr, the density reduced to $0.36g/cm^3$; and it fall down to only $0.22g/cm^3$ when the content of ZnSr was up to 1.6%. These variation trends were similar to the pressure curves (Fig.1).

3.4 Effect of ZnSr on cellular structure

Cellular structure was another important property for crosslinking polyethylene foaming materials. In this experiment, cellular structure was characterized by SEM, Fig.4a, Fig.4b and Fig.4c showed the SEM pictures of 0.2ZnSr, 0.4ZnSr, and 1.6ZnSr, respectively. Cell size was large, the average diameter was about 600μm. And cell shape was irregular and crowd between each other. In order to improve the cell morphology, talcum powder was used as nucleating agent added to the polyethylene mixture. Fig.4d, Fig.4e and Fig.4f showed the SEM pictures of 0.2ZnSr(tal), 0.4ZnSr(tal), and 1.6ZnSr(tal), respectively. The cell became uniform and morphology changed to similar circular structure, and cellular size became small, the average diameters of 0.2ZnSr(tal), 0.4ZnSr(tal) and 1.6ZnSr(tal) were 300μm, 350μm and 450μm, respectively. With the ZnSr content increased, the cellular size appeared the increasing trend, which was similar with the expansion ratio variation trend. In addition, the cell became crowd with the amount of ZnSr increasing. The main reason for this phenomenon was that the foaming reaction preceded by the ZnSr, and cell growth less constricted by crosslinking reaction, finally resulted in the larger cellular size and more crowding cell space[16].

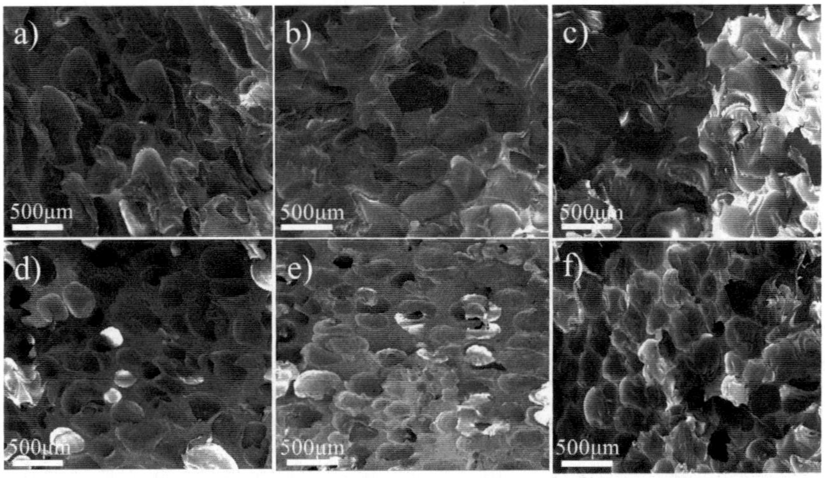

Fig.4 SEM pictures of a) 0.2 ZnSr, b) 0.4 ZnSr, c) 1.6 ZnSr, d) 0.2 ZnSr (tal), e) 0.4 ZnSr (tal) and f) 1.6 ZnSr (tal)

3.5 Effect of ZnSr on compression strength

Due to closely relationship between cellular structure and mechanical property, crosslinking polyethylene foaming materials' mechanical property will be transformed with the cell size and morphology[21,22], which could affect by different amount of ZnSr. In this paper, compression strength was choose as the typical mechanical property for crosslinking polyethylene foaming material, and different content of ZnSr samples' (first group, third group and four group of samples) compression strength and density were also tested.

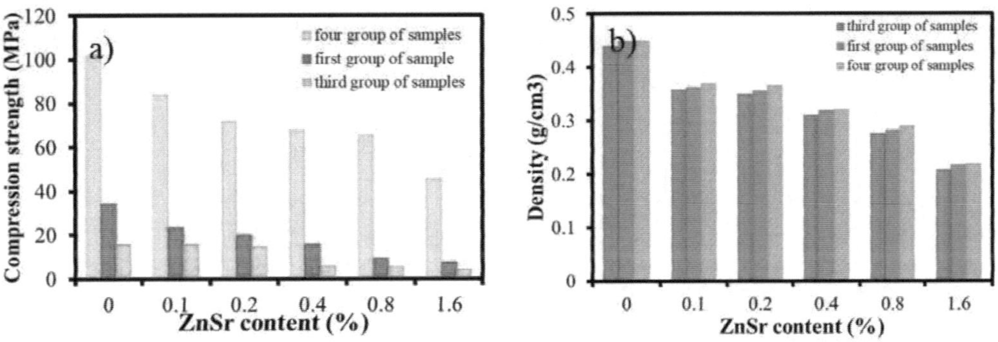

Fig.5 (a) compression strength and (b) density of the first group of samples, the third group of samples and the four group of samples

The compression strength of the first group of samples (0ZnSr, 0.1ZnSr, 0.2ZnSr, 0.4ZnSr, 0.8ZnSr and1.6ZnSr) showed in Fig.5. Compression strength of 0ZnSr was 34.41MPa, after added 0.1% ZnSr, the compression strength dropped to 23.55MPa; when the amount of ZnSr increased to 0.2%, the compression strength reduced to 19.83MPa; when ZnSr content increased to 0.4%, the compression strength decreased to 15.68 sequentially; when ZnSr content increased to 0.8%, the compression strength decreased to 9.49MPa; when ZnSr content increased to 1.6%, the compression strength even decreased to 7.64MPa. These above compression strengths presented decreasing trend with ZnSr content increased. This phenomenon was similar with the density variation trend (Fig. 5b), and it demonstrated that: the more content of ZnSr, the lower density, and the lower compression strength.

In order to demonstrate the mechanical property advantage of this controllable crosslinking polyethylene foaming material, the third group of samples were used as the control samples, which didn't have the crosslinking agent DBPH. Fig.5a also showed that the compression strengths of the third group of samples (0ZnSr(AC), 0.1ZnSr(AC), 0.2ZnSr(AC), 0.4ZnSr(AC), 0.8ZnSr(AC) and 1.6ZnSr(AC)). Test results showed that third group of samples' compression strengths lower than the first group of samples, and their compression strengths also presented similar decreasing trend with ZnSr content increased. This phenomenon demonstrated that crosslinking polyethylene foaming material have better mechanical property[23,24].

Due to the cellular size and morphology have closely relationship with mechanical property, and adding nucleating agent talcum powder, the cellular morphology became more uniform and similar circular structure, and cellular size became smaller, so the compression strengths of mixtures with talcum powder (four group of samples) were also tested in this experiment. Test results showed in Fig.5a, compression strengths of four group of samples were much higher than first group and third group of samples, although these three group of samples with the similar density. This phenomenon main explained by that small cellular size and uniform dispersion foam structure by nucleating agent improved the compression strength[25].

4 Conclusions

In this paper, the controllable crosslinking polyethylene foaming materials were developed by adding foaming agent addition ZnSr. After added ZnSr, foaming reaction realized starting earlier than crosslinking reaction, the foaming window became wider, density became smaller and expansion ratio became larger. On this basis, nucleating agent talcum powder modified crosslinking polyethylene foaming material owned more uniform cellular structure and higher compression strength.

Acknowledgments

This work was financially supported by the Technical Innovation Project of National Institute of Clean and Low-Carbon-Energy (No. CF9300180621).

IOP Conf. Series: Materials Science and Engineering **544** (2019) 012058 doi:10.1088/1757-899X/544/1/012058

References

[1] Wu S He L Zhang C Gong W He Y and Luo Y 2017 *Polym. Test.* **63 367**
[2] Ester L G Javier E and Miguel R P A 2018 *Composites Part B* **148 156**
[3] Sun X and Liang W 2016 *Composites Part B* **87 21**
[4] Laguna-Gutierrez E Escudero J and Rodriguez-Perez M A 2018 *Composites Part B* **148 156**
[5] Pilon L Fedorov A G and Viskanta R 2000 *J Cell Plast* **36 451**
[6] Smedberg A and Hjertberg T 1997 *Polymer* **38 4127**
[7] Lazir M Rado R and Rychli J 1990 *Adv. Poiym. Sci. Polym. Phys.* **95 149**
[8] Adem E Burillo G Dakin V and Vazquez M 1995 *Radiat. Phys. Chem.* **46 937**
[9] Saiz-Arroyo C Rodríguez-Pérez M Velasco J I and Saja J A 2013 *Composites Part B* **48 40**
[10] Hill C and Eastoe J 2017 *Adv. Colloid Interface Sci.* **247 496**
[11] Yamaguchi M and Suzuki K - I 2001 *J. Polym. Sci. Part B: Polym. Phys.* **39 2159**
[12] Abe S and Yamaguchi M 2001 *J. Appl. Polym. Sci.* **79 2146**
[13] Zhai W Park C B Kontopoulou M 2011 *Ind. Eng. Chem. Res.* **50 7282**
[14] Deng F Q Ma J Z Xue C H and Duan Z Y 2013 *Adv. Mater. Res.* **641 355**
[15] Singhal P Rodriguez J N Small W Eagleston S Water J V Maitland D J and Wilson T S 2012 *J Polym Sci Part B: Polym Phys* **50 724**
[16] Rodríguez-Pérez M A Almanza O Ruiz-Herrero J L and Saja J A 2018 *Cellular polymer* **27 179**
[17] Ducháček V and Kuta A 1982 *J. Appl. Polym. Sci.* **27 1549**
[18] Brown H P 1963 Rubber Chem. Technol. **36 931**
[19] Poongavalappil S Svoboda P Theravalappil R 2011 *Eur. Polym. J.* **47 1949**
[20] Reyes-Labarta J A Sempere J Marcilla A 2011 *Ind. Eng. Chem. Res.* **50 7964**
[21] Wang G Zhao J Wang G Mark L H Park C B and Zhao G 2017 *Eur. Polym. J.* **95 382**
[22] Laguna-Gutierrez E Saiz-Arroyo C Velasco J I and Rodriguez-Perez M A 2016 *Eur. Polym J.* **81 173**
[23] Zhu W Zhang G Liu B Chung T C M 2018 *Polymer* **146 101**
[24] Shen L Li J Li R Lin H, Chen J Liao B-Q 2018 *Applied Surface Science* **437 75**
[25] Notario B Pinto J Rodríguez-Perez M A 2015 *Polymer* **63 116**

IOP Conf. Series: Materials Science and Engineering **544** (2019) 012059 doi:10.1088/1757-899X/544/1/012059

Establishment of Constitutive Equation of Hot Deformation of Ti-6Al-4V-0.1B alloy with Widmannstätten structure

Y Yu[1], Z P You[2], H L Hu[3] and W J Yu[3]

[1] State Key Lab of Nonferrous Metals & Processes, GRINMAT Engineering Institute Co,Ltd. Beijing, PR. China,
[2] China International Engineering Consulting Corporation,Beijing,PR.China
[3] Beijing Trillion Metals Co.,Ltd.Beijing,PR.China
E-mail:yuyang@grinm.com

Abstract. In order to study the hot deformation behavior of the Ti-6Al-4V-0.1B alloy with Widmannstätten structure and obtain its constitutive equation,the Ti-6Al-4V-0.1B alloy cylindrical specimens with size of Φ8×12mm were prepared for the hot compression test on the Gleeble-1500 simulation machine,the deformation temperature from 850 to 980°Cand strain rates of 0.01 to 1s-1.The true stress-strain data and curves were obtained from the hot compression test.The results show that the true stress-strain curves of Ti-6Al-4V-0.1B alloy with Widmannstätten structure have the characteristics of flow stress softening and the flow stress is sensitive to temperature and strain rate.At the same temperature, flow stress increase with the strain rate.At the same strain rate, flow stress decrease with the temperature.Through regression analysis of true stress and strain data by Arrhenius hyperbolic sine function, the activation energy Q is calculated, 890.49kJ/mol.Under the range of the experimental deformation temperatures and strain rates, the constitutive equation is $\sigma=11.36\ln\{(Z/e^{88.47})^{1/4.4107}+[(Z/e^{88.47})^{2/4.4107}+1]^{1/2}\}$.

1. Introduction

Ti-6Al-4V alloy is widely used as critical structural materials in aerospace industry, such as landing gears [1-4], due to their high strength-weight ratio and high fracture toughness. Recent research has shown that a trace of Boron addition (~0.1wt %) to Ti-6Al-4V can refine its grain size with nearly an order of magnitude [5-6]. The refined microstructure in the as-cast condition can offer the potential to typically simplify its thermo-mechanical processing, and thus reducing its fabrication cost. [5]. However, the hot deformation behavior has not studied, which is important for its hot working[6]. Constitutive equations for hot deformation have been widely used in many materials[7-9]. It could reveal the hot deformation behavior of materials. It is necessary to obtain constitutive equation for Ti-6Al-4V-0.1B alloy. In this study, by thermal simulation experiment, the stress strain curves are obtained by hot compression test and constitutive equation is then obtained, which can provide a guidance for thermomechanical processing of Ti-6Al-4V-0.1B alloy.

2. Experiment

Ti-6Al-4V-0.1B ingot with 120 mm in diameter was produced by double vacuum arc remelting using sponge titanium, Al-55V, aluminum foil and Al-22Ti-7B boron. The ingots were forged and finally rolled in the $\alpha+\beta$ phase field. The bars were treated by 1020°C/1h,AC+540°C/8h,AC in this study. The Widmannstätten structure is showed in Fig. 1. The hot deformation specimens with size of Φ8×12mm are machined from bars. Hot compression tests were carried out on Gleeble-1500 thermal simulator. Specimens were heated with a rate of 10°C/s to different temperatures from 850 to 980°C and then held for 2 min. the specimens were compressed to 70% height reduction with different strain rates from 0.01 to 1 s^{-1}.

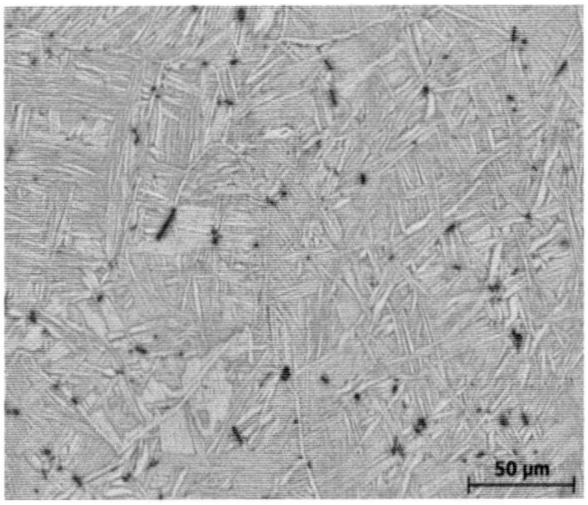

Figure1. Initial microstructure of compressive specimens

3. Results and Discussion

3.1. Flow behavior

Fig.2 shows the true stress-strain curves of Ti-6Al-4V-0.1B alloy deformed at different temperatures and with different strain rates. At the same temperature, the flow stress increase with increasing strain rate, because higher strain rate results in higher work hardening. At high strain rate, a drop of flow stress can be observed after peak stress. At the same strain rate, the flow stress decreases with increasing temperature, because of the atomic bond weakening. At high temperature, the drop of flow stress almost disappears, the stress curves after peak become steady.

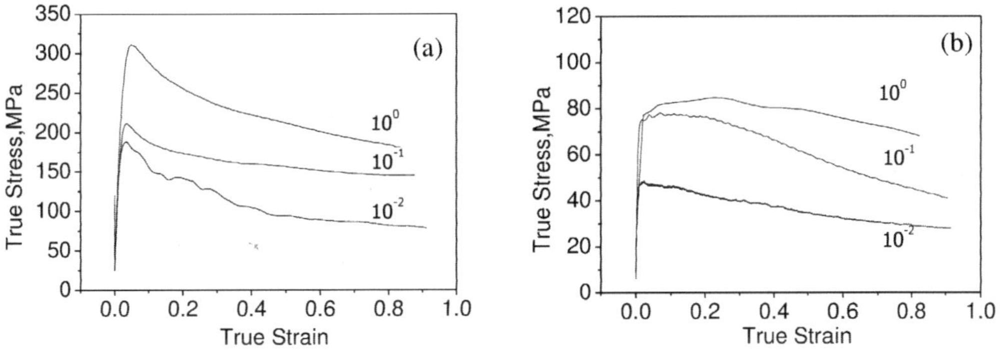

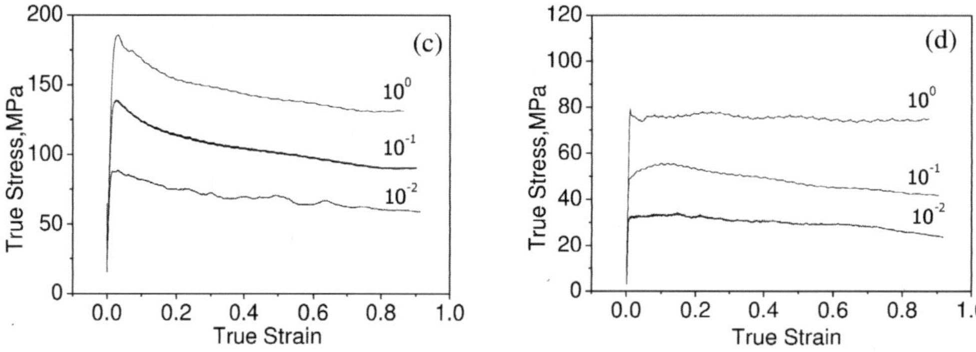

Figure 2. Curve of true stress and true strain at Widmannstätten structure of Ti-6Al-4V-0.1B alloy deformation at (a)850°C,(b)900°C,(c)950 °C,(d)980 °C

3.2. Thermal deformation mechanism

The hot deformation of metals is essentially a thermal activation process. The effect of deformation temperature and strain rate on flow stress can be calculated based on Arrhenius hyperbolic sine function to establish constitutive relations of experimental data [10].

$$\dot{\varepsilon}=A[\sinh(\alpha\sigma)]^{n}\exp(-\frac{Q}{RT}) \quad (1)$$

where A, α, n are constants, $\dot{\varepsilon}$ is strain rate (s-1), σ is flow stress (MPa), Q is activation energy, R is gas constant, T is Thermodynamic temperature.

In different stress state, Equation (1) can be simplified to (2) and (3):

$$\dot{\varepsilon}=A_2\exp(\beta\sigma)\exp(-\frac{Q}{RT}) \qquad \alpha\sigma<0.8 \quad (2)$$

$$\dot{\varepsilon}=A_1\sigma^{n1}\exp(-\frac{Q}{RT}) \qquad \alpha\sigma>1.2 \quad (3)$$

Taking logarithm to eq. (2) and eq. (3), the following equations are obtained:

$$\ln\dot{\varepsilon} = B1+n1\ln\sigma \quad (4)$$

$$\ln\dot{\varepsilon} = B2+\beta\sigma \quad (5)$$

where $n1=\partial\ln(\dot{\varepsilon})/\partial\ln\sigma$, $\beta=\partial\ln(\dot{\varepsilon})/\partial\sigma$. The experimental data are used to plot the relationship between strain rate and peak stress (in Fig. 3 lnσ-ln $\dot{\varepsilon}$ and σ-ln $\dot{\varepsilon}$). According to linear regression, the n1 and β can be obtained, then α can be calculated.

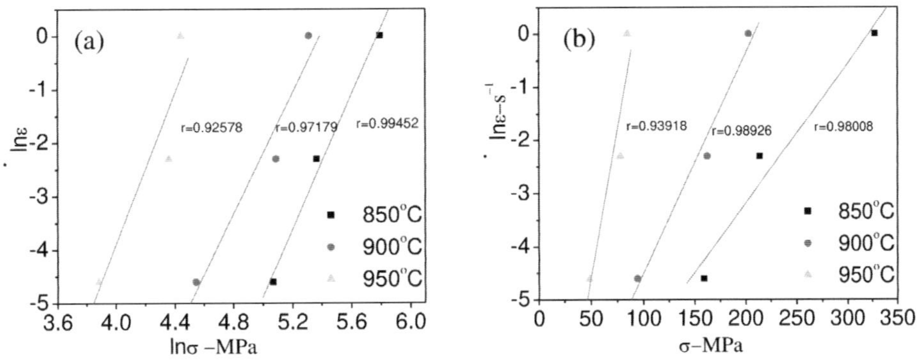

Figure 3. Relationships between strain rate and peak stress: (a)lnσ-lnέ (b)σ-lnέ

Through the inverse of the slope of these straight lines, the results are got: n1 = 6.35, β = 0.05597,α = 0.0088. Taking logarithm to equation(1), the following equations can be obtained:

$$\ln\dot{\varepsilon}= n\ln[\sinh(\alpha\sigma)]+\ln A-Q/RT \tag{6}$$

$$Q=R\{\frac{\delta\ln\dot{\varepsilon}}{\delta\ln[\sinh(\alpha\sigma)]}\}\{\frac{\delta\ln[\sinh(\alpha\sigma)]}{1/T}\} \tag{7}$$

By plotting the relationship between ln[sinh(ασ)]-lnέ and ln[sinh(ασ)]-1/T in Fig.4, activation energy Q can be calculated by equation (7) and the slope of these straight lines. The result is: Q = 890.49 kJ/mol. It reveals the thermal deformation behavior mechanism of Ti-6Al-4V-0.1B alloy.

By plotting the relationship between ln[sinh(ασ)]-lnέ and ln[sinh(ασ)]-1/T in Fig.4, activation energy Q can be calculated by equation (7) and the slope of these straight lines. The result is: Q = 890.49 kJ/mol. It reveals the thermal deformation behavior mechanism of Ti-6Al-4V-0.1B alloy.

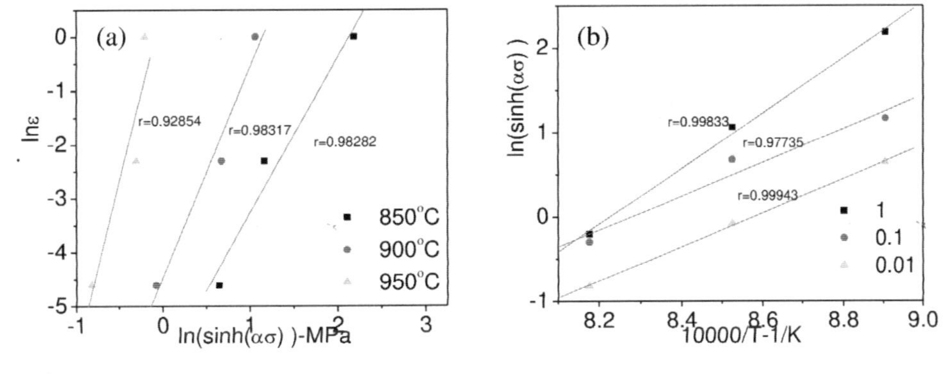

Figure 4. curves of (a) ln-ln(sinh(ασ)) and (b)ln(sinh(ασ))-1/T with Widmannstätten structure

3.3. Constitutive equation

Zener and Hollomon proposed the temperature compensated strain rate [11], where they verified that in high temperature deformation, the relationship between strain rate έ and deformation temperature T can be expressed by Z-parameter like the following equation:

$$Z=\dot{\varepsilon}\exp(\frac{Q}{RT}) = A[\sinh(\alpha\sigma)]^{n} \tag{8}$$

Taking logarithm to eq (8), equation (9) can be obtained:

$$\ln Z = n\ln[\sinh(\alpha\sigma)] + \ln A \quad (9)$$

Plot the relationship between lnZ and ln[sinh(ασ)]. Through linear regression and measuring the slope and intercept of straight line, the constants n and A can be obtained:n=4.4107， A=2.65×1038.Taking all parameters into the equation (1), the constitutive equation is expressed as:

$$\dot{\varepsilon} = 2.65\times1038\sinh4.4107(0.0088\sigma)\exp(-890490/RT) \quad (10)$$

the Z parameter is expressed as:

$$Z= \exp(890490/RT) \quad (11)$$

The flow stress constitutive equation is expressed by Z parameter as:

$$\sigma=11.36\ln\{(Z/e^{88.47})^{1/4.4107}+[(Z/e^{88.47})^{2/4.4107}+1]^{1/2}\} \quad (12)$$

4 Conclusions

1) The flow stress of Ti-6Al-4V-0.1B alloy is sensitive to temperature and strain rate. At the same temperature, flow stress increases with increasing strain rate. At the same strain rate, flow stress decreases with increasing temperature.

2) The activation energy of Ti-6Al-4V-0.1B alloy is 890.49 kJ/mol. The flow stress constitutive equation is expressed by Z parameter as: $\sigma=11.36\ln\{(Z/e^{88.47)1/4.4107}+[(Z/e^{88.47)2/4.4107}+1]^{1/2}\}$

References

[1] Y. Yu, S. Hui and W. Ye 2012 Advanced Materials Research, **428** 185

[2] A.A. Ziaei Moayyed, V. Fallah and A.R. Fotovati 2007 Proceedings of the 11th world conference on titanium. Kyoto: The Japan Institute of Metals 319

[3] Y. Fu, Y. Song, S. Hui 2006 . Rare metals **30**(6) 850

[4] R. Srinivasan, S. Tamirisakandala, Miracle 2007 Proceedings of the 11th world conference on titanium. Kyoto: The Japan Institute of Metals 977

[5] Wang W Q, Yang Y L and Zhang Y Q 2009 Materials Science Forum 173

[6] Jiao L, Li M, and Yu W 2010 Rare Metal Materials & Engineering **39**(8) 1323

[7] Li C W 2017 Rare Metal Materials & Engineering **46** 326 (in Chinese)

[8] Cai J, Wang K and Wen W 2016 Rare Metal Mate-rials & Engineering **45**(10) 2549

[9] Zhang X, Cao F and Yue H 2013 Rare Metal Materials & Engineering **42**(5) 937(in Chinese)

[10] Cote J, Galliard J L and Rossi J 2015 Materials & Design **65** (2) 272

[11] Wang Y, Lin D and Law C C 2000 Journal of Materials Science Letters **19**(13)1185

IOP Conf. Series: Materials Science and Engineering **544** (2019) 012060 doi:10.1088/1757-899X/544/1/012060

Large eddy simulation study on flow characteristics of ultrasonic heat meter at low Reynolds number

Yonghui Liu[1], Fengyan Yi[1*], Changfeng Zhou[1], and Lili Tao[1]

[1] Shandong Jiaotong University, Jinan, China
[*] Corresponding author
E-mail: lyh1981@139.com

Abstract. Ultrasonic heat meter is widely used in heat metering, and its measurement accuracy depends on the flow characteristics of the reflected channel. In this paper, LES (Large Eddy Simulation) method is used to calculate the flow field of the ultrasonic heat meter at low Reynolds number. Through the analysis of the flow field, it is shown that the velocity gradient near the cylindrical reflector is larger, especially at the tail of the first cylindrical reflector; the eddy formed by the flow can scour the reflecting surface and prevent scaling; the average linear velocity over time of the reflecting channel is obtained by the calculation results. The range of velocity variation is 0.609%.
Key words. Ultrasonic heat meter; Large eddy simulation; Low Reynolds number; Cylindrical reflector

1. Introduction

Ultrasonic heat meter has been widely used in heat metering because of its advantages in accuracy and structure. The testing accuracy of ultrasonic heat meter mainly depends on the flow testing accuracy. The principle of flow testing is shown in Figure 1 and 2. Because of the different ultrasonic velocity in the downstream and upstream, the time of the ultrasonic signal received by the two transducers is different, and there is a time difference Δt. The accuracy of flow testing of the ultrasonic heat meter mainly depends on the measurement of time difference Δt.

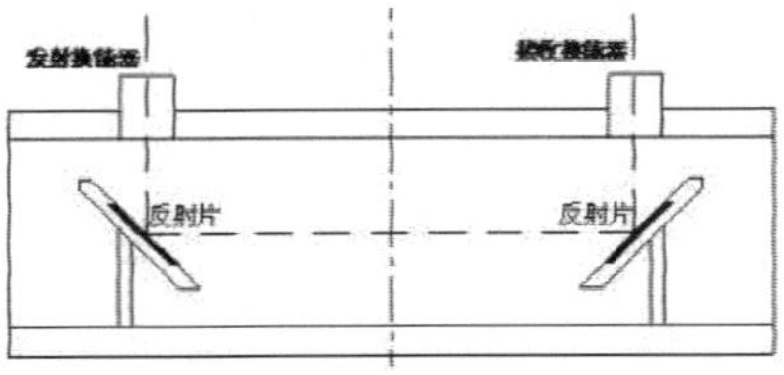

Figure 1. Flow testing principle

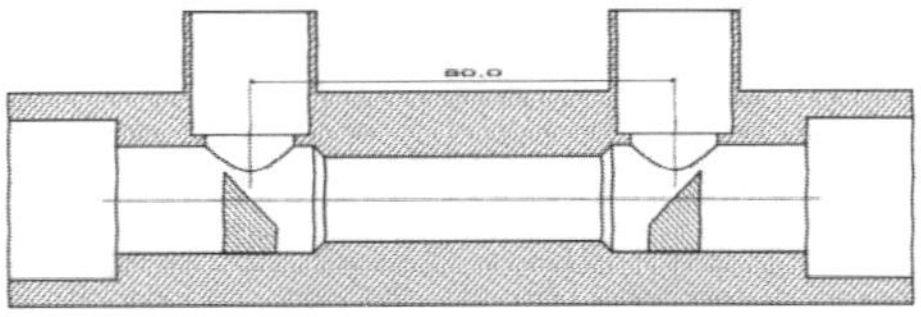

Figure 2. Internal Structure of U-shaped reflector

In order to obtain the flow characteristics more accurately, the large eddy simulation (LES) method is used to study the flow characteristics of the ultrasonic heat meter at low Reynolds number. For the method of LES, Zhang Zhao-shun [1] et al. carried out theoretical analysis of LES; Yao Shi-zhuan [2] et al. used the LES method to study square cylinders with chamfered corners; Ning Tao [3] et al. studied the turbulence of circular tubes with LES method; Wu Yanchao [4] studied the circular cylinders with low Reynolds number via LES method; J. Boudet, Sajjad Mira and so on. N[5,6] et al. used LES method to study the effect of roughness on transition; C.W.H. van Doorne [7] et al. used PIV method to measure the flow in a circular tube; the present author [8,9] also used LES method to study the flow in pipe in the previous study, and used k-e model to study the velocity characteristics of the U-shaped reflecting channel [10]. In summary, there are few studies on the flow characteristics of low Reynolds number in the channel of ultrasonic heat meter, and few studies on the application of LES to the flow in complex small chambers.
In this study, the velocity characteristics in the reflecting channel of the ultrasonic heat meter are studied. The low Reynolds number flow in the reflecting channel is calculated by LES method in order to obtain the relevant laws.

2. Mathematical and Physical Models

2.1 LES Model

Mathematical equation[1] of LES is as follows:

$$\frac{\partial \bar{u}_i}{\partial t}+\frac{\partial \bar{u}_i \partial \bar{u}_j}{\partial x_j}=-\frac{1}{\rho}\frac{\partial \bar{p}}{\partial x_i}+\upsilon\frac{\partial^2 \bar{u}_i}{\partial x_j \partial x_j}+\frac{\partial \bar{\tau}_{ij}}{\partial x_j} \tag{1}$$

Where $\bar{\tau}_{ij}=\bar{u}_i\bar{u}_j-\overline{u_i u_j}$ is defined as Subgrid stress, and Smargorinsky model is used in this paper, which is shown in Equation 2:

$$\bar{\tau}_{ij}=2\mu_t\bar{S}_{ij}+\frac{1}{3}\bar{\tau}_{kk}\delta_{ij} \tag{2}$$

Where μ_t is Subgrid turbulent viscous stress, $\bar{S}_{ij}$ is tensor rotation rate, as shown in Equation 3 and 4.

$$Sij=\frac{1}{2}\left(\frac{\partial \bar{u}_i}{\partial x_j}+\frac{\partial \bar{u}_j}{\partial x_i}\right) \tag{3}$$

$$\mu_t=\rho L_s^{2}\left|\bar{S}\right| \tag{4}$$

Where *Ls* is the mixing length of mesh, and $\left|\bar{S}\right|=\sqrt{2\bar{S}_{ij}\bar{S}_{ij}}C_s$, *Cs*=0.1.

2.2 Reflecting device

U-shaped reflecting device is used in this paper. As shown in Figure 2, the reflecting device is made of stainless steel cylinder and polished by 45 degree reflector. After two reflections, the sound wave is received by the receiving transducer and a measurement is completed. The distance of the ultrasonic reflector on the central axis is 80 mm.

2.3 Grid resolution

According to literature [7] and previous research experience [8,9], the minimum mesh size requirement is as follows:

$$L < 5\, y_0^+ \qquad (5)$$

Where y_0^+ is defined as viscosity scale and its formula is: $y_0^+ = v / u^*$, Where frictional velocity

$u^* = \bar{u}\sqrt{f/2}$, f=0.079$Re^{-1/4}$

When the flow rate is 0.05 m^3/h. According the above formulas, we obtain $u^*=5.7\times10^{-3}$m/s, $y_0^+=1.7\times10^{-4}$m, and the minimum mesh scale $l<5y_0^+=8.5\times10^{-4}$m=0.085mm. Therefore, according the requirement of mesh size, the mesh size is 0.8mm in the middle reflection area, as shown in Figure3.

Figure 3. Mesh distribution in reflection region

Figure 4. Physical model

2.4 Time step resolution

In unsteady calculation, from the previous time to the next time step, the calculation in each time step is equivalent to a quasi-steady state solution. In this paper, the resolution of time step is determined by trial calculation. The initial step is 0.0005s by two orders of magnitude smaller than the ratio of minimum mesh size to average velocity. The time step is gradually increased during the calculation.

2.5 Boundary condition

When the flow rate is 0.05m^3/h, the inlet velocity is 0.01966 m/s, the outlet is free flow, and the temperature is 20℃. The turbulence intensity I = 0.16 $Re^{-1/8}$ = 0.72 is added as the inlet disturbance. The physical model of the calculation is shown in Figure 4.

3. Flow Characteristics of Internal Flow Field in Ultrasonic Heat Meter

3.1 Distribution characteristics of flow field

Figure 5 and 6 are the velocity contour maps of the longitudinal and transverse sections of the flow field respectively at the time t of 7.3235s. From the velocity distribution in the maps, the reflecting device has great influence on flow field turbulence. When the water flows around the first reflecting cylinder, vortices are generated on the upper reflector, which washes away on the reflecting surface. Two symmetrical vortices are formed at the tail. Because of the flow around the reflecting device, the velocity gradient of this region is larger and flow-velocity contour lines are denser. The flow tends to be stable in the middle, and the flow-velocity contour lines are sparse. Then the flow is blocked by the second reflecting cylinder, which leads the denser flow-velocity contour lines and the larger velocity gradient, and the vortex street developed at the tail.

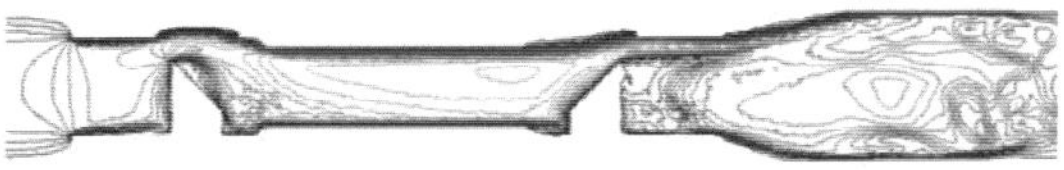

Figure 5. Flow-velocity contour lines of longitudinal sections

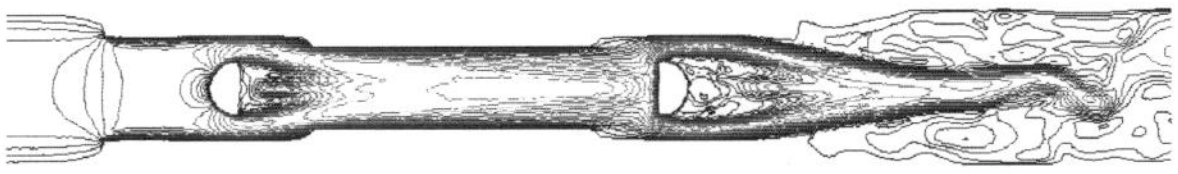

Figure 6. Flow-velocity contour lines of transverse sections

According to the analysis of the flow field, there is a greater velocity gradient near the reflecting cylinders, and the flow is stability in the reflecting device middle. Because of the influence of the second reflecting cylinder, there is no shedding vortex street at the tail of the first reflecting cylinder, and the second reflecting cylinder forms shedding vortex street at the tail and develops backward. The first reflector is washed by the vortex and the second reflector is washed by the fluid in the forward, which can play a self-cleaning role and prevent the reflection surface from scaling.

3.2 The velocity characteristics in the reflecting channel

The reflecting channels of the two reflecting surfaces are shown in Figure 7.

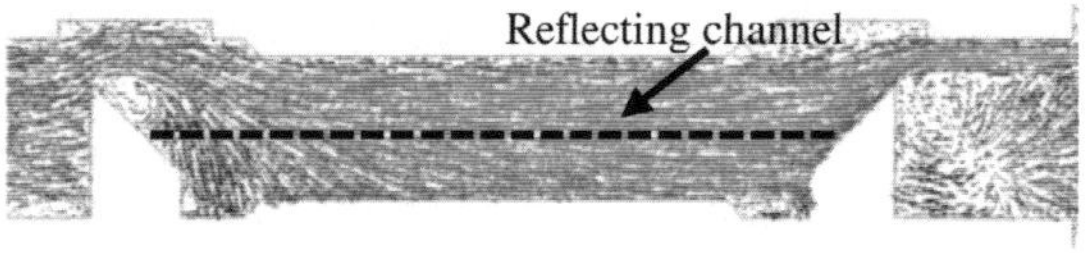

Figure 7. Reflecting channel of heat meter

Figure 8. Linear velocity vector of channel

When measuring the flow rate with an ultrasonic heat meter, the propagation time is mainly influenced by the velocity of the fluid in the reflecting channel, so, it is very important for measuring accuracy of the linear average velocity of the channel. In this paper, the linear average velocity in the reflecting channel is analysed and Figure 9 is a vector diagram of the linear velocity in the reflecting channel. Through the analyzing of the vectors distribution, as shown in Fig.9, the velocity vector direction varies greatly behind the first reflector, and the velocity gradient is larger, then its tends to be stable until the velocity in the area before the second reflector. The flow is blocked by the second reflector, resulting in upward flow direction.

3.3 Analysis of flow velocity characteristics in reflecting channel of ultrasonic heat meter

The stability of the flow measurement of the ultrasonic heat meter is influenced by the stability of the linear average velocity in the time domain. Therefore, this paper studies the variation of the linear average velocity in the reflecting channel calculated at each time step. Fig. 10 shows the linear average velocity in the reflecting channel in the time range of 0-7.323s.

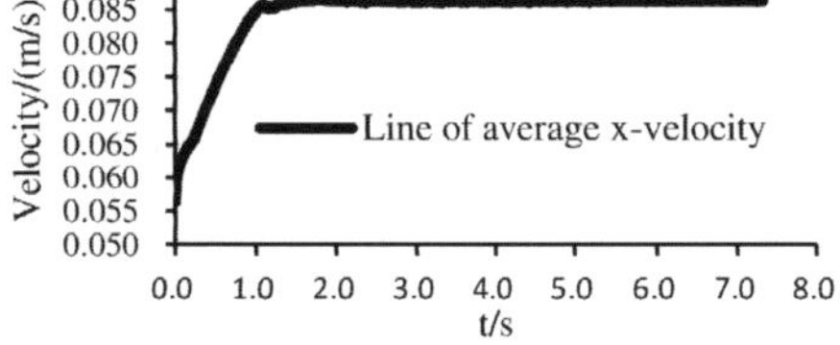

Figure 9. Linear average velocity of 0-7.323s

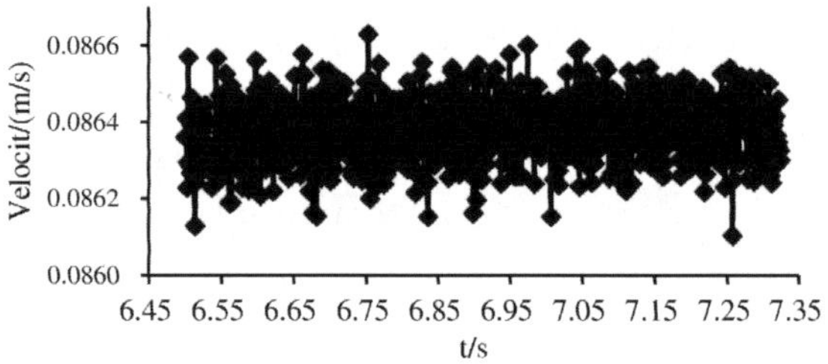

Figure 10. Linear average velocity of 6.5-7.323s

As shown in Figure 9, the velocity distribution increases first and then stabilizes with time. Carve of velocity tends to be straight after 2.5s. Therefore, the velocity at the time step of the last stage of calculation is obtained for analysis, as shown in Figure 10. The average linear velocity in the reflecting channel varies in time domain, ranging from 0.0861 to 0.08663 m/s, with an average value of 0.08638 m/s. And the velocity greatest range is 0.609%. It shows that the velocity variation is relatively small with time of the U-shaped reflector device and can meet the testing requirements.

4. Conclusion

(1) Established the LES calculation method of the complex flow inside the small cavity of the ultrasonic heat meter.

(2) The flow in the middle of the reflecting device is relatively stable, the velocity gradient is small, and the velocity gradient near the reflecting device is large, especially at the tail of the first reflector.

(3) At low Reynolds number flow rate, the average linear velocity of the reflecting channel fluctuates in time domain, and the velocity range is 0.609%.

Acknowledgments

This work is supported by the ***Shandong Province Natural Science Foundation, China.*** The item number is ZR2015EL034.

References

[1] Zhang Zhao-shun, Cui Gui-xiang, Xu Chun-xiao. Theory and application of numerical simulation of turbulent large eddy simulation[M]. *Tsinghua University Press*, 2008

[2] YAO Shi-chuan,SHI Liu-liu, GUO Peng . Large eddy simulation of flow past a square cylinder with round leading corners[J]. *Journal of Hydrodynamics,Ser A*, 2018,33(4):522-529

[3] Ning Tao, Gu Chunwei . Numerical analysis of the drag reduction for turbulent pulsating pipe flows based on large eddy simulation[J]. *J Tsinghua Univ(Sci & Technol)*, 2017,57(05):511-515

[4] WU Yan-chao, LU Xiao-feng. Numerical simulation of flow characteristics past a circular cylinder at low Reynolds numbers[J]. *Journal of Daqing petroleum institute*, 2011,35(05) :73-78

[5] J.Boudet, J.-F.Monier, F.Gao. Implementation of a Roughness Element to Trip Transition in Large-eddy Simulation[J]. *Journal of Thermal Science*. 2015(01)

[6] Sajjad Miran,Chang Hyun Sohn. Numerical study of the rounded corners effect on flow past a square cylinder[J]. *International Journal of Numerical Methods for Heat & Fluid Flow*, 2015 (4)

[7] C.W.H. van Doorne, J. Westerweel. Measurement of laminar, transitional and turbulent pipe flow using Steroscopic-PIV[J]. *Experiments in Fluids*, 2007(42):259-279

[8] Yong-hui Liu, Guang-sheng Du, Zheng-gang Liu, li Li, Li-li Tao. The profile-linear average velocity for the transition in pipes based on the method of LES[J]. *Jouranl of hydrodynamics*, 2010,22(5):366-370

[9] LIU Yong-hui, DU Guang-sheng, TAO Lili, SHEN Fang. The calculation of the profile-linear average velocity in the transition region for ultrasonic heat meter based on the method of LES[J]. *Journal of hydrodynamics*, 2011,23(1):89-94

[10] LIU Yong-hui, DU Guang-sheng, TAO Lili, SHEN Fang. Study of the flow characteristics of the ultrasonic heat meter based on the method of PIV[J]. *Chinese Journal of Scientific Instrument*, 2015,36(10):119-123

IOP Conf. Series: Materials Science and Engineering **544** (2019) 012061 doi:10.1088/1757-899X/544/1/012061

Study on the influence of reflecting channel position on flow velocity characteristics of ultrasonic heat meter

Yonghui Liu[1], Fengyan Yi[1]*, Changfeng Zhou[1] and Lili Tao[1]

1 Shandong Jiaotong University, Jinan, China
* Corresponding author
E-mail: lyh1981@139.com

Abstract. In this paper, LES (Large eddy simulation) method is used to study the influence of different channel positions on the flow velocity of the ultrasonic heat meter, and the variation of the average linear velocity of different channel with time is researched. Through the calculation results, the velocity characteristics of 11 channels of the U-shaped reflecting device ultrasonic heat meter are analysed. The results show that the vortex behind the reflecting cylinder affects the velocity distribution in the channel. Statistical analysis shows that the linear average velocity in different positions of the channel fluctuates with time, but the fluctuation amplitude is basically the same. The average linear velocity asymmetry of reflecting channel far from the vortex zone is smaller, and the deviation of the average velocity is less than that of the average velocity, therefore, the distribution of the velocity is more reasonable.
Key words. Ultrasonic heat meter; Reflecting channel; LES

1. Introduction

It is important of the ultrasonic heat meter for the metering and charging in heating reform. Transit time ultrasonic heat meter is widely used in the market, and its measurement accuracy depends on the structure of the internal reflecting channel. The accuracy and stability of domestic meter are poor Compared with the imported, especially in the case of small flow rate. Therefore, the influence of the reflecting channel position of the ultrasonic heat meter on the flow characteristics at low Reynolds number is studies, which based on the method of LES.

Many scholars have used LES method to carry out applied research, Zhang Zhao-shun [1] et al. expound the theory and application of LES; Wu Ting, Zhao Yue-zhen [2,3] et al. have studied the near wall model of LES and the method of synthesis of imported turbulence; Ai Guo-yuan, Yu Xi, Jang Hyunchul [4-6] et al. have applied LES method to study the wing, circle, the flow around cylinders, horizontal pipes and propellers; the previous research work of the author [7,8] also used the method of

LES and PIV to study pipes and obtain some results. In conclusion, the method of LES is less used to study the ultrasonic reflection channel, but it is feasible to study the internal flow field of the ultrasonic heat meter.

In this paper, LES method is used to study the influence of the position of the ultrasonic reflection channel on the flow velocity characteristics, and the relevant laws are obtained. It provides a theoretical basis for the selection of the ultrasonic heat meter, and has a certain theoretical significance and application value.

2. Physical model

2.1 Internal structure of ultrasonic heat meter

In this paper, the structure of a U-shaped ultrasonic heat meter is studied, which is shown in Figure 1. The reflecting device is composed of two stainless steel cylinders, and a 45 degree reflecting surface located on the center of the flow. After two reflections, the acoustic wave is received by the receiving transducer, and the transmitting process is completed.

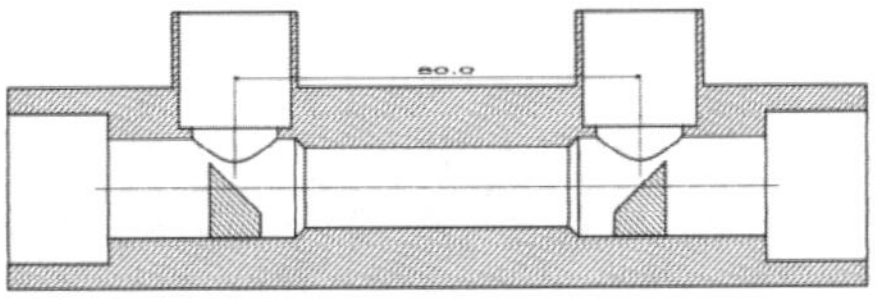

Figure 1. Structure of U-shaped Reflecting device

Figure 2. Computational domain model

2.2 Computational domain model

In order to eliminate the influence of the inlet section, according to the literature [7,8], the 10D-length straight pipe is taken before the inlet, and the 5D-length straight pipe after the outlet of the channel. The physical model of the computational domain is shown in Figure 2.

2.3 Grid resolution

The method of combining structured grid and unstructured grid is used for mesh generation. For the zone with complex flow in the middle, the structured grid method is adopted, and other zone adopted unstructured grid, and the grid close to the wall is refined. The mesh size of the middle reflection area is 0.8mm, and the total number of meshes is 2122545. The meshing is shown in Figure 3.

Figure 3. The meshing of reflecting zone

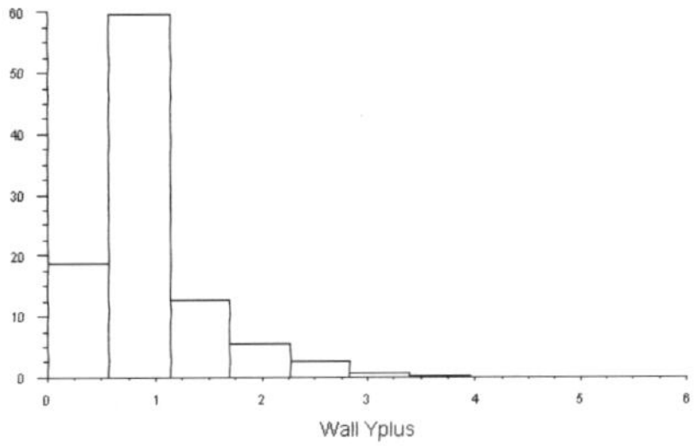

Figure 4. Size distribution of y^+

The LES method requires high resolution grids, and generally requires $y+<5$, where $y+$ [1] is the dimensionless distance from the central mass of the first layer grids to the wall. In this paper, the size of $y+$ is obtained via trial calculation, and the distribution of $y+$ is shown in Figure 4. The horizontal ordinate is the size of $y+$ and longitudinal coordinates is the proportion of size. It can be seen that maximum size are about 1%, which meets the requirements of LES method for grid resolution.

3. Method of numerical simulation

3.1 LES model

The mathematical equations of LES used in this paper are as follows:

$$\frac{\partial \bar{u}_i}{\partial t}+\frac{\partial \bar{u}_i \bar{u}_j}{\partial x_j}=-\frac{1}{\rho}\frac{\partial \bar{p}}{\partial x_i}+\frac{\partial}{\partial x_j}\left(\nu\frac{\partial \bar{u}_i}{\partial x_j}\right)-\frac{\partial \tau_{ij}}{\partial x_j} \quad (1)$$

$$\tau_{ij}=\rho\left(\overline{u_i u_j}-\bar{u}_i\bar{u}_j\right) \quad (2)$$

Where $\bar{\tau}_{ij}$ is defined as Subgrid stress, and Smargorinsky model is used.

3.2 Boundary condition

Inlet conditions: when the flux is 0.05m3/h, the inlet speed is 0.01966m/s;

Inlet disturbance: Inlet turbulence intensity I and hydraulic radius D are used, where $I = 0.16Re^{-1/8} = 0.72$ and $D = 0.003m$.

Outlet conditions: Free outflow is adopted.

3.3 Numerical simulation method

Central difference scheme is used to discretize the diffusion term, and the second-order Euler difference scheme is used to discretize the time term. In this paper, the calculating time step is 0.0005s at the beginning, and then it increases according to the convergence situation. At last, the step time reaches 0.001s after stable flow. A total of 7500 steps were calculated, and the physical time was 7.3235s.

4. Analysis of velocity characteristics of reflecting channel

4.1 Distribution characteristics of flow field

The whole calculation is unsteady. The flow field at the time t of 7.3235s is analyzed in step 7500 of the calculation process.

There are 11 reflecting channels from the two reflection surface, as shown in the Figure 5. It can be seen that there are vortices on the first reflecting cylinder from the velocity distribution in Fig.6, and the action area of the vortices covers the front of channel 7-11. The direction of the velocity vector in the middle is mainly X direction, and the flow tends to be stable. The fluid flow is impeded by the second reflecting cylinder, and velocity vector accumulates on the reflecting surface and flows out from the surface upper forming a trailing vortex area.

Figure 6 is a horizontal cross-sectional velocity vector diagram. Pair of vortices is formed on both sides of the first reflecting cylinder, but they do not develop backwards. The reflecting channels are in the middle of the two vortices. The velocity vector direction of the middle part is basically the flow direction. Two rows of vortices are formed at the back end of the second reflecting cylinder and fall off backwards.

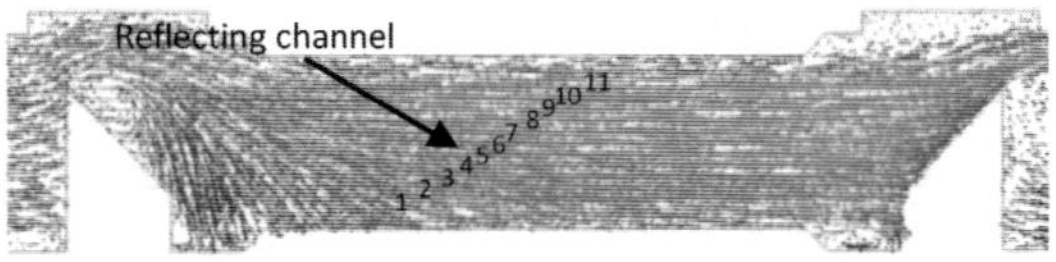

Figure 5. Velocity vector of the longitudinal section

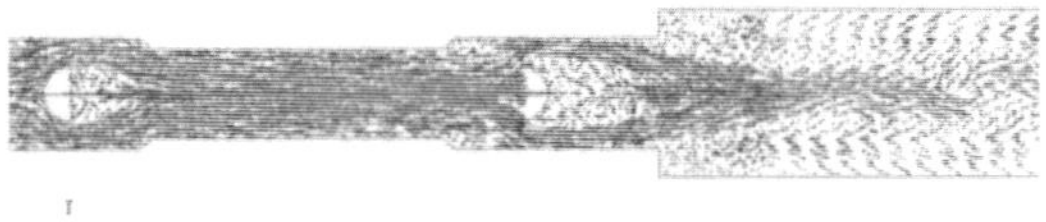

Figure 6. Velocity vector of the horizontal section

4.2 Velocity characteristics of reflecting channel

The propagation speed of ultrasonic wave is influenced by the velocity of fluid in the channel. The distribution of flow field will affect the velocity characteristics in the channel. The linear velocity vectors of 11 channels in the reflector are taken as shown in Figure 7.

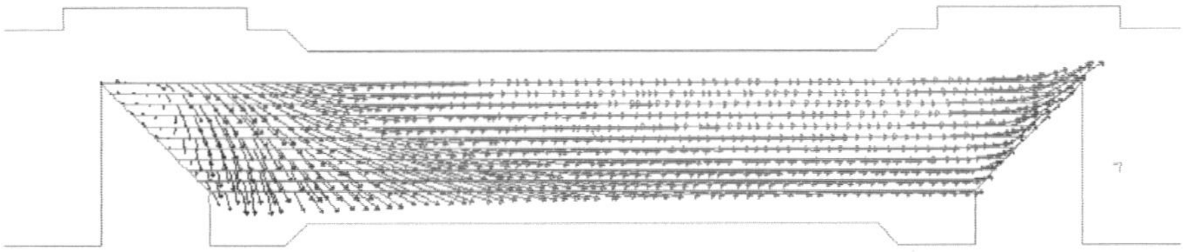

Figure 7. Vector of linear velocity of reflecting channel

From the distribution of vectors, it can be seen that the velocity vector direction of channel 7-11 varies greatly after the first reflector, and then it is impeded by the second reflector, resulting in upward flow direction. The vortices generated at the top of the first reflector influence the velocity characteristics in channel 7-11, and the propagation of ultrasonic is also affected.

4.3 Analysis of the time domain characteristics of velocity in the reflecting channel of ultrasonic heat meter

It is important for the measurement stability of heat meter to the characteristics of linear average velocity in time domain. Therefore, the linear average velocity with time in the reflecting channel is calculated at each time step. Figure 8 shows the linear average velocity in 11 reflecting channels in the time range of 0-7.3235s.

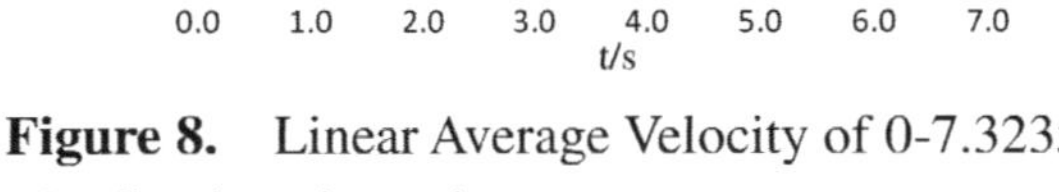

Figure 8. Linear Average Velocity of 0-7.3235s of reflecting channels

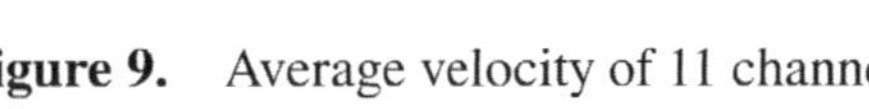

Figure 9. Average velocity of 11 channels

From the velocity curve of Figure 8, it can be seen that the average linear velocity of all reflecting channels increases rapidly at first, and then tends to be stable after about 4 seconds, and the fluctuation amplitude of the average linear velocity in different positions is almost the same. The average velocities of 11 channels are different, and the average velocity of each channel is shown in Figure 9. Six channels are located in the central of the pipeline. The average velocities of 1-6 channels are gradually increasing, which is accord with the distribution of velocities in the pipe. After the seventh channel, it is shown that the velocity becomes irregularly, which increases first, then decreases, and then increases. The main reason is that the influence of the vortex region located the first reflector surface.

5. Conclusion

(1) The vortices generated by the reflecting cylinder of fluid flow will influence the velocity in the channel at different locations, and the velocity distribution law is not consistent with that of in pipe.
(2) The fluctuation amplitude of the average linear velocity in different positions is almost the same with time, which indicates that the reflecting device of U-shaped reflecting device is reasonable.
(3) The average linear velocity in channel 1-6 far from the vortex region is accord with the distribution of velocities in the pipe, and the distribution of the velocity is more reasonable.

Acknowledgments

This work is supported by the ***Shandong Province Natural Science Foundation, China.*** The item number is ZR2015EL034.

References

[1] Zhang Zhao-shun, Cui Gui-xiang, Xu Chun-xiao. Theory and application of numerical simulation of turbulent large eddy simulation[M]. *Tsinghua University Press*, 2008
[2] Wu Ting, Shi Beiji, Wang Shizhao, Zhang Xing, He Guowei. Wall-model for large-eddy simulation and its applications[J]. *Chinese Journal of Theoretical and Applied Mechanics*, 2018, 50 (03),453-466
[3] ZHAO Yuezhen, GUO Jie. Inlet synthetic turbulent methods for large eddy simulation[J]. *Journal of Aerospace Power*, 2018,33(07) 1622-1629
[4] Ai Guoyuan, Ye Jian. Large eddy simulation of low Reynolds number airfoil with different separating flow regime [J]. *ACTA AERODYNAMICA SINICA*, 2017,35(02),299-304
[5] YU Xi, LIU Mei, WEI Hong-yuan. Large Eddy Simulation for Flow Field of Horizontal Pipe[J]. *Journal of Tianjin University*, 2007,40(3):342-345
[6] Jang Hyunchul, Mahesh Krishnan. Large eddy simulation of flow around a reverse rotating propeller[J] . *Journal of Fluid Mechanics* . 2013
[7] LIU Yong-hui, DU Guang-sheng, TAO Lili, SHEN Fang. The calculation of the profile-linear average velocity in the transition region for ultrasonic heat meter based on the method of LES[J]. *Journal of hydrodynamics*, 2011,23(1):89-94
[8] Liu Yonghui. Study of the flow characteristics of the ultrasonic heat meter based on the method of PIV[J]. *Chinese Journal of Scientific Instrument*, 2015,36(10):119-123

Research on the steering gait and error of a novel type of heavy load Automated Guided Vehicle

J Mucklow[1] and A Jansson[2]
Yang Zhang[1 2] , Zhigang Xu[1] and Songkai Liu[1]

1 Shenyang Institute of Automation (SIA), Chinese Academy of Sciences (CAS)
2 University of Chinese Academy of Sciences(UCAS)
E-mail: zhangyang4@sia.cn

Abstract. Automated Guided Vehicle is a kind of mobile robot Guided by vision, magnet, laser or tracking marks and cables on the ground .In the process of changing the driving direction, if the AGV is directly forced to flatten, the wheels will be "locked" in the force and unable to move, which will also cause damage to the driving system. Therefore, we adopt the way of steering in two or three sections to stabilize and overbalance, ensuring flexibility and safety.

1. Introduction
AGV "Automated Guided Vehicle" (as defined by AMHA, America Material Handling Association) is a kind of mobile robot Guided by vision, magnet, laser , tracking marks or cables on the ground.
In this paper, a large AGV is modeled and its turning gait is analyzed to find the main factor affecting the accuracy, and a feasible solution is given.

2. Kinematic Of Automated Guided Vehicle

2.1. Position Description of AGV
Firstly, based on the world coordinate system, a reference coordinate system is fixed for each joint, and then the transformation operator between any two adjacent coordinate systems is determined, and finally the total transformation operator is obtained.
Since the wheel mechanism is parallel, we take a branch road as an example for transformation.
{G}, known to the world coordinate system in the centroid to build A solid AGV coordinate system {A} the x axis parallel AGV length direction parallel to the y axis AGV width direction, {G} in the coordinate system, set up to determine the origin of the {A} vector for ${}^{A}P_{BORG}=[x \quad y \quad z]^T$ connect the coordinate system relative to the rotation of the {G} with ${}^{G}_{A}R$ said, and will coordinate transformation to {A} {G}, first of all, will coordinate {A}.into an origin with {A}, and {G} in the middle of the same coordinate system, By means of the left multiplication attitude transform matrix

$$R=\begin{bmatrix} \overline{e_{xA}}\cdot\overline{e_{xG}} & \overline{e_{yA}}\cdot\overline{e_{xG}} & \overline{e_{zA}}\cdot\overline{e_{xG}} \\ \overline{e_{xA}}\cdot\overline{e_{yG}} & \overline{e_{yA}}\cdot\overline{e_{yG}} & \overline{e_{zA}}\cdot\overline{e_{yG}} \\ \overline{e_{xA}}\cdot\overline{e_{zG}} & \overline{e_{yA}}\cdot\overline{e_{zG}} & \overline{e_{zA}}\cdot\overline{e_{zG}} \end{bmatrix} \tag{1}$$

The vector symbol of A is eliminated, and the attitude of the vector coordinate system is unified. At this point, all vectors can be added. For simplicity' ,we combine ${}_{A}^{G}R$ with ${}^{A}P_{BORG} = [x \quad y \quad z]^{T}$ Using a matrix operator ${}_{A}^{G}T$

Represents the transformation of two coordinate systems

$$\begin{bmatrix} {}^{G}P \\ 1 \end{bmatrix} = {}_{A}^{G}T \begin{bmatrix} {}^{A}P \\ 1 \end{bmatrix} = \begin{bmatrix} {}_{A}^{G}R & {}^{G}P_{AORG} \\ 000 & 1 \end{bmatrix} \begin{bmatrix} {}^{A}P \\ 1 \end{bmatrix} \tag{2}$$

Namely,

$${}^{G}P = {}_{A}^{G}T\,{}^{A}P \tag{3}$$

In general, the motion of the AGV in the vertical direction is very small. Therefore, when researching the overall movement of the AGV, we can reduce the AGV and the world coordinate system to two dimensions.

On the frame, there are four off-centered oriental wheels, and the wheel carrier can rotate freely relative to the car frame. Taking a certain wheel carrierB_i as an example, we can obtain from the above method

$$\begin{bmatrix} {}^{A}P \\ 1 \end{bmatrix} = {}_{B_i}^{A}T \begin{bmatrix} {}^{B_i}P \\ 1 \end{bmatrix} = \begin{bmatrix} {}_{B_i}^{A}R & {}^{A}P_{B_iORG} \\ 000 & 1 \end{bmatrix} \begin{bmatrix} {}^{B_i}P \\ 1 \end{bmatrix} \tag{4}$$

Namely,

$${}^{A}P = {}_{B_i}^{A}T\,{}^{B_i}P \tag{5}$$

$${}^{G}P = {}_{A}^{G}T\,{}^{B_i}P = {}_{B_i}^{A}T\,{}_{B_i}^{A}T\,{}^{B_i}P \tag{6}$$

The wheel The wheel B_i on the rack C_i

$$\begin{bmatrix} {}^{B_i}P \\ 1 \end{bmatrix} = {}_{C_i}^{A}T \begin{bmatrix} {}^{B_i}P \\ 1 \end{bmatrix} = \begin{bmatrix} {}_{C_i}^{B_i}R & {}^{B_i}P_{C_iORG} \\ 000 & 1 \end{bmatrix} \begin{bmatrix} {}^{C_i}P \\ 1 \end{bmatrix} \tag{7}$$

$${}^{B_i}P = {}_{C_i}^{A}T\,{}^{B_i}P \tag{8}$$

$${}^{A}P = {}_{B_i}^{A}T\,{}^{B_i}P = {}_{B_i}^{A}T\,{}_{C_i}^{A}T\,{}^{B_i}P \tag{9}$$

$${}^{G}P = {}_{A}^{G}T\,{}^{B_i}P = {}_{B_i}^{A}T\,{}_{B_i}^{A}T\,{}^{B_i}P = {}_{A}^{G}T\,{}_{B_i}^{A}T\,{}_{C_i}^{A}T\,{}^{B_i}P \tag{10}$$

In order to verify the accuracy of the above operator, the data at the initial position of the AGV were taken and the correct conclusion was drawn. It is proved that the homogeneous transformation operator is correct.

2.2. Kinematics Analysis of AGV

In this research, we regard AGV as a rigid body, whose input is tangential acceleration and angular acceleration.

$$\mathrm{p} = (p_x, p_y, p_z) \tag{11}$$

Is the coordinates of AGV center in geodetic coordinate system {A}.

Three Euler angles （ψ， θ， φ） In the fixed connected coordinate system, the first state equation is:

$$\dot{p} = R(\psi,\ \theta,\ \varphi)\cdot v_r \tag{12}$$

For

$$R(\psi,\ \theta,\ \varphi) = R_\psi R_\theta R_\varphi = \begin{bmatrix} \cos\psi & -\sin\psi & 0 \\ \sin\psi & \cos\psi & 0 \\ 0 & 0 & 1 \end{bmatrix} \begin{bmatrix} \cos\theta & 0 & \sin\theta \\ 0 & 1 & 0 \\ -\sin\theta & 0 & \cos\theta \end{bmatrix} \begin{bmatrix} 1 & 0 & 0 \\ 0 & \cos\varphi & -\sin\varphi \\ 0 & \sin\varphi & \cos\varphi \end{bmatrix}$$

$$= [i|A \quad j|A \quad k|A] = \begin{bmatrix} \cos\theta\cos\psi & -\cos\varphi\sin\psi + \sin\theta\cos\psi\sin\varphi & \sin\psi\sin\varphi + \sin\theta\cos\psi\cos\varphi \\ \cos\theta\sin\psi & \cos\varphi\cos\psi + \sin\theta\sin\psi\sin\varphi & -\cos\psi\sin\varphi + \sin\theta\cos\varphi\sin\psi \\ -\sin\theta & \cos\theta\sin\varphi & \cos\theta\cos\varphi \end{bmatrix} \tag{13}$$

Take the derivative of the above expression, then:

$$\ddot{p} = \dot{R}\cdot v_r + R\cdot \dot{v}_r \tag{14}$$

In another way:

$$\dot{v}_r = R^T\cdot \ddot{p} - R^T\dot{R}\cdot v_r \tag{15}$$

$$R^T\dot{R} = Ad\left(R^T\omega\right) \tag{16}$$

$$R^T\cdot \ddot{p} = \mathrm{a}_r \tag{17}$$

Then

$$\dot{v}_r = \mathrm{a}_r - Ad\left(\omega_{R_1/R_0|R_1}\right)\cdot v_r \tag{18}$$

The above equation is the second equation of state
Finally, we get

$$\omega_{|R_0} = R(\psi,\ \theta,\ \varphi)\cdot\omega_{|R_1} \tag{19}$$

then

$$\begin{bmatrix} 0 & \dfrac{sin\varphi}{cos\theta} & \dfrac{cos\varphi}{cos\theta} \\ 0 & cos\varphi & -sin\varphi \\ 1 & tan\theta sin\varphi & tan\theta cos\varphi \end{bmatrix}\begin{bmatrix}\dot{\psi} \\ \dot{\theta} \\ \dot{\varphi}\end{bmatrix} = R(\psi,\ \theta,\ \varphi)\cdot\omega_r \tag{20}$$

If you extract the vector, you get the third equation of state. There are three simultaneous state equations

$$\dot{p} = R(\psi,\ \theta,\ \varphi)\cdot v_r$$

$$\dot{v}_r = \mathrm{a}_r - \omega_r \wedge v_r$$

$$\begin{bmatrix}\dot{\psi} \\ \dot{\theta} \\ \dot{\varphi}\end{bmatrix} = \begin{bmatrix} 0 & \dfrac{sin\varphi}{cos\theta} & \dfrac{cos\varphi}{cos\theta} \\ 0 & cos\varphi & -sin\varphi \\ 1 & tan\theta sin\varphi & tan\theta cos\varphi \end{bmatrix}\cdot\omega_r \tag{21}$$

Since the objects in this research only move on the horizontal surface, there are $\psi = \theta = 0$

From the model

$$\dot{\psi} = \omega_{r3} \tag{22}$$

$$\dot{\theta} = \omega_{r2} \tag{23}$$

$$\dot{\varphi} = \omega_{r1} \tag{24}$$

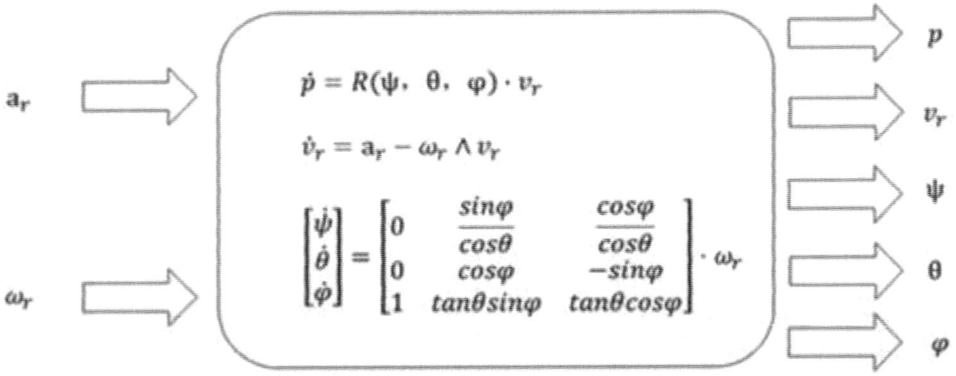

2.3. Dynamic modeling

In this research, we derive the dynamic equation based on the kinematics model, taking into account the angular acceleration and tangential acceleration generated by the force and dynamic characteristics. The relation between acceleration and force is derived from Newton's second law or the basic principle of dynamics.

The mass of AGV is denoted by f and denotes the resultant force in the inertial coordinate system.

$$\mathrm{m}\,\ddot{p} = f \tag{28}$$

$$\mathrm{m}\left(\dot{R}\cdot v_r + R\cdot \dot{v}_r\right) = f \tag{26}$$

$$R^T\cdot \ddot{p} = R^T\left(\dot{R}\cdot v_r + R\cdot \dot{v}_r\right) = \mathrm{a}_r \tag{27}$$

$$mR^T\cdot \ddot{p} = mR^T\left(\dot{R}\cdot v_r + R\cdot \dot{v}_r\right) = m\mathrm{a}_r \tag{28}$$

$$m\mathrm{a}_r = R^T f \tag{29}$$

From this, the tangential acceleration (which is used as an input to the kinetic model) is an algebraic equation for the forces applied to the robot.

3. Error Analysis And Accuracy Research Of Steering Gait

In the translation process, since the off-centered oriental wheel rotation is a process of constant change in the direction of speed, and the transverse deviation is what we care about, and most affects the parameters of the docking accuracy, we take the first-order integral process as an example to calculate the maximum error.

The formula is as follows:

Taking a single off-centered oriental wheel as the research object, the displacement generated when it rotates around the rotation center is studied:

When the off-centered oriental wheel rotates to an arbitrary Angle

$$S_o = \int_0^{\psi} \cos\theta V_o \cdot \frac{\psi e}{V_o} d\theta = \int_0^{\psi} \psi e\cdot \cos\theta d\theta = \varphi e \sin\theta \,|_0^{\varphi} = \varphi e sin\varphi \tag{30}$$

In the range of 0-90 degrees, the migration distance of off-centered oriental wheels keeps increasing. In the range of 90-180 degrees, the displacement direction of off-centered oriental wheels is opposite to that of the previous stage. Finally, it is offset in the position of rotation 180 degrees

$$\mathrm{t} = \frac{\pi e}{2V_o} \; (31)$$

$$S_o = \int_0^{\frac{\pi}{2}} \cos\theta V_o \cdot \frac{\pi e}{2V_o} d\theta = \int_0^{\frac{\pi}{2}} \frac{\pi e}{2}\cos\theta d\theta = \int_0^{\frac{\pi}{2}} \frac{\pi e}{2}\cos\theta d\theta = \frac{\pi e}{2}\sin\theta \,|_0^{\frac{\pi}{2}} = \frac{\pi e}{2} \tag{32}$$

From the integral process, it can be seen that the displacement of the off-centered oriental wheel when it rotates around the center of rotation is independent of its speed, which is only related to the magnitude of the eccentricity e of the off-centered oriental wheel and is independent of its speed.

In terms of the four off-centered oriental wheels of AGV as a whole, the rotation directions of the four off-centered oriental wheels respectively will bring different deviations.

Now, we're going to be rotating 90 degrees with a single wheel

1. The four wheels have the same migration direction and migration path

$$S_F = S_{o1} = S_{o2} = \frac{\pi e}{2} \tag{33}$$

$$S_B = S_{o3} = S_{o4} = \frac{\pi e}{2} \tag{34}$$

So, Angle of deflection is equal to 0

The offset distance is evaluate e/2

2. One turn is opposite to the other three

$$S_F = S_{o1} = S_{o2} = \frac{\pi e}{2} \tag{35}$$

$$S_B = S_{o3} - S_{o4} = 0 \tag{36}$$

$$\varphi = \arctan\frac{\pi e}{2l} \tag{37}$$

or

$$S_F = S_{o1} - S_{o2} = 0 \tag{38}$$

$$S_B = S_{o3} = S_{o4} = \frac{\pi e}{2} \tag{39}$$

$$\varphi = \arctan\frac{\pi e}{2l} \tag{40}$$

3. The two front wheels have the same direction of deflection, and the two rear wheels have the same direction of deflection, but the front and rear wheels have the opposite direction of deflection

$$S_F = S_{o1} = S_{o2} = \frac{\pi e}{2} \tag{41}$$

$$S_B = S_{o3} = S_{o4} = \frac{\pi e}{2} \tag{42}$$

$$\varphi = \arctan\frac{\pi e}{l} \tag{43}$$

4. The deflection direction of the two front wheels is opposite, and the deflection direction of the two rear wheels is opposite

$$S_F = S_{o1} - S_{o2} = 0 \tag{44}$$

$$S_B = S_{o3} - S_{o4} = 0 \tag{45}$$

But the two deflection distances are in opposite directions
The offset distance is 0
From the above process, we can see that in the case of unknown disturbance, there are various possibilities for the lateral error of the docking AGV, but all of them are independent of the speed, and only related to the eccentricity of the off-centered oriental wheel e and the distance L between the two sets of off-centered oriental wheels. Among them, the eccentricity of the off-centered oriental wheel e has the greatest influence on the car frame. However, the greater the eccentricity distance of off-centered oriental wheel, the better its orientation.

4. Mechanics Analysis

The designed product tends to run on flat road, so the performance of the suspension system is not required. We use the simplest method, namely, the wheel itself is designed to be flexible, and the rubber deformable tire is used on the small casters to make an active suspension [autonomous mobile robot introduction p29]. In general, in rolling friction, we assume that the wheels and the ground are point contact, that is, the wheels and the ground have no deformation, but in fact, the wheels and the ground have a small deformation. At the same time, the wheels actually produce a contact surface between the wheels and the ground, and the wheels are subject to distributed forces, which we can simplify to a force and a torque. The introduction of this theory has answered some of our questions, such as:
In general modeling simulation, some simulation results are different from the actual situation. For one thing, the external environment such as the ground is not ideal; for another, there is deformation between the wheel and the ground.
So here we introduce the rolling friction coefficient
The force analysis of ten thousand star wheels in AGV motion (static analysis) is as follows

$$\sum M_A(F) = 0 \tag{46}$$

Namely,

$$-\mathrm{f} \cdot R + M_{max} = 0 \tag{47}$$

For

$$M_{max} = \delta F_N \tag{48}$$

Combine the two

$$f \cdot R = \delta F_N \tag{49}$$

Consequently ,we've got

$$f = \frac{\delta}{R} F_N \tag{50}$$

Let $\mu_r = \frac{\delta}{R}$

The coefficient of static friction is greater than the coefficient of rolling friction

By type, we noticed a problem: when the direction of the driving wheel and castor wheels into 90 degrees, underpowered would happen to drive the AGV investigate its reason, driving wheel power provided by the ground reaction force in the face of the wheel, and at this point, the point of view of universal wheel makes the does not exist between the scroll with the ground, the stress for the static friction force, its theoretical calculation is as follows:

When the drive wheel does not roll

$$F_M = \mu_s F_N \times \frac{2S_M}{2S_M + 4S_o} \tag{51}$$

$$F_f = \mu_s F_N \times \frac{4S_o}{2S_M + 4S_o} \tag{52}$$

Because drive wheel and off-centered oriental wheel size difference is not big, contact surface area is similar, then have apparently

$$\frac{2S_M}{2S_M + 4S_o} < \frac{4S_o}{2S_M + 4S_o} \tag{53}$$

So, obviously, when the drive wheel is rolling (idling)

$$F_M < F_f \tag{54}$$

$$F_M = \mu_r F_N \times \frac{2S_M}{2S_M + 4S_o} \tag{55}$$

$$F_f = \mu_s F_N \times \frac{4S_o}{2S_M + 4S_o} \tag{56}$$

It is worth mentioning that in the engineering practice to process, it is a large demand for power directly to turn around in 90 degrees, easy to cause the drive wheels of the motor overload and instability caused by instantaneous electric current too large, big horsepower driving wheels would increase the cost, the oscillation of the driven wheel will be too large, affect the accuracy, therefore, we turn to its gait is optimized, the first rotating driving wheel, turn 45 degrees in the direction of translation, travel distance and then rotate 90 degrees to the direction, can from straight moving to traverse, the gait optimization, save power and reduce the cost. At this time, the friction is calculated as

$$F_M = \mu_s F_N \times \frac{2S_M}{2S_M + 4S_o} \tag{57}$$

$$F_f = \mu_s F_N \times \frac{4S_o}{2S_M + 4S_o} \times \frac{\sqrt{2}}{2} \tag{58}$$

At this point, the calculation could not be carried out by approximate method. We adopted Adams simulation to verify the simulation results, and after gait improvement, AGV could successfully achieve the purpose of lateral movement.

However, although its demand for power has declined, it is still considerable. In particular, the power of this AGV is provided by batteries, and frequent charging is not conducive to the improvement of production efficiency. Therefore, in the following research, we plan to further optimize the turning gait, and install the steering motor at the turning shaft above the off-centered oriental wheel. Before the steering, the off-centered oriental wheel is rotated to the direction of motion, that is, each off-centered oriental wheel Oi is rotated under the action of torque output by the motor. However, when the power is saved, the torque caused by the motor also causes the rotation of the AGV, and its calculation formula is T=I spark.

5. Conclusion

From this research , we can draw the following conclusions:
1. The offset Angle of the car frame is inversely proportional to the distance L between the front and rear off-centered oriental wheels, and is directly proportional to the eccentricity distance e of off-centered oriental wheels .
2. The best way to reduce the deviation error is to make the deviation errors generated by each wheel offset each other, which is also the aspect that we need to plan after installing the steering motor .
(train of thought motor is used to add disturbance, only to guide the steering, while driving.)
In the process of changing the driving direction, if the AGV is directly forced to flatten, the wheels will be "locked" in the force and unable to move, which will also cause damage to the driving system. Therefore, we adopt the way of steering in two or three sections to stabilize and overbalance, ensuring flexibility and safety.
In terms of turning accuracy, we have another solution: namely, starting from the control system, to establish a feedback control rate of constant correction.

6. References

[1] Chih -Lyang Huang , Chang-Chia Yang , and John .Y.Huang .Path Tracking of an Autonomous Ground Vechile With Different Paylosds by Hierarchical Improved Fuzzy Dynamic Sliding-Mode Control [J]. IEEE TRANSACTION ON FUZZY SYSTEM,VOL.26 , NO.2,APRIL 2018.
[2] K.R.S.Kodagoda , W.S.Wijesoma , and E.K.Teoh .Fuzzy Speed and Steering Control of an AGV [J].IEEE Transaction Control System Technology ,VOL.10,NO.1,January2002
[3] Sok-Yong Lee ,Hai-Won Yang .Navigation of automated guided vehicles using magnet spot guidance method[J].Robotics and Computer-Integrated Manufacturing,28(2012)425-436
[4] Dynamic Model With Slip for Wheeled Omnidirectional Robots. Robert L.Williams , Brian E.Carter , Paolo Gallina , and Giulio Rosati
[5] H.Martínez-Barberá ,D.Herrero-Pérez .Autonomous navigation of an automated guided vehicle in inindustrial environments. Robotics and Computer-Integrated Manufacturing,26(2010)296-311
[6] R.Siegwart, I.R.Nourbakhsh, D.Scaramuzza. Introduction to Autonomous Mobile Robots[M].America, The MIT Press,2012
[7] Luc.Jaulin.Mobile Robotics[M].ISTE Press.2015
[8] Saeed B.Niku.Introduction to Robotics: Analysis, Control, Applications, Second Edition[M].John Willey & Sons， Inc.2011
[9] John J.Craig.Introduction to Robotics:Mechanics and Control, Fourth Edition[M].Pearson Education. 2018
[10] Campion G , Bastin G B, D' Andrea-Novel. Structural properties and classification of kinematic and dynamic models of w heeled mobile robots[J] .IEEE Transactions on Robotics and Automation.1996, 12(1):47-62.
[11] Muir P F, Neuman C P .Kinematic modeling for feedback control of an omnidirectional wheeled mobile robot[A] .Proceedings of the 1987 IEEE International Conference On Robotics and Automation [C] .1987.1772-1778.
[12] West M , Asada H H .Design and control of ball wheel omnidirectional vehicles[A] .Proceedings of the 1995 IEEE International Conference on Robotics and Automation[C] .1995.1931-1938.
[13] Tahboub K A , Asada H H .Dynamics analysis and control of a holonomic vehicle with a continuously variable transmission[J] .ASME Journal of Dynamic Systems, Measurement , and Control.2002, 124(3):118-126.

IOP Conf. Series: Materials Science and Engineering **544** (2019) 012062 doi:10.1088/1757-899X/544/1/012062

[14] Wilson L , Williams C , Yance J, etc. Design and modeling of a redundant omnidirectional RoboCup Goalie [A] . Proceedings RoboCup 2001 International Symposium [C]. Seattle :2001.
[15] Carter B , Good M , Dorohoff M , etc. Mechanical design and modeling of an omnidirectional RoboCup player[A] .RoboCup AI Conference[C] .Seattle W A:2001.1-10.
[16] Byun K S, Kim S J, Song J B .Design of continuous alternate wheels for omnidirectional mobile robots[J] .Proceedings of the 2001 IEEE International Conference on Robotics and Automation. 2001, 1:767-772.
[17] Paromtchik I E , Rembold U.A motion generation approach for an omnidirectional vehicle[A] .Proceedings of the 2000 IEEE International Conference on Robotics and Automation[C] .San Francisco : 2000.1213-1218.
[18] Pin F G, Killough S M .A new family of omnidirectional and holonomic wheeled platforms for mobile robots [J] .IEEE Transactions on Robotics and Automation.1994, 10(4):480-489.
[19] Holmberg R, Khatib O .Development and control of a holonomic mobile robot for mobile manipulation tasks [J] .International Journal of Robotics Research .2000, 19(11):1066-1074.
[20] Park T B, Lee J H , Yi B J, etc.Optimal design and actuator sizing of redundantly actuated omni-directional mobile robots[A] .Proceedings of the 2002 IEEE International Conference on Robotics and Automation [C] .San Francisco :2002.
[21] Moore K L , Flann N S.A six-wheeled omnidirectional autonomous mobile robot [J] .IEEE Control System Magazine .2000, 20(6): 53-66.

IOP Conf. Series: Materials Science and Engineering **544** (2019) 012063 doi:10.1088/1757-899X/544/1/012063

Magnetic graphene oxide for efficient solid phase extraction of DEHP

L Ma, J Huang and M Zhou*

Tianhong Measurement Institute, Zhuzhou 412036, China
E-mail:ljweisz@163.com

Abstract. DEHP (di-2-ethylhexyl phthalate) can cause negative impacts on human health, thus development of reliable analytical methods for the determination of PAEs is important.In this work, magnetic graphene oxide was synthesized and used as the sorbent for solid phase extraction of DEHP. Factors that influence the extraction efficiency of DEHP were studied. The optimum conditions, such as solution pH, extraction time, desorption time, and eluent type and volume were obtained. The analytical performance of the developed method was investigated, and it was found that the limit of detection (LOD) for DEHP was 0.35μg/L and the related standard deviation (RSD) was 4.3 %. The proposed method was applied in the determination of DEHP in environmental water samples with recoveries of 91.6-106.5%.

1. Introduction

Phthalic acid esters (PAEs) are widely used in food packaging materials, personal care products, paints, and building materials. DEHP (di-2-ethylhexyl phthalate) is the most often used one of the PAEs. However, DEHP can cause negative impacts on human health [1-3]. Nowadays, DEHP has been found in various environmental matrices and food samples. Consequently, it is necessary to develop suitable analytical methods for DEHP. Solid-phase extraction (SPE) can be applied for separation and enrichment of pollutants from various matrices, and has been widely used [4-16].

Graphene oxide (GO) is the oxidized derivative of graphene, which has an ultrahigh-specific surface area, superior chemical property and physical stability. Consequently, graphene oxide can be used as the sorbent for SPE [16-22].

Traditional SPE is always a tedious process. However, magnetically separation technology can provide an easy and rapid way for separation of magnetic particles from solution by applying a magnetic field [23-26].

In this work, magnetic graphene oxide was synthesized and used as the sorbent for solid phase extraction of DEHP. Factors that influence the extraction efficiency of DEHP were studied. The optimum conditions, such as solution pH, extraction time, desorption time, and eluent type and volume were obtained. The analytical performance of the developed method was examined. Finally, the developed method was applied for the detection of DEHP in environmental water samples, and satisfactory results were acquired.

2. Materials and methods

2.1. Chemicals

IOP Conf. Series: Materials Science and Engineering **544** (2019) 012063 doi:10.1088/1757-899X/544/1/012063

Nature graphite powder (99.95%) was purchased from Sinopharm Chemical Reagent Co., Ltd (Shanghai, China). Other chemicals of analytical grade were purchased from Tianjin Damao Chemical Reagent Factory (Tianjin, China).

2.2. Apparatus

A HPLC system (SHIMADZU, Kyoto, Japan) equipped with diode array detection (DAD, SPD-M20A) and a Shim-pack VP-ODS column (150 × 4.6 mm I.D.) was utilized for analysis. The data obtained were analyzed by the LabSolutions software. A gradient program with a mobile phase system consisting of two parts as eluate A (0.2 M KH_2PO_4 buffer, pH 2.6) and eluate B (MeOH/ACN 50:50, v/v) was established for the elution of DEHP.

2.3. Preparation of MGO

Graphite oxide (GO) was prepared from nature graphite powders by a modified Hummers method, and magnetic graphite oxide (MGO) was synthesized by a one-pot solvothermal reaction [19].

2.4. Solid phase extraction

50 mg of MGO was added to the 200 mL DEHP containing water sample, and the solution pH was adjusted to pH 7. After shaking for 25 min, the MGO was isolated from solution using a magnetic field. The adsorbed DEHP was eluted with 2 mL of methanol.

3. Results and discussion

3.1. Selection of sample pH

The influence of sample pH on the extraction recovery of DEHP was examined by varying pH value from 4 to 10, and the results were shown in Fig.1. As can be seen, the recovery enhanced with the increase of pH from 4 to 6, and then kept almost unchanged up to pH 8. When the pH value was larger than 9, the recovery decreased. Generally, effect of pH was not great in the pH range of 4-10, and pH 6-8 was optimal.

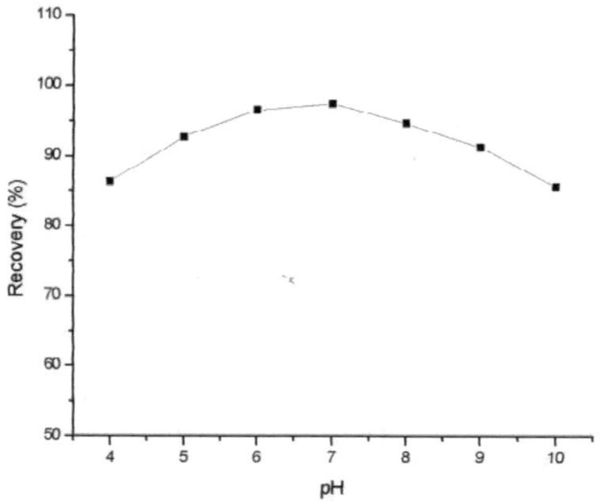

Figure 1. Effect of solution pH.

3.2. Selection of extraction time

The effect of extraction time on extraction recovery of DEHP was investigated. As seen from Fig. 2, the recovery increased significantly at the initial adsorption stage from 1 to 15 min. After 25 min, the recovery showed very little changes, indicating the sorption equilibrium was achieved. Based on these results, 25 min extraction time was employed in the subsequent experiments.

3.3. Selection of eluent type and volume

Methanol, acetone and acetonitrile were used as the eluents for stripping of DEHP from the MGO. It was found that methanol gave the highest recovery (97.3%) for DEHP (Fig.3). Thus, methanol was adopted as the eluent solvent. To examine the effect of methanol volume on the extraction recovery of DEHP, different volumes of methanol (1, 2, 3 and 5 mL) were tested. The results indicated that 2 mL of methanol could completely elute the adsorbed DEHP.

3.4. Selection of desorption time

To obtain the optimal desorption time, various time periods were tested in the range of 1 to 20 min. The results were presented in Fig.4. According to Fig.4, 8 min was enough for quantitative elution of DEHP from magnetic graphite oxide.

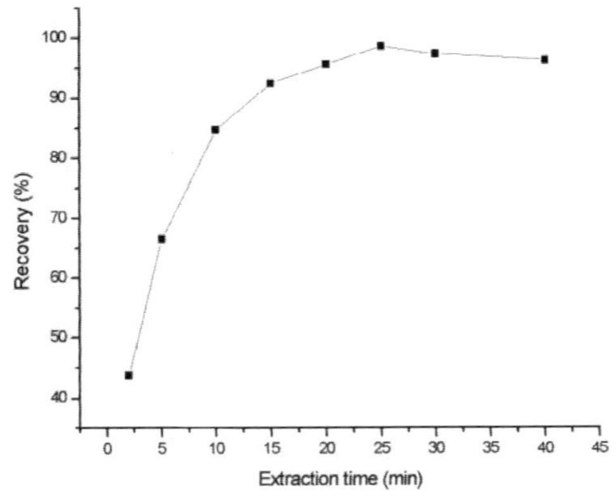

Figure 2. Effect of extraction time.

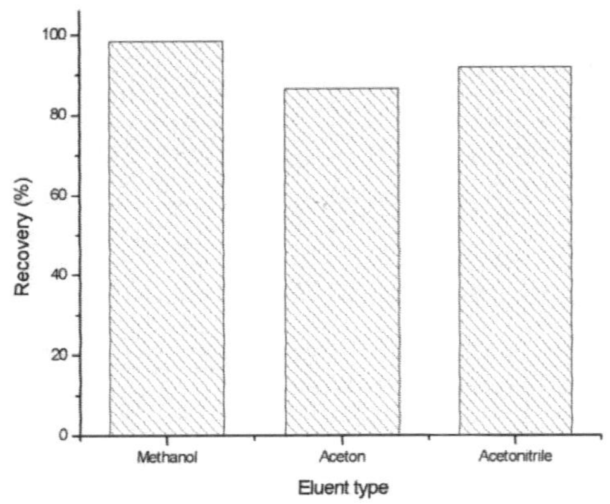

Figure 3. Effect of eluent type.

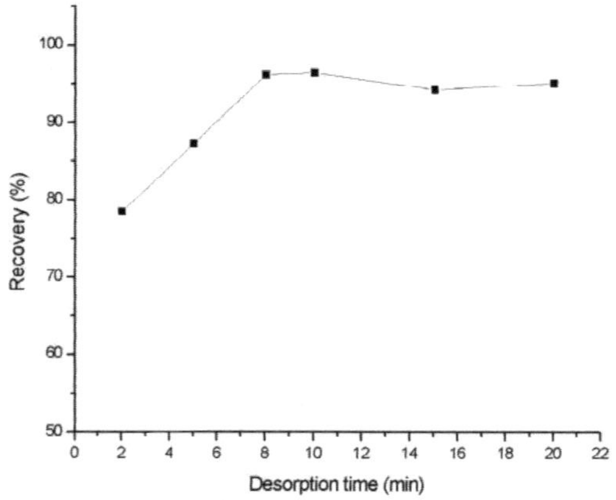

Figure 4. Effect of desorption time.

3.5 Analytical performance

Under the optimum experimental conditions, the calibration curve of DEHP was linear in the concentration range from 1.0 to 100 μg/L (R^2=0.9972).The limit of detection (LOD) for DEHP was 0.35 μg/L The relative standard deviation (RSD) for 6 replicate measurements of 10.0 μg/L of DEHP was 4.3%.

3.6. Water sample analysis

To evaluate the capability of the proposed method for analysis of practical samples with different matrices containing various amounts of DEHP, three real environmental water samples were analyzed, and the analytical results were given in Table 1. It is obvious that the spiked recoveries of these samples were in the range of 91.6–106.5%, with RSD of 2.3-5.4%. Therefore, the developed method was sensitive, accurate and precise.

Table 1 Analytical results of DEHP in water samples.

Sample	Add/(μg/L)	Found/(μg/L)	Recovery(%)	RSD(%)
Water sample 1	0	---	---	5.4
	5.0	4.58	91.6	3.7
Water sample 2	0	2.97	---	3.5
	5.0	7.73	95.2	2.8
Water sample 3	0	6.42	---	3.9
	10.0	17.07	106.5	2.3

4. Conclusions

Magnetic graphene oxide was synthesized and applied as the sorbent for solid phase extraction of DEHP from solutions. Then a new analytical method for trace detection of DEHP was proposed, using a combination of SPE with HPLC-DAD. Results showed that this method had good performance in terms of limit of detection, linearity, accuracy and reproducibility. The proposed method was used to real environmental water samples, and the obtained recoveries of DEHP were in the range of 91.6–106.5%

with RSDs of 2.3-5.4%, indicating the suitability of the developed method, due to its sensitivity, rapidity, and reliability.

5. References

[1] Yılmaz P, Ertaş A and Kolak U 2014 *J. Sep. Sci.* **37** 2111.
[2] Li C, Ma X, Zhang X, Wang R, Li X and Liu Q 2017 *J. Sep. Sci.* **40** 1621.
[3] Lafleur A and Schug K 2011 *Anal. Chim. Acta* **696** 6.
[4] Zou J, Ma X, Dang Y and Chen Y 2014 *J. Anal. At. Spectrom.* **29** 1692.
[5] Chen J, Ma X and Wu Y 2014 *Food Anal. Methods* **7** 1083.
[6] Heidari N,Ghiasvand A and Abdolhossaini S 2017*Anal. Chim. Acta.* **975** 11.
[7] Ma X, Huang B and Chen M 2007 *Rare Metals* **26** 541.
[8] Chen Y, Ma X, Huang M, Peng J and Li C 2016 *Anal. Methods* **8** 824.
[9] Barbosa A, Segatelli M and Pereira A 2007 *Talanta* **71** 1512.
[10] Nong Y, Ma X, Fan S and Yu Y 2014 *Anal. Methods* **6** 4124.
[11] Zhang Y, Ma X and Fan Y 2014 *Food Anal. Methods* **7** 1763.
[12] Fan Y, Ma X and Li Z 2016 *Food Anal. Methods* **9** 16.
[13] Ma X and Li Y 2006 *Anal. Chim. Acta* **579** 47.
[14] Heidari N,Ghiasvand A and Abdolhossaini S 2017 *Anal. Chim. Acta* **975** 11.
[15] Ma X and Zhang Z 2003 *Anal. Chim. Acta* **485** 233.
[16] Ma X and Zhang Z 2004 *J. Anal. At. Spectrom.* **19** 738.
[17] Li D, Ma X, Wang R and Yu Y 2017 *Anal. Bioanal. Chem.* **409** 1165.
[18] Faghihi A, Vakili M and Hosseinzadeh G 2016 *Desalin. Water Treat.* **57** 22655.
[19] Zhang X, Ma X, Li X and Chen M 2018 *Water Air Soil Pollut* **229** 270.
[20] Huang R, Ma X and Li X 2018 *J. Colloid Interf. Sci.* **514** 544.
[21] Xie X, Ma X, Guo L, Fan Y, Zeng G, Zhang M and Li J 2019 *Chem. Eng. J.* **357** 56.
[22] Li M, Liu Y and Liu S 2018 J. *Colloid Interf. Sci.* **521** 150.
[23] Li C, Ma X, Zhang X., Wang R, ChenY and Li Z 2016 *Anal. Bioanal. Chem.* **408** 7857.
[24] Zhao B, He M, Chen, B and Hu B 2015 *Spectrochim. Acta* B **107** 115.
[25] Li X, Ma X, Huang R, Xie X, Guo Li and Zhang M 2018 *J. Sep. Sci.* **41** 2837.
[26] Kazemi E, Shabani A, Dadfarnia S, et al. 2016 *Anal. Chim. Acta* **905** 85.

IOP Conf. Series: Materials Science and Engineering **544** (2019) 012064 doi:10.1088/1757-899X/544/1/012064

Research on the sintered microspheres of high purity molybdenum powders

Jingbo Fu *, Fei Zhuang, Weicheng Cao, Xiao Zhang

JDC Molybdenum Industry Co., Ltd, Xi'an 710077, China
E-mail: jingbo_fu@126.com

Abstract. The process of preparation of the sintered microspheres of high purity molybdenum powders was investigated. The results indicate that the process include spray granulation, degreasing and sintering(spray granulation- degreasing -sintering). And in the process, spray granulation is the key step of deciding the morphology and size of the powders, degreasing temperature is the important parameter to control the purity of the powder and sintering is the necessary process to control the morphology and physical properties of the final microsphere products. The higher of the sintering temperature is, the higher of the bulk density and flowability of sintered microspheres are. And the higher of the sintering temperature is, the lower of the O content of the powder is. The sintered microspheres of high purity molybdenum can be used not only in coating industry, but also in manufacturing industry of precision parts.

1. Introduction

With the development of modern technology, such as electronic information and astronautics aviation industry, some miniature or special-shaped molybdenum parts are in great needs. For example, LCD backlight source of molybdenum electrode, microwave oven magnetron cathode assembly, high performance and molybdenum alloy nozzle and throat with a new generation of aerospace engine. All these parts have complex shape, processing and difficult. Powder (near) net forming is a new technology for preparation of those parts is complicated, long processing cycle. In powder net forming, powder molding is the key technology.

Conventional hydrogen reduction of molybdenum powders are featured with irregular morphology, small bulk density, poor mobility, which are difficult to fufill powder uniform filling, and can not meet the powder (nearly) net shaping demand. Therefore, the preparation of microspheres molybdenum powder with round morphology, large bulk density, good flowability has become an urgent problem.

At present, the main methods to improve the propertied of powders are plasma spherical, sol-gel method, spray drying and spray drying and sintering [1-11]. Plasma spherical has the defect of high cost and low yield, sol-gel process is complex and long production cycle, spray granulation technology can improve the bulk density and flow velocity to some extent, but introduce high impurity. Spray granulation+sintering can not only improve the bulk density and flow velocity greatly, but also lower the impurity of the powder, and more importantly, spray granulation+sintering has the advantages of simple process, low cost, and easy to achieve industrialization.

2. Experimental Results and discussion

The molybdenum powders used are hydrogen reduced by nomal procedures, their Fisher particle sizes are about 3.1μm, and bulk density are 1.0g/cm^3. The morphology of original molybdenum powders are as Figure 1, and their chemical compositions are as table 1. The binder used is PVA(Polyvinyl alcohol) with molecular weight of about 15000~25000.

The procedures of the experiment are as follows: Firstly, PVA is dissolved in the pure water completely, and molybdenum powders are added into the PVA solution as 68% solid content and stirred well to make slurry. Secondly the slurry is spray dried by centrifugal spray dryer with the rotation rate of atomizing disc about 12000r/min. Thirdly, the spray dried molybdenum powders are degreased in the mono pipe hydrogen furnace at 800~900 ℃ . Finally, the degreased molybdenum powers are sintered at 1100℃- 1500℃ in a muffle furnace with hydrogen atmosphere. Bulk density of the powders are determinate by funnel method, their flowability are detected by hall flowmeter, and the morphologies of the powders are observed by scanning electron microscope(SEM). The content of C in the powders is detected by infrared carbon and sulphur analyser, and the content of O is detected by oxygen analyzer.

Figure 1. Morphology of original molybdenum powders

Table 1 Chemical composition of original molybdenum powders (wt%)

Element	Fe	Ni	Ca	Mg	Si
Content（wt%）	0.0040	0.0014	≤0.0008	0.0001	≤0.0015
Element	Al	Cu	Cr	Mn	Pb
Content（wt%）	≤0.0010	0.0003	≤0.0012	≤0.0001	0.002
Element	Sn	K	C	O	Mo
Content（wt%）	0.002	0.0018	0.005	0.08	balance

3. Results and discussion

3.1. Morphology and physical properties of the microspheres of molybdenum powders

Morphology of the microspheres of molybdenum powders at spray granulation, degreasing and sintering states are shown in Figure 2. It can be seen that the morphology of the microspheres at

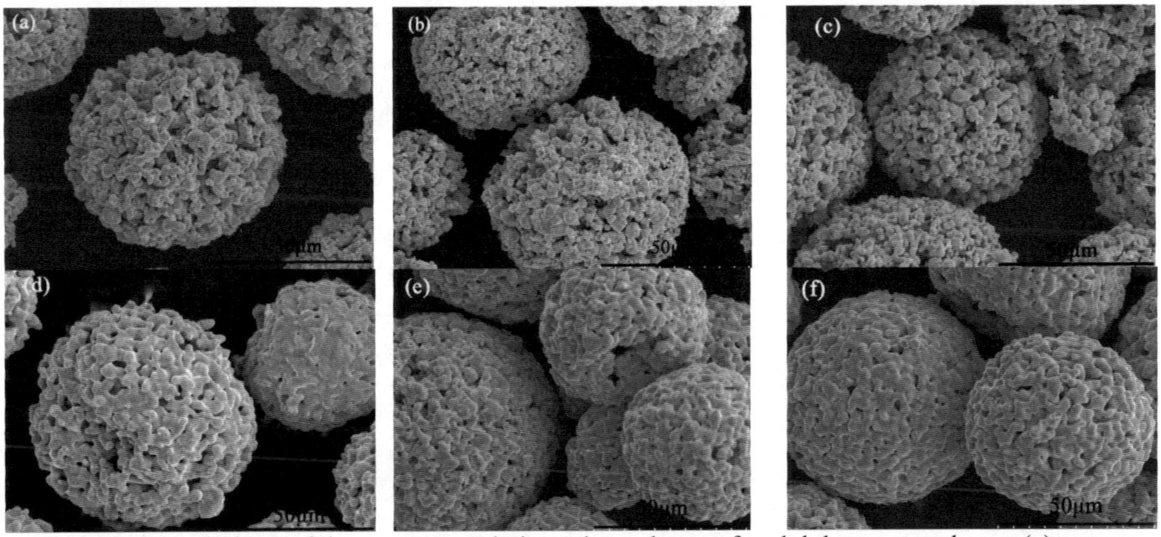

Figure 2. Morphology of the spray granulation microspheres of molybdenum powders at (a) spray granulation and sintering temperature (b)1100°C, (c)1200°C,(d)1300°C, (e)1400°C, (f)1500°C

different states are relatively ideal ball type. Spray granulation powders are smooth spherical which are bonded by fine molybdenum powders as raw material (Figure 2(a)), and spray granulation powders after degreasing are rough loose microspheres (Figure 2(b)). Spray granulation powders after sintering at 1100 ℃, 1200 ℃ are rough but a little shrink (Figure 2(c),(d)) compared with the degreasing powders (Figure 2(b)), and spray granulation powders after sintering at 1300℃, 1400℃, 1500℃ are compacted microspheres with much sintering neck between the original raw powders (Figure. 2(e),(f),(g)).

The physical properties of the spray granulation microspheres of molybdenum powders after degreasing and sintering at different temperatures are shown in Figure.3. It can be seen that the physical properties of the spray granulation microspheres of powders after degreasing will be worse than that of spray granulation microspheres, while it will be better after sintering. And the higher the sintering temperature is, the higher of bulk density and flowability of the microspheres powders are. But at differnet sintering temperature rages, the degree of improvment of the physical properties is diferrent as well. At the sintering temperare between 1100℃-1200℃, the bulk density and flowability of the microsphere powders are improved a little, while at the sintering temperare between 1300℃-1500℃, the bulk density and flowability of the microsphere powders are improved a lot.

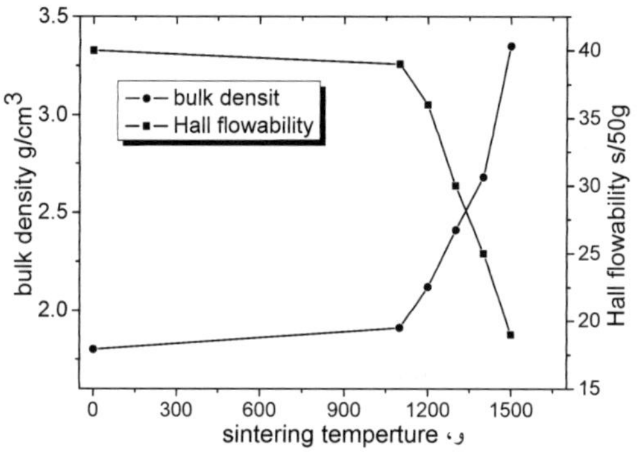

Figure 3. Relationship between the physical properties of microspheres and sintering temperatures

The process of sintering is the migration of substance, and to fill the pore between the particles. The sintering of molybdenum powders can be explained by diffusion and mass transfer[11]. Self-diffusion coefficient expression of the atomic is described as by formula(1)

$$D = D_0 \exp\left(-\frac{\Delta G}{RT}\right) \tag{1}$$

where D denotes the diffusion coefficient, D_0 is the so called pre-exponential factors, R the gas constant and T the thermodynamic temperature. From formula (1), it can be seen that the diffusion coefficient in the particles increases according to the index law with the rising of the sintering temperature. Therefore, with the rising of the sintering temperature, thermal diffusion velocity is

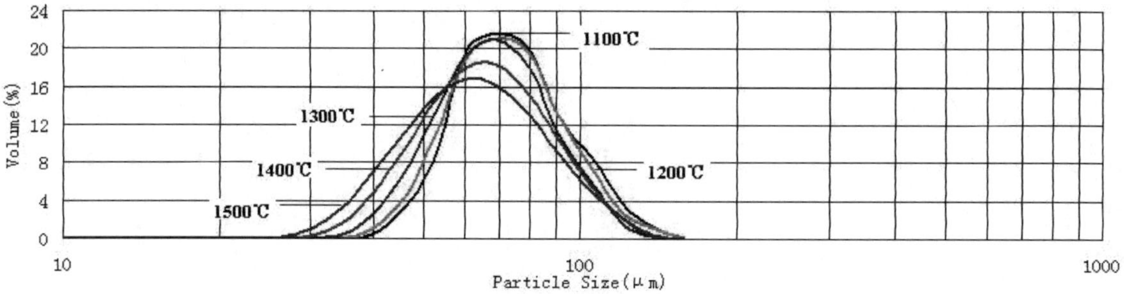

Figure 4. Particle size distribution curves of microsphere molybdenum powders sintered at different temperature

increased, the molybdenum powders are bonded to each other and formed necking. And at the same time, hollows between particles are decreased, the size of the microsphere particles reduced, which lead to the density of the microsphere particles improved, thus the bulk density of the microspheres of powders is increased.

Generally speaking, the flowability is proportional to the bulk density of the powders. It can be explained as follows: powders with large bulk density are more conducive to slip created by the gravity on itself, which conquer the frictions and adhesion between the partices of the powder. Hence with the rising of the sintering temperature, the flowability of the powders is increased. Size distribution curves of the microsphere powders sintered at different temperatures are shown in Figure 4. It can be seen that with the increasing of the sintering temperature, the peak size of the microspheres of powders decreases. Which are consistent with the conclusion from formula (1).

3.2. The chemical constitution of the agglomerated microspheres of molybdenum powders

During our research, in addition of PVA, no other impurities are introduced. As it is known, PVA is a kind of high molecular polymer with chemical formula of $(C_2H_4O)_n$, so C, O are the main impurities elements in the molybdenum powders we should consider. Figure 5 shows C content of the powders at different states. From Figure 5, we can see that the content of C in the spray drying powders is much higher than that in the original powders(as table 1 illustrated), and after degreasing, the content of C are tend to be the same. The high content of C in the spray drying powders is attribute to the adding of PVA, and the lower content of C in the degreased and sintered powders is due to the decomposing of PVA. Generally speaking, PVA begain to decompose into water, acetic acid, acetaldehyde and crotonaldehyde at about 300 ℃,and between 700 ℃ - 800 ℃,it decompose completely. Consequently, after degreasing at 800-900℃, PVA in the spray drying powders decomposed completely, and the contents of C in the powders is as much as in the oringinal molybdenum powders.

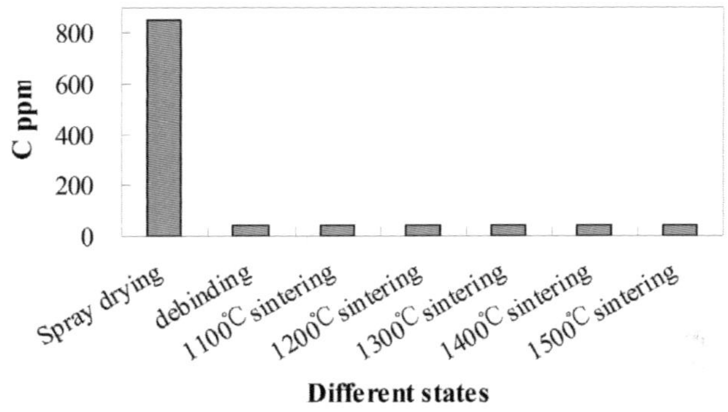

Figure 5. C content of the powders at different states

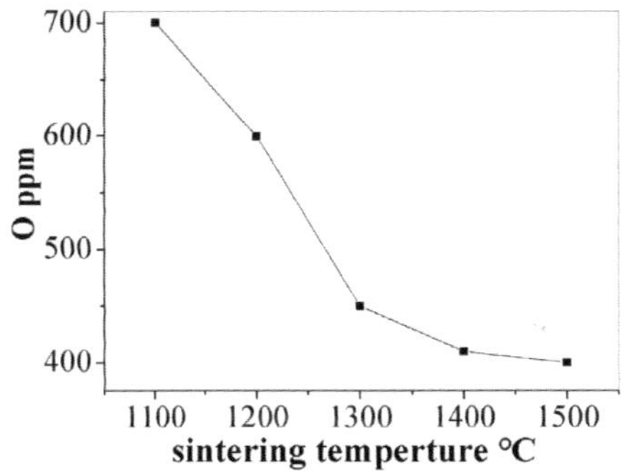

Figure 6. Relationship between O content and sintering temperatures

Relationship between O content and sintering temperature of microsphere molybdenum powders are illustrated in Figure 6. From Figure 6,we find that the higher of the sintering temperature is, the lower of the O centent of the powder is. Which can be explained as following[12]: Some O exsit deeply in the molybdenum particles wich are bounded firmly after hydrogen reduction (the nomal reduction temperature of molybdenum powder is about 1000 ℃), and it will be released only at a higher temperature. accordingly after heating at 1100℃～1500℃, the contents of O is lower than that of original powders.

4. Summary

The sintered microspheres of high purity molybdenum powders can be prepared by spray drying-degreasing- sintering process. And the physical properties and O centent of the microsphere powders are various when the sintering temperature changed. The higher of the sintering temperature is, the higher of the bulk density and the flowability of the powders are. And the higher of the sintering temperature is, the lower of the O centent of the powders is.

IOP Conf. Series: Materials Science and Engineering **544** (2019) 012064 doi:10.1088/1757-899X/544/1/012064

Acknowledgements

This project is supported by the national key research and developmentprogram of China (No. SQ2017YFGX010031-02).

References

[1] V.N. Korzhyk, L.D. Kulak, V.E. Shevchenko, V.V. Kvasnitskiy, N.N. Kuzmenko, X. Liu, Yi Xiang Cai, L. Wang, H.W. Xie and L.M. Zou 2017 New equipment for production of puper hard spherical tungsten carbide and other high-melting compounds using the method of plasma atomization of rotating billet. Materials Science Forum (Volume 898): 1485-97

[2] D Yan Wei Sheng, Zhi Meng Guo and Jun Jie Hao 2012 Characterization of spherical molybdenum powders prepared by RF plasma processing. Advanced Materials Research (Volumes 482-484): 2563-67

[3] Yan Wei Sheng, Jun Jie Hao and Zhi Meng Guo 2011 Study on spheroidization of tungsten powders by RF plasma processing. Advanced Materials Research (Volumes 295-297) P135-9

[4] Romain Vert, Remy Pontone, Richard Dolbec, Luc Dionne and M.I. Boulos 2016 Induction plasma pechnology ppplied to powder manufacturing: example of titanium-based materials. Key Engineering Materials (Volume 704):282-6

[5] P. Sobharaksha, R. Indranupakorn and Manee Luangtana-Anan 2012 Influence of diluents on physical properties of soybean extracted powder prepared by spray drying method. Advanced Materials Research (Volume 506):529-32

[6] Zong Lin Zuo, Jin Wang, Jian Qiu Deng, Qing Rong Yao, Zhong Min Wang and Huai Ying Zhou 2017 Electrochemical performance of spherical Li3V2(PO4)3/C synthesized by spray drying method. Key Engineering Materials (Volume 727):738-43

[7] Jian Hua Liu, Tao Huang, Huan Ran Zhang and Rui Xiang Wang 2011 Prepare uniformity superfine WO_3 by spray drying. Advanced Materials Research (Volumes 399-401): 731-5

[8] Helena Bruncková, Ľubomír Medvecký and Ján Mihalik 2009 Effect of sintering time on the phase composition in PFN ceramics prepared by Sol-Gel process. Key Engineering Materials (Volume 409):317-21

[9] X. L. Jiang, M. Boulos. Transaction of Nonferrous Metals Society of China, 2006,16(1)13-17

[10] H. S. Huang, Y. C. Lin, K. S. International Journal of Refractory Metals and Hard Materials, 2002, 20(3) 175-180

[11] S. J. Guo 2002 Sintering Theory of Powder Metallurgy(Metallurgical Industry Press, China, 2002):11-12

[12] F. Wang and H. Z. Yang 2003 Impulse infrared determination of oxygen in pure molybdenum powder and its make-up. Chinese journal of Rare Metals, 27(1):205-6

Institute of Physics
Dirac House, Temple Back
Bristol BS1 6BE UK

ISSN: 1757-8981
ISBN 978-1-5108-9381-8

9 781510 893818

12th International Workshop on Low Temperature Electronics 2016

Journal of Physics: Conference Series Volume 834

Tempe, Arizona, USA
18 – 21 September 2016

IOP
Institute of Physics